PRAISE FOR *MILK INTO CHEESE*

"The biological, historical, and ecological principles of good cheesemaking are as much a wonder as the cheese itself. It's all connected, and it's all here in *Milk Into Cheese*. An inspired how-to guide for the greatest of culinary transformations."

— **Dan Barber**, chef/co-owner, Blue Hill and Blue Hill at Stone Barns; author of *The Third Plate*

"David Asher has continued to deepen his understanding of cheese in the years since *The Art of Natural Cheesemaking* was published. His new book, *Milk Into Cheese*, is no less than a cheesemaking textbook, with thorough treatment of basics such as the biochemistry of milk and its transformations, rennets and other coagulants, salting, microbial development, aging environments and protocols, tools, and many other variables. These sections are followed by a survey of cheese styles, detailing methods and parameters for each. David explains in detail and with great passion the hows and whys of making cheese in traditional and simple ways, based on extensive and broad experience with the incredibly varied ways in which people have, do, and can work with milk."

— **Sandor Ellix Katz**, fermentation revivalist; author of *The Art of Fermentation* and other fermentation bestsellers

"*Milk Into Cheese* is a near-encyclopedic deep dive into the world of fermented dairy. Having traveled the world, David Asher's intimate and diverse first-hand knowledge of cheesemaking (and fermentation in general) makes the stories, explanations, and recipes within this book feel less like the work of a cookbook author and more like that of an insatiably curious natural philosopher. This book will stand alongside works like Katz's *The Art of Fermentation* and Steinkraus's *Handbook of Indigenous Fermented Foods* as essential reading for anyone wanting to learn all they can about the ways in which humans, and our domesticates, share this world with microbes."

— **David Zilber**, chef and food scientist; coauthor of *The Noma Guide to Fermentation*

"David's first book was life-changing for me, so much so that his methods are at the center of our restaurant's cheesemaking program. As I pore through the pages of this book, the inspiration is abundant. He has leveled up again, delving deeper into natural cheesemaking methods and answering any niggling questions that may be had. As we seek to better understand microbiology and its importance to our own health, this book is leaving a legacy that will become evident in years to come. Set to become a classic of our time."

— **Rodney Dunn**, cofounder, The Agrarian Kitchen

"An incomparable summary of cheese knowledge, containing a lot of different aspects and a very practical and open-minded approach to the subject of milk. Definitely a must-have educational tool for any cheesemaker. Thank you for this gift to the world of cheese."

— **Martin Grüner**, Goas Dairy & Farm Shop

"In this book, David Asher answers the questions I have been wondering and the ones I hadn't yet thought to wonder. It is a deep dive into everything interesting and extraordinary about natural cheesemaking and milk."

— **Robyn Jackson**, owner, Cheese From Scratch

"David Asher's *Milk Into Cheese* is a revolutionary homage to the art of traditional cheesemaking. Through engaging prose, Asher leads us on a journey back to natural and sustainable methods, proving himself a master at blending tradition with innovation. A treasure for those seeking to rediscover authenticity and joy in cheesemaking."

— **Samuel Lai**, Sinnos Formaggi

"A book that's the most complete guide to making and maturing real cheese. A book that defends the traditional ways and justifies the use of raw milk and natural starters for any cheesemaker, commercial or not. If you are a cheesemaker, I guarantee this book will leave you with the desire to test new cheeses and modify techniques to see the different results."

— **Martín Rosberg**, cheesemaker

"An indispensable book for understanding natural cheesemaking and putting it into practice in a simpler, safer, small-scale, sustainable way, referencing traditional practices and valuing the fresh milk, the cultures, the rennet, and the skill of the cheesemaker himself. It creates a new way of understanding the relationship between milk and cheese, building faith in milk's natural ability to make all styles of cheese anywhere and in all cultural circumstances."

— **Malen Sarasua**, director and researcher, Leartiker Dairy Centre

"David Asher has created the world's most comprehensive and empowering guide to natural cheesemaking. Through the history, science, and cultural nuances of making cheese, he illustrates how communities around the world developed the hundreds of different cheeses that we love today. He dives into the nuts and bolts of natural cheesemaking and teaches the reader how to replicate these processes so that they can make many of these cheeses, using natural processes, in their own home kitchens. This book is a must read for cheesemakers, cheese lovers, and anyone looking to reconnect with the most basic, foundational human food."

— **Dr. Bill Schindler**, author of *Eat Like a Human*

"To be a great cheesemaker, one has to understand and respect the milk, and David's book provides the key to both. *Milk Into Cheese* is not just a deep dive into natural cheesemaking techniques but also the three key pillars—milk, cultures, and rennet. The book has an exhaustive list of cheeses and methods to make them, but before you get there, the preceding chapters help you understand a cheese, its provenance, and what makes it unique. For a cheesemaker, it's probably the only book you'll ever need to make better cheese."

— **Namrata Sundaresan**, cheesemaker

MILK INTO CHEESE

MILK INTO CHEESE

The Foundations of Natural Cheesemaking
Using Traditional Concepts, Tools, and Techniques

DAVID ASHER

Chelsea Green Publishing
White River Junction, Vermont
London, UK

Copyright © 2024 by David Asher Rotsztain.
All rights reserved.

Unless otherwise noted, all photographs copyright © 2024 by David Asher Rotsztain.
Unless otherwise noted, all illustrations copyright © 2024 by David Asher Rotsztain.

No part of this book may be transmitted or reproduced in any form by any means without permission in writing from the publisher.

Project Manager: Rebecca Springer
Developmental Editor: Matthew Derr
Copy Editor: Laura Jorstad
Proofreader: Diane Durrett
Indexer: Shana Milkie
Designer: Melissa Jacobson

Printed in the United States of America.
First printing July 2024.
10 9 8 7 6 5 4 3 2 1   24 25 26 27 28

Our Commitment to Green Publishing
Chelsea Green sees publishing as a tool for cultural change and ecological stewardship. We strive to align our book manufacturing practices with our editorial mission and to reduce the impact of our business enterprise in the environment. We print our books using vegetable-based inks whenever possible. This book may cost slightly more because it was printed on paper from responsibly managed forests, and we hope you'll agree that it's worth it. *Milk Into Cheese* was printed on paper supplied by Versa that is certified by the Forest Stewardship Council.®

Library of Congress Cataloging-in-Publication Data
Names: Asher, David, 1980– author.
Title: Milk into cheese : the foundations of natural cheesemaking using traditional concepts, tools, and techniques / David Asher.
Description: White River Junction, Vermont : Chelsea Green Publishing, [2024] | Includes bibliographical references and index.
Identifiers: LCCN 2024008917 | ISBN 9781603588874 (hardcover) | ISBN 9781603588881 (ebook)
Subjects: LCSH: Cheesemaking.
Classification: LCC SF271 .A8434 2024 | DDC 637/.3—dc23/eng/20240326
LC record available at https://lccn.loc.gov/2024008917

Chelsea Green Publishing
White River Junction, Vermont, USA
London, UK
www.chelseagreen.com

This book is dedicated to my mother, provider of my milk, which, naturally, I turned into cheese.

PHOTO COURTESY OF MAX JONES

CONTENTS

Acknowledgments — xi
Preface — xiii
Introduction — 1

PART I THE ELEMENTS

1 Natural Cheese — 15
2 Milk — 27
3 Culture — 51
4 Rennet — 77

PART II MAKING AND AGING

5 The Make — 95
6 The Tools — 123
7 The Salt — 135
8 The Cave — 149
9 Affinage Treatments — 163
10 Ripening Ecologies — 177

PART III THE TECHNIQUES

11 The Families of Cheese — 203
12 Spring — 221

Clabber 222 • Kefir 224 • Junket 226 • Cuajada 228 • Faisselle 231 • Chèvre and Brebis 232 • Crottin 234 • Valençay 236 • Le Sein 238 • Saint-Maure 241 • Saint-Marcellin 242 • Primo Sale 244 • Queso Fresco 246 • Halloumi 248 • Stracchino / Crescenza 250 • Tomme Vaudoise 252 • Traditional Camembert 254 • Modern Camembert 257 • Brie 260 • Pont l'Évêque 262

13 Summer — 265

Mozzarella 266 • Queso Oaxaca 269 • Burrata 272 • Sour Pickles 274 • Feta 276 • Sirene 278 • Tomme de Savoie 280 • Tomme de Chèvre 282 • Tomme de Montagne 284 • Raclette 286 • Tomme Crayeuse 288 • Comté 291 • Emmentaler 294 • Grana 296 • Saint-Nectaire 298 • Gouda 300 • Lancashire 303 • Traditional Cheddar 305 • Modern Cheddar 308 • Caciocavallo 311

14 Fall — 313

Cultured Cream 314 • Long-Cultured Cream 316 • Butter 317 • Skyr and Smjör 320 • Smetana and Tvorog 322 • Goat and Sheep Butter 325 • Clabbered Cottage Cheese 327 • Handkäse 328 • Graukäse 330 • Sourdough Rye Bread 332 • Blue Mold 334 • Camembert Bleu 336 • Fourme d'Ambert 338 • Roquefort 340 • Gorgonzola 342 • Stilton 345 • Natural Wine 347 • Amasi 350 • Milkbeer 352 • Ambarees 354 • Cheese on Lees 356

15 Winter — 359

Yogurt 360 • Gros Lait 362 • Ryazhenka 364 • Kaymak 366 • Yogurt Cheese 369 • Shankleesh 370 • Cream Cheese 372 • Munster 373 • Vacherin 376 • Époisses 378 • Torta 380 • Fresh Whey Ricotta 382 • Ricotta Salata 384 • Ghee 385 • Lactic Cheese Sausage 386 • Salami 389 • Milk Vinegar 391 • Mysost 392 • Paneer 395 • Kalvdans 396

Appendix A	On Clabber Culture	397
Appendix B	On Whey Culture	400
Appendix C	On Kefir Culture	402
Appendix D	On Keeping a Wooden Vat	406
Appendix E	Thermophilic Starters	409
Appendix F	Commercial Considerations	411
Appendix G	Sourdough Starter	417
Appendix H	Keeping a Wine/Beer/Cider Starter	421
Appendix I	Cultivating *Mucor* for Rinds and Rennet	423
Appendix J	Flaws in Cheese	426

Index — 435

ACKNOWLEDGMENTS

I have many to thank for supporting the good work that put this tome together.

To all my fellow farmers who supplied me with my milk over the years, especially the reverends Rob and Ellen Willingham for providing the goat's milk that became my first cheese—it was a religious experience!

To the many champions of natural cheesemaking for working toward acceptance and understanding of this most important food in their locales and beyond, including Peter Dixon, Rachel Fritz-Schall, Trevor Warmedahl, Sister Noella Marcelino, Martin Gott, Marie-Christine Montel, Mateo Kehler, Martin Rosberg, Carolina Bittencourt, Doetie Trinks, Piero Sardo, and more.

To those who offered me a safe house during the years of writing and making this book: Melanie and Tyler Webb of Stony Pond Farm; Ashley and Fañch Guillou and the whole Guillou crew at Ferme Enez Raden; the Svanholm Community; the Carranza family at Rancho Avellanas; and Michael and Eliza Schmidt of Glencolton Farm (your fight for raw milk access in Canada is worth acknowledging too!).

To all those who have hosted us and our classes over the years and helped to get the word out about good cheese in your communities.

To all the cheesemakers who helped refine my cheesemaking skills and deepen my understanding in specific techniques, especially Jennifer Kast and Joe Schneider for insights into Stilton-making methods; Dag Kyrre Lygre for insights into Norwegian mysost traditions; the late Brother Alberic, Canada's last Trappist cheesemaker, for insights into monastic cheesemaking practices; to Sophie and Alexander of Secret Lands Farm for introducing me to the traditions of Russian cheesemaking; Matthäus Rest for your thoughts on alpine cheesemaking and buttermaking traditions in Austria and Switzerland; Trevor Warmedahl for insights and observations into lesser-known cheese traditions of Europe and Asia, and the Daphne Zepos Teaching Endowment (DZTE) that helped fund your research trips; Ana Brito at Projecto Sazonal (and her mother

Manuela Ramos) for instructing on the harvest of cardoon flowers for rennet; and Maryann Borch, for weaving ideas about weaving into natural cheesemaking.

To those in the world of natural fermentation who helped open my eyes to the fundamental fermentations that link these foundational foods: the natural bakers Paul Albert, Blair Marvin, and Sarah Owens, for helping me understand the facets of sourdough bread baking and its relationship to natural cheesemaking; Vasilios Gletsos of Wunderkammer Biermanufaktur for our conversations about natural brewing; John Cox at Quercus Cooperage for fashioning my first gerle and elaborating the importance of wood for the evolutions of so many fermentations; Silene DeCiucies for our conversations about clay (and all those beautiful wood-fired pots!); Esteban Yepes of El Taller de los Fermentos for gifting me my first gourd—its effect was truly shamanic; and, of course, Sandor Katz, for being a guiding light in the fog of industrial fermentation.

To the cheesemakers who contributed photos of their work in these pages, including Doetie's Geiten, Fromagerie au Gré des Champs, Queijaria BelaFazenda, Foradori Alimentari, Ballymaloe, Dag Kyrre Lygre, Charly Ilpide, and Benoît Huyghe.

To all the photographers who took the excellent photos for the book: Aliza Eliazarov, Virginie Gosselin, Chloé Gire, Max Jones, Larry Lopez, Chloé Savard, Mercurio Studio, Merlin Delens, and Kathryn Scarfone.

All the Jerseys, Brown Swiss, Canadiennes, Pie Noirs, Nubians, Alpines, and British Milk Sheep that provided the milk for my cheeses featured in these pages.

To all those at Chelsea Green who assembled and polished this work—especially my friend and editor Matthew Derr.

And finally, to my wife, Kathryn, for putting up with all my stinky cheeses!

PREFACE

July 2023
Brittany, France

Vive le fromage libre!

Here I am, in France, teaching the French to make cheese.

Now, I know what you're thinking—I'm selling sand to the Sahara—but even the French are beginning to realize that something's missing in their most important food.

I've been invited by the country's paysans (politically motivated small farmers in France have reclaimed their title of "peasants") to teach techniques that are exceedingly rare, even here, in this land of raw milk cheese. It's been a shock to find copies of my *Art of Natural Cheesemaking*—translated *en français*, dog-eared, and stained with whey—in the dairies of the makers I've visited. But they've almost nowhere else to turn to for these ideas that have largely been lost.

For nearly ten years—since the release of my first book—I've taught nearly nonstop, from Uruguay to Ukraine, Switzerland to Sweden, Austria to Australia. The ideas I share have struck a chord among cheesemakers in places I never would have expected, and invitations to teach in so many extraordinary places have been hard to decline. And though I'm not inclined to travel so relentlessly, I feel these ideas need to be spread far and wide. To add fuel to my fire, I'm forced to travel to teach, for the good work I do isn't welcome at home.

In Canada, natural cheesemaking practices are all but illegal, making this important food out of reach to most. The sale and transport of raw milk is strictly forbidden there, and the regulations surrounding cheese production—though they don't implicitly forbid the natural—render it almost impossible to make for just about everyone, including myself.

As a dairy revolutionary, I cannot make a living making cheese at home. Instead, I'm focusing on education and advocacy to change the circumstances we find ourselves in today, and show people a better way. I have become a fully nomadic cheese educator; missing my home, my garden, my goats; but planting different seeds, establishing a new culture of cheesemaking in communities all around the world.

Though the rules and regulations regarding raw milk cheesemaking are different in every region, the same natural laws apply everywhere to milk—for milk transforms into its cheese universally. And regardless of whether what I teach is legal to practice in a particular locale, it is of utmost importance for this message to be heard everywhere. For natural cheesemaking needs to be more widely understood and practiced.

Like the right to clean water and access to good food, the ability to make, sell, and eat natural cheese should be seen as a universal human right. This is a significant cultural issue, with the future of our food and farming at stake, for laws that prohibit or severely restrict the sale of raw milk and its cheeses are laws that restrict small-scale, sustainable agriculture. Though natural cheese may seem trivial to some, in many regions, this is the most important food.

In the alpine regions of Europe, for example, cheese is the most nourishing food that can be reaped from the mountainous land, and natural cheesemaking techniques support the working landscape and a way of life that's existed almost unchanged for hundreds of years, despite the industrialization of cheesemaking in the valleys below. In other regions, like Mexico, where the sale of raw milk cheese is now officially banned (industry, government, and health authorities worked toward this hand in hand), it is made and sold nevertheless. Enforcement of the ban would shut down significant foodways in many rural communities; peasants would lose their livelihoods, and might be driven to cities, or even forced out as refugees. It's not uncommon to hear stories like these, for this is the history of raw milk cheese.

———

Though the situations in each region are different, in my travels I've felt this need for true cheese everywhere—and France is no exception. Even here, dairy is increasingly dominated by the industrial. I see inroads of it everywhere: the plowing up of permanent pastures, replaced with fields of corn for silage; the massive tanker trucks picking up milk every three days; the rise of the supermarket—the "hypermarché"; and the predominance of pasteurization.

Early in my cheesemaking career, I considered coming here to learn about natural cheese. But I couldn't abandon my garden, couldn't walk out on my goats. And even if I could have left the farm, I wasn't sure if I could find cheesemakers here who practiced what I was seeking. I had come to understand that natural techniques had nearly disappeared, even in places with longstanding cheese traditions. And even if I did find someone practicing a true cheesemaking, I wasn't sure they'd be eager to share their traditions with this curious Canadian.

Cheesemaking today is almost always a clinical practice, and it's shocking how hospital-like the average dairy is. Cheese was one of the first realms of food production to fully modernize and adapt to the science of its time, and it is the food that we produce today that has most fully embraced industrial standards. But what about France, you might ask? Surely it mustn't be that way there? Well, France isn't just the land of raw milk cheese, it's also the land of Pasteur; and the country set the standard for industrial-artisanal practice that is largely the norm in dairies today all around the world. And even if some aspects of natural practices are in use in local practice here—say, goat's cheeses made with whey starters—the starters aren't understood from a natural perspective, and cheesemakers have most often adopted the industrial approach in most other facets of their production.

The modern doctrine of cheesemaking, with its dismissal of its origins and its disrespect of milk, never sat well with me. When I made my first batch of cheese, I couldn't pasteurize my milk, I couldn't bring myself to open a package of freeze-dried starter culture. I wanted to practice something different—more in line with the natural way I farmed and fermented.

Having a sense that there was a better way, I poured some kefir into my fresh goat's milk, added some calf rennet; let the milk ferment and curdle overnight; and then drained the curd in cloth to make a soft fresh cheese. Little did I know, I was stumbling across ancient methods of cheesemaking that had largely been lost, guided by a different ideal, looking to reconnect this food with its agricultural origins. Every success I had confirmed my intuition and helped me better understand milk's sacred nature. Every cheese I made was the best I'd ever tasted—and every cheese I made contradicted the modern modes of its production.

I found a hole in our understanding of cheese, and so, I decided to fill it. And here I am now, teaching the world to make cheese naturally, sharing what I've learned through careful observation and intuition; adapting techniques and ideas from other realms of fermentation; and feeling my way through the fog.

———

Natural techniques have been used by cheesemakers since cheese was first made many millennia ago, and are the original way that all of the cheeses that we know and love were first created and defined. But rarely today is cheesemaking practiced this way.

Some makers around the world still practice this original cheesemaking, having passed their techniques from generation to generation unchanged—these include some isolated European and Latin American makers, cheesemakers on islands in the Mediterranean and Atlantic, many central Asian and African cheesemakers, and makers of some rare DOP/PDO cheeses in which natural methods are codified. But their practice remains under threat by modernizing food regulations and the continued growth and consolidation of industrial cheesemaking, with its labor- and cost-saving tools and technologies, and its provision of cheap cheese.

Fortunately, an increasing number of cheesemakers are having success reclaiming these old but new ideas. Though hardly discussed or defined even a decade ago, natural cheesemaking is at last a growing concern, joining the ranks of sourdough baking, natural winemaking, and vegetable fermentation, all of which are intimately related, and employ similar communities of microbes, as well as communities of like-minded, revolutionary food producers.

Natural cheesemaking is now practiced by many commercial and home-scale producers, and very nearly everywhere. Champions of natural methods have risen up in many regions; helping support a global movement in their corners of the world. Those who have adopted these methods include established commercial producers who realize that a connection is missing in their production, as well as start-up makers who have only ever envisioned making cheese this way. They include homesteaders who want to make cheese independently, as they raise their animals and as they grow their vegetables; and folks in the city who see making natural cheese at home as a way forward toward a more just, ecological, and nourishing food system.

I imagine that natural cheesemaking will be even more widely adopted within the next decade, for while established cheesemakers are often challenged by the dogma of their current practice (and the perceived risk of losing a vat-full of milk) the next generation of makers is more welcoming of these natural ideas, and more eager to make them work in their circumstances. I know this, as I am one of them: a maker not wanting to follow the flock.

As an outsider, many see me as a disruptor, and my ideas as a danger to the industry. But the true danger lies in the damage we do when we industrialize our food and farming systems to an inhuman degree. There aren't good enough reasons why natural methods should not be adopted by commercial producers. For these techniques are the right way to work with milk.

I'm now dedicated to making real change in cheese, from the ground up. To reflect this shift from the fringe to the cutting edge, I've changed the name of my traveling cheese school to Milklab. This rebrand represents a migration to the mainstream, and the idea that these ideas are for everyone, everywhere . . . for we are all natural cheesemakers!

Included in this tome are techniques for producing many diverse dairy products from around the world, including amasi, an alcoholic milk fermented in a gourd (*Lagenaria siceraria*) all across Africa.

Introduction

*E*verything begins with milk. You are holding in your hands a gift, offered to you like milk from mother to child. In the pages of this book you will find all you need to learn to make the most exceptional cheese. *Milk Into Cheese* defines the theory and practice of natural cheesemaking, providing ideas and tools for makers to transform their milk into an expansive array of delicious and nourishing dairy products in their most original and respectful way.

This book explores in depth the four foundations of natural cheesemaking, the philosophies that allow cheese to take its truest, most delicious form: using methods based on traditional cultural practices; using the freshest possible, least processed milk; coagulating with natural rennet from the stomachs of young animals; and inducing a natural fermentation with milk's own community of microbes. In other words, the cultural (the hand) and the agricultural (the milk); the biological (the rennet) and the microbiological (the ferment).

Natural cheese is transformed from milk most fresh, preferably still animal-warm, and with minimal treatment. It is ideally made from the milk of ancient breeds (though any ruminant's milk will work just fine) nourished on fresh pasture or seasonal browse. And that milk should evolve from an agriculture that involves regenerative, organic, or biodynamic practices, and that assures the most ethical and sustainable treatment of its animals, as well as the sanctity of the soil and life.

Natural cheese invokes the use of traditional starters like whey, clabber, kefir, and wood. These cultures cause milk's lactose sugars to ferment to lactic acid, souring it, developing its complex flavors, and protecting it from unwanted microbes. The starters are carried forward from one day's cheesemaking to the next, allowing milk to best express its innate and intricate microbiologies. It is a cheesemaking allied with other traditionally fermented foods such as sourdough bread, natural wine, and naturally soured vegetables like pickles, kimchi, and sauerkraut.

Natural cheesemaking is curdling milk with naturally sourced rennet, derived from the stomachs of young animals of the herd, or certain plants or fungi that also cause milk to gel. It's a realization of the biological processes that are the foundation for all cheese's many transformations. This rennet is synchronous with the dynamic of traditional dairying practices and assures the most ethical and sustainable treatment of young animals born into the dairy.

And natural cheesemaking is employing techniques that are time-honored and respectful of traditional practices from cultures around the world. It is using natural tools that enable every cheese's best evolution. It is cultivating the right ripening organisms through a naturally minded affinage, and aging cheeses in a more natural cave. And it is trusting your milk, your culture, your rennet, and your skill as a cheesemaker.

Natural cheesemaking should inspire awe and wonder like other extraordinary natural phenomena: powerful thunderstorms, meandering rivers, old-growth trees, and distant galaxies. It's like working with the forces of both life and death simultaneously—for natural cheesemaking encompasses remarkable forms of biological magic.

This book's aim is to prove the worth of natural methods in an uplifting and spirited exploration of our beautiful world's diverse cheeses and dairy ferments. *Milk Into Cheese* is a celebration of milk, in all its forms.

The infinite possibilities of natural cheese: a fermented whey ricotta coalesces into curds much like distant nebulae birth stars. PHOTO BY CHLOE GIRE

A Second Book

This book is a deeper dive into milk—a more expansive and explorative look into the subject of natural cheesemaking than my first book, *The Art of Natural Cheesemaking*—showing how all styles of cheese are interrelated, and how they can all evolve from the same milk, with the help of its exceptional community of microbes and its most appropriate enzymes. It better represents the scope of what I practice and what I teach today.

This time I wanted to take the opportunity to write a definitive guide on the subject of cheesemaking and dairy fermentation, taking the natural philosophy of cheese to an all-encompassing completeness, with the goal of proving the rightness of these techniques for all styles of cheese in all circumstances. It offers a global, universalist perspective on the ways to make this ancient and important food.

Almost no cheese is left unturned in the pages of this book, from the simplest cultured butter, cottage cheese, and kefir, to the most challenging Brie, Comté, and caciocavallo. It includes the most popular cheddar, mozzarella, and ricotta, as well as the rare amasi, tomme crayeuse, and cuajada. From the softest and pudgiest to the hardest and most crystalline, the freshest and most heaven-sent to the most decayed and putrescent, this book will help you make them all.

There is a great diversity of techniques within each class of cheese, taking a detailed look at the craft and representing a greater picture of milk's many possibilities. For example, I include eight or more different alpine cheesemaking techniques (tomme de Savoie, tomme crayeuse, tomme de montagne, tomme de chèvre, raclette, Comté, Emmentaler, and grana).

The techniques are also more streamlined. The philosophies of natural cheese have continued to inspire and improve my approach, and I've realized many ways in which the cheesemaking process can be made even simpler and more elegant with these ideas. Go ahead and add your natural starter and rennet to milk at the same time, for example, and save yourself an hour in each and every make.

On the natural side, I've now worked extensively with wooden cheesemaking vats, copper kettles, calf and kid stomachs, and many of the trappings of traditional cheesemaking that I only briefly explored previously or had not even considered. I've discussed the nuances of wood and other natural materials with respect to fermentation with coopers, vintners, and cider makers. And I've worked with potters and weavers to better understand the nature of clay and reed and their relation to natural techniques.

I've spent many of these intervening years, since my last book, milk-jamming with many other practitioners around the world (see the acknowledgments) who have helped me understand the nature and nuance of the cheeses they make. I've experimented endlessly with new techniques to understand more diverse cheeses and redefine their makes in a natural way. And I've learned about (and tasted) many more, often obscure styles of cheese, and come to understand what each one teaches us about the nature of milk.

In this book, I cast light on an entity I call clabber culture, which is derived from raw milk's incredible microbiology and can be kept, like sourdough, by regularly feeding it fresh milk. In partnership with kefir, clabber makes this art most accessible to all (kefir can be, for many, harder to obtain than raw milk, while for many others the opposite is true). It is my goal to be as inclusive as possible with these philosophies, and to offer opportunities for every cheesemaker, anywhere, who wishes to make good cheese.

In *Milk Into Cheese* I focus less on kefir as a "wunderkultur" (there's a certain magic-bean-like tone to my praise of it in *The Art*) and instead talk more broadly about the nature of milk's microbes. For all of these different fermented milks are elaborations on the same community of milk-based microorganisms. And though each one is cared for differently, each one functions more or less the same when it ferments milk, and each one can be used more or less interchangeably in a cheese.

For that matter I also include information for successful dairy fermentation in wood, clay, and even gourds. I've even elaborated techniques of creating your own rennet microbiologically with the help of friendly fungi akin to koji, for those who are into the DIY approach but remain averse to using the stomachs of young animals.

There are even techniques that push the boundaries of what's possible in meat curing, natural winemaking, and cider making, borrowing from the philosophy from natural cheesemaking, which of course borrowed many concepts from these same realms. I've come to believe that on the exceptionally wide spectrum of natural cheesemaking, wine, bread, and fermented meats should all be included!

For commercial producers, there's also much more information (in the appendices) on how to have success with these natural methods in your dairies, based on my and others' experiences, including how you can prove the safety of keeping natural starters, and even how to use traditional rennet stomachs.

What This Book Is and Is Not

This book aims to establish a new narrative in cheesemaking and show unequivocally that natural methods are the best (indeed the only!) approach for making all styles of cheese. I have not written a scientific treatise, for many of these new/old methods have not been approved by government or certified by peer-reviewed science. Rather, this book is an experiential work, formed by my farm-based research, and confirmed by other practitioners of natural techniques.

However, this book isn't necessarily unscientific; it elaborates on much of the most recent research into milk and cheese and is likely in line with the current microbiological understanding of these foods (cheesemaking as generally practiced today is defined by a microbiological understanding from the 1960s!). But this book does not use these scientific ideas as a foundation for its philosophy, for the techniques of cheesemaking expounded upon in here are based fully on traditional technologies, and not upon the latest findings from the DNA profiling of milk's microbes. Such research can be important to helping us prove the veracity and effectiveness of traditional techniques, and to understanding how they work from our present perspective, but it should not necessarily inform our daily practice. The specific microbiology of your milk, for example, does not matter as much as you might think (it may not even matter at all), and the exercise of analyzing it may actually shake your confidence in your materia prima.

This book aims to create a new way of understanding the relationship between milk and cheese. The reductionist breaking down of milk into its component parts can confuse the cheesemaking practitioner. Milk, just like anything else of any complexity, does not function the same when its specific parts are isolated and analyzed. Our scientific understanding of the processes and players at hand will always fall short, and the more we dig, the more we'll discover what we do not know. But traditional techniques will always work and have been proven by millennia of practice. Milk functions as a complex whole, and we must have faith in something beyond the established science or a trained microbiologist to fully grasp its inherent capacity for cheesemaking.

This book is a practical guide to working with milk, to building faith in its natural ability to make cheese and recognizing that you as a cheesemaker can be in complete control of your milk's transformation. Natural cheesemaking's complexity can be overwhelming as a novice practitioner (or one transitioning from industrial practice). But as a keen observer you can come to grasp the subtlety of the craft—the little things you do that make the big differences—and this book will help you observe these changes for yourself and help your cheeses evolve in their most extraordinary, most natural way.

For those in old cheesemaking regions, where cheesemaking culture is established and remains part of the recognized pathway for milk, this book is meant to be a reminder of original ways that have by and large been forgotten or suppressed. In many European and Latin American countries, traditional cheese culture has been forcibly oppressed and only remains intact on the remotest islands or tallest mountains. Cheesemaking practice in what we often consider original cheesemaking regions of the world has been devastated by industry and governmental decree. Even with a Protected Designation of Origin (PDO), cheeses made in these regions today may resemble their progenitors in name only. Old-world cheeses were once all made in this truer way, and this book will help your cheeses achieve their original character.

For those in new cheesemaking regions, in English-speaking realms, or in those places without an established cheesemaking culture, this book can be a guide to establishing a new aspect of your agriculture that helps milk realize its fullest potential.

Transforming milk into cheese isn't often a part of our culture; if it is, it's most often done on the factory floor. And if it is performed on a farm, it's typically executed with the same industrial approach. Dairy farmers are selling themselves short by selling fluid milk, and we may be doing ourselves a disservice by drinking glasses of the stuff. This book will show you the many different ways that milk can be manipulated and help you realize a more exalted purpose for milk.

As you might have realized, this is not a book about milk drinking. I don't believe that milk (either pasteurized or raw) should be drunk by adults, or even children for that matter, aside from those that are breastfeeding. Milk is made for cheesemaking. And even the simple act of fermenting it into kefir or yogurt brings about a miraculous transformation in its flavor, texture, and nutrition that makes that glass of milk seem mundane and unappetizing. Indeed, milk's fermentation and its coagulation combine in a sort of pre-digestion that

Any milk, anywhere, is capable of developing a universal fermentation that promotes the growth of beneficial microbes like *Geotrichum candidum*, shown here manifesting atop a naturally made Saint-Marcellin.

makes milk most nourishing and delicious, mimicking the stages of breakdown that take place in a milk-drinking infant's digestive system.

There is very little history of drinking fresh milk. Most if not all traditional dairying societies transformed their milk instead to cheese, a food of significantly greater purpose. Fresh milk or cream likely couldn't even leave the dairy historically, for without the modern industrial technologies of pasteurization, refrigeration, and sterile stainless steel, milk would ferment by its handling, and curdle within hours of leaving the udder. The idea of drinking a glass of milk may be a wholly modern Anglo phenomenon, one without foundation, and one that may be leading us into nutritional voids and ecological ruins. We should instead take inspiration from the rest of the world, for the rest of the world makes cheese.

This book is meant to chart another course for milk. Our scientific-industrial dairying complex is focused on bringing liquid milk unchanged to the table, and this direction has led to the closure of countless small family farms across continents, and the growth of dairy farms of unimaginable scale. Milk has become our most industrialized and most politicized food. It's also become our most divisive—no other food is more revered and more reviled. But with a different consideration—that milk is made for cheesemaking—a whole new way of dairying can be conceived that can restore the honor and respect of this most perfect food.

This book isn't meant to shame established cheesemakers that practice the standard model of production. The modern approach is what most cheesemakers have learned and experienced and what they have had their own success with. And it has been this way for generations, most cheesemakers having learned their modern techniques from other cheesemakers who learned these ways from their predecessors. For those who practice this other sort of cheesemaking, consider this book not as an affront, but as a source of ideas and inspiration to help improve the various aspects of your cheesemaking and make it your own.

This book aims to democratize and decolonize cheesemaking; to make any cheese accessible to anyone, anywhere. It is not meant to be an elitist or even a Eurocentric tome, for cheese is peasant food, made around the world, in Asia, South America, and Africa in simple circumstances, often with animal-warm milk poured straight into a bucket and handled in a most straightforward way. Raw milk cheeses are often seen as a luxury, one of the most expensive foods in our stores, but they should be understood as a good food for all that could be most affordable and accessible if simpler production methods were the norm and the cheeses were locally produced instead of imported.

This book places no region's cheeses on a pedestal; all dairying regions of the world are equally capable of making great cheese. Judging the intoxicating amasi of southern Africa against the richest-tasting alpine cheeses of Europe is a pointless exercise—both are expressions of cultural practices in particular parts of the world that create products that are the most delicious and appreciable of their milieux. And either could be created in any other region with equal results, given a respect for original cheesemaking

This book takes a defiant stance against the industrialization of dairying and aims to reclaim the most ecological form of agriculture: natural cheesemaking. PHOTO COURTESY OF QUEIJARIA BELAFAZENDA

INTRODUCTION • 5

practices—for anyone anywhere can re-ferment milk in a gourd to make amasi, and anyone anywhere can cook their Comté curds in a copper kettle. It should not be considered cultural appropriation to re-create another culture's cheese, but rather cultural respect and appreciation; a realization and celebration of the infinite potential of milk.

Milk Into Cheese is intended to build the capacity of local food systems, to help cheesemakers and dairy farmers everywhere produce the best cheese and put an end to its costly and environmentally damaging international import. The reigning European cheese hegemony is an expression of cultural imperialism and a relic of colonization; it diminishes the cheeses produced in other parts of the world, which has led to a vast industrialization and export-driven development of agriculture in those most famous cheesemaking regions—and has by and large degraded the quality of their once esteemed production. To state that cheese is better when produced in Europe is misguided and damaging to local and small-scale cheese producers everywhere, who are capable of making better cheese than is most often made today in a cheese's place of origin!

This book is not meant to make cheesemaking inaccessible to those who do not have access to raw milk; it is the domination of the milk processing industry and the industrialization of agriculture (and its consequent unchecked urbanization) that have made good milk inaccessible in the first place. Like apples plucked from a tree, raw milk fresh from the udder should be available to all. This sort of cheesemaking is remarkably simple to practice—it truly makes itself, so long as the right ingredients can be sourced; if, however, they cannot, I recognize that this may be the most impossible cheesemaking book that has ever been written.

This is not a book about raw milk or raw milk cheese, but about the transformation of *milk*, a most healthful gift made by mothers, into its *cheese*, the most healthful food. I have a concern with the use of the terms *raw milk* and *raw milk cheese*, for the well-practiced process of transforming milk into cheese constitutes an effective sort of pasteurization, enabling a more effective and long-standing microbial control than heat treatment alone. (And if pressure treatment can constitute a legitimate form of pasteurization, as it is in Australia, for example, why can't fermentation?) I hesitate to use the term *raw milk* in my language, due to the negative connotations, as I'll discuss in chapter 2.

That said, a natural cheese can be made with pasteurization, if required by local authorities, for what is really unnatural about the cooking of milk? Many natural cheese makes involve a cooking step, including the boiling of milk for yogurt, the hot-water stretching of pasta filata styles, and the high-temperature curd treatments of grana cheeses, which are nearly equivalent to the temperature of thermization, a common preventive microbiological treatment for the industrially produced raw milk cheeses of Europe. What's unnatural about pasteurization is the practices it engenders—in essence the industrialization and consolidation of our dairying, which would never have happened without it.

This book's intent is to show that the practices of natural cheesemaking make any milk safe to consume. Pasteurization should not be seen as the only pathway toward safely produced dairy. There is another way, and that is small-scale, well-informed, and well-practiced natural cheesemaking!

This book aims to rebuild our capacity for making cheese in a safer way, according to the enlightened culinary practices of our ancestors. When the food system is industrialized, there is an increased food safety risk. In our modern times that risk has been mitigated through the use of pasteurization. But our cheesemaking predecessors knew how to properly produce this most important food, and fermented dairy products would not have become so prevalent in so many societies were they a danger to eat! Food sovereignty, food security, and food safety are intimately interrelated; controlling the means of our food production and deregulation (the simple methods of producing natural cheese should be understood as the most effective and easily implemented safeguards against unwanted microbes) are essential for robust, resilient, and healthful food systems that can feed all those in need.

How to Use This Book

And so, to further the cause and improve the practice of natural cheesemaking, I offer the world this book. But before we begin, I'd like to outline how to use it.

This is not a recipe book, but a philosophy book. And so I recommend you don't just jump to the cheesemaking techniques, for there is a more fundamental understanding that should first be imparted.

It's very important that you first build your assurances in milk and its microbes and their inherent capacity to make good cheese, that you expand your understanding of the elements at play, and that you strengthen your grasp of the methods before you attempt your first batch. So much of this philosophy is built on faith and feel, and so I believe you should first acknowledge the integrity of traditional methods of cheesemaking, trust the nature of your ingredients, and build your understanding before you attempt the techniques.

I therefore encourage you to first read the introductory chapters, especially "Natural Cheese," "Milk," "Culture," and "Rennet," before making the cheeses of this book; for this is an approach to the subject that can require a deeper familiarity, and not just the following of a recipe. Having an early failure with the methods, most likely due to a simple misstep or misunderstanding, can cause you to lose faith with the philosophy, which can cause your natural cheeses to continue to fail.

However, you can also learn another way: by doing. The cheeses have been put in a certain order in the final chapters to build your understanding from the ground up. Following the techniques one by one teaches about cheesemaking from a practical angle, showing how each new cheese is elaborated and what makes it unique from the previous ones explored. Each cheese tells a story about milk and how it responds to the way we handle it. Often what distinguishes one cheese from its kin is just one slight difference in how we handle it, either during its make or during its affinage.

Exploring all of these styles can help you make each cheese better. And though I don't necessarily advocate for making each and every style in the book, making them all will give you a perspective that can help you fully understand how milk evolves into each and every cheese, and how each particular one requires a series of often careful steps to be made just right.

Techniques in this book can be exceedingly challenging or remarkably simple. Some of them are more complicated and involved, require hard-to-find tools like copper vats, and can demand the greatest endurance and hardest work of any food transformations within the entire realm of gastronomy. Many require acquiring one of the most restricted and difficult-to-obtain of all ingredients: milk still warm from the udder. Yet other techniques are among the most easily realized transformations in food, if you can get the ingredients—simply leaving milk at room temperature with a bit of backslop to make clabber, for instance.

Some cheesemaking techniques I share on these pages are very strict; others are quite laissez-faire, reflecting the fascinating diversity of cheeses. There's a cheese in these pages for every persuasion and every character. (Some, like aged lactic cheeses, are good for folks with OCD; others, like alpine cheeses, are best for folks with anger management issues!) But in general, the harder the cheese, the more challenging and demanding the cheese make. Nevertheless, don't be dismayed by the complexity and time span of some of the techniques. Just make cheese—for with your help, cheese really does make itself!

The Chapters

Now on to the chapters of the book, fifteen in number, plus the appendices. The first eleven define the theory and practice of natural cheesemaking, while the final four are devoted to eighty or so cheese makes. The appendices explore detailed techniques for keeping natural starters and other odd bits I couldn't find a place for in the main sections of the book.

Chapter 1, "Natural Cheese," sets the philosophical tone for the book, helping the reader understand the medium from a more natural perspective. The chapter explores the biological, microbiological, cultural, and agricultural origins of this incredible food.

The next set of chapters gives background information on the ingredients of primary importance to cheese: milk, culture, and rennet. **Chapter 2, "Milk,"** explores this most fundamental fluid from which all cheeses and dairy ferments, indeed even our lives, evolve; and helps us to understand its many nuances, and why it can evolve in so many diverse directions. **Chapter 3, "Culture,"** looks at the fermentation philosophy that defines natural cheesemaking, and how to best cultivate the exceptional communities of microorganisms that make cheese evolve most naturally. **Chapter 4, "Rennet,"** focuses on the enzymatic aspect of milk's transformation, how and why rennet works its magic on milk, and the various ways to prepare the ingredient that is key to milk's physical transformation into cheese.

The next suite of chapters explores how the primary ingredients come together to become cheese, with chapters exploring the various aspects of cheese's makes and the tools used by cheesemakers to transform their milk. **Chapter 5, "The Make,"** focuses on the various aspects of the diverse cheese makes and how each act of the cheesemaker helps bring about the ultimate transformation. **Chapter 6, "The Tools,"** explores how the tools used define the cheese's make and how a cheesemaker can best use their tools to their cheese's advantage. **Chapter 7, "The Salt,"** addresses various aspects of salt and salting cheeses—how this integral ingredient preserves cheese and enables the art of affinage.

The following three chapters explore the various aspects of preserving cheese after the make: the cheese cave, cheeses' ripening microorganisms, and affinage techniques. **Chapter 8, "The Cave,"** outlines the requirements for effective cheese-ripening spaces and discusses how to create the right circumstances for cheese's best degradations. **Chapter 9, "Affinage Treatments,"** explores the various options for handling cheeses as they age. And **Chapter 10, "Ripening Ecologies,"** looks at how cheeses are transformed by microorganisms as they ripen—essentially exploring the second fermentation that happens in cheeses as they age.

Chapter 11, "The Families of Cheese," outlines the various relations between the styles of cheese, breaking them down into categories defined by differences and similarities in their makes, helping set the stage for the individual techniques.

Finally, you'll find the techniques, organized by the season in which each style of cheese is best made. **Chapter 12, "Spring,"** features fresh fermented milks and small cheeses with white rinds, those aged cheeses that ripen most quickly. **Chapter 13, "Summer,"** explores the cheeses made from milk in abundance, which can be preserved through the year, and those made in warmer weather around the world. **Chapter 14, "Fall,"** shares cheeses ripened with *Penicillium roqueforti* and other styles best made in cooler weather with the fattiest milk; and **Chapter 15, "Winter,"** is filled with cheeses involving cooking milk, and with washed-rind styles.

The **appendices** include various practicalities of keeping the different cheesemaking cultures and their close kin. I cannot include enough information about these starters in the book, for successful natural cheesemaking most depends on successful fermentation, and so my advice has spilled out into these extra chapters: **A**, keeping clabber culture; **B**, a whey starter; **C**, kefir culture; and **D**, how to use wooden vats; as well as **E**, advice on thermophilic starters. Appendix **F** explores commercial considerations for practicing these natural methods. Appendix **G** is dedicated to the care and keeping of a sourdough mother, as this culture relates strongly to cheese's natural side. Appendix **H** explores the keeping of a natural yeast starter. Appendix **I** is dedicated to growing the fungus *Mucor*, as a source of ripening culture for certain cheeses, as well as a source of microbial rennet. And appendix **J** investigates the flaws that can happen in cheesemaking, and how to solve them in a natural way.

Notes on the Techniques and Ingredients

I must admit, I am not a follower of recipes, so writing techniques for eighty cheeses has indeed been a painstaking task. I've done my best to test every technique in the book and assure its fidelity, and have also enlisted the help of other cheesemakers to test many of the techniques.

I've also done my best to make these methods as accurate and authentic to the original styles of each cheese as possible. In many cases these styles have been re-imagined, for cheeses are no longer produced today with natural techniques.

That being said, there is a remarkable amount of flexibility within the many methodologies; each cheese, after all, is a relative of all the others, and often changing just one aspect of the make changes the cheese entirely. And to rigidly define a certain style falls in line with the specialization and commodification of dairying in the twentieth century. There was once significantly greater diversity within individual styles of cheese, but many developments (including PDOs) have brought many cheeses that were previously only vaguely defined into standardized production. The diverse ways that milk can be handled, even within a certain style, contribute to the exceptional diversity of cheeses in the world—so please don't feel beholden to the techniques, and develop them according to your experiences and circumstances. It is these details that will make your cheeses your own!

Cheesemaking to Scale

One of the more important aspects of this adherence to tradition is that the amount of milk used in a batch matters to the evolution of each cheese. It should be understood that, in general, softer cheeses evolve best with smaller quantities of milk; harder cheeses need more milk to turn out right. I highly recommend keeping to these relative scales when making your cheeses, and not attempting their makes if you haven't nearly the right amount of milk. There are important reasons, for example, that I recommend using 100 L milk to make a single clothbound cheddar or 200 L for an Emmentaler. In the making of these cheeses, the massiveness of the cheese is a part of its proper evolution, and making it smaller will result in a cheese that doesn't evolve right at all.

I've put the techniques into different volume classes based on how much fresh milk is recommended to make each well, which you can also imagine as how many goats or cows are needed for each technique. These are the different quantities:

250 ml		1 cup milk
1 L	1 bottle of milk	1 quart
4 L	about 2 goats' worth of milk from 1 milking	1 gallon
10 L	about 1 cow's worth of milk from 1 milking	2.5 gallons
20 L	about 2 cows' worth or 10 goats' worth	5 gallons
40 L	about 4 cows' worth of milk	10 gallons
100 L	about 10 cows' worth	25 gallons
200 L	about 20 cows' worth	50 gallons

So if you're only milking one cow, I don't recommend making those cheeses in which I call for ten cows' worth, because the cheeses won't evolve well at the different scale. There is some wiggle room—you can make a cheddar with 20 L cow's milk, for example—but I'd recommend that you make a Lancashire with that quantity of milk, a similar recipe that yields a similar cheese, but that evolves much better, and with much less work, with two cows' worth of milk. Best, I'd say, is not to go up or down too many volume classes for any particular technique.

And please, don't save up small quantities of milk in the fridge to make a larger batch at the end of the week. The cheese will be greatly degraded by the milk's refrigeration. Consider instead pooling fresh milk with your neighbors to make a more significant wheel—cheesemaking is often a community endeavor all around the world!

As for the Milk

The milk for all cheesemaking techniques should preferably be used still warm from the udder. If I could distill a single definition for milk that makes the best cheese, this would be it.

Milk is degraded by any processing other than cheesemaking. To make the best cheese, any milk treatment, especially its refrigeration, should be avoided. For a farmer-cheesemaker I therefore recommend twice-daily cheesemaking, always with the freshest milk, preferably still warm from the udder. You will make a better cheese (flavor-wise, nutrition-wise, efficiency-wise, and food-safety-wise) if you make smaller batches of cheese more often, using the freshest milk with each batch. Some styles like lactic cheeses and barrel butter are effortless and make regular cheesemaking easier—easier even than putting that milk into the fridge!

Chilled milk, or mixed milk out of a bulk tank, can be used (though I personally choose not to), but doing so will cause problems in the evolution of a cheese, resulting in off flavors, unwanted microbiologies, and possible risks to human health. For reasons that I'll explain in chapter 2, the longer it's refrigerated, the worse milk and its cheese gets. Milk refrigerated longer than 12 hours should probably be pasteurized, a technology that's best avoided. I'm not going to get into the details about it in this book, as it's entirely unnecessary when working with the freshest milk.

Milk, still warm from the udder, is as good as it gets, and though its sale off the farm is largely illegal (if raw milk can be sold it is generally mandated to be chilled to 4°C (39°F) immediately after milking, damaging that milk), it is worth seeking out for its many benefits. Refrigeration is such a norm on dairy farms today that milking systems immediately chill milk on its exit from the animals, providing no option for a farmer to take out warm milk. Producers typically have to work around their built-in systems to do so, often no easy task. Find warm milk if you can, by bypassing the industrial system that largely destroys the best qualities of our milk and making connections to get it directly from your local farmers. Be sure to use the milk right away, or within 2 to 3 hours of milking, for best effect.

Any unprocessed milk from any style of farming will work for these techniques: from pastured cows to cows inside on hay and even those fed baleage, silage, or total mixed ration (TMR). Though I hold the practice of feeding fermented corn silage to be among the most destructive inroads of industrialization into dairying, corn-silage-fed cow's milk still makes remarkably delicious cheese when transformed in a natural way. When working with it I have been surprised to find it respond and ferment the same way as other, more naturally produced milks. It seems to me to be a misconception that milk made with fermented feeds causes cheeses to bloat; the problem is more likely a systemic one. Use what milk you can get access to and uphold your own ideals. Fresh milk direct from an animal is still milk and will work for all of these techniques.

And for those that aren't dairy farmers, or cannot source the freshest milk, pasteurized and unhomogenized milk will work well enough for most techniques in these pages. However, you should avoid High Temperature Short Time (HTST) pasteurized milk (72°C [162°F] for 15 seconds) and ultra-high-temperature-treated (UHT) milk (110°C [230°F] for 10 seconds). Bulk pasteurized milk (62°C [144°F] for 30 minutes) does respond to most cheesemaking methods, but the curds are less responsive and often need more delicate handling and a prolonged stirring to achieve the same result as a raw milk version.

Some techniques work specifically for cow's milk and others specifically for goat. I do recommend in general that you stick to those particular species for those particular makes, switching things up only by making appropriate modifications. Every species' milk works slightly differently than every other, and in the milk chapter we'll explore the different tendencies of each milk.

When It Comes to Culture

Please keep your culture natural! I hold natural fermentation to be the strongest pillar of natural cheesemaking. Good cheese can still be made with low-temperature-pasteurized (62°C [144°F]) milk so long as it's as fresh as possible, and genetically modified rennets still form curd (I can't deny that they work). But natural starters are not easily replaced by freeze-dried starters, which alter the ecology of cheese and do not allow the food to be so effortlessly and fantastically fermented. I cannot with any good conscience recommend the use of industrial starter cultures or ripening adjuncts; makes are significantly changed by their frailties and shortcomings. The convenience of freeze-dried starters comes with a large cost—they are the most complicating and misconceived aspect of modern cheesemaking practice.

The techniques to cultivate starters include naturally re-fermenting milk as a clabber culture; using a kefir culture; saving whey from batch to batch; and making cheese daily in a natural material such as wood. All of these starters require regular care and feeding to remain in good order, for the activity of these cultures rises and falls much like a sourdough starter. The success of a natural cheese make depends on their health. These starters will be explored in chapter 4, as well as in specific appendices at the end of the book. Understanding their evolutions is essential to having success in natural cheesemaking.

As for Rennet

The milk-coagulating enzyme known as rennet is probably the most controversial and misunderstood ingredient in our food system. Now, I don't expect everyone to sacrifice a calf to source this enzyme from its stomach in what I believe is the most natural and ethical way, but I also do not recommend just buying any seemingly innocent brand of "vegetarian" rennet (often genetically modified) off a cheesemaking supply website. Every rennet works to coagulate milk, but each works slightly differently from every other, and each comes from a different source whose origins I believe should be considered (and advertised) when making cheese.

Unfortunately, today the vast majority of cheesemakers in many regions of the world (especially the United States and Canada) use genetically modified rennets. But few advertise their origin. Misunderstanding or ignorance on behalf of the cheesemaker is one reason; unfortunately, it can also be the result of the mistaken and insensitive belief that the public doesn't know what's best for them. Manufacturers, of course, contribute to misunderstanding by obfuscating that their rennets are genetically modified.

Though I personally hold the slaughtering of young animals not needed on dairies and the use of their stomachs for rennet to be the highest ideal in natural cheesemaking, for many, such a practice is completely impractical. And for many in the world, the commercial

Natural cheesemaking techniques can be, indeed should be, incorporated into commercial cheesemaking operations.
PHOTO COURTESY OF DOETIE'S GEITEN

rennets I would recommend simply aren't available. So go ahead and use the rennet you can find in your locale.

In my previous book I offered advice on rennet brands to purchase, but I've chosen not to do so in this work—in part because the manufacturer of the natural rennet that I recommended, Hundsbichler in Austria, which formerly produced the Walcoren tablets, was bought out by the world's largest manufacturer of enzymes, and the provider of enzymes and microbes used in at least 50 percent of the world's cheeses, the Christian Hansen company (Chr. Hansen) of Denmark. As well, in light of the different availabilities of rennets to different makers around the world, and the constantly shifting and consolidating nature of the industry, I've simply advised that you use the recommended rennet dosage for a standard amount of milk for the rennet that you've purchased (commercial rennet should come with operating instructions and a recommended dosage). You can then do the math to find out how much rennet is needed for the amount of milk you're working with.

Almost all the techniques in the book (aside from lactic goat cheeses) call for the same standard amount of rennet, which fully curdles milk at 35°C (95°F) in about 45 minutes. If you can't find that information on the package, contact the manufacturer, and they can give you advice, or you can test the strength of the rennet on a small quantity of milk kept warm at 35°C to determine its dosage for a larger batch of cheese. Lactic goat cheeses use a quarter of the standard dose.

In chapter 4 I have provided more directions on how to prepare and use rennet stomachs if you choose to do so. In some regions these stomachs can still be purchased from cheesemaking supply shops or found in local markets, but in general they're hard to find and their production is restricted or even taboo. Typically, the only way to procure one is to take it from an animal yourself and transform it in-house, which can violate the regulations of commercial dairying. (Though there's nothing inherently illegal about using animals' stomachs to curdle milk, ingredients used in modern dairies need to come from certifiable sources, not just from a backyard-slaughtered buckling.) However, anywhere in the world that animals are being milked, making animal rennet is a realizable possibility. And so to keep with the theme of universality, this is the technique I will focus most attention on in the book.

The Units

Though I do have a fondness for imperial units (a gallon of milk makes about a pound of cheese, how perfect!), I've chosen to use metric units (aside from temperature) in my techniques to simplify the text, leaving more room on a page for more important information, and avoiding confusion that might result from having two conflicting numbers. I understand that many readers will not be familiar with liters and kilograms, but given the universal nature of the lessons in this book (and the tendency of even commercial cheesemakers in the imperial kingdom of the United States to use metric), I thought it would be more in line with the philosophy to unify the units.

We should also realize that a cheese doesn't need measured degrees to evolve right, and that cheesemakers originally made cheese without the aid of Celsius/Fahrenheit or milliliters/ounces, simply being comfortable and familiar with the capacity of their rennet, culture, and salt to do the work they do. Rest assured, I don't measure anything in my cheesemaking (aside from the amount of milk, and the corresponding amount of rennet needed, and the pH, but only with more traditional tests). And I've done my best to avoid unnecessary scrutiny in measuring, by providing alternative ways of measuring temperature, acidity, and humidity that are more intuitive and that measure just as effectively (if not more effectively) as more technologically advanced instruments. See chapter 5 for more detailed explorations on natural measurements.

Enjoy!

What else can I say without digging into the meat of the matter—just jump in and enjoy making the most of your milk! You can look at cheesemaking not as a chore, but rather as a practice to be enjoyed and savored as it is performed, a meditation on the transformation of your milk into its most beautiful, enlightened form. Consider cheesemaking and dairy fermentation a regular ritual to look forward to, acts that feed your soul as much as they feed you, your family, and your community.

PART I
THE ELEMENTS

Cheese is biological: It is a phenomenon that originates in the stomachs and digestive tracts of young milk-drinking mammals. PHOTO COURTESY OF QUEIJARIA BELAFAZENDA

CHAPTER 1

Natural Cheese

I'll start this book off officially with a question: What is cheese?

Though cheese can be many different things to many different people—Cheese is work! Cheese is money! Cheese is the most flavorful food on the planet!—for me, cheese is milk's *destiny*. From a natural cheesemaking perspective, there are four interconnected philosophies that make this imperative: (1) Cheese is a food that reflects the culture and practices that make it. (2) Cheese is a most significant product of our agriculture. (3) Cheese is curdled milk, evolving biologically with the aid of rennet. And (4) cheese is transformed from milk by an inherent fermentation. These fundamentals can be elaborated into four interrelated identities of cheese: Cheese is cultural, cheese is agricultural, cheese is biological, and cheese is microbiological.

Milk is an incomplete food; transforming it into cheese completes it. These four fundamental identities of cheese help us recognize this. They tell the truest story of this most important food, cast light on its intricacies, and enable a beginner or established cheesemaker to best grasp the most traditional and natural methods of its production.

In all the chapters of this book, we will return again and again to these four fundamentals, for to have success with a natural cheesemaking, we must understand the purpose of this most incredible food.

Cheese Is Biological

Cheesemaking is a biological process that is an inextricable and exceptionally important part of a most significant life force: the feeding of young mammals by their mother's milk. This is particularly apparent in the manner in which milk curdles, for this is the biological precedent of all cheeses, everywhere.

Milk coagulates into cheese naturally. It forms curds, a separation of its solid fat and protein from its whey, with the assistance of enzymes that cause it to rearrange itself into a semi-solid gel. And milk curdles in this way for a very good reason.

When a young mammal, be it a calf, kid, or lamb, drinks their mother's milk, they transform that milk into cheese within the belly to get a better handle on the nutrition it provides. Within the lining of the young animal's stomach, digestive enzymes and acids are produced that naturally curdle the mother's milk, turning it from a liquid to a solid and giving it a more tenable form. Thus the stomach is better able to hold on to and churn the now-solid milk and transform it before passing it along through the intestines, so the young animal is able to make the most of this milk.

But it's not just young cows, goats, and sheep that transform their mothers' milk into cheese. Most likely every young mammal, including our own infants, curdles their milk into cheese. Evidence of cheesemaking that happens inside our young can be seen when they spit up on our shoulder: The curdly bits are a natural human cheese that coagulated in the little cheesemakers' stomachs! We've all made cheese before; grasping this concept of the craft should therefore be intuitive. Indeed, if we take inspiration from the natural circumstances in which milk transforms into cheese, we'll have much greater success, and make a much more delicious wheel.

So if we wish to practice a successful natural cheesemaking, it's best to consider the original biological instance in which milk transforms into cheese: when a calf drinks its mother's milk. When a young calf suckles, they get rather excited about the upcoming meal, about their connection with their mother. With this anticipation, the alignment of their gastrointestinal tract (specifically the esophageal groove) shifts, and the milk goes right down the throat and directly into the fourth stomach (known as the abomasum) in which the cheesemaking happens. In so doing, the milk bypasses the first three stomachs (which are in essence fermentation chambers for digesting grasses that will be important when the animal is older) and instead goes directly from the udder and into the cheesemaking vat still warm from the mother.

When I make cheese, I often consider the cheesemaking pot to be that young calf's stomach, and I recommend you do the same. To simplify and improve the make, use milk that's as fresh as possible, ideally still warm from the udder. Add to the milk some starter culture, a preparation that's carried forward to ferment milk's sugars to lactic acid and that aids its coagulation. That culture is a stand-in for the acid produced by the calf's stomach and provides the microbes that help the calf digest the otherwise indigestible sugars in their mother's milk. Then add a preparation of rennet derived from the stomach of a young calf, which contributes the biological enzymes that make milk curdle.

We humans didn't invent cheese; we simply stumbled across the biological phenomenon and adopted it, even perfected it, into our cultural practices through the development of dairying technologies. All of the peculiar behaviors of milk and cheese can be explained through milk and cheese's biological connection. For example, the curdling of milk when it ferments to a pH of 4.5 is related to its cheesemaking capacity. Infants' stomachs (both cow and human) produce acids that bring their semi-firm cheesy contents toward this same ultimate pH; at this point the calcium dissolves from the cheese's proteins, softening its texture (into a lactic-cheese-like paste) and enabling the creamy-textured curd to be passed along into the intestines, where it ultimately nourishes the infant. This dissolving of casein's calcium is the same phenomenon that causes milk to curdle when it ferments.

Because of this biological origin of cheese, its history is likely pre-agricultural (despite what I say in the next fundamental philosophy). Ancient peoples, who appreciate nearly every part of an animal, would have relished the nutritious and delicious cheese found inside the stomachs of young milk-fed mammals, and probably even fermented the milk that flowed from a butchered animal's udder. It's likely that we've been consuming this important food for much longer than the beginning of agriculture 10,000 years ago.

Cheese Is Agricultural

Cheese is an extension of its agriculture, and cheese is best made from milk if it happens in a continuum of that milk's evolution from the animals and the land. Separating cheesemaking from its agriculture has numerous negative consequences. Cheese can be produced as an industrial process instead of an agricultural one, but the manner in which it is made taints the cheese, robs it of its best nutrition, and makes a poorer-quality processed food with a much dirtier environmental footprint. Technologies like refrigeration, pasteurization, homogenization, and milking robots, as well as the ingredients of modern cheesemaking—its freeze-dried cultures and transgenic rennet—all are efforts to extricate cheesemaking from its agriculture in order to scale up and industrialize the process. All of these technologies generally benefit neither the farmer nor the cheesemaker but rather the agricultural industrialists; and all cheeses made with them reflect the changing landscape of farming that has by and large degraded our soils, emptied our rural communities, and decayed the roots of our culture.

In traditional agricultural systems there is no divide between the milking animals and their cheese—cows are put out to pasture, eat fresh grass, ruminate, and

Cheese is agricultural: It is a natural extension of pastoralist practices. PHOTO COURTESY OF SINNOS FORMAGGI

produce milk. That milk is curdled, still warm from the udder, with stomachs derived from the milking animals' calves, then fermented with culture saved from the previous day's cheese make that originally derives from animal origins. The cheesemaker then cares for the cheeses with the same love and attention they reserve for their animals, and sells their cheeses to eat in place of their meat. With industrial or conventional cheesemaking today, a literal wall is required to be put up between animals and their cheese, with the dairy being shut off completely from the cheesemaking space. Only milk may pass between the two, certainly not the milk's microbes, and most definitely not the animal's enzymes.

Making cheese with milk is a choice made by farmers, to produce something better with their milk. Instead of raising calves on their mothers, the young are taken away and sacrificed, and elements of the calves (their enzymes) are then used to curdle their mothers' milk. In so doing, farmer-cheesemakers procure more nourishment from that milk—making cheese with it produces almost twice as much food for us as does raising young animals on that same milk and then eating those animals. With this extra effort—the twice daily dairying and cheesemaking—we're able to procure more nourishment from our land. Producing food this way may be the most ecological and ethical form of animal agriculture, indeed of any agriculture on earth (more on this in chapter 2). To make cheese is a most exceptional agricultural effort, akin to planting seeds and growing a garden—the most fundamental and important of our human activities.

Milk is a fluid in flux, in transition between the cow and calf, doe and kid, ewe and lamb, that turns to cheese in the young animal's body. Historically, the dairying wasn't done until the milk was properly transformed into its cheese. Much as many crops need daily water and livestock need daily care, cheesemaking was and should again be practiced daily. Just as a farmer would never consider neglecting their milking goats for a day, so cheesemaking should be considered a daily ritual and responsibility, a final agricultural act of the morning or evening milking's routine.

Even aged cheeses are tended to in underground spaces that encourage carefully cultivated rind microbiologies including many bacteria and fungi that originally evolved from the same soil that grows the grass that supports the animals that give the milk. The cheeses become living, breathing entities that need air to thrive, that support ecologies of their own, and that

NATURAL CHEESE • 17

reflect the nature of their handling (rind washing can be considered equivalent to weeding a garden!). And those cheeses, if neglected, ultimately return to the soil, which of course was once their cradle.

Naturally, if you change aspects of the agriculture, you change aspects of the cheese.

Keeping historical breeds of animals in traditional shepherding systems that move over the landscape from day to day improves the animals' health and well-being, and, when produced daily on the farm, makes a most sought-after cheese with extraordinary harmonies. There are added benefits for the soil, preserving biodiversity and producing food without contributing to climate change. In fact such a cheese preserves grassland ecologies and their abilities to build soil from otherwise detrimental greenhouse gases. These traditional approaches to livestock husbandry stand in stark contrast with industrial methods where confined high-yielding Holsteins are fed a total mixed ration of monocultured genetically modified corn silage. Their milk is then chilled for days and trucked to central processing facilities to produce a cheese that reflects our modern agriculture's misguided attempt to control nature and to produce as cheap a food as possible with the fewest farmers standing in their way. Such a cheese tastes of and contributes to a broken food system.

Even if they aren't on the farm where the milk is produced, cheesemakers should approach cheesemaking as if they were, creating links with farmers to source milk as fresh as possible, preferably still warm from the animals, and caring for their cultures with a daily feeding as if they were bottle-fed lambs. If a naturally produced milk is processed with the tools of industrial agriculture, it will suffer from the same detriments to its taste as any industrially produced cheese. No milk is immune to the many mistreatments that it can receive once it leaves the udder.

Cheese is not a fancy food, but an agrarian one. It is a way of life, defined and refined by farmers and pastoralists, that has traveled around the world along with other primary agricultural techniques, domesticated animals, grain crops, and diverse vegetables. It was likely part of the original Neolithic package—a suite of tools and techniques that enabled agricultural civilizations to take hold all around the world, providing the foundations for our diverse cultures to evolve their own unique ways. Cheeses and fermented dairy are fundamental foods that ought to be considered first and foremost when feeding ourselves, and not just something to eat on significant occasions. For every day is a significant occasion when we nourish ourselves with the best food our land provides.

Cheese Is Microbiological

What would cheese be without culture? Simply sweet milk, coagulated. But because of certain practices carried out by the cheesemaker, extraordinary microbial communities are carried forward from one batch of cheese to the next that cause their best evolution and preservation. These fermentations cause cascades of transformation that reach a crescendo of flavors, aromas, and textures, and ultimately define the character of very nearly every cheese, both fresh and aged.

Cheese, like chocolate, wine, beer, cider, and sourdough breads, holds a place among the world's most flavorful foods, yet all cheese evolves from a very simple flavor profile: mildly sweet and creamy milk. That milk contains multitudes of potential masterpieces, but its flavor out of the udder hardly hints at these possibilities. Only when it is broken down microbiologically is milk given the capacity to express itself fully. Cheeses are among the most delicious foods because our bodies recognize their exceptional capacity to nourish, thanks to fermentation.

It is fermentation, for the most part, that makes cheese so expressive. Fermentation breaks down the highly nutritious components of milk, allowing that milk to express itself in most extraordinary ways, and in ways that we respond to almost emotionally. Nearly all that we recognize flavor-wise in cheese comes from the many pathways of fermentation that cheesemakers create in handling their milk in so many different ways. Mildly sweet lactose is broken down by lactofermenting bacteria into tangy lactic acid, which in turn is metabolized by yeasts and fungi on the rind into expressive esters, ketones, and alcohols. Its fats are broken down or lipolyzed by microbes into more flavorful and nutritious fatty acids; the proteins are proteolyzed into the richest umami notes, indicating an equally rich and incomparable source of bioavailable nutrition. And it

Cheese is microbiological: Its production is made possible thanks to a most phenomenal form of fermentation. PHOTO COURTESY OF FORADORI ALIMENTARI

is microbes that by and large lipolyze, metabolize, and proteolyze all these components of milk into something much more dramatic and delicious, each different microbe transforming each different milky component in its own particular way.

An exceptional diversity of microorganisms enables cheeses to develop their particular textures, colors, and flavors. And though many hundreds or thousands of different microorganisms may be involved in milk's many transformations, natural cheesemakers can be in complete orchestral control of this microbial evolution by creating just the right circumstances during their makes and affinage for the right microorganisms to thrive—exactly those groups of microbes that allow a cheese to develop just the right character.

The different communities of microbes encouraged by the different conditions created by the cheesemaker give cheeses vastly different identities: Different guilds of rind-ripening bacteria, yeasts, and fungi can enable cheeses to develop beautiful blue-green veins, bright orange-pink rinds, gorgeous cream-colored wrinkles, or fluffy white crusts. Cheeses can be made to swell from the production of gas-producing bacteria within their paste; milk can even be turned mildly alcoholic from the growth of often unwanted, but sometimes sought-after gas-producing yeasts. These microbes work in partnership with one other and with us, growing in the right conditions that the cheesemaker helps to create, and protecting the cheese from unwanted microbiologies that can alter its tastes and textures, and impact its safety. The beneficial microbial communities encouraged by a cheesemaker limit many foodborne pathogens that could otherwise cause great harm.

And these protective microorganisms that thrive in milk do so for an important, biological reason—to protect the health and well-being of young animals that drink their mothers' milk! The same microbes that create cheese thrive in the digestive tract of young mammals and offer protection from pathogens. They help to transform milk's lactose into protective compounds in the intestine, the same transformations that enable milk to take its most fundamental first step into cheese in the vat. This microbiological aspect of cheesemaking is thus also interwoven with the biological, for the community of microbes in cheese reflects the inherent microbial community of mothers and their milk-drinking offspring.

Our attraction to cheese's many flavors is elemental to our biological nature. The aromatics of fatty acids, the umami of amino acids, and the tangs of lactic acid are expressions of cheese's exceptional capacity to nourish. We are drawn to this expressive food as our tongues recognize it to be a most nutrient-dense source of nutrition, rich in the elements our bodies most need, in their most available and assimilable forms. All of these diverse flavors evolve from the simplest sweet-tasting milk, as expressions of its capacity to provide the most significant nourishment from mother to child.

The incredible expression within cheese is representative of the partnership between milk and the many microbes that it is meant to harbor. Cheese is thus as much microbiological as it is biological, as much about fermentation as it is about farming.

Cheese Is Cultural

But this most delicious craft is also cultural. The evolution of a cheese reflects the cultures and customs of its traditional makers: the particulars of their day-to-day practice, the tools they use to transform milk, and the cultural context of their makes. And to recognize how a particular cheese is to be best made, it's important to understand how and why it's made the way it is.

Cheese is cultural: Its making is informed by a material culture, such as the use of copper kettles. PHOTO BY CHLOE GIRE

Cheese is the meaning of milk. It is misrepresentative even to describe cheese as preserved milk or a value-added product, for most often, in traditional dairying societies, milk is made to be made into cheese. Many cultures did not even consider milk to be a food until it was transformed, and milk could have hardly, historically, left the farm without fermenting or curdling of its own accord.

One significant philosophical difference between North American and European cheesemaking is that in Europe there is a cultural imperative to make cheese. In many regions cheese is simply what is done with milk. Whereas in North America (at least the United States and Canada), the reason for a farmer choosing to transform their milk into cheese is largely an economic choice to add value to their production (which of course is also a cultural decision that influences their cheese!). Many European cheesemakers recognize that milk doesn't have its true value until it's transformed into cheese, and they make their cheeses with warm milk according to their agricultural and cultural practices, often defined by their material culture. North American cheesemakers, however, typically follow a recipe for a cheese from afar, purchase the tools they think they need to make it from afar, and refrigerate their milk, to the cheese's detriment, until they're ready to transform it into the cheese they've chosen to make.

To create a cheese in its truest sense, we must consider the cultural conditions in which it's made. If we consider these origins in our make, we can create a cheese that is a truer version of the original. Realizing the how and why of every aspect of the make allows a cheese to take on its truest taste and form.

All cheeses originate from shepherds', cowherds', and goatherds' ever-present need to transform the milk of their animals into more significant, useful, and durable products. And it is the particular way farmers handle their milk that defines the evolution and flavor development of their cheeses. But the way the makers handle their milk during its transformation reflects the culture they are steeped in. All the diverse ways that milk is handled allow cheeses to take on an incredible diversity of flavors and forms.

Milk and its cheese are incredibly adaptable, amenable to the hundreds or even thousands of different ways they can be handled. The incredible diversity of human culture—our diverse ways of interacting with animals and their milk—has contributed to the incredible diversity of cheeses around the world.

These human cultures define our cheesemaking culture. All styles of cheese reflect the way in which their original makers related to their landscape, their climate, their agriculture, their economy, even their religions. South Asia, for instance, with its Hindu majority and reverence for the cow and calf, transforms its milk almost entirely without rennet, and the region's cheeses all reflect this difference in culture. And those mountain cheeses made in the Alps don't taste necessarily of alpine pastures where the animals graze, but of the whole mountain culture that makes those massive wheels of alpine cheese—the fresh warm milk, the backslopped whey, the calf stomachs, and the material culture that includes massive copper kettles used over fire.

But it is not just a regional culture that makes cheese evolve a certain way; it can be a personal culture, too. Korean kimchi makers often refer to the flavor found within their own products as *sson mhat* (손 맛), the flavor of the hand. This is not literally the taste of the maker's hand or its microbes, but rather the sum of all the personal acts that go into the making of their product; the acts that are different from their neighbors' and make their particular ferment evolve in a distinct way. And as with kimchi, one cheesemaker can handle or transform their milk in hundreds of different ways and make hundreds of different cheeses.

Everything done differently to a cheese changes it; even slight differences in milk and cheese's handling can encourage different microbiological breakdowns, different flavor expressions. Every different approach to making a cheese, every different material used to handle the curd, every different type and breed of animal makes a difference to the evolution of a cheese. And all of these differences represent human relation to land, to animals, and to other humans; they represent our culture. It's not just one of these aspects that makes a certain cheese taste a certain way, but a combination and a culmination of them all. This is where the exceptional diversity of flavors in cheese comes from: Cheese is not necessarily a taste of the land, but a taste of the hand.

The Story of a Cheese

To help with this four-way understanding of natural cheese, let's examine the make of one particular cheese within these contexts—a cheese made in southern Italy called caciocavallo, translated into English as "horse cheese" (though it's not at all made with horse milk). This peculiar name reflects the cheese's cultural heritage, which created and continues to create the unique conditions that allow the cheese's unique characteristics to evolve.

Caciocavallo is a most beautiful spun-curd cheese, shaped like a baby or a hanging gourd. Its flesh varies with age and origin from gold to white. It has a thin waxy skin that peels like a grape and a taste as rich and flavorful as any aged alpine cheese, combined with the complexity of a freshly made mozzarella. To understand what it takes to create this cheese, let's consider the cultural context in which it was and (to a dwindling extent) still is made.

This exceptional cheese is a part of an important cultural tradition known as the seasonal transhumance, a practice still carried out in many parts of the world. The transhumance, from the Latin roots of "across the land"—*trans humus*—involves seasonal shepherds taking animals up from their valley farms into the mountains in the summer months to feast on the lush grasses that grow there, while preserving the valleys' hayfields for the preparation of winter forage. On the mountain, their milk is transformed daily (even twice daily) by the shepherds into caciocavallos, which are then aged in the cellars of their summer cheesemaking huts.

During the caciocavallo-making season, milk is taken still warm from the udders and made into cheese after every milking in the traditional wooden cheesemaking vat. The original makers would not have refrigerated their milk, but worked instead with the inherent warmth of the milk from their animals. The wood preserves this necessary warmth and helps the cheese evolve best. The twice-daily milking therefore calls for a twice-daily cheesemaking, and doing the same will create a truer version of the cheese, one not affected by rancidification and psychrotrophic (capable of growth in cold environments) microbes that result from milk's refrigeration.

The milk used for caciocavallo is often a mixture of milks, some cow's, some sheep's, and often some goat's, reflecting the traditional agricultural practice of mixed-animal husbandry in the region, a more balanced ecological agriculture. But that's not the only reason for the mix of milks. Making the cheese with just cow's milk results in a cheese that's missing a certain nuance, for the aromatics of the caciocavallo are especially expressive when sheep's and goat's milk are included. However, the cheese cannot be made exclusively with sheep's and/or goat's milk, as these two milks lack the capacity to respond to the pasta filata method that births the caciocavallo. And so the cheese *must* be made with a mixture of milks to be made right!

Culture is added to the warm milk. This would have been second nature for traditional caciocavallo makers, who made their cheeses in the wooden vat known in southern Italy as a *Tina*. The unwashed Tina carries culture forward in its slats from one day's cheesemaking to the next without fail in an effective fermentation. In the absence of wood, whey saved from the previous day's cheese, or a daily-fed clabber or kefir culture, could be used as a starter to add the ever-important microorganisms to the milk. But still, something would be missing from the cheese without the flavor from the chestnut wood, a strong but lightweight wood most often used to make the portable Tina.

Rennet is then mixed into the milk to form its curd. To create the right effect, consider adding the rennet of southern Italy, typically a lamb or goat rennet derived from a young suckling animal, and prepared in a way that values the cheese found in the young animal's stomach as described in chapter 4. The inclusion of this first cheese in the rennet results in an extra enzymatic effect, a lipase-rich flavor typical of many Southern European cheeses that use this particular ingredient, originating from the very same biological origins as the milk they use. A caciocavallo won't taste quite right with a more modern rennet, lacking the necessary diversity of enzymes!

Upon setting, the curd is cut with a traditional wooden spino (a branch cut from a tree) into small curds and stirred by hand to remove as much whey as possible, making a more solid, long-preserving cheese. The mass of curd would be brought together under its whey and left to ferment for half a day to develop its acidity. The right handling of the curd is essential for making a cheese of the firmness that enables

Caciocavallo, a traditional cheese of southern Italy, can be made in its best way anywhere, if its maker respects and re-creates the culture from which it evolves. PHOTO BY CHLOE GIRE

caciocavallo to take on its most beautiful form, its desirable texture, and its excellent preservability.

The fermentation of the curd allows cheesemakers to melt it. In what is a hallmark of the pasta filata cheesemaking process, cheeses like caciocavallo are brought with microbiological processes to just the right acidity to yield a curd that melts when heated into a dough (pasta) that can be spun (filata) into almost any shape. In the case of caciocavallo, the curd is melted with the steaming-hot whey left over from making ricotta (with ecological, energetic, and organoleptic improvements). The melted curd is kneaded and then shaped into a sort of gourd, with a body, a head, and a distinctive neck, a shape that reflects the cultural origins of this cheese.

For in the south of Italy, where the transhumance is as old as the hills, milking animals are escorted up the mountains by their shepherds on horseback, and the cheesemaker-shepherds spend the summer months tending their flocks and preparing their caciocavallos with their milk. And then, at the end of summer, the season's worth of cheeses are tied together, two by two, from their necks, and taken back down the mountain, slung over the backs of their horses.

So if you really want to make a caciocavallo taste just right, you'll also have to put it on the back of a horse and take it down a mountain! But of course, it's not just the horse that makes the caciocavallo; you'll have to re-create all the right conditions, each as essential as the next for being faithful to the cultural context. You'll have to use the right mixture of goat, sheep, and cow's milk; you'll have to use the milk fresh from the udder, still warm; you'll want to make the cheese in a Tina, or at least save culture from batch to batch; the milk must be curdled with a traditional lamb or kid rennet; the curd must be cut to just the right size with a spino, then fermented to the right degree of acidity. It is then melted with whey still hot from ricotta making, and then stretched in just the right way, into just the right shape; before being salted, and then aged. Every one of these steps together is what really makes that cheese on the horse a true caciocavallo.

If any of those traditional steps are omitted, the cheese will be missing something, it will not evolve right; but if all of those steps are followed, the cheese will taste as it should. For what makes caciocavallo taste the right way is not necessarily that it is made in southern Italy; it's that it is made according to certain cultural traditions that are practiced there—and perhaps originated even farther afield! The origin of caciocavallo is actually quite mysterious; it may originally be from Greece, Persia, or anywhere in between. This cosmopolitan cheesemaking tradition can be re-created with fidelity and celebrated anywhere in the world. Indeed, this can be the case for each and every cheese.

Many would argue that the particular flavor of a particular cheese is inextricably linked with the place of its origin. That the particular breeds, microbes, plant profiles, climate, and soil are what make cheeses taste their particular way. I disagree with this concept of terroir, preferring instead a variation on the theme—a cultural terroir. For it should be recognized that cows are not of a place; they've traveled around the world since they were domesticated in the Neolithic. Nor are the grasses in a pasture that support them—most pasture plants and grasses are Central Asian in origin as well, having hitched a ride with cows around the continents with the spread of agriculture. And the microbial communities of these creatures came along for the ride. Our milking animals, their microbes, and their feeds represent a sort of cultural "backslop" (more on that term in chapter 3) that creates the circumstances behind the evolution of every cheese.

I have also had a falling-out with the concept of the PDO (or AOC or DOP), Protected Denomination of Origin. I believe that milk is capable of becoming any cheese anywhere, and I do not believe that any particular place should have a monopoly on any particular cheese's production or control over its name. The name of a cheese's place of origin is often its common identity, and a cheese's identity is created not by making it in a certain place but by making it in a certain way. We are often forced, outside a cheese's place of origin, to adopt meaningless names for what we make; but imagine if we couldn't call a crescent-shaped laminated pastry a croissant outside of France, a fermented sprouted barley drink a beer outside of Belgium, or a nixtamalized flattened corn bread a tortilla outside of Centroamerica! I believe we should be reclaiming the names of our cheeses, despite international trade laws that dispute this. Such laws do not typically benefit the small traditional producers of these cheeses but instead larger commercial producers who have formed marketing organizations to protect and promote their products (that's all that PDOs really are). Many PDOs have become cash cows for the industrial producers that now control the production standards for their benefit—infamously, the PDO for Stilton prevents the cheese from being made with raw milk—and do not typically promote natural techniques, often suppressing small-scale production. PDOs should be understood as a form of neocolonialism, and I'd personally prefer to see them abolished, instead creating some other recognition for cheese that's linked with the traditional, natural methods of its production and not with any particular place. Or perhaps, as the natural wine movement (no © there!) has collectively decided, it's best to avoid the standardization that comes along with certification.

The Universality of Cheese

Interestingly, all of these possibilities of cheese have a common ancestry, a historical relationship, that we'll see over and over again in the many lessons of this book. Cheeses thought of as distinctly British, for example, have close cousins in France, Italy, and Spain; many of them are Roman or Celtic in origin, and were probably produced by various cultures across the continents for millennia. For milk has the capacity to evolve its extraordinary transformations everywhere in the world, and all the flavor possibilities of diverse cheeses can be created from a single milk, or any other one, by following the many possible pathways of transformation.

All cheeses reflect the culture of their making and are an extension or continuum of their makers' relation to the landscape. The caciocavallo is a continuum of its particular Puglian culture—the Mediterranean seasonal transhumance; the crottin is a continuum of the small-scale French farmstead culture; Comté is a continuum of the traditional Central European alpine transhumance or *alpage* (much as an American vacuum-sealed and pasteurized cheddar is a continuum of its industrialized American culture). The cheese you make is a taste of your culture, the particular way that you relate to your cheese, and if you re-create the culture of caciocavallo, you can taste that in your cheese, too!

Like music, language, or wine, cheeses and dairy ferments are among the most appreciable and valuable elements of our cultures. And like any of the greatest of these human creations, traditionally made cheeses are timeless and transcend their place; they can be, should be, made or enjoyed by anyone anywhere.

Natural Cheese as Religion

Folks often debate whether cheese is an art or a science. I'd prefer to say cheese is a way of life, a worldview, or even a religion: a set of beliefs and (agricultural) practices upon which we cheesemakers live our daily lives. And there are many different belief systems within the world of cheese—many different sets of codes and practices that cheesemakers ritualize—as well as a belief in a higher authority (Pasteur) that defines the way we interact with this most important food. Is any one more right than any other? I wouldn't say so with regard to a discussion of the general question of religion, but I certainly believe that natural cheesemaking is the *right* way to perform this ancient and foundational practice!

Natural cheesemaking seems to solve all the problems of industrial cheese production; the same can't be said of the reverse, for natural cheesemaking is flawless, seemingly preordained. Cheesemakers have been indoctrinated for over a century into a set of fallacies: a near-complete misunderstanding of milk, a belief that sterility is desirable, and the idea that cheese must be separated from the animal.

With a different understanding, a different belief system, we can break down these barriers and open up cheesemaking to a higher power; not just a human-engineered system to be dominated by top-down control, but one governed by a sacred biology that is meant to be and can be practiced in a remarkably simple and straightforward way. But it takes faith in this different way of cheese to have success with this method. For if we do not believe, we do not practice; if we do not practice what ultimately needs to be practiced, achieving success with these ideas becomes unattainable.

Making cheese takes faith, either faith in the industrial system of food production or faith in the natural approach. It can take time to build a belief in the natural way, especially if you've lived your life and made your cheese (it's hard to easily separate these) according to a different set of principles and standards. You must practice the craft daily, observe and meditate upon it, and recognize the reasons why a cheese may not evolve as expected—it's a question of imperfect practice and not imperfect natural methods. (I can't emphasize enough how right natural cheesemaking techniques are!)

Fermented milks, butters, and cheeses (along with kefir grains) traveled the path of many expanding cultures all around the world, probably even propelled them. These original naturally made dairy products feature strongly in the symbolism of the organized religions of our agricultural era. For instance Zoroaster, the founder of Zoroastrianism, supposedly spent years meditating in a cave, feeding himself entirely from a massive wheel of cheese, while Hindu deities and sadhus are often bathed in yogurt. I daresay, industrialized dairy products could never inspire such devotion!

Western civilization is a milk-based civilization, and our modern Western way of life may well have begun with the first milking of a cow (or goat), not with the first planting of a seed. The technologies of butter, fermented milks, and cheeses enabled us to produce more food from our animals and our land (without the destructive and intensive practices of plowing and harrowing), allowing towns and cities to grow, and for our respective cultures to flourish. And we should understand that it is natural cheesemaking practices that are the foundation for all of this.

We've gotten this far without pasteurization and its attendant technologies; it's uncertain how far we'll make it with its prevalence. Arguably, cheese and fermented dairy products are our most important foods, not just because of their foremost place in the global larder—they are perhaps our most consumed food, allowing us to feed ourselves a perfect food since the dawn of agriculture—but most importantly because cheese is how our infants digest their mothers' milk.

The name of our economic and political system, our modern religion of capitalism, is derived from the Latin *caput*, head of cattle—the original owned property. But perhaps a new milk-based religion is in order, given the circumstances we find ourselves in today, one based not on the heads of cows but on the udder, the motherly organ that nourishes and sustains, and that provides us the material with which to make our cheese.

The modern practice of cheesemaking is remarkably out of touch with many realities of the world around us and driving us recklessly toward uncertain ends. A new cheesemaking is needed to reestablish our most important connection with the earth. This radical rethinking of our dairying practices is essential if we are to be able to continue practicing animal agriculture, and to continue to create this most exceptional of foods. The time is right to take the leap—so will you join us in this church of cheese?

CHAPTER 2

Milk

Milk is the materia prima of both mammals and cheese. It is a phenomenal fluid with exceptional capacities for transformation and nourishment. And its unique composition and extraordinary qualities make it uniquely capable of supporting life and becoming what is probably the most flavorful and nutritious of foods created by our many cultures.

Milk provides specific nutrition for the young animal, but it is also a perfectly designed delivery vehicle for that nutrition from the mother. It is a complex liquid love, providing almost everything the young animal needs to thrive, short of sunlight, air, warmth, and care. For milk doesn't just provide food; every element of milk also has many functions within the infants that drink it, which enables the nutritional capacity of that milk to be fully realized and for the infants to have their best chance at life. And these essential functions are remarkably interwoven with milk's cheesemaking capacity.

Simply put, milk is made for cheesemaking! When a young mammal drinks their mothers' milk, their stomach produces enzymes and acids that naturally curdle that milk into cheese. This transformation of milk into a more solid form aids its handling and digestion within the gastrointestinal tracts and helps assure the animal's best chance at life.

Milk is multifunctional, much like an organ, like the blood from which it's made. Its proteins, fats, sugar, minerals, enzymes, microbes, and myriad other necessities each have multiple important roles to play as they are passed along from mother to child to assure the offspring's success and well-being.

As milk is consumed by the young mammal, its complex parts are broken down and reassembled in ways, both biological and microbiological, that assure its best assimilation into the body. Milk proteins are cleaved by specific enzymes the infant makes into more digestible forms and then into more assimilable amino acids, which are reassembled into various functional proteins within the young animal; milk's sugars feed both the young animal and the microbes that milk provides, helping to establish an essential and protective microbial ecology in the intestines. The fats are broken down into more digestible fatty acids, and various enzymes help assure the continued growth of beneficial microorganisms and prohibit the development of pathogens. These many functions are mirrored when milk is instead transformed by our hands into cheese.

If butchery is seen as the breaking down of animals, cheesemaking can be understood as the building up of flesh from milk. The various methods of making cheese elaborate upon milk's ability to nourish and build the bodies of young mammals—so much so that it seems more than coincidental that the cheeses made from milk can come to resemble the flesh and blood of animals. Cheeses can be made hard and brittle as bone; soft and tender as liver; rich as fat; or textured like muscle (queso Oaxaca is commonly called *pechuga de pollo*—chicken flesh—for its textural and flavor resemblance to meat). They can be as thin and lustrous as a strand of hair, or as strong and elastic as a tendon. Milk's many facets provide the component parts that help build all of these elements and also allow each cheese to take on its particular character.

Milk's capacity to transform into cheese should thus be understood as sacred; but any industrial processing that milk goes through (i.e., refrigeration, pasteurization, homogenization) desecrates it, destroying its ability to become its cheese. Such "milk abuse" diminishes milk's capacity to nourish and sustain, and degrades our best food into what may be the most damaging one.

To help us understand these various cheesemaking aspects of milk, it helps to recognize milk's intrinsic biological purpose. But the opposite is also true: It's surprising how little is understood about how milk works, and natural cheesemaking can help inform us about milk's incredible nature.

Milk's Origins

Milk is generated from the blood of its mothers, within the udder or breasts, also known as mammary glands, producing milk from elements they receive through the blood in the milk vein (500 L blood must pass through the udder to make 1 L milk!). Though we don't typically think of milk this way today, in some historical Christian societies, milk was considered white blood, and consumption of milk and other dairy products is thus restricted during times of penitence like Lent, much like meat.

Milk's particular components are synthesized from selected components in a mother's blood within special cells of the mammary glands, known as epithelial cells. Those components—sugars, amino acids, fatty acids, and so on—make their way into the blood through their ruminant mother's fermentation, enzymatic transformation, and digestion of browse in her four stomachs and intestine. Ultimately these parts of milk originate altogether from the feed of the animals, their pasture, and the sunlight, air, water, and soil that support it.

The epithelial cells grab the particular elements they need from blood that flows through the mammary glands and build from them the component parts of the milk according to formulae that are species-specific, providing for nearly every need. Milk's many types of sugars (lactose and other oligosaccharides) are created from glucose in the blood. The caseins and other proteins are generated from the foundational amino acids in the bloodstream. The milk fats are built from sugars and particular free fatty acids found in the blood from digested food. The milk fats are assembled by the epithelial cells into their butterfat globules, in a complete form, and then exuded along with the other components into milk storage sacs known as the alveoli.

The alveoli hold the milk as it is generated over time and keep it stable (milk is prevented from fermentation in the alveoli with various chemical controls). When an infant suckles on the teat or nipple, a resulting hormonal release of oxytocin (the love hormone) causes the alveoli to contract, driving their milk into

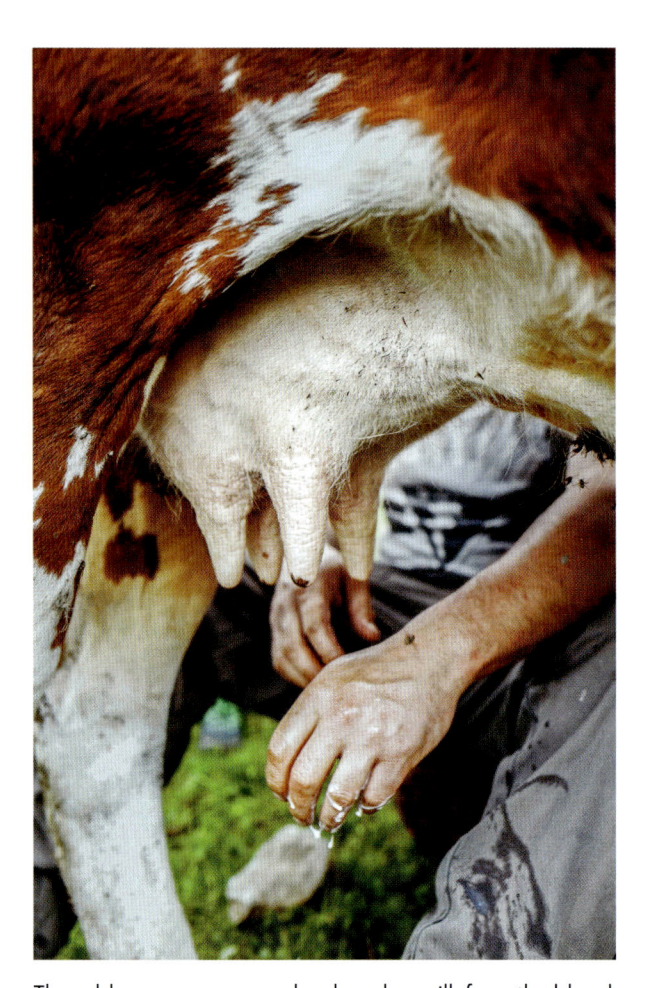

The udder, or mammary gland, makes milk from the blood that courses through it. That milk becomes cheese in the infant's stomach, nourishing the infant and sustaining their growth in much the same way that the placenta nourishes the animal in utero. PHOTO COURTESY OF MAX JONES

the milk ducts and the udders' cisterns, where it is then forced by the action of the suckling from the teat into their mouth.

Milk is made of the mother via her diet. But even if a mammal does not receive enough nutrition to support her young's milk needs, she will "milk off her back"—losing her essential fat stores and muscles in order to generate milk. Every lactating mother essentially produces milk of herself, for rather than simply producing milk from the elements of her digested meals, she builds their milk how she must according to a precise but dynamic formula (it changes over time) that provides for an infant's every need.

The components and functional qualities of an animal's milk therefore are mostly similar across individual species, and only slightly (if at all) influenced by the mother's particular diet. As such, the feed of an animal does not significantly influence the way her milk transforms into cheese. But more significantly, a poor diet is detrimental to the health of the mother, causing her body to make up for the debt of nutrition in her feed.

For the continued health and therefore the continued production of lactating animals and the good quality of milk that goes along with that, it is essential to assure good nutrition through abundant, biologically appropriate feeds, healthy pasture for cows and sheep when seasonally appropriate, and browse for goats; but the milk provided by milking animals functions much as necessary to nourish and support the young drinkers regardless (mothers are selfless in their making of milk!). And though it may only have a marginal effect on the flavor and evolution of a cheese, certainly from a sustainability perspective, from an animal welfare perspective, from a lens of rumen and animal health, the benefit of the soil, and the future of agriculture, feeding animals grass is exceedingly important!

Milk is made of many parts that function as an extraordinarily complex whole. Within milk are diverse proteins that take a unique and purposeful shape; fats in the form of fat globules that rise to create a layer of cream when milk sits; and an array of natural sugars that feed both the infant and their microbes. Each of these components plays an indispensable role in giving life to young animals, as well as in milk's transformation into various styles of cheese and fermented dairy products.

The following are explorations of these various facets of the milk, including how they interact with one another, from both a cheesemaking and a digestive perspective.

Casein Proteins

Milk contains a number of proteins that provide essential amino acids to young animals. But the proteins themselves function in important ways so that their perfect nutrition is best assimilated. The primary proteins are divided into two groups: caseins, which curdle from the milk; and whey proteins, which remain in the whey after the curd has formed.

Milk's primary proteins, caseins, are exquisitely designed, and carry an incomparable form of nutrition that's exceedingly rich in calcium and phosphorus and many other micronutrients, and that give skim milk its whitish color. The nourishing proteins themselves are not easily passed from mother to infant, as they are most easily digested in a solid form, but in milk, the form these proteins take allows them to exist in a liquid state, which permits their transfer from mother to offspring. It's only when the caseins find their way into the infant's stomach that milk comes into contact with enzymes that cause it to curdle, enabling its proper digestion as cheese.

Caseins, as they are produced by the mammary glands, exist temporarily in a fluid state in milk because of their particular form. There are several types of casein protein in milk, and most of them are water-repellent, so on their own, they do not remain dissolved in the liquid. Kappa-casein, however, is water-soluble because of hairs that protrude from the surface of the protein that are attracted to water. And it's thanks to kappa-casein that the primary milk proteins can be easily transferred in a liquid form from mother to infant.

Kappa-casein forms a protective wrap around the less soluble caseins (alpha1-, alpha2-, beta-, and gamma-caseins), which are all twisted around inside the kappa-casein covering, keeping them all in a more stable form known as micelles. The micelles themselves are water-soluble, free flowing, and self-repellent in milk, so they do not curdle out spontaneously. And the key to the micelles' proper functioning, their

Milk's proteins are made to curdle. Its casein proteins coagulate into cheese with the addition of rennet, an enzyme produced in the fourth stomach of young ruminating mammals, that enables its proper digestion.

proper digestion, is the kappa-casein's sensitivity to the chymosin enzyme in rennet, produced by the stomachs of young mammals. This chymosin cuts the protruding hairs of the kappa casein envelope, making the casein micelle insoluble, which causes the micelles to attract one another, forming a new structure in the milk. This new structure turns the milk to a gel in the stomach, which traps the fat globules within. This newly formed curd allows the proteins and fats to be held within the stomach for several hours and be slowly churned to firm them, as their proper digestion requires.

With the acidification (down to a pH of 4.5 from the stomach's hydrochloric acid) that then happens in the infant stomach, the formerly firm casein micelle is broken down by the dissolving of the calcium from the bonds between its casein proteins. The curd is ultimately turned into a soft paste (the same effect is seen with lactic cheeses) that permits the passage of the now water-soluble proteins into the intestine, where they are readily assimilated. This concept of curdling and its breakdown will be explored in more detail in chapter 4.

Whey Proteins

Milk's whey proteins are those that are not coagulated out with the help of rennet. They are composed primarily of lactalbumins, closely related in form to albumin, a family of proteins commonly found in egg whites as well as blood; and immunity-transferring proteins known as lactoglobulins.

These secondary proteins in milk serve several functions. The lactoglobulins are immune factors, which transfer the mothers' immune response to their young; this constitutes a sort of defensive and protective aspect of the milk. The lactalbumins contain certain amino acids that are essential to the health and well-being of milk-fed mammals. They are very easily digested and complement the amino acids provided by the diverse casein proteins.

Because these whey proteins are not affected by rennet coagulation, they're left behind by the rennet cheesemaking process and do not become a part of the cheese. Like the albumin proteins in egg whites, they are clear and colorless, and not easily noticeable in the whey. They are also heat-sensitive, and heating the whey in slightly acid conditions causes the whey proteins to denature and curdle out into ricotta curds. The same whey proteins in milk cause yogurt to thicken more significantly with the cooking of milk above 80°C (176°F), which denatures them, resulting in a more significant yield of cheese if the yogurt is then drained.

Whey proteins consist of almost a fifth of the protein content of milk. Making whey cheeses after a rennet cheese helps cheesemakers realize a greater amount of food from their milk than making rennet cheeses alone. It's no wonder that many dairying cultures make a second whey cheese after the first rennet cheese (the name *ricotta*—"recooked" in Italian—refers to this second cheesemaking, quite common across the Mediterranean region), and that many dairying cultures regularly practice milk boiling. Boiling milk prior to its fermentation, often seen as a traditional public health practice equivalent to pasteurization, is actually a traditional milk handling practice that allows for the denaturation and coagulation of the whey proteins into the curd, which makes the cheeses made from it more bountiful, more delicious, and a greater source of nutrition.

Milk's Fat

Milk fat provides the primary source of energy to young milk-fed ruminants (and our young as well). It contains essential fatty acids that they can only derive from their diet and conveys many fat-soluble vitamins that are essential to milk-drinking-animal health.

But it's not just the nutrition within the fat that matters, though, but also the form that the fat takes within the milk. The fat in milk forms round globules, surrounded by a membrane that keeps the fat soluble in the water portion of the milk, so that the fat doesn't spontaneously separate out into butter (which would complicate the feeding and digestion of milk, and ruin its prospects of cheesemaking!). Because of their high fat content, the globules themselves are lighter than the liquid milk and slowly rise over time, forming a layer of cream at the top of the milk. This cream is composed primarily of the milk fat globules, but it also contains some caseins and other milk proteins. The cream is commonly considered a fat-in-water emulsion, like mayonnaise, but I think that misses the point because it neglects the complexity of the form of the milk fat.

Cream rises to the top of milk as it sits, but when it's transformed into cheese in the stomach (or vat), that cream does not have a chance to separate. It instead becomes purposefully trapped in the casein curd that quickly forms from the milk's interaction with rennet. Separation of milk's cream is a sign of its age and handling, because fresh from the udder, regardless of the milk's season or composition or species, the cream is integrated fully within the milk. This is significant because the fat of the milk is not held or digested slowly in the stomach if it is not first curdled along with the milk proteins into cheese. And the form of the globules themselves permits them to be held within the curd, and thus be more fully digested with the help of the animal's lipase enzymes.

The milk globules themselves are composed of two portions: the inner fat and the outer milk fat globule membrane. The fat globules, formed by the epithelial cells within the mammary gland, are variable in size, depending on the species—smaller for goats, larger for cows. And the globules comprise different fats that are reflective of the particular biochemistry of each species, not necessarily their particular diet, though it may be their source.

When cream is churned into buttermilk, the buttermilk may show a slight brownish color, revealing that it isn't really milk at all, but rather is composed mainly of the milk fat globule membranes now broken up and separated from the globules. As such, the flavor and texture and color are distinct from skim milk, for example, which mainly consists of milk's water, proteins, and sugars.

The milk fat globule membrane, which includes cholesterol and other lipoproteins, helps young animals digest and transport the fat in their bodies. These lipoproteins are also integral to the proper functioning of infants' cell membranes, and consist of many bioactive compounds, so the milk fat globule membranes therefore are an essential part of milk and infant nutrition. Butter transferred from the mother on its own would not be as readily assimilated. It's possible that butter should be eaten with its partner, buttermilk, as it often is in traditional dairying societies.

The milk fat within the globule, like all fats, is made up of triglycerides—three fatty acids connected to a glycerol molecule. The lipase enzyme produced by infants (in glands beneath their tongues) helps break apart the triglycerides in milk into their component parts, which are much more readily assimilated through the gut than is the whole fat molecule.

This process of lipid breakdown, or lipolysis, is mirrored in cheesemaking, first by the addition of lipases from some traditional rennets containing this enzyme in the stomach contents, which impart particular flavors to cheeses made with it; but also by bacterial lipolysis, for the complex communities of microbes in cheese feed on fats and ferment them, breaking them down into their more digestible component parts.

Milk fat is made of particular types of mostly saturated fat, which is now understood again to be

MILK • 31

beneficial to human health (how could it not be if it's the main fat in milk!). The composition and makeup of these fats has a strong effect on the flavor development of cheese, for particular fats contain different fatty acids that help define the flavor of cheese as they're fermented and aged. The fatty acid makeup is primarily related to the species of milk-giving animal.

The short-chain fatty acids that give cheeses their most elemental flavors are built according to species-specific formulae from simple fatty acids in the blood, most significantly in ruminants the very short chains of acetate and butyrate that result from the fermentation of grasses' long-chain carbohydrates in the rumen. The longer-chain fatty acids are sometimes feed-specific and relate to the original state of the feed, be it grass or corn (or even industrial feeds such as palm kernel fat). Grass feeding has been found to improve the quality of these long-chain fatty acids, both nutritionally (in terms of relationship of omega-3 to omega-6) and, to a mild extent, organoleptically. But the vast majority of the components and flavors in a particular milk's fat exist independent of the form of feeding.

The cream is the most savory portion of the milk, giving it its most flavorful character (the fats and their fatty acids provide the most expressive aspects of cheese's many tastes) as well as its rich texture. The globular form that the cream takes permits cheesemakers to make the most luscious and flavorful cheeses with their fullest quantity of fat.

Milk's Sugars

Milk is an abundant source of sugars including lactose and other oligosaccharides that are essential to a young animal's health and development. But these milk sugars don't just provide an essential source of energy to milk-fed animals; they also help to promote the growth of a healthy microbial ecology within an infant's gut.

The primary sugar in milk is lactose, a complex sugar made of two simpler sugar molecules—glucose and galactose. Milk provides other sugars as well, known as oligosaccharides—meaning "made of a few saccharides"—which are slightly more complex than lactose.

These oligosaccharides are known to specifically nourish the development of the milk-drinking mammals' microbial ecology. But lactose may also serve as a source of nourishment for the community of microorganisms provided by milk (more on that in a moment) that enables this beneficial microbial community to take up residence and grow within the gut, with many benefits to the young mammal.

Not much is understood about how lactose is digested within infant and adult gastrointestinal tracts. The lactase enzyme that digests it is theorized to be synthesized in the small intestine, but it may be synthesized by the microbes of the infant's gut, most of which are nourished by lactose and other sugars provided by the milk, breaking them down into simpler sugars, like glucose and lactic acid, that are more easily absorbed into the gut (lactose doesn't flow across the intestinal barrier as readily as glucose or lactic acid). Lactic acid itself is a very important fuel to help infants' brain development and may be an important by-product of infants' digestion of milk.

The presence of lactose-fermenting species within infant guts, and the fact that the intestines host a massive population of these microorganisms (equal to the weight of their brain) that must feed on something, leads me to believe that lactose is indeed being transformed by these complex communities of microbes. There are many theories as to why mammal milks provide lactose, not the glucose or other simpler sugars found in blood. One is that these less easily digested sugars specifically feed the community of microorganisms that is so important to the infant. My own explorations of the microbial community of infant ruminant guts appears to confirm this.

Certainly the fact that the secondary oligosaccharides are digested by the microbes in an infant's gut and not by the infant themselves is well understood. What we haven't fully realized (somehow, we've neglected to consider this) is that milk's main sugar, lactose, also supports the growth and development of bacteria in an infant's lower gastrointestinal tract. (In cows, goats, and sheep, milk's lactose is the main sugar, and their young animals' intestines host a fermentation as well—this is not unique to infant humans and their milk's oligosaccharides.) Milk is thus meant to ferment, as we'll explore more fully later in this book.

But it's not just the milk sugars that fuel fermentation (simple sugars—or more complex carbohydrates for that matter—cannot sustain a fermentation on their own, because microorganisms need other nutrition to grow). Milk contains micronutrients galore, along with macronutrients, abundant fat, and protein, making milk an extraordinary medium for fermentation. Indeed, altogether this makes milk possibly the strongest and most supportive natural medium to support the growth of microorganisms that we know.

No other food ferments quite like milk, in both its readiness to ferment as well as the complex pathways its many fermentations can take. No doubt this is due to the complexity of the incredible nourishing components that milk provides, both for young mammals as well as for their beneficial microbial community. Chapter 3 will explore this idea in much greater detail.

Milk's Other Proteins and Enzymes

Milk is provided to its drinker along with many enzymes that help them make the most of it. Though infants produce most of the enzymes needed to digest the various component of milk, there are still a significant number of enzymes gifted by mothers in their milk that are known to have various important roles, including various proteases, lipases, and more. The particular nature of these enzymes allows untreated milk to be changed in its most authentic way, the same way that it is transformed within the gut.

The lactoferrins and lysozymes provided in milk control against unwanted fermentations in the transfer of milk from mother to offspring and seem to restrict bacteria like salmonella and coliforms from developing. These enzymes keep milk as milk in the mother's alveoli, preventing microbial growth and milk's degradation. After milking, the lactoferrins degrade over time, slowly losing their ability to combat microbial development; but for several hours after leaving the udder, these proteins seem to limit the spontaneous fermentation of milk. They do not, however, seem to impede milk's fermentations during active cheesemaking—indeed, they may encourage its finest fermentations. Regardless of their presence in the freshest milk, it is still best to make cheese with cow-warm milk.

Milk's Microbes

A most essential element of a mother's milk is the microbial community it contains, which seems to be intentionally taken from the mother's beneficial microbial ecology and passed along to the infant. This maternal microbiological transfer, also known as the entero-mammary pathway, is recognized as an important contribution to the health and well-being of infants and should be understood to play the same role in young ruminants.

The establishment of an infant's microbiome is fundamental to their development. It's now known that many organs—including the skin, the lungs, and most importantly the intestines—support beneficial microbial communities that protect the health and proper functioning of these essential biological systems. These microbial communities seem to help with the development of the infant's immune response; with the proper functioning of the gut as well as many other internal organs; and with the production of various compounds that affect an infant's mood and well-being (many appear to be synthesized by the microbes themselves!).

Infants are first colonized by microbes on birth, with the first main exposure to their mother's microbiology as they pass through her vaginal canal. Skin-to-skin contact is also understood as an important source of beneficial microorganisms for an infant's development. But perhaps the most significant and sustained source of beneficial microorganisms for young mammals is mother's milk. Drunk multiple times daily (every 4 hours or so), mother's milk provides millions of beneficial bacteria daily, every time infants suckle at their mother's breast.

The main microbes found in mothers' milk include all of the usual characters used by cheesemakers to ferment their milk—and this is no coincidence. Milk contains a wide diversity of beneficial microorganisms, including microbes in the genera of *Lactobacillus*, *Lactococcus*, *Streptococcus*, *Corynebacterium*, *Propionibacterium*, and *Bifidobacterium*. Most of the research into the milk microbiome has been focused on breast milk, but many of these ideas and microbes cross over into the world of ruminant milk. This microbiology carries forward in the gut of milk-drinking infants. In the guts of healthy milk-drinking infants born to term, *Bifidobacterium*, *Lactobacillus*, and *Streptococcus* dominate; these microbes develop thanks to the lactose, oligosaccharides, and many other compounds that encourage fermentation provided by the milk that forms this community into the best version of itself.

The microbiology of milk comes from both within and without—through biological pathways that bring beneficial microbes from mother to infant in the milk, but also through exposure on the skin that can bring both good microbes and harmful ones from the skin's microbiome and the environment. Sterilizing the udder with iodine can have significant consequences on the microbial character of a milk, and should not be seen as good practice in dairying, as the natural biological balance of milk depends on skin's microbiome.

Milk's microbiology is exceptionally diverse and contains many species from across numerous families of bacteria, fungi, and yeasts. It's important to understand that it contains common cheesemaking species, but it can also contain microorganisms known for their pathogenic relationships. The balance of these species in milk and the gut is dependent on maternal health, exposure to microorganisms in the environment, and exposure to chemicals (especially antibiotics) that can cause dysbiosis.

But regardless of what's present in a particular milk, the microbiology of the milk isn't strong (the numbers are relatively low), and milk will not typically ferment well on its own. Milk's microbiological diversity may be what it is in part as a way to initiate fermentation within the guts of infants that first drink their mothers' milk; only re-fermenting with milk will cause it to develop its safest most stable microbial community. Because milk is both probiotic (it contains beneficial microbes) and prebiotic (it contains elements that feed beneficial microbes), it is completely capable of developing an exceptionally beneficial and stable culture.

Milk's microbiology can be developed into an incredibly complex, diverse, and stable yet manipulable culture through a simple process of re-fermentation that selects for the most appropriate species of this community and inhibits the unwanted players. This involves leaving milk to spontaneously ferment with its indigenous microbes until it acidifies and curdles, then re-fermenting this first curdled milk with fresh milk daily until a reliable fermentation is achieved.

This process of developing a cheesemaking starter likely mirrors the microbiological processes that occur in the gut of a young animal, regularly drinking their mother's milk every 4 hours, constantly feeding and re-fermenting the essential microbial community that exists within them. The complex microbiota of this milk-based microbial community sows the seeds of the appropriate ecologies that allow for the proper evolution of nearly all styles of cheese, and includes rind-ripening fungi like *Geotrichum candidum* (*Geotrichum* or *Geo* for short), which can help cheeses develop beautiful and protective natural rinds. Chapter 3 will expound on the origins and development of this natural starter.

Milk's Warmth

It should also be understood that an inherent quality of milk is its animal warmth. Milk is delivered from the udder to the young animal's stomach at between 37 and 39°C (98.6–102.2°F), depending on the species. Take away this warmth, and the milk fails to do what it's meant to do: it fails to make its cheese.

I often borrow a term from German—*kuhwarm*, cow-warm—to define the concept of milk with its own animal warmth. Milk that is cow-warm, its animal warmth still intact, supports the young mammal in the best way, and evolves into its best cheese. Milk that is chilled, then warmed again, does not function the same (more on the detriments of refrigeration below). It is more ecological, more biological, and more delectable for cheesemakers to work with cow-warm milk.

For the milk proteins to transform to cheese in their preordained way requires warmth. Casein proteins transform into a most digestible cheese in the right amount of time (again, about 4 hours) only when they are at the appropriate temperature in the stomach. The warmth of the animal's body supports the growth of the best milk-fermenting and -digesting microbes (this is one of the reasons we keep ourselves

Milk is meant to ferment; it can develop into an effective starter culture known as clabber simply with its inherent microbiology. PHOTO COURTESY OF MAX JONES

Though milk is assembled from elements of an animal's diet, it is created according to species-specific formulae that make it and its cheese remarkably consistent around the world. PHOTO COURTESY OF QUEIJARIA BELAFAZENDA

at temperatures of approximately 37°C [98.6°F]). And refrigerating milk is damaging to its capacity to curdle as well as to ferment.

All dairy farmers know to feed their young animals milk at mother temperature, for this helps assure the best coagulation and assimilation of the milk in the stomach and intestines (it's an important reason why we warm the milk in a baby's bottle, too!). Cold milk also lowers the body temperature of infants, and warming themselves again depletes their energy.

But one of the first things cheesemakers and dairy farmers typically do with milk is strip it of its warmth by chilling immediately after milking, collecting milk from several milkings together, then warming it again to body temperature to make cheese. This is detrimental to the milk in myriad ways, as is every other processing milk is most often subject to (which all originate from the negative effects of refrigeration, as described below); every other processing, of course, aside from cheesemaking!

Milk Is Milk

All milks straight from the udder respond well to the cheesemaking process—but there are some slight differences in the ways different species and breeds of milking animals' milks respond to the same cheesemaking and handling. Seasonal differences in the evolution of milk through its lactation can also change the way that a certain milk responds to a certain cheesemaking technique, but not that significantly.

I do not necessarily believe that individual breeds make a significant difference to the quality of a cheese; or that climate, feed, pasture ecologies, or underlying soil makeup or geologies are significant factors dictating how a milk behaves or its cheeses taste. (These are, however, important considerations for sustainability, animal health and welfare, and the heritage and culture of a region; they also ultimately relate to the inherent goodness of a cheese—see below.)

Milk is made for cheesemaking, and regardless of the circumstances of its production, or the breed being milked, all milks respond to the digestive process and cheesemaking process in remarkably similar ways. For me, the most significant differences among milks are the particular qualities that come from the particular species of animal that produces it.

The four globally predominant species of milking animals for dairying are cows, goats, sheep, and water buffalo. These domesticated ruminants' milks are also the most suited for cheesemaking, due to the excellent way their high-casein-content milks curdle with rennet. The following is a rundown of the ruminants, their milk, and the best cheeses and ferments made with each of them.

All animals' milks are made for cheesemaking; but certain animals' milks are best suited for certain cheeses and ferments. Nearly all cheeses can be made with any milk, but the character and quality of certain milks makes them more suited to particular styles.

The most significant consideration in natural cheesemaking is that the milk used is fresh and unprocessed, so that it behaves as it is supposed to. With that in mind, cow's milk behaves like cow's milk and makes cheeses that taste of the cow's particular chemistry, while goat makes cheeses taste deliciously of goat, sheep of sheep, and buffalo of buffalo, regardless of what they are eating. Remember, milk is made by the animal itself, and not necessarily from its feeds. All are made in a certain way according to a prescribed formula within the mammary gland and behave in a certain way, to provide for every need of their young.

Cow's Milk

Cow's milk is the gold standard from which all other milks are measured. Though cows are excellent milk producers per animal (in particular Holstein-Friesian), and produce milk of great cheesemaking quality, their cheeses generally pale in flavor (though not in color) when compared with their counterparts made with sheep's, goat's, or buffalo milk.

Cows are the workhorse of the dairy world, providing the majority of modern Western farms' milk. Their milk is suited to all styles of cheese, but best for those aged for over 2 months, as cow's milk's simpler fatty acids require a longer affinage to develop the milk's best flavors. Fresher cheeses made with goat's or sheep's or buffalo milk are generally better tasting. As an example, a fresh lactic goat cheese, chèvre, is a sought-after ingredient in chef circles, whereas the cow equivalent, vache, is less flavorful by comparison. But aged cow's milk cheeses stand their ground against these other animals.

Cow curd is firm and strong, holds its shape, drains well, and performs as it should under normal cheesemaking circumstances. In this book, most techniques are written specifically for cow's milk, and may need to be slightly altered if other animals' milks are used in replacement. The goat, sheep, and water buffalo sections below provide information on how to adapt cow's milk methods to these other milks. In general, other milks' curds need less handling (cutting, stirring, cooking, and so on) to achieve their goal characters.

Cow's milk and cow's milk cheeses show a beautiful golden color. This particular hue is due to the unique way that cows metabolize the carotene in the grass they eat. Carotene, a pigment that contributes to the photosynthesis of grass and other plants, is the precursor for vitamin A, an essential vitamin for animal health. When cows consume the carotene in their grass, they pass it along in their milk to their calves, and the fat-soluble pigment remains in the fat portion of the milk, making the cream colorful. Other ruminants—goats, sheep, and buffalo—transform the carotene into the colorless vitamin A and then pass that vitamin on in their milk to their young, making their milks much whiter than cow's.

Cheesemaking and buttermaking concentrate those carotene pigments in the fat-rich cow curd, making cow cheeses and butter extra golden. This is especially the case in summer, when the animals are eating fresh grass and extra carotene is to be had (it diminishes when grass is dried, and changes form when grass is fermented). This is seen even more significantly with breeds like Jerseys that produce more fat in their milk, and that also seem to pass along more carotene in that fat. Once you've seen Jersey butter or cheese from animals on plentiful pasture, all others pale in comparison.

On that same note, cow's milk is best of all the beasts' for butter production. It has large fat globules that rise relatively quickly, forming a thick layer of cream that can be much more easily skimmed than goat's or sheep's milk. This natural separation of cream allows for a straightforward buttermaking process, especially when compared with goat and sheep, whose butters can be produced only in more roundabout ways. Cow's butter is gorgeous and golden, especially when the cows are on grass, the only ruminant butter to show such a color.

Some important cattle breeds to consider: The most common milking breeds, like Holsteins, Brown Swiss, and even rarer ones like my favorite, Canadiennes, give milk that is well balanced in protein and fat, and therefore very well suited to cheesemaking. Though Holsteins produce milk that has lower solid content than most and yields less cheese per liter, its high protein-to-fat ratio makes a strong curd that responds well to heavy-handed cheesemaking techniques like alpine, stirred-curd, or milled-curd makes.

Breeds like Jersey and Guernsey (bred to give more easily separated cream for buttermaking) provide milk with a higher fat-to-protein ratio, which can make very rich cheese. And though their high-fat milk can make exceptionally creamy lactic and half-lactic cheeses, much of that golden fat is lost to the whey when making harder alpine styles. However, the high level of solids in their milks means that the curds typically need less handling to achieve their goal firmness. Their milks are also ideally suited for buttermaking, as their larger butterfat globules rise readily and yield a thicker cream line than cheesemaking breeds. This is especially apparent when their milks are used to make traditional cream-top yogurts.

In hotter climates, zebu cow breeds like the Gyr provide excellent milk. They are used for cheese- and buttermaking across South America, Africa, and South Asia.

Goat's Milk

Goat's milk is extraordinary for both fresh cheeses and hard aged styles but shines brightest when its cheese is long fermented. It is a milk that's sensitive to its handling compared with cow, but it develops into cheese of an exceptional flavor early on and becomes even more nuanced as it ages. Goat's milk lacks the structure necessary for proper stretching with the pasta filata method (though you can mix it 50–50 with cow and make a most amazing mozzarella). But for just about any other style of cheese, goat's milk is magnificent, providing greater flavor with less aging, and often less curd work, to achieve the same firmness.

Excellent animals for smaller farmstead producers, or those without access to pasture or developed farmland, goats are great animals to keep, and easy animals to hand-milk. And for cheesemaking on a small scale, no milk compares! Their flavorful cheeses always attract greater attention and can bring greater income than a cow's milk equivalent.

Goat's milk and its cheeses are strikingly white, especially when compared side by side with equivalent cow's milk cheeses. Goats transform the carotene in their diet into the colorless vitamin A, passing along the nutrient but not the color into their milk and their cheeses. That vitamin A is likely more assimilable, though, than the more colorful carotene.

Their milk makes excellent fresh lactic cheeses, extremely flavorful on their second day, and indescribably delicious aged lactic cheeses. Goat's milk contains particularly flavorful fatty acids that make exceptionally good lactic cheeses (especially compared with cow's milk) as a result of their longer fermentation. But goat's milk also makes sublime alpine cheeses with very little effort (like tomme de chèvre).

The goat-esque (and sometimes grotesque) flavors often associated with goat's milk cheese are not inherent to goats, but rather result from their flavorful milk's mistreatment before cheesemaking. Goat's milk's particularly flavorful fatty acids are broken down into particularly unsavory ones through the rancidification resulting from refrigeration (and its stirring), pasteurization, and homogenization; most industrially produced goat's milk cheeses therefore have a very offensive stink. Don't blame the buck for an unsavory goat's milk cheese—primarily "milk abuse" that leads to such "cheese trauma"!

Cherished cheeses made with goat's milk include fresh lactic cheeses like faisselle and aged lactic cheeses like crottin. Half-lactic Camemberts and half-lactic tommes (tomme crayeuse) are also excellent candidates for goat's milk. But less primary fermented cheeses like Camemberts, washed rinds, and alpines also develop unique and delicious characters when made with goat's milk. And no milk beats goat's milk for making kefir when it comes to flavor and drinkability.

Goat's milk's curds are more delicate and brittle than other milks and require a gentler handling to assure they don't shatter. With rennet cheeses made with goat's milk, it's always wise to wait longer between cutting the curd and stirring it to allow the curds to firm and strengthen, and then to stir very slowly to ensure that the curds remain intact.

With lactic goat cheeses, a smaller rennet dose can be added to ensure that the curd forms slowly and doesn't crack; the longer coagulation time doesn't lead to cream loss as it does with cow's milk, because the smaller goat fat globules don't rise as readily.

When you're making cooked curd cheeses like a tomme, it is also important to recognize that goat's milk is more sensitive to high temperatures than cow's and need not be cooked at all to achieve the same goal curd firmness; simply cutting the curd small and stirring for several minutes at 35°C (95°F) achieves the same textured tomme as a cow curd cut and stirred and cooked to 45°C (113°F)! If the goat curd were cooked to that temperature, it would take on a grana-like character.

Goat's milk's cream does not rise as readily as cow's, so it cannot be separated easily for buttermaking. And though there are traditional methods of making goat butter, they are not as simple as those used for cow butter that rely on gravity to separate cream (see the technique for goat and sheep butter in chapter 14). Once churned, though, goat butter is beyond compare.

And though it is exceedingly flavorful, goat yogurt never thickens quite like cow; it can, however, be made thicker with the addition of slight amount of rennet, making the sought-after dairy product in Spain known as cuajada.

All goat breeds produce milk of relatively similar qualities, and all can be used nearly interchangeably in the makes. For the most part, goat's milk is goat's milk. The slight difference in fat-to-protein ratio between a Nubian milk and a Saanen milk will yield slight differences in a cheese's creaminess, workability, and yield. Don't, however, expect a different flavor profile from one breed's cheese to the next, all else being equal.

Sheep's Milk

Sheep's, or ewes', milk, though rare in the Western world, is considered by many to be the finest milk for cheesemaking: What sheep lack in quantity of milk, they make up for in quality. Though the animals yield a trickle of milk, on the order of 1 to 2 L per animal per day, their milk gives twice the cheese yield of cow's or goat's milk. Thus a single sheep's cheese yield is similar to that of a goat. Their milk is full of cheese but also full of flavor.

Sheep's milk is exceptional for cheesemaking, because of its satisfying cheese yield and its flavor capacity. In many ways sheep's milk is like goat's: whiter than cow's, full of comparably flavorful fatty acids, but more satisfying to work with because of its high yield. When I make the same cheeses side by side with fresh cow's, goat's and sheep's milk, sheep typically wins. Like lamb over veal, there's just something about these ovines that makes all of their parts sought after by cheesemakers, butchers, and even knitters, who love the lanolin of their wool.

Sheep's milk is ideally suited to softer, higher-moisture makes—those styles that preserve and celebrate sheep's milk's high fat levels. High-moisture rennet cheeses like Camemberts, and lactic cheeses, especially, are very well suited to sheep's milk. Fresh brebis (sheep chèvre), sheep faisselle, and aged lactic sheep cheeses are rare, but they make most flavorful and bountiful cheeses when used for these styles.

Sheep's milk is also extraordinary for making simple fermented milks like clabber, yogurt, or cuajada. Its thickness and composition result in sumptuous and flavorful curds when fermented, with nearly twice the nutrition of goat's or cow's equivalent in a spoonful.

Stirred-curd blue cheeses like Roquefort are also extraordinary when made with sheep's milk, and represent the best of their class. The milk's extra-high fat content makes for especially creamy blues, but their

Sheep make milk that's most extraordinary for cheesemaking, with flavorful fatty acids, high yields, and an ivory-white color. PHOTO COURTESY OF SINNOS FORMAGGI

curd has to be very delicately handled to preserve its high fat. Of course, alpine sheep's milk cheeses, from regions like the Basque country, are also best in their class and are considered among the most flavorful of all cheeses.

When making any rennet cheese with sheep's milk, the curd, when set, is about twice as dense as that of cows or goats. It should therefore be understood that to achieve the same goal firmness in a curd of a certain technique for calling for goat's or cow's milk, the sheep's milk must be handled less heavily, if at all. For example, with a traditional Camembert, sheep's milk should not be cut before it is ladled; doing so would release too much whey and make too firm a cheese for the method. And when you're making cooked alpine cheeses, sheep's milk must be handled differently, as the curd when cut small behaves like a cow curd cooked to 45°C (113°F) when it isn't cooked at all.

Sheep milk's proteins are unfortunately, like goat, not suited to the pasta filata methods, as their curd does not stretch intensely as cow and buffalo do. There are some rare spun-curd cheeses made with sheep's milk, including the smoky and alluring oscypek from the Carpathian Mountains in Poland. But that unique cheese may be made as part of a technique for extracting as much fat from the milk as possible, to get the maximum yield of sheep butter, which cannot be made from sheep's milk the same way as from cow's milk—sheep's milk's delectable cream rises only very slowly.

Sheep's milk cheeses, color-wise, exist in a space between goat's milk and cow's milk. When first made, the cheeses have an ivory whiteness almost like goat cheese, but on exposure to air, their fats yellow and take on the color of cow's milk. This appears to be a result of the oxidation and degradation of the vitamin A in the cheese back into a more colorful form (this phenomenon occurs in other milks as well, but cannot be so easily observed by the naked eye). Their butter is white in color, and delicately lamb-like in flavor.

Buffalo Milk

Water buffalo milk may be the most versatile and richest of milks in the group. The animals are, however, the rarest of the milked ruminants (outside of India,

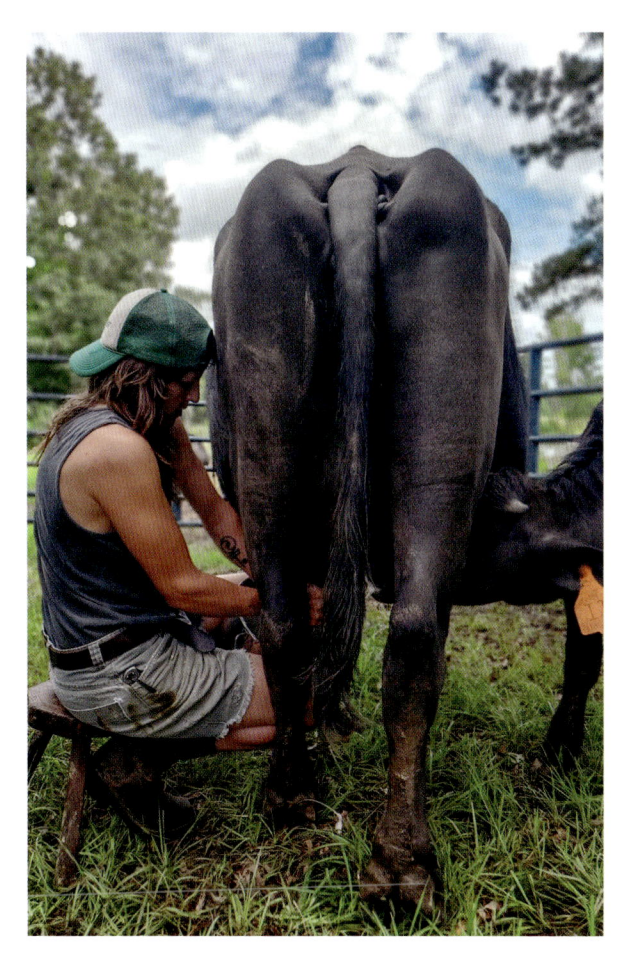

Water buffalo are remarkably calm and friendly milking animals whose milk makes a highly sought-after cheese.
PHOTO COURTESY OF BAYOU SARAH FARMS

Egypt, and central Italy, where they predominate). They deserve to be more common.

Originating in either South or Southeast Asia, water buffalo (not to be confused with American bison, often mistakenly called buffalo) have been raised for meat, milk, and draft purposes across Asia and Africa for millennia, and are only recently gaining traction as farm animals in North America and Europe. They are best adapted to warm and wet climates, with plenty of water for wallowing, though many herds of these beautiful black animals are kept in drier, cooler climates including Canada.

As lovable and responsive as giant dogs, buffalo form strong relationships with their keepers. Their surprisingly white milk (the animals are nearly pure black)

has many characteristics that make it one of the most sought-after milks for cheesemaking.

Buffalo milk is fantastic for all styles of cheese; it has the workability of cow's milk, the whiteness of goat's milk, the high fat-to-protein ratio of sheep's milk, and a delicate flavor all its own. The only problem is that this milk can be hard to get. While they're exceedingly friendly creatures, buffalo are the most hesitant of all the species to let down their milk. Often dairy farmers have to keep their young by their side and treat the animals as well as possible to convince their mothers to allow milking. A similar effect (resulting from the release of oxytocin) can come from the tactile stimulation of the udders before milking.

Like sheep's milk, buffalo milk begins its transformation into cheese twice as solid as cow. As such, most buffalo cheeses do not need as much cutting and stirring to achieve their goal firmness. An especially light touch is needed to make a soft and tender mozzarella di bufala—too much curd handling can easily fault the cheese into a firmer caciocavallo di bufala.

Buffalo milk, like goat's milk, is pure white, as the animals convert the carotene in their diet into vitamin A in their milk. Their cheeses (most significantly, their mozzarella) are therefore white as snow. Yogurt, perhaps the finest product to come from buffalo milk, firms even more fully than cow's milk yogurt, and has the thickest, most strikingly white cream line.

Buffalo milk has an exceptionally high fat content—up to 9 percent—which results in very high yields and especially creamy cheeses. Though known mainly for its mozzarella, lactic cheeses are especially creamy when made with bufala, as are Camemberts, blue cheeses, and tommes. Though it works for the method, water buffalo milk wasn't traditionally used for mountain cheeses, as the warmth- and water-loving animals aren't as well suited to mountain agriculture.

Other Milks

Yak's milk is a common milk in the Himalayan Mountains, provided by animals especially suited to the harsh climate of the region. Their milk is exceptionally high in fat and protein and makes delicious and bountiful cheese.

Horse's, or mare's milk, is common in the Asian steppes, but horse milk cheesemaking is not known to exist, perhaps because this milk has low casein content. It is, however, high in lactose, so much so that it's sweet on the tongue, and is exceptionally suited to making koumiss, an alcoholic milk ferment from Mongolia. The rich sugar content allows for a high lactic acid content after primary fermentation; with continued agitation and aeration, as in the technique for amasi or milkbeer, those lactic acids continue their transformation into a rather powerful alcohol.

Camel's milk, white and mineral-rich like goat's milk, responds well to cheesemaking and fermentation, and has a long history of use in the Middle East and other desert regions in Asia, Africa, and now Australia and America. The milk, however, does not respond to bovine rennet and must be curdled with camel rennet. Its curd makes flavorful lactic cheeses and creamy "Camelberts," but will not stretch for mozzarella.

A final milk I feel must be mentioned is human milk. Mother's milk makes cheese much like any other! Rich in lactose and other oligosaccharides that fuel microbiological development in infants, mother's milk ferments both in a jar and in the gut. However, its behavior inside infant stomachs is still not well understood, and certainly not from a cheesemaking perspective. I believe mother's milk's cheesemaking character merits further study, for understanding breast milk's function during its digestion is fundamental to the argument against the use of infant formula, which is almost certainly an impediment to proper infant nutrition and an opportunity for enormous profit from some of the largest players in the food industry.

Wading into the breastfeeding-versus-formula-feeding debate is not my intended goal with this book, but natural cheesemaking provides a unique contribution to the conversation. When milk is understood to have a greater purpose in its transformation into cheese, this significance to mother's milk is something that cannot be overlooked, for processed formula cannot behave the same as mother's milk, as it's not produced with the same goals whatsoever. The choice to feed an infant formula should not be taken lightly. And mothers need all of the support that they can get to nourish their children the right way. If mothers are unable to provide, milk from donor mothers should be considered the best alternative. But though wet nurses are currently culturally objectionable (due partly to the current dominance of formula), breast milk banks almost always pasteurize and depend on refrigeration and freezing, preservation technologies that diminish breast milk's functionality and nutrition much as they do in ruminant milk's transformation into cheese.

The character and properties of every milk are specific to the needs of the young, including the balances of fat to protein to sugar and the various components of each of those—the particular amino acids and fatty acids. Cow's milk is the way it is to provide the perfect nutrition to calves; ewe's milk is the way it is to provide for the developmental needs of lambs. Any overprocessed milk or milk replacer (for calf or human) will forever fall short of their mother's milk warm from the udder (or breast). Milk doesn't just provide nutrition, it has functions that depend on a connection between mother and infant.

From this perspective, I'll acknowledge that taking milk away from a young animal and then sacrificing the young one to use their stomach to transform their mother's milk into cheese may seem barbaric to many; but I'll do my best to address the ethics of this act in chapter 4. Today the dairy industry is able to salvage its unneeded, formerly sacrificed calves and produce meat from them—but only thanks to industrially produced milk replacers and confined animal feedlots. The replacers are typically fabricated from processed whey and casein proteins, supplemented with unhealthful soybean oils, and are an unfortunate development that has allowed dairy calves to be raised without their mothers' milk. Really, nothing compares to fresh warm raw milk, right from the udder, for raising calves, goats, and sheep, because only such milk responds best to the complex biological processes of a young animal's stomach.

Happy Cows, Goats, Sheep, and Humans

Now, how bold of me to state that the feeding regimen and lifestyle of an animal has little if any influence over the flavor of a cheese produced from their milk! Of course we cannot simply gloss over the needs of our milk animals; their health and well-being are the foundation of all of this. Good husbandry helps to assure our animals remain happy, stay healthy, and produce milk for many seasons. And we cannot talk about milk without a discussion of its own materia prima, the pasture grasses, leaves, and other browse that support the life of cows, goats, and sheep.

Our ruminants need minerals to lick, and other animals to lick, too! They need a harmonious herd hierarchy and horns to make their positions known within it, among other important reasons. They need grandmothers in the herd (they take care of the little ones) and a father or two, too (just like the rooster in a flock, they protect as they inseminate!).

Milking animals need attention during labor and their calves, kids, or lambs by their side after birth. The young animals, if they are to be raised, need access to as much fresh, warm milk as they can drink, as often as they can drink it, preferably straight from the udder, still warm. It's not common to raise young heifers or doelings entirely by their mother's sides in dairies today, but likely it's best for the babies to have their fill—generally they're not fed enough milk, and don't grow as well as young animals raised for meat.

They need good food, preferably fresh grass and browse in summer, and dry hay or silage in winter, readily accessible both day and night. Ideally, they are able to walk out to their pastures from the parlor in the spring, summer, and fall to forage for themselves—it's ecologically responsible, energy-efficient (low in fossil fuel usage), and good for their welfare for them to feed themselves in the grasses, hedges, and forests. They need space, to run, to be free, to be animals.

Their pastures and hayfields should be managed with regular manure spreading, supplementing with additional composts and natural fertilizers. Interspecies control of weeds helps. Goats do the best job weeding and offer other benefits that come from mixing animals in pastures. Such benefits of mixed grazing includes a reduced parasite load (parasites of sheep don't affect cows and vice versa) and mixed-milk cheeses, which are extraordinary. However impractical it might seem given our current agriculture's focus on monoculture, this is the way our farm animals are meant to be kept.

warmth of the vat for firmer cheeses, while the cheeses traditionally made in winter often prefer cooler weather. Even the aging conditions for certain styles of cheese are best during the periods they were traditionally made. For example, heat and flies in summer make it challenging to produce softer, higher-fat cheeses. And the summer season of abundant milk would have provided plenty of long-aging cheese to last through the hungry months of winter until the spring freshening, when the first fresh-eating cheeses would have been made. More on the suitability of each season's cheeses will be discussed in part 3.

Colostrum

During the first few days after birth, milk is provided in a special form called colostrum, which has a unique character very much unlike milk. As the young animal's stomach develops its capacity to digest milk, around day 3 of their life, the product of their mother's lactations transitions from colostrum to true milk.

Particularly rich in diverse milk albumins and immunoglobulins, and not in casein, colostrum does not curdle as milk does. This change can be understood from the perspective of the developmental stages of the infant's stomach: In the first days, the young mammal's primary stomach (the abomasum) is small and undeveloped, and not capable of curdling milk with its own enzymes. Colostrum does not need to be digested in the same enzymatic manner as milk; the albumin in this first milk passes on into the small intestine, where the simpler protein is much more easily assimilable without an enzymatic contribution from the young animal. Indeed, the enzymes necessary for its digestion may be present in the milk itself. Immunoglobulins provided by the mother to help build the infant's budding immune system are directly absorbed into the still-porous intestine.

Colostrum, because it lacks casein, does not turn to cheese with rennet. However, on cooking, it turns to an egg-like custard without the addition of eggs, for its predominant albumin proteins are denatured into a solid form by heat. Kalvdans pudding, in chapter 15, is a historically universal recipe that curdles colostrum through its peculiar sensitivity to heat. This is a product that is celebrated in dairying cultures around the world, for it comes with the birth of the calves and the arrival of first milk: the beginning of the season of cheesemaking.

The Problems of Milk Processing

Only milk that is in its original form performs the way it is meant to, as described above, both in the young animal and in the vat. Milk is remarkably sensitive to its handling: in nature it passes from the udder directly into the cheesemaking stomach still warm and without any mistreatment. Any divergence from this path takes away from milk's essential function, for any mishandling damages milk's many facets, making it misbehave.

Milk isn't simply a liquid; it has an exceptional capacity for transformation and nourishment that transcends its unremarkable white appearance. It should be recognized for its complexity and diverse functionality, its exceptional and irreplaceable ability to support life, as an extension of the mother toward her offspring that gives them the greatest capacity to thrive. Identifying milk as a maternal organ (like a placenta that nourishes the fetus) can give milk the significance—the reverence, really—that it needs. If milk is seen as such, we should then expect pasteurization or other processing to negatively affect its function. If a liver were pasteurized or blood refrigerated, would you expect it to maintain its proper functioning? Shouldn't the same be realized of milk?

As such, I'd prefer not to use the term *raw* to define milk in its original state, though I often fall into using it by habit when I'm differentiating it from pasteurized milk. *Raw* conjures danger or disgust (raw wounds, raw meat, raw sewage—a medium I fermented when I was once a sewage treatment engineer) that need not be introduced into the language (likely this was a weaponized descriptor conjured up by pasteurized milk producers, processors, and scientists in the ongoing era of the milk wars). Milk is meant to be the way it is made by mothers and need not be modified or pasteurized to make it safe.

Thus, when I write of milk in this book, invariably I shall be writing of so-called raw milk. If milk's purpose is to make cheese, then pasteurized milk should not be called milk, for it doesn't turn to cheese in the same way. In turn I will always call out pasteurized milk by its full name.

And milk, as I define it, should be understood to be unhomogenized as well as unrefrigerated, two of many processes that negatively impact milk's cheesemaking ability. It's hard to consider your typical "milk" on the grocery store shelf as milk, considering its lack of responsiveness to the cheesemaking process. If an infant cow, goat, sheep (or human) were fed the stuff, their stomach wouldn't know what to do with it. The milk would fail to curdle in the stomach, and the young animal would have a serious case of the scours (aka the runs) and, ultimately, fail to thrive.

Milk is most often subject to various handlings that each cause significant damage. Refrigeration, pasteurization, and homogenization are the most common, and are addressed in this chapter. If what we called milk was simply milk fresh from the udder, our choice for cheesemaking would be so clear and easy to make: goat or cow or sheep or buffalo!

Because of its unfathomable complexity, milk possesses a remarkable memory. Everything leaves its mark on milk; this is seen with the beneficial ways milk transforms into cheese and also in the damage done by milk's unneeded, indeed unwanted processing.

Refrigerating milk should be recognized as the most damaging way of handling, as it is in some cheesemaking regions. Some PDOs require that milk not be subjected to chilling before cheesemaking, necessitating a twice-daily production, for farmer-cheesemaking and even for industrial-scale production. These cheeses are often considered the best in the world, and it is likely that the use of the freshest milk is the greatest contributing factor to their high quality. Yet for the most part, refrigeration is a nearly unavoidable technology that's applied to raw milk to supposedly preserve it. Unfortunately, it does not significantly slow its decay, and instead shifts its microbiology toward less favorable species than the beneficial milk microbes that prefer body warmth to grow. It is probably more dangerous to keep milk cold than to leave it at animal temperature. The only problem with leaving it animal-warm is that the milk will soon turn to cheese on its own!

Refrigeration causes a number of significant changes to milk, both biological and biochemical.

Fresh milk that's unrefrigerated, unpasteurized, and unhomogenized is best for cheesemaking. Any effort to keep milk as milk takes away from its natural ability to respond to the cheesemaking process.

Psychrotrophic (or cold-tolerant) microbes like *E. coli*, salmonella, and listeria will grow when the milk is kept cold and unfermented (the best microbes instead grow when it's kept warm!). Among other unwanted microbes, using refrigerated raw milk also encourages the growth of *Pseudomonas*, which can cause aged cheeses to glow green and develop horrible tastes! This is an unavoidable consequence of refrigerating milk. Any amount of refrigeration is detrimental, but the longer the refrigeration, the worse the effect, and invariably it leads to the need to pasteurize (which, of course, further damages the milk).

Refrigeration (along with the stirring needed to keep milk from freezing and separating in bulk tanks)

MILK • 47

also causes sensitive milk fats to rancidify due to their exposure to air, which can cause significant problems with off tastes in aged cheeses made with it (and possibly also significant health problems associated with rancid fats—something we don't talk about at all in the world of dairy). I limit refrigeration as much as possible, preferring always to use raw milk still warm from the udder for cheesemaking (this also saves energy and time and makes for a most sustainable cheesemaking). The near-constant stirring in bulk milk tanks also causes cream to churn to butter, which can almost always be seen floating atop a vat of milk that has been overhanded, causing a loss of cream and lost yields, flavor, and quality in cheesemaking.

I generally recommend using only animal-warm milk for cheesemaking or fermentation; keeping milk refrigerated for a maximum of 12 hours is marginally acceptable, so long as it's mixed with the freshest milk (which, if you're milking animals twice a day, means at least a daily cheesemaking). But the less the milk is exposed to a cooling treatment, the better it performs, and the healthier its cheese will be. Milk held for longer than 12 hours should probably be pasteurized for commercial or home production, if it's used at all for cheesemaking. Some farmer cheesemakers, both cow and goat, only milk once a day, which the animals tolerate, and which causes a reduction in milk production of around 30 percent, but this saves significant time and enables makers to make the best possible cheese with just a once-daily make (ultimately, the udder is a better milk storage vessel than a bulk milk tank—it preserves the milk with its biological protections).

If you are purchasing raw milk from a farmer or a store, it's of utmost importance that you source that milk as fresh as possible. This understanding of raw milk's degradation under refrigeration is not commonly held (aside from among raw milk cheesemakers in France, Italy, and Switzerland), and though it sometimes is sold in a store, raw milk doesn't sit well on a shelf. If you are purchasing it from a store (some states and countries allow it, but I personally don't support this—I feel raw milk should only be sold on the day of its production, as is the rule in New Zealand), be sure to pick up the milk the day it is delivered (hopefully it's delivered the day it's produced, but this isn't often

the case) and transform it hastily into cheese! Better, I'd say, to source it direct from a farmer, preferably still warm (which is, unfortunately, almost always illegal—raw milk sales, when permitted, always mandate the chilling of milk before sale).

Even the milk produced under the best circumstances from the healthiest animals out on the greenest grass degrades significantly under refrigeration and can be rendered dangerous or unsuitable for raw milk cheese. I've tasted plenty of cheeses made with the most exalted milk from heritage breeds, on perfect pasture, that were too offensive in flavor to eat (and possibly dangerous to our health from many perspectives), because the milk had been damaged by just 2 days of refrigeration.

Freezing milk, even more damaging than refrigeration, causes a severe and irreversible separation of cream, and a degradation of milk's microbial ecology as well as its fats, due to long periods involved in freezing and subsequent thawing. By the time a milk is frozen and thawed, it's been effectively refrigerated almost 24 hours! Though frozen milk can be used for soapmaking, it should not be used in any circumstances for a natural cheesemaking.

Pre- and post-dipping with iodine is damaging to milk's microbial ecology, as well as the protective microbiome of the teat and udder. Better to wash udders with 1-day-old fermented whey, I'd say, though I recognize this increases the microbial loads in milk well beyond acceptable limits for standard dairy operations. But for cheesemakers practicing natural methods, whey washing improves udder health and induces beneficial fermentations in a milk as soon as possible, resulting in the best-quality cheese.

Milking systems damage both milk and farming culture to different degrees, the most destructive being milking robots. Machine milking systems require treatments with acid, and sterilization to reduce microbiological development and biofilms in the system. The physical pumping of the milk damages its proteins (pipes and pumps get covered in milkstone from the damage) and changes its cheesemaking character. To work effectively these machines are washed in place after every milking with a collection of chemicals, which leave residues in milk and affect its capacity for beneficial fermentation. Not washing these systems results in growth of biofilms that clog the pipes, while washing with just water washes out the protective lactose and lactic acid, and results in dangerous microbiological developments.

Standard milking systems, though seen as essential to farm businesses today, are damaging to milk and also to rural economies and culture. They result in consolidation in the industry toward those that have the capacity to invest in the most advanced systems, which reduce labor needs and ultimately the cost to produce milk, forcing those who can't invest in the latest technology out of the market. This is the long and unfortunate history of nearly all dairying technologies—including modern cattle genetics, artificial insemination, cream separators, and more. Machine milking systems can be adapted to a twice-daily, animal-warm cheesemaking, to make extraordinary natural cheese. But typical milking systems are set up to refrigerate milk immediately upon exit from the udder, making the withdrawal of animal-warm milk a challenge.

Robot milkers, the latest technology to disrupt the barnyard, milk animals all day and all night long, refrigerate the milk as soon as it leaves the udder, and send it immediately to the bulk tank, making working with animal-warm milk impossible. Robots also change the habits of milk cows—they milk themselves every 4 hours and refuse to leave the barn to go to pasture because they all want to be together near their milking machine. If you've got a milking robot and you want to realize best cheesemaking practice, you should probably start by tearing it out!

Warming milk back to cheesemaking temperature (worse yet, to pasteurization temperature) from refrigeration also causes damage. Cooking milk causes it to lose calcium from its proteins at the point of heat contact where the milk is heated above its normal operating temperatures of 37°C (98.6°F). A deposit of calcium known as milkstone will build up on cookware regularly used to warm milk. This buildup can be dissolved with the use of a mild acid like vinegar. This loss of calcium is surprising to many—isn't pasteurized milk supposed to be a good source of calcium? This hinders milk's ability to coagulate, a process that is calcium-dependent. Many cheesemakers add calcium back to milk to supposedly undo this damage, but from what I understand and have experienced, this calcium does not reenter casein proteins and does not restore milk's original capacity for coagulation. Cooking milk before fermenting it is acceptable natural cheesemaking practice, and in my opinion typically improves the quality of fermented milks like yogurt; but milk's ability to respond to rennet in cheesemaking is severely damaged by any cooking. The traditional practice of cooking milk, which is not performed for the purposes of pasteurization, will be explored in chapter 5.

Pasteurization

Now, how about pasteurization? Pasteurization is a completely unnecessary technology for natural cheesemaking, and significantly challenges the practices of ecological, small-scale, and sustainable dairying. It should be a technology that's required only with the processing of industrially derived, cold-stored milk.

There are several types of pasteurization, each one of which eliminates milk's beneficial microbiologies as well as possible pathogenic species. However, some types of pasteurization are more damaging to cheesemaking than others. Bulk pasteurization, heating a vat full of milk to 62°C (144°F) for 30 minutes, is the only acceptable pasteurization for cheese milk, and should probably be employed if that milk is refrigerated for longer than 12 hours (and perhaps any refrigeration at all). This low-temperature pasteurization, sometimes referred to as a gentle pasteurization, is much less damaging to milk and still permits its transformation into cheese, albeit a bit more stubbornly than raw milk. Bulk pasteurized curd develops more slowly and has less strength than raw. Typically, only small farmstead creameries still employ bulk pasteurization—it is an older, less expensive, but also less scalable technology that is now often considered inefficient, ineffective, and obsolete.

High Temperature Short Time pasteurization, also known as in-line pasteurization, is a quick heating of milk to 72°C (162°F) for 15 seconds. This more costly form of pasteurization can be more quickly and efficiently applied to milk and is the preferred pasteurization for larger industrial producers. HTST is the standard pasteurized milk of commerce, with most milk sold in grocery stores now being HTST. Unfortunately, it renders a milk ineffective for cheesemaking by causing excessive damage to its proteins due to the high temperatures involved. HTST curd will be very slow to firm and give off whey, and if made into pasta filata cheeses will fail to stretch.

Ultra-high-temperature milk (UHT) is cooked under pressure to above 135°C (275°F) for several seconds, rendering it completely lifeless and unresponsive to cheesemaking. Such "milk" retains none of its cheesemaking qualities, for its proteins are completely denatured by the heat. Common in hotter climates where milk spoils quickly, UHT milk is becoming increasingly common in supermarkets, especially, unfortunately, with large brands of certified organic milk in the United States to better preserve milk when distributed nationally. UHT milk can still be used to feed a clabber or kefir culture, although such clabber or kefir will not taste as fine or curdle as strongly as a fresher, more delicately handled milk.

Homogenization is an aesthetic milk processing technology that stops milk's cream from rising. It is also a completely unnecessary milk treatment, but its ubiquity sheds light on the reasons for milk's nearly universal processing. Nearly all milk sold in grocery stores is homogenized but need not be for any supposed health reasons. However, milk that is sold on a supermarket shelf needs to be stabilized to stop its cream from separating. Ironically, it separates even more substantially when milk is cooked to high pasteurization temperatures, necessitating a vicious cycle of milk processing! Prior to the evolution of the supermarket, there was no such need for homogenization, for milk was delivered daily, direct from the dairy to the home, often unpasteurized.

Only with the advent of the supermarket and its First-In-First-Out philosophy was milk remade to sit on the shelf for a week or longer. Pasteurization was most widely implemented during this era, not as a public health measure (though it was certainly sold as such!) but to stop milk from spontaneously fermenting and curdling in the store; and homogenization became equally required to stop that typically high-temperature-pasteurized milk from separating its cream as it sat, forming a butter-like plug that prevented its pouring from the container!

Pasteurization is seen as one of the greatest public health achievements of the nineteenth century, but its implementation has been mostly of benefit to industrialized dairying and permitted milking animals and their milk to be mistreated, without harming the health of consumers (at least in the short term). It is a technology that has vastly altered the landscape of agriculture around the world. Formerly, dairy animals and their products were the cornerstone of rural economies. Now only those largest producers that can afford milk handling technologies dominate the industry. Its aura of modernity and near-universal adherence have completely changed our relationship with milk—a food that was rarely drunk by adults or even children before—and damaged the reputation of traditionally produced cheeses all around the world.

Milk was never made to sit on a shelf, yet all of these technologies enable "milk" to be held for long periods without decay. Unfortunately, any efforts to keep milk as milk are in vain, for milk is remarkably sensitive to its handling, and the overprocessed versions sold in the supermarket are rendered more difficult to digest and possibly damaging to drinkers by their excessive handling. Whether raw or pasteurized, milk may not do a body good—what likely makes milk more digestible and safer (and certainly more delicious, health promoting, nutritious, and sustainable) is respecting its destiny: using it still warm from the animal and transforming it naturally into cheese.

CHAPTER 3

Culture

Cheese is a fermented food—it would not exist without culture.

To begin nearly every cheese, cheesemakers start by adding a ferment. This culture, or starter, breathes life into milk; it is the seed that brings about the microbiological developments that are necessary for milk's most effective preservations and its most flavorful expressions.

There are four ways of cultivating the microbial community necessary for milk's transformation into cheese:

1. In the culture absorbed into wood and clay vessels
2. Backslopping whey from batch to batch
3. Maintaining a milk-based starter (aka clabber)
4. Keeping a kefir culture

Though different in their form, each one of these starter regimes carries forward a complex community of milk-fermenting microorganisms that evolves from raw milk, each method assuring that this essential population of microbes is in its best form though a regular cyclical feeding.

The natural approach to the fermentation of milk (and other foods) marries traditional methodologies with a more modern microbial understanding. The approach rejects nearly 150 years of what is considered by many to be dark age of food production and fermentation—the era of Pasteur. It should be understood that current culturing practices in cheesemaking are based on a microbiological understanding from the 1960s and are now largely out of touch with the reality of cheese.

The regulatory industrial regimen does not recognize cheese as a fermented food in part because the standardized definition of fermentation is the transformation of sugar into alcohol in the absence of oxygen by yeasts, common to bread baking and beer brewing. In a traditionally fermented beer or bread, however, there is a remarkable diversity of microbial interactions, both bacterial and fungal (and both aerobic and anaerobic), which cause the beer to brew and the dough to rise (also, though they don't need air to ferment, yeast need air to grow). A more realistic and representative definition of fermentation encompasses these many microbiological breakdowns of any foods that happen with or without air. As a diverse process that elevates foods as it simultaneously causes their decay, fermentation must also include the many microbiological transformations included in cheese; indeed, cheeses and fermented milks are among the most diverse and delicious of the fermented foods we eat.

All methods of effectively fermenting milk were created before science recognized that microorganisms were responsible for milk's transformation into cheese. Since microbes and their effects upon our foods were discovered under the microscope, no improvements have been made in the effectiveness of milk's fermentation; likely our modern cheese production methods and the sterile practices that have arisen from a misunderstanding of milk's microbes impede the possibilities of making our best cheese.

A French cheesemaker once said to me, "Cleanliness is the mother of cheese." To her I responded,

"Well, if cleanliness is the mother, then filthiness must be the father!" For a small amount of measured messiness is a must for the best evolution of any cheese. It is that unwashed wooden vat, the backslop of whey, or the addition of a fermented milk or kefir that assures a harmonization with the sacred biological cycles of cheesemaking, indeed of all the different forms of fermentation.

Fermentation— A Cyclical Philosophy

No sourdough bread baker would ever bake their bread simply by getting their wheat wet. Instead, they carefully cultivate and recultivate the starter that assures their bread's rise. Similarly, no cheesemaker should ever make cheese without adding culture, intentionally or not.

Fermentation is not a linear art, but a circular and cyclical one. This cycle is influenced by and a part of the sacred cycles of birth, life, and death; and these cycles of fermentation perpetuate themselves, so long as the continued circumstances they need are provided for. Most fermentations carried out industrially today, be they bread, wine, fermented vegetables, or cheese, rely on a set of ingredients and controls that assures the fermentation moves forward in a linear fashion, much like our modern Western society strives for in all of its ideals. This linearity, however, always misses the mark, neglecting the diverse cultures responsible for many beneficial transformations that assure a successful fermentation and are propagated through a more cyclical approach. Even with the goal of achieving food safety through eliminating the transfer of possible pathogens from one batch to another, the safety of the individual isolated batch is compromised by the lack of microbial diversity in this sterile approach.

Wine, beer, and cider are all traditionally and best fermented in a similar way to fermented milks and cheese, through a process sometimes called backslopping. It's an aptly named and effective food preservation strategy, for it is a certain sloppiness that carries forward the effective community of microbes that is responsible for a food's finest fermentation. There is no word more specific in the English language to describe this act; some prefer, however, to use the less messy *carry forward*. I'll often use the term *re-ferment* to explain the intended goal of backslopping, but indeed I also like to reclaim the slur of *backslop* as an effective word that inspires successful fermentation. In French, the term *repiquage*, transplant, describes this same act. In natural winemaking circles, this act is often described as *pied de cuve*, which can be translated

Natural cheese and naturally fermented bread, alcohol, and vegetables all inspire and inform one another; indeed, they exist altogether on a continuum of fermentation. LEFT PHOTO COURTESY OF DOETIE TRINKS

roughly as "bottom of the barrel"—another unsavory but effective moniker for fermentation.

Whatever you might call it, backslopping is bountiful! Traditional cultures can be cultivated endlessly through re-fermentation. If not regularly, cyclically fermented, the cultures grow, then plateau, and ultimately decline; but by regularly re-fermenting the culture at its peak, it will grow again, then reach its peak again, endlessly, never entering a state of decline unless long neglected (wherefrom it can almost always, nevertheless, be resurrected). And the cultures will retain the same microbial communities, because the microbes create conditions during their fermentation that encourage them and their counterparts to thrive while discouraging the development of unwelcome guests. Ultimately through re-fermentation, the most appropriate and safest microbes that ferment a medium are cultivated, for the medium itself may well be made to support this essential fermentation.

But where do these cycles involved in backslop originate? In the case of a sourdough starter (the most understood, enlightened, and universally respected form of natural fermentation), the diverse community of microbes in the starter comes from the microbiome of the grain—be it wheat, rye, spelt—that originates in the soil. The microbiome of the grain plays an important part in the immune function of the mother plant, protecting it from unwanted microbes—like pathogenic fungi such as rusts—as well as deepening its relationship with the soil that supports it. These essential microbes that exist upon the grain are passed down from generation to generation, from seed to seed to seed. Once the seeds ripen and fall to the soil to grow again, any ungerminated seeds are returned to the soil by the same community of protective microorganisms. Ultimately these protective/destructive microbes are those that become the microbial community of the sourdough starter, or nearly any other natural ferment. This is the same sourdough community, for example, that causes the breakdown of the grain, and also evolves into the essential grass-digesting microbial community within the rumen of cows, goats, and sheep.

These cycles of fermentation are universal, and work in more or less any fresh food that contains sugar or starch, including fruit, vegetables, grains, and milk. Indeed the fermentation of sourdough is equivalent to the fermentation of milk and may well be the same microbiological phenomenon that makes our foods most digestible and desirable. It may well even be that the rising of a sourdough bread is the same fermentation as the growth of certain types of cheese rinds—in particular *Geotrichum candidum* rinds. And all living animals—but specifically mammals—depend on these same cyclically fermented beneficial microbes.

The Biological Origins of Cheesemaking Culture

Where do milk's microbes come from? And how and why does this incredibly stable community of milk-fermenting microorganisms exist? These microorganisms have a much greater purpose than simply to make cheese; they may well be necessary for the health of all mammalian-kind.

The microbes in milk are an inextricable part of the milk itself, forming an essential aspect of its microbial ecology and contributing toward its proper functioning within young mammals' digestive tracts and beyond. Milk provides these microorganisms through biological pathways that carry beneficial bacteria from the gut to the infant via mother's milk. And milk also seems to provide nourishment for this microbial development within the infant gut in the form of lactose and other milk sugars that, in being broken down into simpler, more digestible sugars by their microbes in their gut, feed both infants and their integral microbial ecologies.

The microbial ecology gifted by mothers to their young through their milk provides many important benefits. These microbes form an essential part of infant mammals' developing immune system—essentially a microbiological defense that complements and cooperates with the body's own immune function. They contribute an essential part of an infant's digestive tract functioning, and seem to help even with the development and function of its other internal organs. These microbes may well be passed on from mother to daughter endlessly over countless generations, much like a natural starter.

A milk-fermenting culture must be fed regularly to perform at its best. This mimics the gastrointestinal tract ecology of milk-feeding mammals that typically feed on their mother's milk every 3 or 4 hours, continuously nourishing in their guts the microbial community that they are completely dependent upon for survival. The sacred microbiological cycle is one of about 3 to 4 hours in a calf (about as long as it takes for yogurt to ferment at the same temperatures). The intestine is linear, but a universal dead end exists in the intestines of all animals—the appendix in our case, the cecum in ruminants—which is believed to be a carrier of microorganisms and could be the digestive equivalent of backslop. Our internal microbiological ecologies, and the proper digestion of our food, may well depend on this re-fermentation.

From my experiments on the internal microbial ecology of young ruminants, young mammals' intestines host a milk-based fermentation that must aid the digestion and absorption of lactose in their guts. By regularly consuming milk every few hours, the milk sugars feed a continual microbial fermentation in the hindgut that aids lactose's transformation into glucose, and perhaps on to lactic acid, both of which are much more readily digestible. When contents of the hindgut of a milk-fed ruminant are put into milk, the milk ferments almost the way it does when a yogurt culture is added (it's also acknowledged that the microbes in yogurt originate in the guts of young milk-drinking ruminants)!

An important question of course arises from all of this: Does the infant or adult human gut function in the same way when it digests milk? This is not something that we can easily study; but from the understanding of guts as fermentation vessels, and because of milk's propensity for fermentation, especially in the warm animal environment, it's hard to believe that lactose is digested in any other way in our guts. Though lactose digestion is seen by medical science as an enzymatic process, infants may well digest the lactose in their mother's milk with the aid of milk-derived and milk-sustained gut microorganisms. We host at least 2 kg of microbes in our gut as adults, and probably 0.5 kg in milk-fed infants. Something's got to be feeding those essential intestinal microbial communities! It's currently understood that HMOs—human milk oligosaccharides—are the sugars that feed these microbes, and not lactose. In cow's milk and goat's milk (and human milk), however, lactose is the primary sugar (bovine oligosaccharides are very low in concentration) and therefore lactose must fuel microbiological development in calf and kid intestines, and probably in milk-drinking infants as well.

We speak of lactase persistence in adults as a genetic condition that permits the continued production of lactase enzymes that aid the digestion of lactose into adulthood, and one that gave Northern Europeans a genetic advantage over other groups in their newfound ability to drink milk as adults. But is it instead a community of milk microorganisms involved in fermenting dairy that allows proper lactose digestion? Interestingly, diagnoses for lactose intolerance do not rely on any genetic testing, but on the production of hydrogen in the gut resulting from microbial activity—mainly coliforms that release hydrogen when they consume lactose. A strong milk-fermenting culture in the gut that would be sustained by regular milk drinking would keep coliforms at bay and reduce gas production. Studies also seem to point out that symptoms of lactose intolerance fade with regular consumption of dairy products. And many lactose-intolerant people regularly eat fermented cheeses and fermented milks. Could the theory of lactose intolerance soon be disproved? To me it's a racist theory whose time has passed. I believe that what gave early pastoralists everywhere an advantage was the adoption of cheesemaking and dairy fermentation.

This cyclical fermentation of dairy in the gut must be the source of the beneficial community of microorganisms that ferments milk into cheese. It is the regular feeding of the culture—equivalent to the feeding of a calf—and the regular nourishing of its intestinal cultures that keeps the ferment active and assures a proper cheese make. This cyclical microbial ecology is simply meant to be.

Our commensal (the term means "dining at the same table") microorganisms are most likely meant to have a share of our first and most important food. If this is true, this understanding of the microbial purpose of milk is much more nuanced than the Pasteurian view of the role of microorganisms in milk, mainly that they are contaminants that cause milk's degradation.

Ultimately, however, the community of microbes in milk may not just be responsible for the building of life; it may help usher in the end of life. As in milk transformed into a cheese, and just like if a grain of wheat falls to the ground and fails to sprout, the milk-derived microbial community that resides within us returns the body of the deceased mammal to the soil by hastening its decay and decomposition when life ends.

Initiating Milk's Cyclical Fermentation

Milk can develop itself into a starter culture for cheesemaking through a process of cyclical fermentation that controls and selects for the most life-giving microorganisms that it contains, in a process that mimics its young drinkers' biology.

Milk will ferment spontaneously when left at a warm temperature (between 20 and 45°C [68–113°F]) because it contains a community of milk-fermenting microorganisms that's essential to the well-being of the infants that drink it. These microbes break down lactose into lactic acid, which causes milk to eventually sour.

But this first fermentation is a linear one, one that doesn't always proceed well because the seed of microorganisms in milk is not strong. Maternal microbes aren't transferred in milk en masse, and so the first fermentation is slow to progress. Unwanted microbes may grow when the fermentation is slow, so that first linear fermentation is rarely perfect. The slower the fermentation progresses, the stranger, gassier, and less savory the result will be, as unwanted microbes are given a chance to develop. Milk will typically thicken on its own in 2 to 3 days—longer in colder weather, quicker in warmer temperatures; longer in cleaner dairies, quicker in "dirtier" ones. (This first fermentation of milk functions best if the milk is taken from a milking machine rather than taking milk from the udder in a sterile, linear manner. Fermentation happens in the pipes and pumps of milking machines—the French refer to them as *nids à bactéries*, bacteria nests—because the plastic parts have lots of connections and tend to grow biofilms and hold on to beneficial milk-fermenting microbes due to their regular use, even when dairy farmers make their best efforts to clean and sterilize them.)

Backslopping strengthens the fermentation. Once that first milk sours from lactic acid development, it curdles, and regardless of its quality, its first fermentation can be considered complete. To backslop or re-ferment it, a small amount of that curdled milk can be transferred to more fresh milk. With the addition of the previous ferment, the seed of beneficial microorganisms is stronger, and the good fermentation that progresses will be faster. This second fermentation, the first cyclical one, will inevitably result in a better curdle in less time. And the third fermentation, made with a bit of backslop from the second, will bring the culture into perfect balance.

Milk has an inherent microbiology that is integral to its function and assists its fermentation, but milk cannot be assured to ferment well entirely on its own. For the microbiology of any given animal's milk is a crapshoot! The spontaneous fermentations of a dozen different cow's milks from a single herd will evolve in a dozen different ways, some say due to an individual animal's diet, the cleanliness of their udder, their age, their stage of lactation, et cetera. This variability, however, is more likely due to the milk's microbiology not being fully developed when it departs the udder and its generally low microbial count when backslop is absent, which results in a chaotic microbial development, especially in clean modern dairies. It is fermentation, specifically re-fermentation, that allows milk's microbiology to develop and to best express itself, and this expression is a universal one, with remarkable consistency worldwide, given the same conditions of fermentation. For the repeated feeding of milk selects for microbes that are best suited to fermenting its lactose, oligosaccharides, and so forth, and keeps unwanted opportunistic microbes from developing.

There are simply never enough active microorganisms, even in fresh warm raw milk, when milked in a modern manner (without wood, and with the use of sterile processes), to assure its best fermentation. When making cheese or fermented dairy products with a modern milk, it is therefore always best to add an active starter culture, which brings about an effective fermentation. It should also be understood that this starter overrides the inherent microbiology of any given milk, because it develops much faster and suppresses the milk's spontaneous fermentation. But the microbiology of that fermentation *is* a development of the culture inherent in milk and should not be seen as an affront to the milk's microbiological character.

The microbes in the re-fermented milk culture are the ones you want to cultivate from your milk, and your cheeses should not lose any character as a result of the addition. Many natural fermenters are upset by the concept of adding a natural starter, because it conjures ideas of commercial ferments, and prefer instead to ferment spontaneously. But this is throwing the baby out with its dirty bathwater; if anything, consistency, good flavor, and control of unwanted microbes all come about through the use of the refined, cyclically fermented cultures. The beneficial microbes exist universally in raw milks around the world, contribute to the health and welfare of young mammals that drink their mothers' milk, and have helped traditional cheesemakers make their best possible cheeses since the beginning of animal agriculture. Under no circumstances should we truly desire to make cheese without them.

Fermentation in Cheesemaking

Fermentation can be carried out in cheesemaking the same way as in traditional bread baking, the same way as in beer brewing: through carrying a culture forward from one day's production to the next and cultivating the diverse species responsible for the medium's many diverse fermentations.

The four traditional ways of culturing these magnificent milk-transforming microbes are using wooden or other natural materials in the dairy, saving whey, cultivating a milk-based clabber culture, and keeping kefir. Though each of these methods is different in its form, each one cultivates the microbes responsible for a safe and successful cheesemaking in much the same way. Each technique involves saving culture from one batch of cheese to the next, encouraging an active fermentation by refreshing the culture with milk on a daily basis. Each propagates many diverse species of milk-fermenting bacteria and fungi that help cheeses evolve their many forms and myriad flavors. And each more or less makes cheeses that evolve in much the same way as all the others (if each one is handled and fermented in the same way).

I generally recommend keeping just one of these starters, so as not to have to keep or feed too many cultures, which complicates the daily feeding and makes it less likely to be regularly done. Stick to one, and treat it right, and you'll have the best results . . . but which one should you choose? For beginner cheesemakers who have only access to pasteurized milk, kefir is most easily sourced and managed, and can help familiarize a cheesemaker with milk's fermentations. For a small cheesemaker with access to good milk, and practicing an irregular or infrequent cheesemaking, a clabber culture is easiest to manage. For those who are committed commercial producers, a whey starter is the easiest to care for; and only for the most serious producers looking to practice a most simplified and elemental cheesemaking, and working with the freshest warmest milk daily, should a wooden vat be used as a sole starter.

There are only a couple of cheeses made in this book without an added starter culture. These cheeses are made specifically without a culture to induce developments or the growth of microorganisms that aren't generally appreciated in cheese. If cheeses are made without starters, you can expect odd fungi like *Mucor* to grow on their rinds, or for the cheeses to ripen prematurely; these odd cheeses teach important lessons about the importance of fermentation in cheesemaking, but even with their oddness they are still delicious when made with the best milk. In combination with poor-quality milk, however, such cheese makes are dangerous, for in making them without a starter, the cheeses are susceptible to the development of opportunistic or even pathogenic microbes that may be present in poorly sourced or handled milk.

Historically, and in many regions of the world still today (across Latin America and Eastern Europe), cheese was made without added starters; but the cheeses made in these regions still ferment, and an added starter culture helps assure that their best fermentations are realized. The microbial quality of fresh milk in these places, where dairying is smaller and less industrialized, with less sanitation, is typically better—in my mad fermenter's opinion—than milk in the West. I can rely on it more to ferment itself than milk from more sterile dairies, but still, it's not generally good enough. For though its microbial counts are higher, they're not high enough or of a good enough quality, and the spontaneous fermentation that often arises quickly isn't perfect. Equipment in these regions is usually washed out only with water, which doesn't stop milk's fermentation but instead shifts it, diluting lactose and lactic acid, resulting in the growth of some unwanted species. From a natural cheesemaking and a public health perspective, it would simply be better not to wash out those buckets, so they carry forward a better fermentation, instead just covering them with cloth to keep out flies and other contaminants. I'm always in a hurry to add my starter to raw milk when I make cheese in these regions, to stop the strong spontaneous fermentations that often arise. Pasteurization isn't always possible in such places; fortunately, a good starter culture added in a timely fashion is always a more effective microbiological control.

But why do these methods of saving culture even work, and why should they be valued? To be convinced of the effectiveness and rightness of traditional methods, it's essential that you understand how our cheesemaking predecessors' techniques worked. So how do these traditional methods of saving culture work? This chapter provides an introduction to the various ways that milk culture can be carried forward. And for details on how to care for and manage each of these fermentations, clabber, whey, kefir and wood, see appendices A through D.

Clabber Culture

Clabber is the Irish term for fermented milk, meaning "mud" or "thick." The word was brought to America, where many Irish immigrated (along with their culture of fermenting milk), and was adopted into American English. (British English doesn't really recognize this word or concept, for in their culture milk is not left to ferment!) Nearly every dairying culture on earth has a word for this fermented milk, because everywhere on earth animals are milked, their milk ferments! And that fermented milk was appreciated historically for its flavor, its healthfulness, and its ability to make cheese and butter.

Before the days of refrigeration, it was the norm for milk to be left at ambient temperatures. When that milk, rich in milk-fermenting microbes, before sterile modern dairying practices, fermented, the milk inevitably curdled into clabber.

And while clabber is milk, fermented, a clabber culture is milk, *re*-fermented. To begin a clabber culture (you can also call it a milk mother or a fermented milk culture), milk is left at 20°C (68°F) to ferment spontaneously into clabber within several days; a small amount of that fermented milk is then fed fresh milk, much like a sourdough starter; within 12 to 24 hours at 20°C (68°F) that milk will ferment into clabber once again; and the process can be repeated, preferably on a daily basis, endlessly. This endless raw-milk-initiated fermentation can even be carried forward endlessly with pasteurized milk. For the elements of the milk are still mostly intact, and still nourish and sustain this essential culture.

The simplest form of the universal milk starter is this clabber culture. Milk curdles when it acidifies (to a pH of 4.5), due to the unique behavior of its casein proteins. Once the milk has curdled, its first fermentation is at its peak. The majority of its lactose is transformed to lactic acid, and the clabber can be refreshed with fresh milk to develop the starter again for the following day's fermentations or cheese make.

Clabber is the most assured of the four cultures a cheesemaker can keep. It evolves simply through the re-fermentation of milk, and can be judged for its strength and quality more easily than whey. It also does not need to be managed in the same way as kefir (the kefir grain can complicate things); and it does not overferment and develop gas as readily as a wooden vat.

Clabber culture is what I have come to rely upon in recent years, due to its simplicity and the straightforward carrying-forward method. Every day I take a spoonful of clabber that I prepared the previous day, put it in a jar, and pour a cup of fresh milk overtop, in a ratio of about 1 part clabber to 50 parts milk, which can be scaled up as needed. I feed the culture warm milk in the morning, leave the milk to ferment overnight, and the following day, the clabber is ready to use as a starter. But before I use it as a starter, I feed it once again, to be sure my starter is ready for the following day's cheese. Kept this way, my clabber culture has never let me down. I can, and you can, have complete faith in its capacity for fermentation.

Every day, any cheesemaker can prepare their clabber the day before their make in the appropriate quantities (again a ratio of 1:50 to 1:100); and it will be ready to add to the milk the following morning, day after day after day. But how do you know when it's ready and active? A re-fermented clabber will curdle when it's ready, due to the transformation of its lactose into lactic acid, typically 12 hours after feeding at 20°C (68°F). Once the pH has reached 4.5, that milk curdles, indicating that a majority of the fermentation is complete and the ferment is full of active milk-fermenting microbes. After 24 hours at 20°C that culture is usually at the end of its active phase; if left much longer, it won't be effective when used as a starter.

Milk can curdle at a higher pH for other reasons, such as due to the growth of protein-digesting fungi like *Mucor*, which can happen in a cheesemaking environment that is washed with water (not whey) and not sterilized. So simply observing its curdling is not necessarily enough to know that milk is fermented—a slight sour taste and a pleasant, fermented smell are essential as well. However, in a well-fed and balanced clabber culture, fed good milk, the good fermentation will keep *Mucor* and other unwanted microbes from growing, and the milk will only curdle due to its acidification. *Mucor* can also grow if milk is long refrigerated before it is fermented, whether pasteurized or not.

The thickness of the curd will be different depending on the milk—cow clabber will be strong and gelatin-y, goat clabber will have a more liquid curd, and it can be hard to tell when it's curdled. Sheep curd will thicken more substantially than cow but be softer and creamier. And atop the curdled milk, especially cow's milk, there will be a thick and luscious layer of fermented cream, the best part. The cream will rise to the top and thicken before the skim milk curdles below due to the cream straining out the fermenting microbes as it rises. It shouldn't be considered active until the whole pot of milk has curdled. I usually keep this culture in a glass jar to be able to see the development of the curd most easily.

The flavors and aromas will be pleasing, but will depend on the species of milk and its degree of fermentation. Clabber should have a mildly sour nature and mild but enjoyable animal notes, as well as a surprisingly floral character (resulting from the fermented fatty acids).

It will also have a slight yeastiness from the growth of *Geotrichum* at the surface. A strong sour flavor indicates overfermentation, which is usually achieved after 24 hours at 20°C (68°F), or 12 hours at 30°C (86°F). If the flavors are unpleasant, that's usually a sign that either the milk is older or the fermentation is irregularly practiced. Atop the clabber culture, a light velvety layer of *Geotrichum candidum*, which feeds on lactic acid and is milk's most readily fermenting fungus, will be visible after 24 hours at 20°C. This growth of surface fungi obscures the reflection of light from a window or lightbulb (you'll see fuzzy islands floating on top!).

Finally, a well-fed clabber culture's curd will be free from gas and without whey separation, showing an absence of coliforms (they produce unwanted gas—hydrogen—that can cause a cheese to blow, and a ferment to be full of bubbles). In essence the ecology of cheese in a glass, what happens in this cultured milk will happen in a cheese prepared in the same manner, and the clabber's character (it should always be absolutely delicious when prepared right) can provide a window into what the cheese will become when it ripens.

Wood

Wood was the original cheesemaking starter culture, much as it was the first sourdough starter and beer-wine-and-cider starter. Before the days of sourdough starters being kept intentionally by bakers, unwashed wooden dough troughs, in many traditions, carried culture forward from one day's bread baking to the next, and a daily bread baking would keep the starter in its active phase. All the original natural vessels for milk, including wooden buckets and vats, clay pots, gourds, and animal stomachs and skins, served as carriers of culture that assured a predictable fermentation. Cheesemaking's departure from traditional starters is mirrored in nearly every form of food fermentation; where once an active culture was fed daily by the fermenter, knowingly or not, now a packaged culture replaces this important relationship with a manufactured convenience.

After every milking, cheesemakers would pour the fresh, warm milk into a wooden vat. Following the addition of rennet to the still-warm milk, the milk would set to curd. The makers would cut the curd to

Clabber culture can be re-fermented by carrying forward and feeding the culture fresh milk daily. This universal milk starter culture can evolve from any raw milk anywhere in the world, and will come to develop the same character if treated and handled the same.

smaller pieces and stir them gently to separate the whey. Once the curds had reached the right consistency, they'd transfer them to forms to make their cheeses. The whey-filled vat would then be emptied (though not washed), or left to allow the cream that rises from the whey be skimmed, until the next milking, when the vat would be filled with fresh milk yet again and the cheese made anew. This cyclical practice of cheesemaking successfully propagated the microbes that cause cheese to evolve its best.

But the ferment may have initiated even before that. The wooden milking bucket that first receives the milk is also a carrier of culture; and no doubt it had a role in culturing cheeses in early-industrial cheesemaking practices involving steel vats. Recipes from the early twentieth century that called for making cheese in a modern clean dairy without adding a starter neglected to realize that the milk used at the time would have been pre-cultured by its handling in natural materials on the farm.

The way the wood functions in a cheesemaking context is relatively simple: It holds on to whey (or fermented milk) from one batch of cheese and carries it forward to serve as the starter for the next. The beneficial microbes in the whey continue fermenting after the cheese is made. By the next cheese make, the culture in the wood is active and ready to ferment milk again. The milk, when poured into the vat, feeds the culture carried forward in the wood, and can keep this culture active indefinitely if regularly practiced.

Traditional makers that use wood well do not wash their vats with anything but whey; washing with water dilutes the activity of the beneficial microbes and results in a delayed fermentation, with consequent loss of flavor and possible faults in the cheese. Cheesemakers who use wood make cheese on a regular (preferably daily or twice-daily) basis, assuring that the culture within the wood is always in top form. Cheeses made in other natural materials such as clay carry culture forward from one batch to the next in essentially the same way thanks to the porosity of all these materials.

A wooden vat must, however, be used right to assure that its embedded culture doesn't overferment and shift into a yeasty realm. If a barrel is emptied of whey after the make, the culture fermenting within the wood is exposed to air and can develop a rind of yeast, especially if exposed to air for days. Leaving the whey to ferment in the vat after the make can reduce air exposure and yeast growth, keeping the culture in its best condition. But if the wood vat isn't treated right, it can make curd and cheese that fail to ferment well and are full of yeast and/or coliform bubbles.

But some gas development is inherent to the culture that develops in a wooden vat. Due to the porosity of the natural material, small amounts of air pass through into the cheese that's made within, changing the microbial ecology. When lactic cheeses are long fermented in wood, they are subject to an aeration that changes their evolution compared with an impermeable vat, which encourages the growth of yeasts, resulting in a small amount of gas, and adds to the complexity of a cheese. Of course, additional nuance and flavor will come from the wood.

A new or long-unused wooden vat will lack the necessary microbes required to effectively ferment milk. So how can a cheesemaker initiate fermentation in their vat? By making the first batch of cheese with an added starter, like kefir, clabber, or whey. The vat will then carry the universal starter forward when used day after day. More information on managing wooden vessels for fermentation is provided in appendix D.

Whey Starters

Whey from a cheese can easily be saved at the end of a make, left to ferment and develop its microbial community, then used for the next day's make, day after day after day. Whey has been used as an inoculation for successful fermentation for hundreds if not thousands of years, and has a long history of use in Italy and the alpine regions of Europe, those regions practicing cheesemaking in copper kettles, where presumably the practice of saving whey began.

Unlike wooden vats, the correct use of copper kettles calls for cleaning and drying them between use to prevent verdigris—thus halting fermentation—and then intentionally adding a starter, typically in the form of fermented whey from the previous make, often with calf stomachs steeped in them to extract their rennet. Even if the practitioners didn't know they were adding a culture or microorganisms in their whey per se, the careful and regular practice of preparing stomachs by leaving them in fermenting whey overnight led to a successful fermentation in each and every make.

Whey is the yellow-green liquid that runs from the curd after it sets. It consists of the milk's water, the milk's water-soluble proteins unaffected by the rennet coagulation, and the milk's lactose, among other sugars. When the whey is first taken off the cheese, it's sweet, because most of its lactose is as yet unfermented. That sweet whey is then set aside overnight to develop its activity through the fermentation. The now sour-tasting whey (it will be lemonade-like with a pH of 4.5 or less) can then be used as a starter for making the next day's cheese. A cheesemaker can simply add the whey to the milk, then the rennet, and continue with their make. A whey starter, unfortunately, doesn't offer the same assurances of its activity as kefir or clabber, as it doesn't curdle at a pH of 4.5.

Once saved at the end of the make the whey can be fermented either mesophilically or thermophilically, depending on the desired microbiologies for a particular cheese. For making most styles of cheese, those prepared at temperatures below 35°C (95°F), leaving the whey to ferment at 20°C (68°F) for 12 to 24 hours will suffice; but for those that are cooked above 35°C—high-temperature-fermented mozzarella, washed-curd cheeses, semi-alpine, and full-alpine cheeses—it is better to incubate the whey as it continues fermenting at 40°C (104°F) for 6 to 12 hours until it's sour to select for the most appropriate thermophilic microbes.

Whey is the most fluid and effective method of carrying culture forward from one batch to the next, if cheese is made regularly. No extra milk needs to be set aside for the process—the culture is fully recycled and free. Whey is also the quickest of cultures to mix into milk. No shaking or stirring of the whey is required, and no inconsistencies result in the curd from adding curdled milk as a starter. Sometimes bits of clabber, kefir, or yogurt are found in the final cheese, but this is never the case with a whey starter. And the microbiology propagated by its regular use is most in line with the microbiology of the cheese being made. The exact conditions of the cheese make encourage the development of its most appropriate microbial ecology. You simply have to remember to save it at the end of the make.

Though some experienced cheesemakers might argue saving whey at the end of the make is microbiologically dirtier than keeping clabber or kefir starters, as hands have been in the vat, the effective fermentation of a healthy whey culture means no incidental microbes have the opportunity to develop. The easier a culture is kept, the safer its practice, as simpler practices are more likely to be carried out by their keepers, and at the end of the cheesemaking day, nothing's easier than saving whey.

This straightforward carrying-forward practice (a true cheese backslop) is capable of yielding incredibly consistent and excellent-tasting cheese—if it's practiced consistently. It is of utmost importance that cheese be made regularly to maintain the activity of a whey starter (though it can also be refrigerated once active, for up to 1 week, or frozen for months). It's the regular refreshment of the culture with milk in the vat that allows the microbes to flourish and maintain their activity.

All you really need to make cheese is whey. But where does that first whey used to make that first cheese come from? This chicken-or-egg conundrum can be solved quite easily. A first cheese can be begun with clabber, or even kefir. Whey from the first clabber- or kefir-made cheese can then be handled and propagated as a whey starter. Whey, microbiologically speaking, equates to clabber or kefir; texturally speaking, it is clabber or kefir absent its curd.

Kefir

Kefir culture takes the most intriguing form of all milk's fermentations, for it is a culture that emanates from a mysterious entity known as the kefir grain. Though the ferment is associated historically with semi-nomadic peoples of Central Asia and the cultures of the Caucasus, evidence of dairy fermentation with kefir grains goes back many generations in Europe and Latin America and may have been a part of dairying cultures there for ages, perhaps since dairying began.

Unlike grains such as wheat or barley, kefir grains contain a community of symbiotic microorganisms including many diverse species of bacteria and yeast that form touchable colonies in any milk and reliably initiate its fermentation. Kefir grains will ferment raw milk,

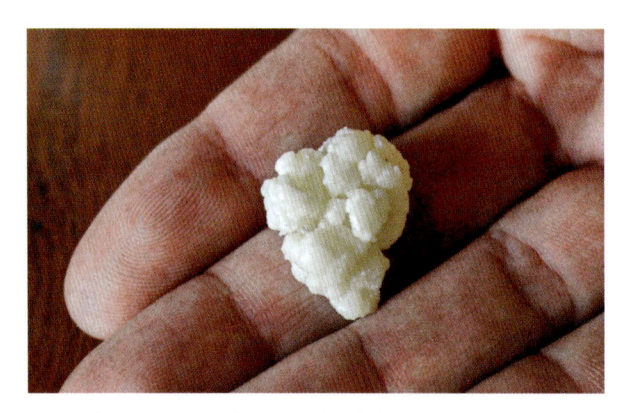

The kefir grain is a vehicle for backslop. It is an extraordinary source of beneficial milk-fermenting microbes, especially for those without access to raw milk.

pasteurized milk, even UHT milk; they'll ferment goat's milk, sheep's milk, cow's milk, buffalo milk, even human milk. But kefir grains do not grow in plant-based "milk," though they will cause them and other foods to ferment reliably if regularly fed these drinks.

No one knows exactly where kefir culture comes from or how it was first created, and as far as I understand, no kefir grains have been created from scratch in modern times (I certainly have tried!). Only kefir grains beget kefir grains, and to source the culture, you must first find a kefir keeper to share their grains. Fortunately many keepers are generous with their grains, as they grow and cannot all be kept. Once obtained, the grains can be kept indefinitely by regularly feeding them milk and making kefir, and the culture you keep can be passed down to your grandchildren, who can pass it on to theirs.

Trace any kefir grain on earth back to its origins, and you're likely to find that it originated from the same first kefir grain. However, very little is truly known about kefir's origins. Not much is written about kefir culture in antiquity; and very little archaeological evidence of its history has been found, save for one tantalizing example: A mummy (the famous and perfectly preserved Beauty of Xiaohe) was found in the desert regions of western China who was laid to rest with a necklace of dried kefir grains! Thousands of years ago this dairy culture was clearly essential to the survival of the nomadic people that lived in that region and subsisted off the fermented milk of their animals. The culture was so important to them in the present life that it was preserved in necklaces to be carried forward to the next life, where it would certainly be just as sacred!

The Beauty, as she's known, though she was found in China, was European. DNA analysis found her to be Caucasian. So what was this European woman doing keeping sheep and goats and making kefir in the far western parts of China? It may well be that she was part of an early human migration of pastoralists that brought agricultural technologies like seeds, cows, goats, cheese, and kefir grains to the far reaches of Europe and Asia. Intriguingly this woman may have been a part of an early European culture, remnants of which exist today across Europe and Asia. What makes me believe this is that her hat, with its rope and feathers, is nearly identical to the felted wool hats worn today by alpine cheesemakers in Tyrol (a region straddling Austria and Italy), who make cheeses (such as graukäse) that are more like the sour milk cheeses of Tibet and Mongolia, and less like the big wheels of rennet cheese made in the western Alps.

It may be that the early human pastoralists who first milked their animals created kefir grains unknowingly through their regular fermentation of raw milk in animal skins or stomachs. When the grains were discovered, they were recognized as an embodiment of the force that was so important to the fermentation and subsequent preservation of their milk, and were valued, perhaps even worshipped. But it could also be that kefir grains (like everything else in cheese) predate their human discovery and existed out there in the wild as part of the microbial community of raw milk within the gut or udder of a young milk-drinking mammal. Intestines, specifically the appendix, seem to support the growth of biofilms. Perhaps the natural formation of the grain allowed the culture to adhere to the lining of the intestine and aid with the organ's continual internal milk fermentation. It could also be that the community only took its tangible form once milk was re-fermented in the cooler, aerated conditions of an animal skin. Again, no one knows the origins of kefir; many cultures around the world rightfully consider the grains to be gifts from gods.

According to my experience and the understanding I've formed, kefir grains act as a vehicle for fermented milk backslop. Their purpose seems to be as a carrier of culture, rather than a source of it. If kefir grains are washed with water of the culture within their folds, they fail to ferment milk well. The cauliflower-like, fractal-esque architecture of the kefir grain maximizes its surface area as it grows; in so doing, the kefir grain becomes optimized for holding as much culture as possible when transferred from one batch of kefir to the next, assuring the successful growth and propagation of the microbes that it carries. If kefir grains are washed, the backslop is rinsed away and the fermentation time nearly doubles, resulting in the probable growth of unwanted and possibly dangerous microbes! The grain itself may just be a secretion of polysaccharides by certain species within the culture that grows over time in milk, and the tortuous shape and the ability of the grain to carry culture forward would be essential for the continued success of the entire microbiological community of kefir.

But kefir grains may be a living creature, or perhaps more precisely a holobiont: a life form consisting of a host and symbiotic species that live upon it. If the grains are cooked to temperatures above 50°C (122°F), they disintegrate and fail to grow in subsequent feedings. The grain itself may be a manifestation of an integral milk microbial community: one that is sensitive to heat, one that feeds, breathes, and grows, if it's treated right.

Kefir grains grow and grow and grow, doubling in size every 10 days or so, if fed once a day. As they grow, their structure expands itself outward, increasing its surface area and leaving a hollow core, much like an old-growth tree. If you consider that kefir grains all have a common origin, and that they are a continuously growing organism, it may well be that kefir is among the world's oldest living creatures, having been passed down from milk fermenter to milk fermenter for possibly as long as 10,000 years. And this same kefir grain has successfully fermented milk, batch after batch, for eons.

I often liken kefir grains to lichen, themselves mutualistic holobiont relationships between fungi and algae wherein different species from across the microbial kingdom don't just coexist but depend on one another for sustenance and survival. The individual cultures within kefir do not propagate without their kin, each providing essential contributions to the health and function of the community. I believe that the culture should be recognized like other organisms with a scientific name to protect the identity of kefir, which is often produced industrially with freeze-dried starters, and to give credibility to the commercial use of this natural starter, which is indeed rare (there are still only a handful of commercial producers making true kefir in North America, or any other region for that matter).

Most commercial "kefir" is cultured with freeze-dried "kefir" culture, which is a selection of laboratory-raised microbes chosen by microbiologists to ferment milk. These selected starters do not contain the great diversity of kefir grains but are a neutered version, missing the majority of the significant species, as well as being devoid of beneficial yeasts; including them and their gas production would cause plastic packaging to burst! Such "kefir" is lacking flavor, texture, probiotics, and beneficial yeasts (as well as their alcohol), cannot be propagated or used as a cheesemaking starter, and leaves its producers dependent on the same culture companies as standard cheesemaking starters.

Like seeds of beneficial raw milk microorganisms, kefir grains add active culture to milk—it's almost as if the kefir culture can unpasteurize milk by reestablishing the community of beneficial milk-fermenting microorganisms that's destroyed by pasteurization. And though

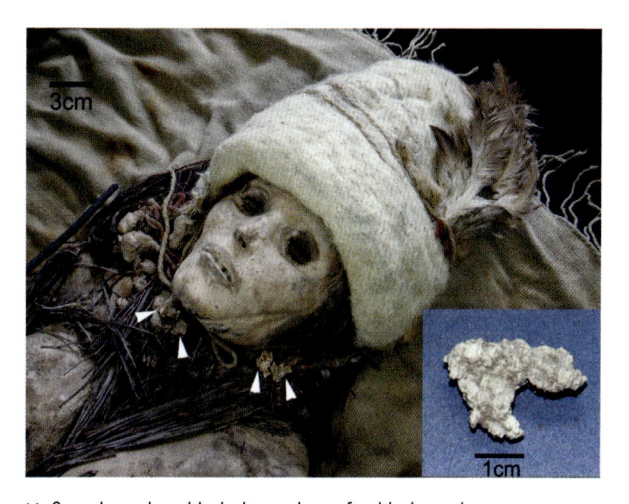

Kefir culture has likely been kept for likely as long as milking animals have been milked. But the only archaeological evidence of its ancient origins was found on a mummy named the Beauty of Xiaohe, who was buried along with kefir grains in the deserts of western China over 3,500 years ago. IMAGE COURTESY OF YIMIN YANG

the individual species in kefir may not be exactly the same as clabber made with just raw milk's microbes, the two fermented milks behave the same, taste the same, and make cheese the same way.

So how is this mystery of milky mysteries kept? The grains themselves can be easily fed by placing them in milk to ferment (a ratio of 1 part kefir grain to 50 parts milk is best); after 12 to 24 hours of fermentation at 20°C (68°F), the milk curdles as its lactose is converted into lactic acid by the microorganisms carried forward by the grain. Once a pH of 4.5 is reached, milk's sensitivity to acid causes a soft curd to form that is equivalent to clabber. The kefir grains are then strained from the curd by gently passing the kefir through a sieve (shake the kefir first in a jar to break up the curd first and facilitate the straining), and the grains are placed anew into fresh milk (without rinsing them—ever!) to ferment once again. And the liquid kefir, strained of its grains, can then be drunk, preserved in the fridge, or used as a cheesemaking starter culture. In this way, kefir culture can be placed into milk in the morning, strained of its kefir the next day (it's often thickened and fermented in 8 to 12 hours, especially in warm weather), and prepared fresh as part of a daily ritual. More information on keeping kefir can be found in its technique in chapter 12, and in appendix C.

Confounding (or perhaps enlightening) our understanding of this odd organism, kefir culture can be propagated without the kefir grain. So, what happens then when you culture kefir with just the liquid milk kefir? Are you making kefir, or are you making clabber? This question is a sort of dairy koan, for it has no answer, so long as you consider kefir and clabber to be two separate things. However, if you consider kefir and clabber to both be fermented milks (and just two of the four ways of fermenting milk), the truth of kefir comes to light.

Kefir is a fermented milk, just like clabber, that consists of a community of microorganisms that is nourished by milk, and that likely is provided by milk, but that takes a fascinating form. In essence, the kefir grain ferments milk the same way as backslopping a spoonful of clabber into milk, the same way as whey saved from one cheese to the next, and in the same manner as making cheese in a wooden vat, hosting a stable fermentation that feeds on milk's various components and can ferment indefinitely so long as the culture is regularly fed fresh milk.

But where does this community of milk microorganisms come from; why do these cultures take such stable forms; and how can we trust that this fermentation will carry on every single time? To answer to these questions, we must understand one of milk's most important purposes: Milk is meant to ferment.

A Microbiological Protection

If the microorganisms in milk form an essential part of the infant's developing immune system, can the microorganisms in a traditional cheese culture be seen as an immune system? In my understanding and experience this seems to be the case. A properly fermented culture protects fresh milk and natural cheese on many levels and may constitute a better protection than pasteurization (for milk can still be contaminated post-pasteurization).

Good fermentation quality eliminates gas development from coliforms, indicators of microbiological contamination. A naturally made lactic cheese like crottin only shows the growth of the fermentation-derived *Geotrichum candidum* on its rind, and not wild fungi, which are protected against by the *Geo*'s protective mycotoxins. Only the desired microorganisms are visible on cheeses that are made right, and believe it or not, this visual assurance is a good microbiological assurance of the microbial quality of a dairy product.

The mere presence of these beneficial microorganisms constitutes a defense against unwanted ones. And if the pathogenic microorganisms of concern in raw milk are mainly bacteriological, good fermentation practices will take care of all of them, by eliminating their food source and creating chemical compounds that impede their development. A proper fermentation in a traditional cheese make is known to be the most effective control of *E. coli*, salmonella, and listeria. It should be understood to be a control of *all* microorganisms of concern in raw milk.

But it's not just bacteria (and unwanted fungi) that seem to be controlled by these communities of milk microbes; viruses seem to be limited in their reach when natural dairy products are a part of the diet.

This may be another aspect of milk's inherent capacity for fermentation and an important aspect of these microbes' benefits that is interwoven with our immune system's function. The antiviral properties of kefir (and probably of other naturally fermented dairy products containing similar microbiologies) have been very well documented. Many reputable studies over many years point toward the effectiveness of probiotics in reducing symptoms of viral illnesses. Their regular consumption appears to bolster immunity and reduce the severity of viral illnesses. (During the early days of the pandemic, a flurry of research even identified that consumption of kefir limited the fury of the cytokine storm—a severe inflammatory reaction that many had to Covid-19.)

Furthermore, bacteriophages—viruses that can destroy an entire vat's worth of bacteria, which are of great concern for industrial cheesemakers who work with monocultures of microbes—seem to present no issues to natural producers (kefir after all has been kept for possibly 10,000 years without succumbing to them). It may well be that natural starters prevent them from developing, that perhaps they have their own virome (the viral equivalent to the microbiome) that prevents an overgrowth of unwanted viruses, or that bacteriophages are an important part of the balanced ecology of these starters and aid in the progression of microbiologies as the cultures are propagated.

Fermentation Quality

We should speak of fermentation quality the same way we address milk quality. It's something I've come to understand as one of the two strongest pillars of good cheesemaking practice. It almost doesn't matter how your rennet curdles, how you make your cheese, or how you age it; if the milk is as fresh as possible, cow-warm, and the fermentation in its best form, any cheese or fermented dairy product you make will be its best. Of course, good cheesemaking practice is important in assuring the cheese follows its proper path. But if the fermentation quality is off (or the milk quality is off), no skilled cheesemaker can save a cheese.

So how do I define *fermentation quality*? It is the power of the culture to ferment milk or cheese in an effective and timely manner and the capacity of the culture to re-ferment repeatedly at a peak fermentation time at a certain temperature (12 hours at 20°C [68°F]; 3 to 4 hours at 40°C [104°F]). If a culture is lagging because it has not been regularly fed, it cannot be considered active but should be referred to as overfermented.

A starter culture can be either inactive or active, depending on the state of its fermentation. It can be young and growing, and not yet active, which makes it an ineffective starter; it can be at its peak activity and quality, at its perfect point for making a cheese; or it can be overfermented, its activity having passed its peak, at which point it will again ineffectively ferment a cheese.

To help understand this concept, it's important to have a grasp of how the microbial communities in cultures like clabber or kefir change over time as they're regularly fed or neglected.

The Dynamics of Microbial Communities

The microbial communities in kefir, clabber, whey, or wood are as much a living entity as you or me. They eat food and create waste; they grow, sleep, and die. And though they may not dream (they may have ambitions to make a natural cheese!), they respond to love and affection. Conversely, they suffer if neglected, if not given what they need to thrive.

The activity of these cultures rises and falls over time as they consume lactose and other sugars in the milk, converting them into lactic acid and other metabolites. If neglected and left unfed, eventually a culture runs out of food and drowns in its wastes, losing its activity. To keep the culture in top form, the microbial community must be regularly refreshed with fresh milk; the infusion of food sustains the continued growth and activity of the culture.

When the natural starter is fed milk at a ratio of 1:50 or so, the culture first enters a phase of rapid growth, doubling in number about every half hour with the abundant nourishment and warm temperatures in the milk. But the exponential growth of microorganisms cannot be sustained forever on the finite amount of milk they are fed. The lactose and other sugars in the

a natural starter at 20°C

growth of lactofermenting bacteria

milk curdles at pH 4.5

culture leaves active zone after 24 hours

growth of *Geotrichum candidum*

number of microorganisms

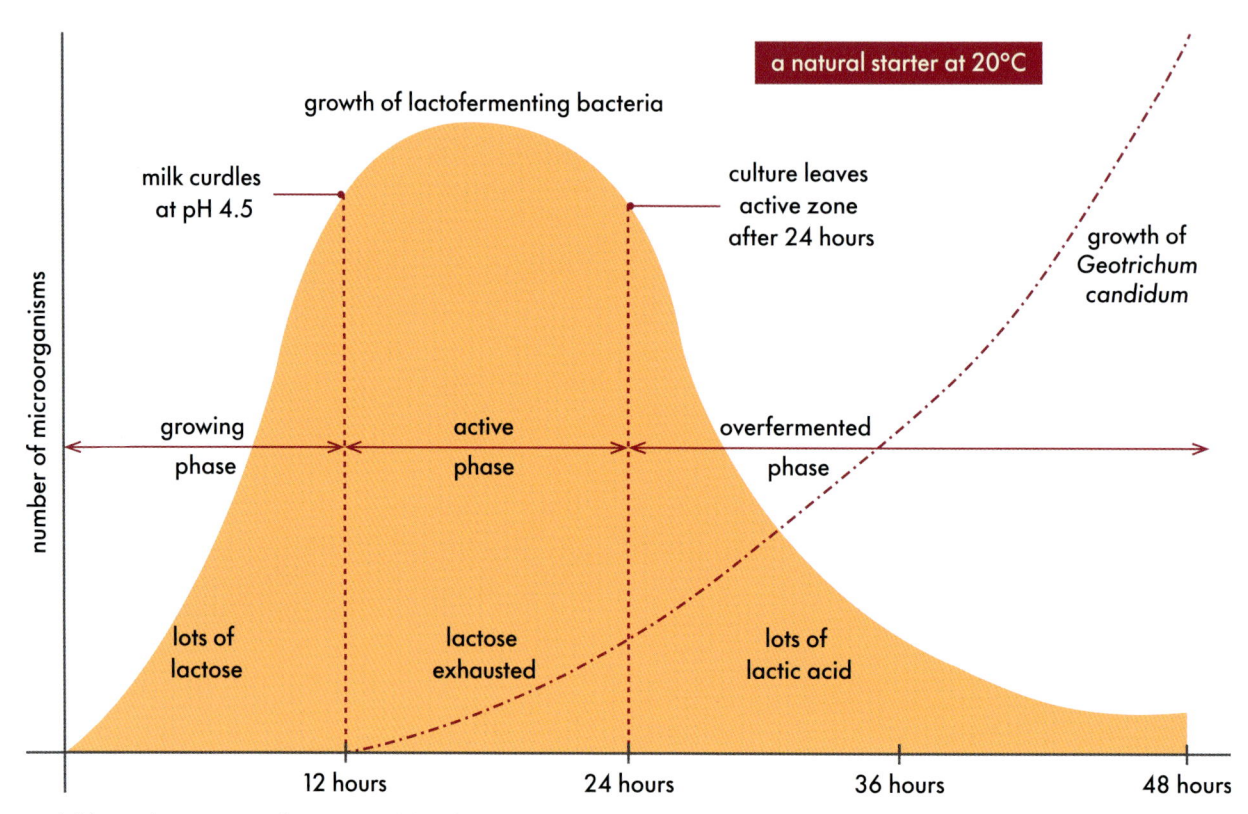

growing phase | active phase | overfermented phase

lots of lactose | lactose exhausted | lots of lactic acid

12 hours | 24 hours | 36 hours | 48 hours

A clabber culture grows, plateaus, and then begins to decline. It should be fed or used as a starter once the culture has reached its peak and thickened, usually after 12 hours at 20°C (68°F). If it's left to overferment, beyond 24 hours, the culture will become yeast-dominated and will cause off fermentations in a cheese. The same holds true for kefir or a whey culture.

milk soon begins to run out, and the metabolites of that lactose—lactic acid, alcohols, and esters—build up in the environment. As the culture begins to run out of food and is affected by its waste products all around it, its activity starts to peak. The growth of microorganisms eventually flattens, typically once most of the lactose has been transformed to lactic acid. This is when the number of microorganisms in the milk reaches its maximum (the culture is now as it was before it was fed), and conveniently it occurs once the acidity of the milk reaches a pH of around 4.5, when milk serendipitously (it seems meant to be) curdles due to its sensitivity to acid.

This peak is the time at which the culture is at its most active. And it's easy to judge, at least in a kefir or clabber culture, when the milk sets to a soft gel. This is when the culture is at its best; when it is ideal to use as a starter for a cheese, and also when it is best to reculture or feed. And this stage is usually reached about 12 hours

at 20°C (68°F) after feeding the culture, depending on the temperature and the amount of starter added.

When a small amount of this active culture is back-slopped again into fresh milk (or when that culture is used as a cheesemaking starter) the active culture finds itself again in an abundance of food and grows once again exponentially, reaching its peak activity again in about 12 hours. This peak activity can be sustained by regularly feeding the culture milk, waiting until the culture ferments before each feeding. Every day or so this culture can be fed, and every 12 hours later it will again be ready for cheesemaking or reculturing. This cyclical process of feeding and fermenting can be repeated ad infinitum.

But what if the culture isn't refed? What if it isn't refueled with fresh milk and doesn't get the love and attention it needs? The culture remains at its peak of activity for about as long as it took to ferment. If the culture took 12 hours to ferment at 20°C (68°F), it will remain in its active zone for another 12 hours at that

a natural starter at 30°C

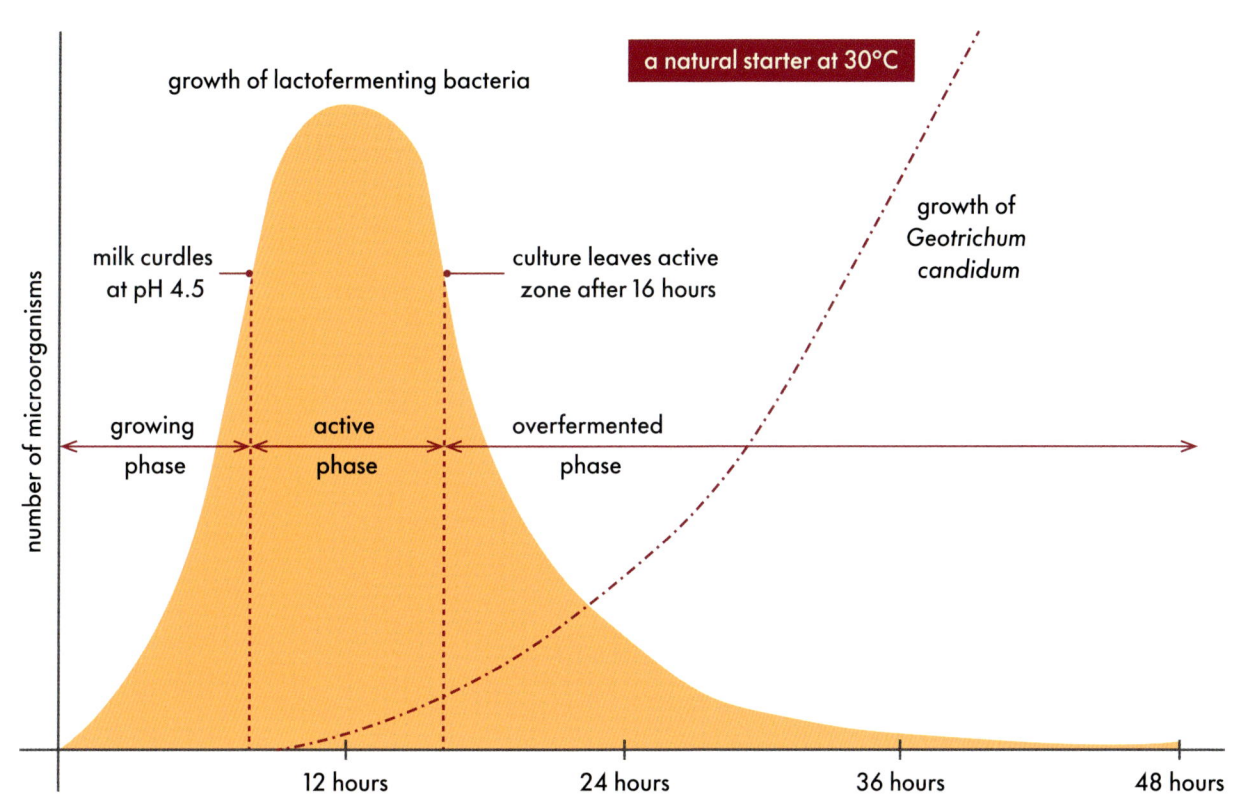

- growth of lactofermenting bacteria
- milk curdles at pH 4.5
- culture leaves active zone after 16 hours
- growth of *Geotrichum candidum*
- growing phase | active phase | overfermented phase
- 12 hours, 24 hours, 36 hours, 48 hours
- y-axis: number of microorganisms

At 30°C (86°F) a clabber culture ferments much more rapidly, reaching its peak within 8 hours, and becoming overfermented within 16 hours. The culture must be fed twice daily, or refrigerated after 8 to 12 hours, to keep it from overfermenting. Similarly, a smaller amount of starter can be added, increasing the fermentation time, or the culture can be kept in a cooler environment.

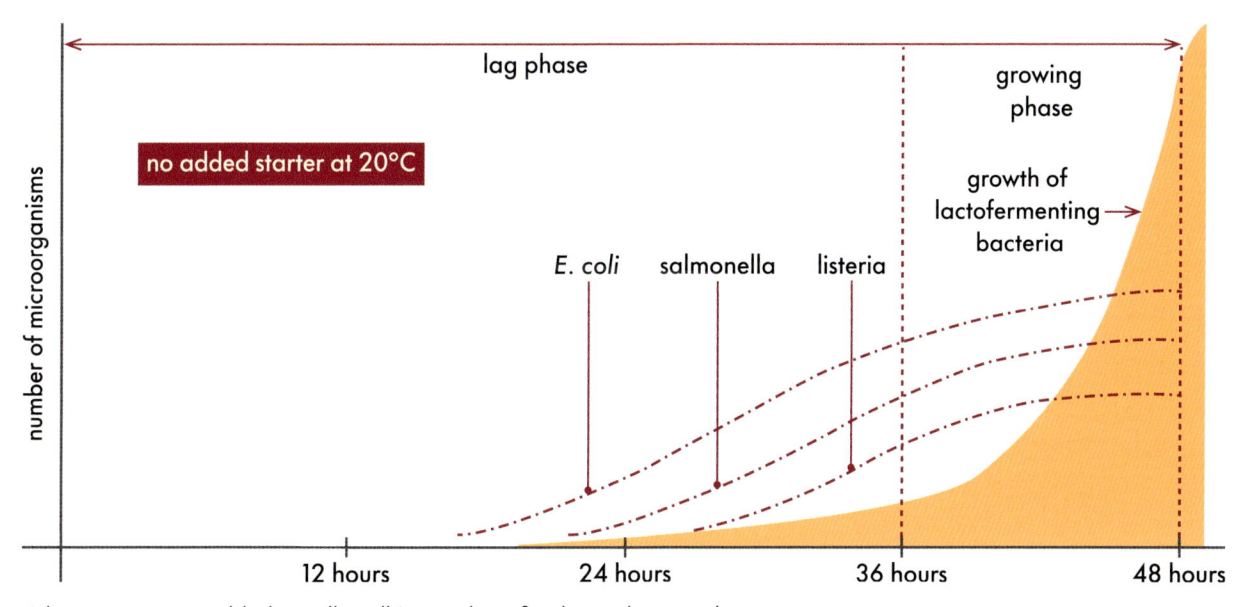

no added starter at 20°C

- lag phase | growing phase
- growth of lactofermenting bacteria
- *E. coli*, salmonella, listeria
- 12 hours, 24 hours, 36 hours, 48 hours
- y-axis: number of microorganisms

When no starter is added to milk, milk's most beneficial microbes are slow to grow, giving an opportunity to opportunistic microbes like salmonella, listeria, and *E. coli*. Similar circumstances occur when milk is long refrigerated, or when an overfermented starter is added. None of these circumstances result in cheese of a good quality.

CULTURE • 67

same temperature until, inevitably, it begins to decline. If at 30°C (86°F) it took 6 hours to ferment, the culture will leave its active phase 12 hours after being fed. Eventually the microorganisms become overwhelmed by their wastes; as the acidity declines below a pH of 4.5 and the supplies of lactose are exhausted, the culture finds itself in dire straits. By day 2, the culture is working at half its capacity; by day 4 it's down to around an eighth. Much as the culture began its life with an exponential growth, the microorganisms experience an exponential decay, slowly, inevitably losing their activity.

But when the lactose-fermenting microorganisms reach their peak, a new regime of organisms finds itself in a preferred position. Yeasts that feed on lactic acid and that need air to thrive begin to develop at the top of the culture. *Geotrichum candidum*, a fast-growing yeast that's readily propagated by backslop, reproduces itself upon the surface of these cultures (if they're kept at below body temperature, or mesophilically). The *Geo* finds a cornucopia in the wastes of the primary fermenters, ushering in a stage of secondary fermentation.

Geo will continue fermenting the lactic acid from the surface of the culture as long as the circumstances that support its growth last. So long as there is enough lactic acid, so long as the culture is kept in a warm enough environment, so long as there is enough moisture, and so long as there is sufficient air for the fungus to breathe, *Geo* will continue to grow. The fungus will metabolize that lactic acid into alcohol and other secondary metabolites, slowly consuming the contents of the ferment, including proteins and fats. Interestingly, no other fungus will grow atop the ferment, because the *Geo* protects it and ultimately consumes it.

Active versus Inactive Cultures

For a culture to ferment milk in its finest way, it must be in its active state when used as a cheesemaking starter. If the culture is used when it is inactive, because it either hasn't yet fully fermented or has overfermented, the subsequent ferment will likely fail.

If the culture is added to milk in its growing phase it will be slow to ferment the milk, due to the low numbers of active lactofermenting microbes. The culture in this phase will still be liquid milk, having not yet been acidified sufficiently to curdle. A lag time will result in the development of unwanted microbes from the milk and can cause problems in a cheese's development. It is always best to wait for a culture to curdle before using it as a starter. Unfortunately whey does not curdle when it reaches its active phase, but it can be tested for a slightly sour taste indicating a pH of 4.5 or less.

If the culture in its declining phase is added, the fermentation will also be slow, resulting in a lag phase where unwanted microorganisms will grow. And the longer a culture is neglected, the further into the declining phase it will descend, and the slower and more unpredictable the subsequent fermentation will be. Furthermore, the declining culture will be dominated by yeasty species that will cause the fermentation to progress in a yeasty manner, featuring gas development (CO_2), rising curds, and odd tastes. This unwanted gas can also be a sign of coliform development, which can grow with a lag time, allowing milk's unwanted microorganisms to develop unimpeded.

How to tell if a culture is its declining phase? It can be hard to judge by its appearance, unlike a sourdough starter, which rises and falls with its activity. Time typically tells me; if the culture was fed 24 hours ago at 20°C (68°F) (12 hours at 30°C [86°F]) and it curdled within 12 hours, the culture is most likely entering into its declining phase, though it will still work as a culture. If the culture is 48 hours old, it certainly will be a poor fermenter. You can often see the development of excessive yeasts on the surface of the older culture, indicated by the telltale wrinkling of the *Geotrichum candidum* fungus. And the culture will often taste excessively sour at this point, due to the buildup of lactic acid.

But despite their poor fermentation quality, overfermented cultures have not perished. The lactofermenting microbes mostly die when neglected, but they retain very small numbers of viable organisms. They can always be brought back from the brink through one or two successive feedings. Though the first ferment will be slow and unpredictable due to the low numbers of live microbes, the second ferment will proceed as if the culture had never been neglected in the first place. Luckily these cultures don't hold a grudge!

Mesophilic and Thermophilic Cultures

Milk's microbes tend to have a preference for either higher or lower temperatures. The terms *mesophilic* and *thermophilic* refer to a culture's preference for fermenting at medium temperatures (meso is about 20 to 35°C [68–95°F]) or warm temperatures (thermo, about 37 to 45°C [99–113°F]). The dividing line between the two different states of fermentation is just below body temperature; and our bodies therefore best support the growth and development of thermophilic microbes.

It's important to acknowledge that a traditional culture prepared at a certain temperature zone ferments a cheese best at that same zone. For example, a mesophilic culture kept at 20°C (68°F) will not ferment a cheese well at 40°C (104°F). The lower-temperature species dominating the starter are not as effective at fermenting at higher temperature; the species needed for the higher-temperature fermentation are not in abundance in the lower-temperature starter. As such, I'll include a synopsis of mesophilic and thermophilic cultures here to help folks understand the best culture to use for a cheese.

Mesophilic, or medium-temperature-loving, starters are the go-to for most of the cheeses in this book. Fermented at temperatures lower than the udder, between 20 to 35°C (68–95°F), clabber culture, kefir, and wooden vats are generally considered mesophilic starters. Whey can also be meso if it is saved from a cheese made at temperatures at or below 35°C.

Thermophilic starters prefer warmer temperatures at or above those of the udder but have an upper limit of where they will grow well. They perform best at temperatures between 37 and 45°C (99–113°F). Thermo cultures include yogurt, which ferments best at around 40°C (104°F), and whey saved from batch to batch in the making of mozzarella or high-temperature alpine cheeses like grana or Comté. Similarly, a clabber culture can be re-fermented at 40°C and will become a thermophilic clabber (different from yogurt in that the milk isn't cooked first) suitable for using as a starter for high-temperature cheese makes.

Thermophilic starters are not independent from other forms of milk fermentations; rather, high-temperature culture can take the form of clabber (fermented milk), whey (fermented milk serum), or yogurt (fermented cooked milk). Each one of these, when prepared thermophilically, is simply fermented at a higher temperature than its mesophilic counterpart. Wooden vats, however, are not amenable to a thermophilic fermentation. While kefir grains do not ferment well and may not persist at temperatures above body temperature, at high temperatures their yeasts don't thrive and the grains may not grow.

A mesophilic natural starter can be adapted into a thermophilic starter natural starter and vice versa when re-fermented two or three times at a new temperature regimen. For though they are dominated by species that prefer the temperature regimen they are kept in, they still contain other species that prefer other temperatures, though not in such high numbers.

In general, there is more microbiological diversity in a mesophilic starter than a thermophilic one, and the diversity of microorganisms seems to produce a diversity of flavors in a cheese. In microbiology this is referred to as heterofermentation. The few microbial species best adapted to a thermophilic temperature regimen tend to ferment milk in a more homogeneous way, producing primarily lactic acid from the milk's lactose. In microbiology this is referred to as homofermentation.

Mesophilic cultures are more diverse microbiologically, and hence develop more diverse products of lactose fermentation, not just lactic acid but also alcohols, ketones, and other aromatic compounds. This results in greater flavor development, not just acidity. Mesophilic starters also tend to contain, significantly, more surface yeasts and fungi, which don't grow well at body temperature or above. All cultures, be they clabber, kefir culture, whey, or even a wooden barrel, kept at 20 to 35°C (68–95°F) and fed daily will grow a white veil of *Geotrichum* after 24 hours of fermentation, indicating the establishment of a healthy surface ecology, which can help a cheese achieve rapid rind development.

There are fewer microbes adapted to growing in warmer weather, and those that thrive in thermophilic conditions tend to develop more acidity than meso ones, resulting in thermophilic cultures having lower pHs. This is why thermo yogurt is typically tangier than a mesophilic yogurt like gros lait, which is instead more flavorful than yogurt. The higher acidity further constricts the growth and development of many species, permitting only the most acid- and heat-tolerant microbes to grow. Few if any fungal or yeast species will develop in the high-temperature and high-acid conditions within a thermo starter. It has been proposed that the high temperature of our bodies is a mechanism used to control or limit the growth of fungus among us! This, unfortunately, leaves cheeses made with only high temperature starters susceptible to the growth of unwanted surface flora, as they lack a protective microbial ecology on their rinds.

With most cheeses, I'd say the choice of a meso or thermo starter is an easy one to make: Most cheeses are made at temperatures lower than the udder, and thus a mesophilic starter should be used.

Lactic cheeses, like chèvre, are long fermented at lower temperatures most suited to the use of a mesophilic culture. Rennet cheeses, like Camembert, ferment mostly between just below body temperature and ambient temperature and so are best made with mesophilic starters as well. Most styles of cheese aside from alpine cheeses follow similar fermentation rhythms and temperatures, beginning with milk at near udder temperature, and slowly lowering over time. Thus they do best with a mesophilic starter.

I recommend using a thermophilic starter for only three types of dairy products: high-temperature-fermented milks like yogurt; modern cheddar; and high-temperature-cooked alpine cheeses brought to temperatures of 40 to 55°C (104–131°F), though for alpine cheeses, I usually also recommend adding a mesophilic starter. Mozzarella and some other pasta filata styles can also be adapted to a higher-temperature fermentation by using a thermophilic starter and fermenting the curd at 40°C (104°F). Yogurt making involves keeping the fermentation temperature around 40°C for the entire make. And the yogurt itself, through backslop, or its whey, becomes the thermophilic starter for the next batch. This yogurt can be used as a starter for other high-temperature-fermented cheeses like mozzarella and full-alpine cheeses.

Alpine cheeses, whose make involves bringing the curd to a temperature over 35°C (95°F), prefer a higher-temperature culture to ensure a proper primary fermentation in their first stages of development before salting. But once the cheese cools after a day, mesophilic cultures will gain the advantage, helping to establish the important and protective rind ecologies. If only a thermophilic culture is added to such a cheese, the cheese will be susceptible to wild yeasts and fungi at the rind. According to my understanding and experience, it is therefore best to add both mesophilic and thermophilic starters to a high-temperature-cooked cheese. Ultimately the microbiological diversity of a mesophilic starter helps these cheeses develop their rind ecologies earlier, and likely therefore improves them.

More information on developing and keeping a thermophilic starter can be found in appendix E.

Primary and Secondary Fermentation

The fermentation in a cheese progresses through two or three different phases as the cheese is made and aged.

Milk, re-fermented in mesophilic conditions, comes to grow a thick veil of *Geotrichum candidum*, an important rind-ripening microorganism, at its surface. The Geo is just visible as a soft velvet coat; with the light right, its presence obscures the reflection on top of the fermented milk. PHOTO COURTESY OF MAX JONES

The two main phases, the make and the aging, can also be termed the primary and secondary fermentations. In some rare styles, the secondary fermentation is followed by a tertiary, or third, fermentation.

The primary fermentation is dominated by a fermentation of lactose into lactic acid, primarily by lactose-fermenting microorganisms; this fermentation is for the most part anaerobic (not dependent on oxygen) and performed by bacteria.

The secondary fermentation is mostly aerobic, progressing from the rind to the interior by rind-ripening microorganisms, usually fungal. These microorganisms feed primarily on lactic acid, as well as the proteins and fats in the cheese. A typical secondary fermenter is *Geotrichum candidum*, which feeds on the products left by the primary fermentation (these secondary microbes won't grow if the primary fermentation doesn't happen). *Penicillium roqueforti* is also in this category, as is the washed-rind ecology, all of which feed on lactic acid resulting from the primary fermentation. Ripening cultures are best propagated onto the ecology of cheeses through backslop (as in the case with *Geo* or the washed-rind ecology, backslopped from the rinds) or the intentional introduction of spores (as in *P. roqueforti*).

For a cheese to be well made and ripened naturally, the cheesemaker must propagate both primary fermenters and secondary fermenters. A thermophilic starter culture will only contain primary fermenters, for all yeast and fungi are impeded by the conditions created in the high-temperature thermophilic culturing process. A mesophilic starter, exposed to air, however, will grow a healthy population of surface-ripening, oxygen-consuming secondary fermenting yeasts and fungi. These will come to dominate the cheese rinds as they age.

The primary fermentations in a cheese are equivalent to the microbiological processes in an infant's gut that cause the digestion of the lactose into more digestible sugars and lactic acid. The secondary fermentations upon the rind that feed on the lactic acid are thus equivalent to the infant's metabolism that depends on the primary fermentation for the sugars and lactic acid. It's almost as if babies are equivalent to *Geotrichum* in their modes of digesting milk.

When we eat a fresh, unripened cheese, we digest the by-products of a primary fermentation, just as the rinds on a cheese consume the microbially digested, primary fermented curd. But if a cheese is left to ripen too long, ultimately the entirety is consumed by the rind-ripening ecology, leaving nothing left for us to consume. On the way, however, the secondary fermentation of the cheese makes it more digestible, breaking down proteins into more delicious and digestible amino acids, and giving the cheese its most desirable qualities.

The Universality of Natural Starters

Any cheesemaker anywhere in the world can have success cultivating the microbes they need for cheesemaking from either the milk they work with or kefir. It is an outdated understanding of milk's microbiology that states only certain cheeses can be made in certain regions; instead, it should be understood that all milks everywhere have all the microbes they need to become any cheese!

It should be recognized that these microorganisms are all a part of an agricultural package comprising their livestock hosts and their supportive pasture grasses that has traveled around the world. These microbial communities came along with cows, goats, and sheep and their favorite forage plants. They've come from the regions where these animals were domesticated, across Asia, to the highest mountains in Europe, to remote islands in the Atlantic, and to the lush dairying regions of South America as a cultural backslop.

The microbial character of starter cultures developed from raw milk is remarkably consistent regardless of where they are cultivated. The particular pathways of cultivating a starter culture select for the most appropriate microbes of milk, which are nourished by the conditions created and supported by the milk itself. And even with milks of supposedly differing microbial ecologies (we cannot fully measure the microbes in milk using even the most modern microbiological methods), creating the right conditions through refermentation selects for the best and most appropriate milk-fermenting microorganisms from this vast microbiological reservoir. Once that culture is first created from raw milk, it can be sustained even on pasteurized milk, for the community of microbes in the culture is fully sustained by the exceptional nourishment of milk, which helps keep the beneficial community in balance, even when it's not provided by the milk. That community is carried forward effectively, regardless of the milk used: The exceptional strength of the fermentation overrides any and all of the original microbiological character of the milk so long as that milk is fresh and hasn't been exposed to conditions that cause it to initiate its own fermentation.

Given the right circumstances, given the same regular re-fermentations, given differences among species, all milks anywhere in the world will come to grow the same communities of beneficial microorganisms that derive from their milk and grow best in those circumstances. All milks when re-fermented in the same circumstances will become dominated by certain lactose-to-lactic-acid-fermenting bacteria; and all milks will grow *Geotrichum candidum* atop their surfaces when re-fermented the right way.

Recognizing the strength, resiliency, and ubiquity of the culture takes a certain faith that comes from experience and understanding. I built my own faith in milk's capacity for fermentation through a combination of recognizing the merits of traditional cheesemaking and my own cheesemaking practice; carrying forward a successful fermentation by daily feeding my milk cultures (kefir and clabber) on my travels all around the world; and realizing that regardless of which continent I fermented my milk on, regardless of which animals' milk I fermented (even human milk), my culture remained remarkably consistent so long as I remained consistent with my practice.

Faith in Fermentation

Historically, the fermentation of all foods required faith because fermenters fermented unaware of the microbiological underpinnings of their efforts. The complex microbiologies of breads, preserved vegetables, meats, and cheeses were not revealed until modern times; microbes themselves were not discovered until the late 1600s, and their involvement in the fermentation of foods was unacknowledged until Pasteur made the link in the 1860s. Now, as a result of the intrusion of scientific simplifications into the production of our food, coupled with unbreakable belief in outdated science and a dependence on the expertise of the professional microbiologist, cheesemakers no longer have faith in this foundational fermentation.

Faith is not just a blind belief; it is trust that the work we do will bring rewards. We have faith that the seeds we plant in our gardens will bear fruit, and they inevitably will, so long as we plant the seeds and do the work we need to do for them to germinate and grow. We have faith in feeding young animals, knowing someday that they will feed us back. And we can have faith in feeding a sourdough or clabber culture, knowing that tomorrow the culture will be ready to bake our bread or make our cheese.

Beginning with Pasteur's postulation that unwanted microbes were responsible for food's degradations, our approach to preserving food has lost its faith. Microbes have been almost universally viewed as agents of decay; only a select few have been declared good, and they and their propagators have profited immensely. The scientific method of creating "pure cultures" has allowed individual species like *Saccharomyces cerivisiae*, brewer's yeast, to dominate beer brewing, bread baking, and winemaking, when previously dozens if not hundreds of diverse species were involved in their production. An incredible homogenization and degradation of culture, of taste, of nutrition, as well as a dependence on the scientific method and an abandonment of tradition, has resulted from this gross oversimplification.

Our belief in the scientific approach to producing food has overtaken our historical faith in fermentation. This chapter's goal is to bring back this trust in the goodness of the microbes all around us. Science is not a rock solid belief system, it's more of a sandbox; its sands shift depending on where we stand, our previous understanding collapsing on itself as we dig deeper. Our scientific understanding is constantly evolving as new research is performed. In the last decade or two a more nuanced view of milk's microbiology has evolved, one that casts a different light on the traditions of natural cheesemaking.

Fortunately a more complex understanding of milk's microbiology is evolving, one that values and confirms the traditions of cheesemaking and that can help build our faith. It turns out that milk is a fountain of beneficial microorganisms, and not sterile in the udder, as science believed a short while ago. Microbes need not be understood to be a contaminating element of milk (contamination is a state of mind); the majority of the microbes that ferment a natural cheese originally derive from the animals that provide the milk.

Not having faith that the starter will ferment, and not doing the work needed to sustain it, will cause its fermentation to fail and these unwanted contaminating microorganisms to develop (such "contamination" is not inherent, but a consequence of missteps that allow unwanted microbes to develop). Instead we can trust these cultures' ability to ferment, and trust ourselves to be able to maintain them and, together, create the best conditions for cheese's many different evolutions.

Faith takes time, experience, and understanding to build. Though an eager cheesemaker can transition completely from industrial practices to natural starters in 1 week, I often recommend keeping a traditional culture for over a month before using it as a starter to assure its success, for lacking faith in the culture's abilities can cause it to fail. Feeding a culture fresh milk daily, and watching it curdle daily, builds faith in its capacity for fermentation. Taste the culture daily, know that it is good, know that it is capable, before you faithfully add it to the vat. Know that cheesemakers have been making cheese this way for millennia, and have faith that milk's fundamental fermentations will make good cheese.

Monastic cheesemakers faithfully made their daily cheese, even on their Lord's day. Taking a day off their responsibilities to feed their animals and themselves was seen as a breach of faith. And taking the day off from cheesemaking would have set back the cultures that sustained the fermentation they unknowingly depended on for their continued good work.

Industrial Starters versus Natural Fermentation

Packaged starters or DVIs (direct vat inoculants) have many downsides and disadvantages that are not regularly discussed among their users. Cheesemakers should not feel they need to use them. It should be understood that cheeses and cheesemaking suffer from their use. And anyone, anywhere can improve their cheesemaking practice through natural starters, regardless of their scale.

Most commercial producers of cheese believe that packaged starters are the only option for cheese's proper production; that milk is deficient in the appropriate microbes and rich in dangerous ones; and that they are incapable of realizing the work that is normally done by trained microbiologists. DVIs (Direct Vat Innoculants—commercial freeze-dried starters) are considered the only acceptable way to safely make cheese, and the most convenient option for producers, big or small.

But all conveniences come with costs, dozens of them in the case of industrial starters. And every fault of these often corporate cultures can be resolved with the use of easily maintained natural starters like kefir, clabber, wood, and whey. Every single downside, both in their practical use and in the questions of their origins, is righted with natural cheesemaking.

Industrial starters are by and large produced by multinational corporations. Danisco, the most prolific starter producer, is based in Denmark and is a subsidiary of DuPont. This corporation and others like it profit off cheesemakers' demand for a product that they do not truly need.

The philosophy behind their use creates an atmosphere of microbial fear. There is no acceptance of the possibility that the microbes of your milk can work in your favor; they cannot be trusted to make good cheese, it is believed, and only microbiologist-approved strains can make cheese as its meant to be made.

Industrial starters are monocultures of microorganisms that have no precedent in nature and need perfectly sterile environments in order to function correctly. They are out of touch with the reality of cheese, which needs dozens if not hundreds of species of microbes to evolve according to their safest and most flavorful pathways. All of this is becoming even clearer through more advanced, but still incomplete, methods of DNA sequencing.

Their production is based on outdated understandings of milk's microorganisms. The practice of their use hasn't changed much since the early 1900s when selected starters were invented, aside from an entrenchment in the 1970s with the development of freeze-dried versions of the same cultures.

Freeze-dried starters should only be used once. Once the package is opened (with sterile scissors!), they should not be put back into the freezer for reuse if they are to function best. The challenges of using them according to proper protocol make them less convenient than natural starters.

These starters also cannot be re-fermented by a cheesemaker—their sensitive simplified ecologies are not propagable without the sterile laboratory environment. Cheesemakers cannot easily (or cheaply) produce a sterile enough space in their dairy to reproduce them.

The cultures have an expiration date and are slow to ferment milk after this point has passed. In fact, their effectiveness of fermentation declines daily after their date of production and leads to inconsistencies in every production of cheese.

Industrial starter monocultures are also susceptible to bacteriophage viruses. This susceptibility then demands a certain code of sterile practice that can be challenging and costly for cheesemakers to follow, and that probably exacerbates the problem. Furthermore, the strains of culture a cheesemaker uses need to be changed weekly to stay one step ahead of the phages.

Freeze-dried starters have an incomplete ecology that leaves a cheese susceptible to the growth and development of unwanted microbes. Milk is remarkably complex in its constituents, and industrial starters do not provide a diverse enough ecology to successfully ferment all of any given milk's parts, allowing opportunistic microbes an opportunity to grow.

Different starters must be bought for nearly every cheese. For a cheesemaker to produce ten different cheeses, standard protocol calls for purchasing (at least) ten different packages of microorganisms. Often only a single strain is contained within a package, and it's recommended for the production of but a single type of cheese. Even if some starters are more diverse, upon the advice of the culture producers, different collections of cultures are recommended for the production of specific cheeses.

Culture companies sell an extraordinarily complex and mind-boggling number of different industrial starters from which to choose. I don't even attempt to understand them. Danisco produces at least fifty different lines of starters, adjuncts, and bio-protecting cultures, each identified by a code. By doing so, they and others add to cheesemaker confusion, create hierarchies of understanding in the industry, and reinforce dependence on their products. It's the culture companies that inform the consultants and suppliers (and other cheesemaking guidebook authors), which are often one and the same.

Many different starters have to be added, often just to make a single cheese. A typical "informed" cheese make today will call for the addition of three different cultures: a starter, a yeast/fungus, and a bio-protector, which, ironically enough, shores up the ineffectiveness of the industrial starter.

There is a delay in their activation, often of 30 minutes or longer, a possibly dangerous period during cheese makes that can allow unwanted microbes to grow in the vat. This also adds time to the daily production.

Their fermentation is ultimately also slower than that of natural starters, which are much stronger in their ability to acidify milk than most cheesemakers expect them to be! This draws out the length of the make of industrial cheeses and contributes to the starters' ineffectiveness.

Industrial ferments require pH meters to work with successfully, another costly, difficult-to-use, inconsistent, and imprecise instrument that cheesemakers must rely on to make safe and consistent cheese with these unnatural starters.

Because they allow undesirable microorganisms to grow with their lag time and their incomplete, slow fermentation, freeze-dried starters do not function well in a raw milk cheese make. It is often recommended to pasteurize or at least thermize a milk if industrial starters are used.

They do not contain appropriate microbes for goat and sheep cheesemaking, but instead are cow-specific. They are propagated only on cow's milk, if they are even propagated on milk to begin with.

Industrial starters are never of organic origin, though as of yet they are at least still not GMOs. However, they are propagated on industrially produced milk (milk powder typically) and even packaged on this questionably sourced dried milk powder for transport. It's not a stretch for me to propose that certified organic and biodynamic cheese should be produced with natural starters. But alas, organic agriculture is often dependent on the products of industry and industrial agriculture.

Freeze-dried starters are an unneeded daily cost for cheesemakers. A typical make according to the instructions of the culture producers, with three packages of culture—two species of bacteria and one fungus—will typically cost twenty to thirty euros/dollars.

They are energy-dependent, requiring freezing and costly overseas transport. And your dairy needs to be plugged in with freezers to use them!

They are not easily accessed outside regions where they are regularly used by producers: Europe, the US, and Canada. Outside these centers, cheesemakers often cannot access the ingredients they are told they need in order to make cheese, thus rendering the practice of cheesemaking out of reach. Even in industrialized regions like Australia, import restrictions limit the availability of selected starters, making it challenging for cheesemakers to follow the industrial status quo.

There are probably even more problems that arise with these starters than I can think of! And with natural starters a cheesemaker need not lose sleep over *any* of these issues, acknowledged or unacknowledged!

The Benefits of Natural Starters

I cannot overstate the significance of the use of natural starters in cheesemaking. They are free for the world's benefit. They function fully, without flaws. And they work as nature intended. Natural starters are available for anyone, anywhere. As long as there are animals being milked, milk's integral microbial community can be propagated with successful results and extraordinary quality.

Cheesemakers can produce their own natural starters independently, enabling them to take back control of their production and have autonomy in their efforts. They empower cheesemakers to be responsible for their own fermentation, and eliminate dependency on microbiologists, their laboratories, and the corporations that typically profit from the use of industrial starters.

Natural starters can be cultured indefinitely. They have no expiration date, and if cultured regularly they can be reused infinitely! They are remarkably predicable, re-fermenting true, day after day after day. So long as the conditions of fermentation are kept consistent, the fermentations will be consistent.

Just two starter cultures, one thermophilic and one mesophilic, can provide almost all the microbes a cheesemaker needs. There is no need to add adjunct ripening cultures, like yeast and fungi, because all of the most sought-after ripening cultures are present in a typical mesophilic starter.

Natural starters allow a cheese to evolve in its most complex way, providing a diversity of microorganisms that helps with both primary and secondary fermentation. Commercial cultures were originally selected from traditional and naturally made cheeses and are now purported to be superior to them. But natural starters are more in line with the original practices that led to the development of all the cheeses of the world.

They are more in touch with the now better-understood microbial ecologies of a cheese and respect the microbial diversity that is known to exist in milk. Still, DNA sequencing cannot truly understand the scale and complexity of the microbial interactions in a natural starter or a cheese or even milk, or identify species that have not yet been identified by science, or show how the complex ecologies evolve over time.

The simple practice of daily feeding a culture on fresh raw milk is remarkably simple, monumentally easier than the significant steps required for effective use of freeze-dried starters. Milk need not be sterilized, pasteurized, or thermized to successfully propagate a natural starter. Simply feeding a culture the freshest, preferably cow-warm milk regularly assures a strong and stable and safe ferment. Starters can be propagated daily, weekly if refrigerated, or monthly if frozen.

They do not need perfectly sterile conditions to grow and be propagated; cheesemakers can easily reproduce their natural starters more safely and effectively through backslop, and unwanted microbes from the environment do not get a chance to grow. You can use more sterile methods to propagate these starters if you feel you have to, though I recommend a simpler propagation procedure on fresh raw milk, which saves time and energy and cultivates a starter specific to milk—and not boiled milk, whose sugars are changed by its cooking.

Natural starters are infinitely propagable. Kefir grains have existed for at least 8,000 years, re-fermented daily throughout modern human history and prehistory, while freeze-dried cultures cannot be reused even once with fidelity!

Natural ferments immediately initiate a fermentation in milk, resulting in zero lag time with their proper use, even dropping the pH slightly to sought-after levels immediately upon addition due to their acidity. Both of these qualities provide significant protection against unwanted species and savings in time.

Natural starters are not susceptible to bacteriophages. Because of their microbiological diversity, they do not suffer die-offs as do selected starters. In fact, the phages may play an important role in the evolution of natural starters.

No pH meters are required for successful use of natural starters. Cheesemakers can have faith in a natural ferment, observing the coagulation of the starter after a certain number of hours. When that properly fermented starter is added, there is no need to verify their effectiveness with pH meters.

You can add natural starters and rennet at the same time, and don't have to wait for measured pH drop (or Dornic degree rise) before adding the ferment. Traditionally, cheesemakers added their starters and rennet at the same time. They often considered them to be one and the same. Natural cheeses can be made more quickly this same way, and you can more readily work with the animal warmth of the milk without having to heat it back up.

Natural starters fully ferment any given milk, providing a safe and effective form of bio-protection against unwanted microbes. They consume all of the fermentable sugars in milk, due to the incomparable microbiological diversity and selection that comes from regular backslop. Unwanted microbes simply don't get a chance to grow if these methods are properly practiced.

Cheesemakers save additional time every day through their use, because goal pHs in cheese makes are typically achieved faster with natural starters thanks to their diversity as well as their ability to immediately begin to ferment a milk upon addition.

Cheesemakers can create goat-, sheep-, and buffalo-specific cultures by fermenting their natural starters on their species-specific milk (each species' milk has different sugars and fats, which support the growth of different microbes). This evolves the absolute finest flavor and safest practices for these remarkable cheeses, which are not as well fermented or flavorful with cow-specific cultures.

Natural starters can be propagated on certified organic and/or biodynamic milk, assuring a good origin for all the ingredients in a cheese.

No plastic packages are required with natural techniques. There is no costly international transport. And no refrigeration/freezing is needed if they are fed daily (twice daily in the tropics or during heat waves).

Natural starters allow microbes to be microbes, without forcing them to submit to our industrial processing needs. Natural fermentation enables the continuing evolution and natural behavior of microorganisms, including natural selection, which can happen remarkably quickly with bacteria, and horizontal DNA transfer, allowing different species to exchange their strengths and thrive.

And finally, natural starters foster a respect and appreciation for the diversity and beauty of beneficial microorganisms in milk and in cheese. They truly let us love our microbes!

CHAPTER 4

Rennet

When a young calf, kid, or lamb drinks their mother's milk, that milk transforms into cheese in their fourth stomach thanks to the effects of its enzymes, known as rennet.

This cheese formed inside the young animal's stomach is milk's greater purpose. For the milk to be properly digested, it must first be transformed into a solid that can be more easily held by the stomach, passed along slowly, and broken down thoroughly as it moves through the gastrointestinal tract.

Nearly all styles of cheese require rennet to catalyze their creation. Rennet causes milk to curdle in a way that makes it exceptionally suited to its best transformation and preservation, enabling the best elements of the milk to be separated from the liquid portion that hastens its fermentation and decay. An extraordinary diversity of cheeses evolves from the manner in which milk curdles, and the many different ways this curd can be handled. And even those cheeses that do not use rennet still evolve from milk due to milk's unique sensitivity to this magic ingredient.

The original rennet is a preparation made from the fourth stomach, also known as the abomasum, of a young ruminant. And to source these essential cheesemaking enzymes with the most natural philosophy, the sacrifice of young animals is called for. But it's important to understand these animals are not just slaughtered for their rennet; most significantly, they're sacrificed for their mothers' milk. For a cow, goat, or sheep to continue to produce milk year after year, the mother must give birth to a calf, kid, or lamb annually, most of which aren't needed in the dairy. To achieve a sustainable ideal of tip-to-tail cheesemaking, it is important to acknowledge the vital role played by the calf and (typically) their rennet.

Though not seen as such today, using animal rennet to transform milk into cheese should be understood as an ecological and ethical act. Sacrificing young animals and using their stomachs to curdle milk in lieu of raising the animals and eating them makes sense on many levels. There are many reasons why we should consider the act of raising cheese, instead of raising calves.

In the English-speaking world this transformation is not generally recognized to be a part of the biology of the animals; however, in other, more cheese-centric cultures, this knowledge is a part of the intrinsic understanding and can be related as such through the words used for the abomasum. In some languages this stomach is often referred to as the cheesemaking stomach, especially in a young animal, such as the French *caillete*, related to *caillé* or curd, or the Spanish *cuajar*, related to *cuajo* (the word for rennet). And, in general, these cultures have a greater respect and appreciation for the biological process at hand in milk's transformation into cheese.

The Rennet Effect

Nourishment from mother to infant is provided in a liquid form, what we know of as milk. Rennet helps makes that milk more digestible by turning it to a solid, which aids the subsequent breakdown and assimilation of what are perhaps the most important portions of the

the fate of the young animals born into dairies is to ignore an important truth about animal agriculture. Acknowledging the crucial role that young animals play in traditional dairying systems is a first step in taking responsibility for all dairy animals' welfare, not just the milking animals.

Once upon a time, before refrigeration, pasteurization, and stainless steel, dairy farmers around the world were also cheesemakers, and before commercial rennets were available, dairy farmers would derive their rennet from stomachs sourced from their herd or their region. The stomachs of their surplus young animals born on the farm would have provided a sustainable as well as an ethical and local source of rennet for their cheesemaking. But for a variety of reasons, as explored in upcoming pages, it's rare to find a farmer today that makes their own.

Farmstead cheesemaking operations can produce all the rennet they need from the animals of their farm. For all dairy farms produce young animals that aren't needed on the farm. To make milk, a cow must give birth to a calf every year. Only one calf is needed to replace their mother at the end of a long milking career. The other animals, especially the males, are considered surplus. To raise them is a liability as they'd drink much of the mother's milk, and therefore the farmers would be raising meat instead of milk and cheese.

These calves, known as bobby calves in some English-speaking countries, would historically be slaughtered at a very young age, their stomachs taken out for rennet making, and their flesh sold as milk-fed veal. The same held true for both goat and sheep dairies. And these sacrificial lambs were once central to our agriculture, even to our religions (the reasons why we celebrate young animals like lambs during the springtime festival of Easter is because we used to sacrifice and eat the young animals just after they were born in the spring, and use their stomachs to transform their mothers' milk to cheese).

But these days few consider the significance of the sacrifice of the calves. Rennet no longer needs to come from the herd, or even from commercially prepared calf rennet. It can be produced in endless amounts with the help of transgenic bacteria, a so-called vegetarian rennet, which should go by the name *genetically modified* as will be discussed below. Calves are no longer slaughtered on farms because we have chosen not to eat them young. Rather they are sold at auction at several days of age, often without having even drunk colostrum, shipped across continents, then raised in concentrated animal feeding operations (CAFOs) without access to pasture, and fed only milk replacer, grain, and antibiotics. The predominant animal at CAFOs in America (this same form of agriculture is growing in Europe) is the surplus dairy calf that was once sacrificed.

With the growth of the animal rights movement that began in the 1960s, we stopped eating veal, but we kept on drinking milk and we kept on eating cheese. And arguably, the most egregious industrial agricultural system has evolved from our denial of the calf's continued existence. The young animals born onto dairies are now considered too cute to eat. This Bambification of agriculture means that calves today are of almost no value to their farmers (especially the small Jersey males, who are of little worth at auction, and are now often shot and discarded at birth). Of course, there are more ethical ways to raise veal, by keeping the animal at their mother's side until slaughter; but to acknowledge the importance of their slaughter, it is exceedingly important to recognize the context into which dairy calves, kids, and lambs are born.

If the milk was taken away from young animals, farmers historically had no choice but to slaughter them. Milk and its cheese were considered the most important product of the farm. And to have the best yield of milk, but also to be able to transform the milk in its best way, with rennet, the young animals of the dairy farm would have to be slaughtered.

It should be noted that there is no gain in food that comes from keeping the calf; indeed it's quite the opposite. Raising young animals on mother's milk (or even on milk replacer) for beef results in a loss of food value compared with sacrificing the young animals and keeping milk cows for milk/cheese (it's an entropy thing). We get twice as much food from our grazing animals if we consume their cheese in place of their meat; and we need to raise and slaughter half as many animals to feed ourselves. Even if we consider pasturing and forage feeding of the young in the equation, raising calves means less pasture and forage crops for milk cows and

The relationship between a cow and calf is sacred. But sacrificing the calf for their mother's milk, their veal, and their rennet should not be seen as a violation of a natural order or an unethical act. In fact, doing so makes more food from our animals and our land, for by transforming milk into cheese we feed ourselves directly, instead of raising animals on their mother's milk and eating the animals. PHOTO COURTESY OF VIRGINIE GOSSELIN

less milk for cheese. Raising animals for milk and transforming that milk naturally to cheese should be understood as one of the most ethical and ecological ways of feeding ourselves. And there is significant precedent and history of this all around the world, for dairying was one of the most enlightened improvements of our agricultural technologies and was fundamental to the development of our modern societies.

In my cheese travels, I've encountered a few cultures that continue to sacrifice their excess young animals and use their stomachs for their cheeses. The state of Nuevo León, just south of Texas but a world away, is a traditional goat cheesemaking region of Mexico that still produces its own rennet. Many *campesinos* there make cheese from their small herds of milking goats, taking them out to pasture by horseback, and bringing the cheeses made that day back home by horseback, too.

The region is also celebrated for its *cabrito*, the meat of a "little goat." In all of the finest restaurants of the state's capital, Monterrey, cabrito features prominently on menus, and nearly all of the animals' parts are served—all the cuts of meat, the liver, kidneys, intestines. But there's one part of the little goat conspicuously missing from the menu, the fourth stomach, which is the reason why the cabrito is celebrated. At markets all around Nuevo León, the cabrito's cuajares, or abomasums, can be bought and sold for cheesemakers to use for their goat cheese. The region is one of the few in the world where young animals are still sacrificed, in part to take their mothers' milk for cheesemaking, but equally important, to have their rennet to preserve the milk as cheese.

Conversely, in Hindu tradition, cheeses are prepared without the slaughter of young animals. Fresh Indian cheeses made without rennet, like paneer and *dahi* (yogurt), can be produced without having to sacrifice calves. But the question of the calf is still one that must be addressed in India by all dairy farms, regardless of religion, regardless of their use of rennet. So what happens to excess dairy calves on dairy farms there? The Muslims of India address the issue by consuming the young animals that are taboo in Hindu society, though they are often persecuted for this symbiotic act.

The question of the calf is largely ignored; covered up, in fact, on most dairy farms and cheesemaking operations, large and small, around the world. Some of the worst practices in animal agriculture evolved from neglecting to address this important aspect of dairy farming and cheesemaking. By acknowledging the surplus calves, valuing their veal, and using their rennet to transform milk into a greater food, we can forge a more sustainable future in dairying and cheesemaking.

On the Age of the Animal

When the young animal is slaughtered, it is best if they are about 2 weeks old, and preferably only drinking mother's milk. Fourteen days seems to be the sweet spot for both a rennet stomach that's most effective for curdling milk, and a young animal that's sizable enough to be worth eating.

At 2 weeks the young animal is, more or less, a walking abomasum. Though the ruminant has four chambers in its stomach, at 2 weeks of age the fourth chamber, the abomasum, is fully developed and dominant, and has its best composition of enzymes for cheesemaking. The first three stomachs, having no purpose when the young animal is drinking only milk, are only buds that will develop their full size as the animal ages.

In the first few days of their life, the animal barely even has a fourth stomach—the abomasum develops quickly, but only after birth. It reaches a significant size within 3 days, harmonized with the transition of colostrum to milk. But once the animal has reached just 1 month of age, the first three stomachs—the rumen, reticulum, and omasum—will already be filled with digesting grasses, bubbling with fermentation. The fourth stomach will then be partly filled with their contents, messily mixed in with cheese.

At 14 days, the ruminant's first three stomachs are empty. This negates the need for cleaning of the abomasum in the making of Northern European rennet and enables goat and sheep cheesemakers to prepare their traditional Southern European rennet containing the stomach contents, primarily cheese. Some rennets are prepared from digested grass-filled older animal stomachs in Italy and consequently are green in color! But these are not as effective at making cheese, and tend to result in unwanted flavors due to their changing enzymatic profile.

To slaughter animals at a younger age also makes a more efficient and effective (and delicious) rennet. The young stomach provides more appropriate milk-curdling enzymes, and more of them. As the animal ages, their stomach also loses its coagulating power and contains less effective and less desirable enzymes, like pepsin. It's as if a young animal is born with enough rennet to coagulate all of the mother's milk that will be produced during that one lactation cycle. As the animal ages, the rennet concentration of the stomach declines and the organ's enzymatic activity become dominated by pepsin, a more generalized proteolytic enzyme that we produce in our own stomachs. But the way in which pepsin breaks down proteins causes cheeses to develop stronger, more unsavory flavors, like those in our stomachs, which take away from the fine flavor of cheese (such cheeses taste of vomit!). And chymosin is likely more efficient and effective at curdling milk—there must be a reason most young milk-drinking mammals produce it. The chymosin-to-pepsin ratio of an abomasum is around 95:5 at 1 week; 50:50 at 2 months; and mostly pepsin after 10 months, once the milk supply dries up.

Slaughtering young animals at a consistent age is part of what makes consistent traditional rennet. To slaughter every animal at 2 weeks will yield a stomach with a certain strength of rennet, whereas to slaughter at 30 days will yield a stomach with a significantly reduced strength. A standardized rennet can be created by assuring a standardized age of slaughter.

Older ruminants' stomachs, and other animals' as well, have historically been used to curdle milk. In many places in South America, the stomachs of older cows are used to produce rennet, though the quality and the yield are greatly reduced. Meat is often the priority in these regions, to the detriment of cheese quality. In Brazil and other Latin American countries, legend tells of early cheesemakers using capybara and armadillo stomachs to curdle milk. And in Italy, some cow's milk cheeses continue to be made with the stomachs of pigs. But though all of these alternative rennets do cause milk to curdle, they do so in a roundabout way and don't make the best cheese: Older animals stomachs, pig stomachs, chicken gizzards, armadillo stomachs, fish stomachs, crushed snails, and even insectivorous plants have all been used historically in different corners of the world. They contain primarily the more general proteolytic pepsin, making a poorer-tasting and -performing cheese than one made with chymosin.

On the Subject of Slaughter

If you choose to produce your own rennet from an animal, I recommend you find help with the slaughter and butchery if you're not comfortable with performing the acts. Being comfortable taking an animal's life is not necessarily a goal to strive for, but in acknowledging the importance of the calf's role in cheesemaking, and in understanding that we can receive more food from our grazing animals if we take and transform their milk ourselves, we can recognize the purposefulness of their life and accept the need for their death.

Never has it been easy for us to swallow this necessary sacrifice of young animals for dairying. Many of our (agricultural-era) religions address this issue, likely because of the duality it has always presented. Humans cherish and nurture our own young, but slaughter and eat the young of our animals. The Jewish prohibition against cooking a calf in their mother's milk (and therefore the mixing of milk and meat) likely evolves from this concern. Most religions also address this question of sacrifice—and advise on the most acceptable methods of animal slaughter to assure the least suffering. And though Christianity doesn't necessarily advise on the best practices of slaughter, some of its central mythology revolves around sacrifice—specifically Jesus's representation as the Lamb of God, sacrificed for his followers' sins.

The animal should be allowed to suckle their mother up until their last moments. The connection with the mother, as well as the drinking of milk, can help soothe the young animal and make the moments before slaughter less traumatic. The soothed animal is also easier to handle and makes less noise. All of these reasons make home slaughter that much more humane. But in many jurisdictions, killing an animal at home or on the farm is illegal; in most countries, only animals slaughtered in approved slaughterhouses can produce meat products for sale, and presumably stomachs for use in cheesemaking. But abattoirs are often distant from the farm and unfamiliar to the animal and present

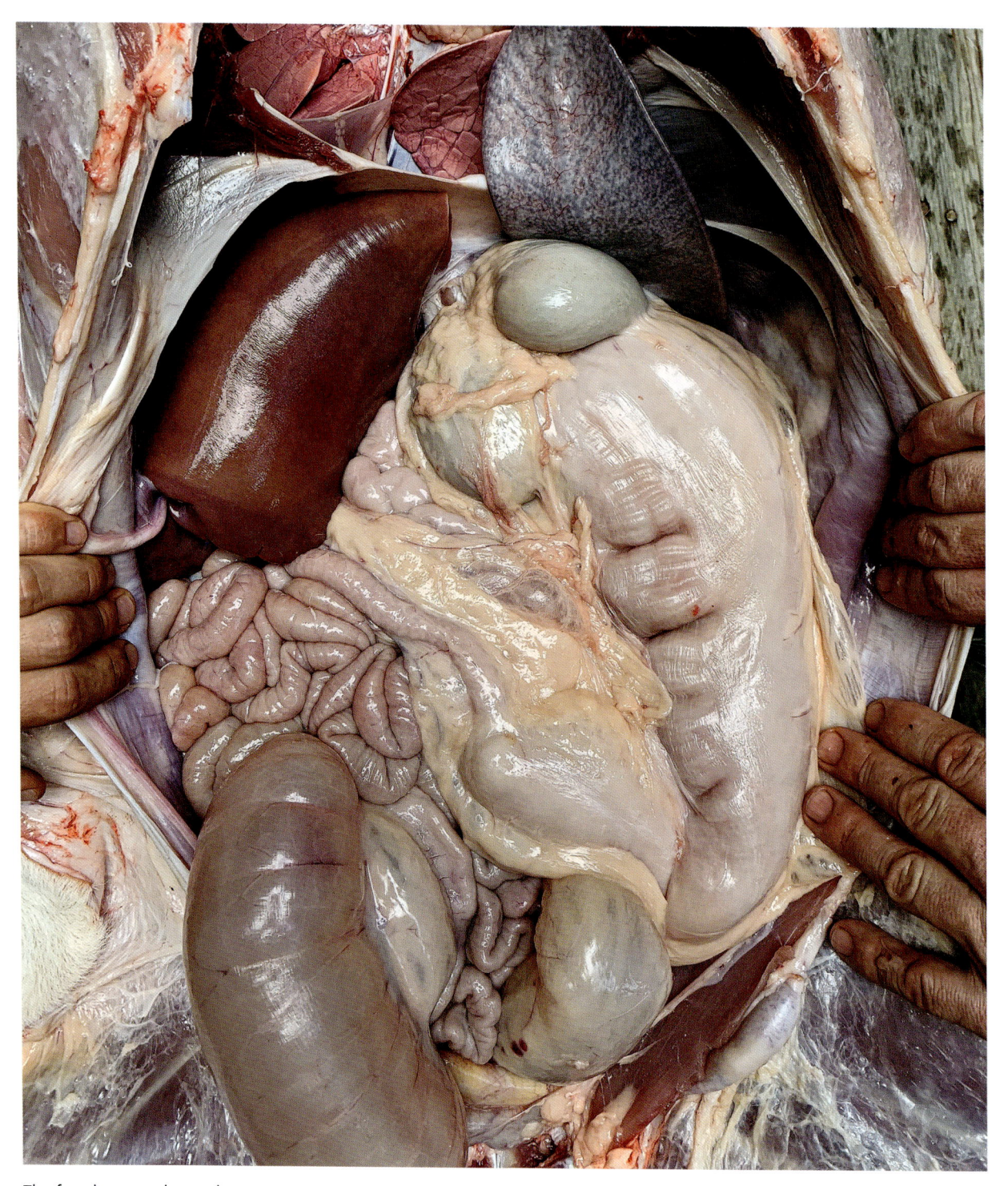

The fourth stomach, or abomasum, is positioned center-right in the abdominal cavity of a young calf. It is harvested by making two cuts, the first through the small intestine, just downstream of the abomasum, and the second, through the intensely folded omasum, just upstream of the abomasum. The stomach can be salted, hung, and dried with its contents of cheese and whey inside, as per the Southern European method of rennet production. Or the stomach can be emptied of its contents, inflated, salted, and dried, as per the Northern European method. PHOTO BY QUINN VEON

many stressful situations. This presents an unfortunate limitation on the humane slaughter of all animals, but especially young ones. Many slaughterhouses won't even consider killing such small animals.

I feel that the Jewish kosher or Islamic halal methods of slaughter, slicing through the neck of an animal with a long, sharp blade, are the most humane and the most accessible to a home slaughterer, though the dominant practice in Western nations, for supposedly ethical reasons, is for the animal to be shot or bolt-stunned before being bled. The young animal can be restrained with extra hands or ropes. Then with one hand, hold their muzzle back toward you, and with the other hand, hold a long, well-sharpened knife. With a swift and firm movement, draw the blade through the throat just below the jaw, severing the carotid artery, the trachea, and the esophagus. Within several seconds, the animal's life ends. But for several minutes after slaughter, the animal's body convulses, as the nervous system tries to make sense of its new reality.

Once the animal is still, the carcass should be hung with a gambrel through the Achilles tendon on the rear legs. The skin should be removed carefully and preserved; it makes for incredibly fine leather or pelts. (Astrakhan comes from very young karakul lambs slaughtered for their rennet, meat, and milk in sheep cheesemaking regions of Central Asia.) The abdominal cavity should be cut open, taking care not to puncture any internal organs. By cutting out the anus and tying off the intestine at that end, you can avoid spilling the intestinal contents into the carcass. If you do, don't worry, it's mostly just fermented milk. (Like a cheesy spaghetti, the cheese-filled intestines of milk fed calves, known as Pajata are cooked and enjoyed in the markets of Rome!) The gastrointestinal tract should come out in one piece, from the anus through to the esophagus, with all the organs—stomachs, liver, intestines—attached. Place it on a table to find and remove the abomasum.

The fourth stomach is easy to find: It's the final stomach in the gastrointestinal tract before the intestines, the one that is fully developed, the others being just buds when the animal is young. The texture on the inside is smooth and only slightly folded; the first three stomachs have much more significant texture (papillae, which increase the surface area and fermentative and digestive capacity of the first three stomachs). And, it should be filled with cheese from the animal's last meal of milk! The cheese will form firm grains, or be just a gentle curd, depending on how recently the animal suckled.

The Preparations of Rennet

There are likely as many ways of preparing rennet as there are of cultivating milk's microbes. I'll explore two different approaches to handling stomachs that are used in two different cheesemaking cultures. The first, from the Swiss Alps, represents the Northern European method; the second, from the island of Sardinia, represents a Southern European method.

The two methods are quite different, each being an adaptation to cheesemaking in the cultures and climates of the region, and each making a cheese that tastes specifically of that area.

The character of a cheese is to a great extent given by the rennet used to make it. And if you wish to make a cheese that tastes of a certain place, it is important that you use the rennet of that place to make its cheeses. This is as much the case with a caciocavallo, prepared with a Southern European rennet method, as it is with an alpine cheese, prepared with the Northern European rennet method. And the flavors given by their respective rennets are one of the most significant differences between the cheeses of Northern and Southern Europe.

The main aspect that distinguishes the two methods is what's done with the contents of the stomach when the abomasum is prepared. In the northern method, the stomach is emptied of its cheese, and often washed out to remove its contents. The primary rennet enzyme of this preparation is chymosin, which gives a clean and simple finish to a cheese. Whereas in the southern method, the cheese contained in the animal's stomach is left in, making a rennet with a strong lipase component, which gives cheeses made with it more flavor. Lipase is a lipolytic enzyme that breaks down fats via lipolysis to more flavorful fatty acids as the curd is broken down and digested in young animals' stomachs. Once trained to the taste of lipase in cheese, it's easy to recognize its effect. And the easiest way to train yourself to taste the difference is to taste a small amount of the remarkably flavorful cheese directly from the ruminants' stomach—it is perhaps the strongest lipase that can possibly be tasted! And it tastes of Parmigiano-Reggiano, pecorino, and caciocavallo, all cheeses made with the southern rennet and its lipase.

Calf, kid, and lamb stomachs prepared in these traditional manners in modern certified facilities are available for purchase commercially in Western Europe (from Cuajos Caporal in Spain), and in the common markets of many countries like Moldova and northern Mexico. If you happen to live in one of these places, it's possible to purchase them and possibly legal to use them in commercial cheesemaking. In many countries, however, there is no legal or even illegal source for the purchase of abomasums. Producing them yourself is likely the only way to obtain one, though the process is far from legal (it's not implicitly forbidden to use an animal stomach to curdle milk, but it's not legal to use the home-prepared stomach from a home-slaughtered animal in a commercially made cheese).

It should, however, be understood that renneting with traditionally prepared, salted, and cured stomachs is an effective, safe, and culturally appropriate way of making cheese; alas, this may not be the case where you're reading this book. But adding a piece of an animal to coagulate milk should not immediately cause alarm bells to ring, for what is milk, after all, but an animal part?

Northern European Rennet

Northern European rennet is prepared by emptying the young animal's abomasum of its contents, then inflating and drying the organ. A stomach prepared in this manner is known as a vell.

Vells are typically prepared with the stomachs of calves, the predominant milking animal of Northern Europe. The stomachs are emptied of the contents likely because the colder, wetter climate made the preservation of a large calf's stomach full of cheese more challenging. It could also be that there is a cultural aversion in the northern cultures to the half digested contents of these stomachs. Of course, the method can also be used on lamb and kid stomachs, and this may well have been done historically in Northern Europe.

To prepare this rennet, the stomach is cut from the digestive tract, including a short length of the intestine and a small part of the previous stomach, the intensely folded omasum, to include the sphincters in the vell.

The stomach is not to be cut open to be cleaned; doing so jeopardizes the process. The contents are instead worked out by hand through the upper sphincter. Once emptied, the stomach can be washed out with water to make a simpler-tasting, almost purely chymosin rennet, or can be left unwashed.

The stomachs are salted like a cheese or dried without salt—each region that prepares the stomachs, and sometimes each farmer in a region, does it differently. This diversity of rennet preparations helps contribute to the diversity of cheesemaking evolutions. If you wish to preserve the stomachs until you are ready to salt and hang many at one time, they can be frozen for several months (though this will cause some degradations in quality) or put in a container and salted with equal parts salt and stomach.

The organ is then fully inflated with your breath, with a straw stuck into its intestinal end and strings tied around the sphincters. The stomach can be left to dry in a warm, fly-free space for about 1 week. The inflated and dried stomachs can then be cleaned of fat and veins and trimmed of the narrower outlet part of the stomach, which lacks enzymes. Such a vell can be preserved for years if kept dry.

To use the vells, cut them into slices and steep in fermented (to pH 4.5—sour) whey (1 L whey per stomach) for 12 to 24 hours at room temperature to extract their enzymes. The whey extraction can then be strained. Salt can be added at around 2 to 5 percent to preserve this precious liquid, which then doesn't need refrigeration. And the liquid can be added to milk in a certain quantity based on its tested potency (see below).

In the traditional process of Switzerland and the alpine regions of France and Italy, cheesemakers combine the preparation of the starter and the rennet together. To do this, they save whey fresh from their cheese make, then add the quantity of stomach they expect to use to curdle the next day's milk. These are fermented together for 8 to 12 hours in incubators at 40°C (104°F)—thermophilically. (However, if you're preparing a rennet liquid to be preserved for months, do not expect the rennet liquid to function as a ferment; a separate starter culture should always be added.)

By my calculations, there is enough rennet in the vell of a 2-week-old calf to curdle 2,000 to 4,000 L

The traditional preparation of rennet—stomachs steeped in whey—can be found in traditional dairies around the world. The preparation is used to both curdle and ferment milk into its cheese. PHOTO BY CHLOE GIRE

milk—approximately all of the mother's milk during a single lactation cycle. But before using the rennet, its strength should be tested (see below).

Southern European Rennet

Southern European rennet is made by leaving the cheese and whey inside the stomach, then salting, curing, and drying the cheese-filled stomach. This rennet style is typically prepared with suckling lamb or kid stomachs, though it can be prepared with calf stomachs as well, as is the case in some parts of Eastern Europe. Leaving the stomach contents within contributes a greater diversity of enzymes to the cheesemaking process, not just chymosin from the stomach but also lipase from the tongue, which contributes to the fine flavor and

The rennet of Southern Europe is the young kid or lamb abomasum, with its cheese still inside, salted and hung to dry. The cheese and its enzymes, including lipase, give much more flavor to and improve the nutrition of cheeses made with these stomachs.

digestibility of cheese made with it. The stomach contents are also rich in chymosin, so wasting them—as in the case of Northern European rennet—is also a less efficient use of the animals' stomachs.

The abomasum is cut from the digestive tract in much the same way as the vell used in Northern European rennet, to include a length of the intestine and a small part of the omasum before it; but in this case the stomach contents—the cheese and its whey and anything else that young animal nibbled upon—are left inside.

The stomachs can be tied together two by two, salted like a cheese, and hung to dry in a warm and fly-free space for several weeks, and then cleaned of excess tissue. This is easier to do in the winter or early spring, when the stomachs are typically harvested, and flies are scarce.

Once dry, the stomachs are remarkably stable, even in warm climates, so long as they are kept dry. They will last for years, but will eventually be consumed by cheese mites if kept too long, or if not sufficiently salted.

To use this Southern European rennet, a dried small ruminant stomach is cut into pieces or ground to a paste. A single stomach is submerged in 1 L fermented, sour whey and left to steep for 12 to 24 hours at room temperature. The enzymes will migrate into the sour whey overnight, and the liquid can then be strained through cloth, squeezing as much as the liquid from the solids as possible. This liquid rennet extract is fairly stable, containing both lactic acid from the whey and salt from the stomachs, but additional salt can be added (up to 5 percent) to keep the enzymes for months at room temperature. *Geotrichum* will likely grow atop the lactic acid–rich liquid rennet, without damaging its strength.

Each stomach from a kid or lamb of about 1 week of age will typically coagulate up to 500 L milk. Dissolved in 1 L whey, you should expect the liquid to have a strength of 1:500; 1 part of the rennet liquid will curdle about 500 L milk. But the rennet should be tested before determining its dosage, as described below.

Cardoon Rennet

Cardoon, a relative of artichokes, has been used as an alternative coagulant to animal rennet for centuries in parts of the Mediterranean where the plant grows wild. In Portugal, Spain, Italy—specifically the islands of Sardinia and Sicily—and parts of North Africa, the plants' purple flower petals and stamens are harvested when they bloom, then dried in an airy space. Older brown cardoon petals lose their effectiveness, so flowers must be harvested and dried continuously as they bloom over the course of the summer. Purple-flowered thistles I've tested in other regions of the world seem to have a similar coagulating effect.

Producing thistle rennet takes a bit of advance preparation. The day before the make, steep the dry stamens in cool water overnight to make a tea, at a rate of about 1 gram per 4 to 8 L of milk to be curdled. Add the thistle tea to the warm milk after the culture, to get a curdle within 40 minutes at 35°C (95°F). Thistle rennet can be used in place of animal rennet for many styles of cheese but is especially suited to the making of torta del Casar (see chapter 15) and Moroccan *raïb*, a fermented curd similar to Spanish cuajada but made with thistle.

Mucor Rennet

I have found that you can even harness the coagulating powers of fungi to derive your own rennet from a microbial source grown easily enough at home. Several fungi produce proteolytic enzymes that can cause caseins to curdle, and one of these, *Mucor miehei*, can be readily grown and harvested for its enzymes. This is the source of the *Mucor* rennet commercially available to cheesemakers as a vegetarian alternative to calf rennet. Because certified organic cheese cannot be made with genetically modified rennets, as described below, and because they prefer not to scare off their conscious customers with the use of an animal rennet, most certified organic cheesemakers use *Mucor* rennet.

Essentially equivalent to koji (aka *Aspergillus oryzae*) in its ecologies, *Mucor* is a fungus used in the traditional production of sake, miso, tempeh, and many other Asian ferments before the modern microbiological interventions akin to cheesemaking settled on *Aspergillus* as the preferred fungus, probably because of its clean whiteness, compared with the gray-black appearance of *Mucor*.

Mucor is a proteolytic fungus that feeds on a wide variety of foods. Its enzymes cause milk to curdle much the way chymosin does. It can be grown at home simply enough on bread that is made without fermentation (commercial yeasted or sweet-dough bread). The bread, lacking lactic acid, is susceptible to the growth of wild fungi, and for whatever reason *Mucor miehei* is the fungus that seems to grow best in this medium when re-fermented. By leaving such bread to go moldy in the right conditions (high humidity, room temperature, backslop of a previous batch of *Mucor* bread), you can grow a pure culture of *Mucor miehei* that can be used to furnish a fungal coagulant for your cheesemaking.

The preparation of *Mucor* rennet isn't that challenging, and can be done by anyone, anywhere. But the process is demanding enough that a longer description is needed to explain its production and extraction. I've included much more information about this fungus so helpful to cheese and other fermented foods in the appendices.

Essentially, one medium-sized slice of *Mucor*-fermented bread can coagulate up to 20 L milk into cheese and a loaf of twenty-five slices can curdle 500 L. And a sustainable supply of this cheesemaking enzyme can be easily conjured.

Testing Natural Rennet

Though you can come to determine the strength of your liquid rennet preparations from experience and expectation of the coagulating capacity of a stomach or a certain number of thistle flower heads or slices of *Mucor* bread, it is best to perform a coagulation test to determine the best dosage of the rennet before using it on a larger batch.

To determine the strength of your rennet, the liquid preparation in question can be tested in various doses on small quantities of milk. It is best to use fresh raw milk in medium-sized jars (1 L or smaller) and recreate the conditions for cheesemaking therein. Warm the milk to 35°C (95°F) and add a small amount of starter (roughly 1:100) to mimic true cheesemaking conditions. Keep the jars of milk in a warm-water bath to preserve the 35°C required for optimal curdling. Then add various amounts of rennet to several jars to see which dosage coagulates the milk softly in 15 to 20 minutes; and fully in 35 to 40 minutes.

To determine the dosages to test, it's best to make an educated guess about the strength of the rennet. For example, one dried kid stomach with its cheese inside from a 1-week-old animal typically coagulates around 500 L milk in 15 to 20 minutes. If that stomach is steeped in 1 L whey, that liter of liquid should curdle 500 L milk.

The estimated dosage of the rennet should therefore be 1:500 (1 part rennet liquid to 500 parts milk). This can be one of the tests, with 1 L milk in the test jar. You'll have to add 2 ml of the rennet liquid to get the 1:500 dosage.

Four test jars can be prepared, one at 1:1000, one at 1:500, one at 1:250, and one at 1:100, spanning an order of magnitude around the estimated dosage. The rennet doses should all be added to the milk at essentially the same time, which should be noted. And the jars should be observed every few minutes to determine the time to a light curdling. The jar that curdles the milk lightly by around 15 to 20 minutes is the desired dosage. The first batch of cheese made with it on a larger scale can help fine-tune the dosage. For larger producers going this route, it helps to prepare a larger batch of rennet with many stomachs and many liters of whey to get enough liquid to last many makes without having to test the strength of the rennet too often.

The rennet liquid should be expected to preserve its strength for months, even at room temperature, if salted.

Rennet-Free Cheeses

A good number of cheeses can be made without rennet; however, rennet is necessary for the proper curd development that allows most styles of cheese to evolve. And the cheeses that are made without a coagulant typically do not preserve as well as rennet cheeses, and don't have tastes or textures as desirable. There are two different styles of rennet-free cheeses: acid-coagulated cheeses and heat acid cheeses.

Milk will curdle on its own without rennet, but to do so, the curd becomes very acidic (pH 4.5) and the soft gel that forms does not exude whey very well. It makes only a soft, sour, and high-moisture cheese. That doesn't mean that the cheese isn't good; some excellent styles derive from this sort of cheesemaking, but the cheeses' options are limited. In this book, I've included a number of rennet-free makes including kefir, clabber, skyr, tvorog, yogurt, ryazhenka, yogurt cheese, shankleesh, handkäse, graukäse, and ambarees.

Nevertheless, these cheeses made without rennet still rely on how milk is affected by rennet. For the physiology of milk is perfectly partnered with its rennet, and even if rennet is not added, the milk still curdles solely with acidity, or a combination of acidity and heat, because of the way its proteins are made to curdle with rennet.

The curdling of milk without rennet results from a sensitivity of the liquidy casein micelles to the growing acidity of the milk, which causes a stripping of calcium from their structure, causing a destabilization and consequent coagulation out of the milk, but not with the same intensity or quality as with rennet. This destabilization and coagulation happens in all milks naturally at around a pH of 4.5 to 4.6, known as milk's isoelectric point.

However, at increasing temperatures, milk is more sensitive to acid and more prone to curdling. This can be seen when heating slightly soured milk, which curdles as the temperature rises and the milk reaches its isoelectric point. Paneer is the most emblematic of cheeses made by heating milk and adding acid.

Finally, we have kalvdans, a Nordic colostrum custard that goes by many different names in many different regions. Though not exactly a cheese, kalvdans is made by slow-cooking colostrum with sugar and spice until its albumin-rich curd coagulates. Interestingly, the colostrum contains little to no casein protein, because it takes several days for the young animal's abomasum to prime itself for curdling milk. In the meantime, the colostrum's albumin is absorbed just fine by the intestines without having to be first transformed into cheese. This albumin-rich first milk is full of this heat-sensitive protein, and simply slow-cooking the colostrum causes it to solidify into a simply prepared, nourishing, and satisfying first food.

Commercial Rennet

Of course, you shouldn't feel obligated to source your own rennet from calf or kid stomachs, especially if you are unable to easily find these. Sourcing and using animals' stomachs in cheesemaking presents often insurmountable challenges, especially to commercial producers who cannot procure them with an official paper trail. It's easy enough for a farmer-cheesemaker to do it themselves, but to use them legally, they will likely have to come from a registered slaughterhouse, which may be difficult to arrange in many regions unaccustomed to the practice.

Commercial rennets work, and you shouldn't feel like you're breaking a code of practice of natural cheesemaking if you use them. I've simply chosen to emphasize the practice of using stomachs to underscore the importance of reclaiming this lost aspect of our agriculture. But before you choose to use the stuff, it's important to understand where it comes from: It's best we not turn a blind eye.

As of the writing of this book, there is no rennet manufactured in North America, and few other countries have maintained a local supply of the ingredient, even though it is entirely possible and sustainable to do so with the surplus calves born into the dairy industry. Every farmer-cheesemaker could theoretically produce all the rennet they need to transform all the milk they take from their animals. But few, if any, anywhere, do so today, because of the seeming convenience and low cost of the commercial stuff. This should be considered a cause for concern for reasons of food sovereignty, food security, and food ecology.

Commercial rennet evolved in the late 1800s with the development of standardized rennet by the Christian Hansen company (Chr. Hansen) in Denmark. Using a process that involves acidification and maceration, the rennet enzymes are extracted from young calf stomachs. The chymosin is then separated and purified, and sold to cheesemakers as ready-to-use rennet in standard strengths, in either liquid or solid form. Liquid rennets generally contain preservatives to keep them shelf-stable, which may adversely affect the microbial ecology of cheese. Powder or tablet rennets consist of enzymes and salt, and are more appropriate for a natural cheesemaking.

Commercial rennet helped to transform cheesemaking from an agricultural craft into an industry in much the same way as commercial cultures. To use them in cheesemaking is a convenience, but conveniences always come with consequences . . .

Commercial calf rennet is one of the most globally integrated of all food ingredients. Though I still recommend it as one of the most readily available natural rennets, the calf rennet whose use I advise in my last book comes mostly from milk-fed veal calves raised and slaughtered in New Zealand (New Zealand is one of the few nations that has maintained its milk-fed veal industry and continues to slaughter very young calves). Their stomachs are frozen and shipped to Austria, where they are transformed into rennet; the rennet is then shipped to North America, where it is distributed to cheesemakers around the world.

Often rennet has been on no less than three continents before it is used—quite a few food miles for an absolutely essential ingredient for making cheese, a product of the land that's considered to be the most emblematic of local foods! And how can a cheese made with such a globe-trotting ingredient possibly express a terroir?

The production of rennet is now consolidated by a small number of firms, some of which are subsidiaries of multinational agricultural-chemical-pharmaceutical companies like Danisco. As with the culture of cheese, rennet is no longer produced by farmers or cheesemakers, who now rely entirely on cheesemaking supply companies for this essential ingredient. Farmers' and cheesemakers' independence has slowly been whittled away by agricultural corporations, a trajectory paralleled in nearly every sector of food production.

Another disincentive for farmers to fashion their own is that most rennet used today is not of animal origin, and most consumers, mostly for their lack of understanding of how dairying works, prefer it this way. The alternatives are also considerably cheaper. Cheesemakers around the world now have a very affordable and attractive alternative: rennet produced in bioreactors with the superficial "vegetarian" label. But let's consider the context of these other rennets.

In the 1970s, when the Bambification of agriculture forced milk-fed veal to disappear from our dining tables, the market for young calves in America bottomed out, and many dairy farmers had no choice but to slaughter their surplus calves on farm and dispose of them. Because the veal wasn't valued, they weren't being slaughtered, and their stomachs weren't being harvested. The result was a shortage in vells and a steep rise in the price of rennet.

The first-generation microbial rennet derived from *Mucor miehei* arrived on the scene shortly thereafter, much to cheesemakers' relief. The product became widely used and made satisfactory cheese, though supposedly with several downsides including bitterness. I have not seen this effect myself, but an equivalent cheese made with animals' stomachs is certainly more savory.

In the early 1990s a second-generation microbial rennet came on the market that worked better and gave cheese comparable flavor because it contained the same enzyme found in calf rennet: chymosin. This rennet has the supposed advantage of not being derived from a slaughtered calf, but it has the unmentioned disadvantage that it derives from a genetically modified source. Produced in bioreactors with transgenic strains of bacteria modified to produce the enzyme normally made only in the stomachs of mammals, this rennet is known as Fermentation Produced Chymosin (FPC). I do not appreciate the use of the term *fermentation* in regard to this genetically modified enzyme, as it is more a biotechnological process than a microbiological one.

Its manufacturers acknowledge that their FPC cannot be used in organic agriculture because it is a product of genetic engineering; but they nevertheless deny that the enzyme itself is genetically modified. Genetically modified food labeling laws typically focus on GMOs, organisms that are genetically modified, and ingredients that are derived directly from genetically modified plants or animals. But because the genetically modified chymosin is chemically the same calf chymosin, and it is separated from the genetically modified organisms, it itself is not considered a GMO and therefore is not required to be labeled. Most of its consumers eat their cheese without knowing that it contains a genetically modified enzyme; and many cheesemakers use the product without knowing its true origin.

FPC rennet was the first genetically engineered ingredient to be approved for human consumption in the USA by the FDA in 1990. This technology was brought to us by none other than Pfizer, a company in the habit of using gene-editing technologies to commodify things that need not be commodified. It was passed through with very little, if any, testing, oversight, or opposition—it was seen as good for calves, simply by showing that it was chemically equivalent to bovine chymosin, thus paving the path for the approval of many other genetically modified foods.

The development of GM rennet is often heralded by vegans and climate campaigners as a positive step toward eliminating animals from agriculture (or even eliminating agriculture altogether!). For if we can engineer this milk-curdling enzyme, why can't we simply engineer milk? This bleak vision for the future will find us in a world fed by precision "fermentation" (genetically modified bacteria grown in bioreactors fed by corn, soy, and electrically generated hydrogen) and dominated by industry behemoths that already dominate our food systems and would love nothing more than to eliminate farms from our landscape.

Do we need any more reasons to consider making our own?

PART II
MAKING AND AGING

and the many different uses of the whey left over from the make.

To begin the chapter and help us better understand all that we do to milk to transform it into cheese, we'll first explore the origins of the cheese make, both in the infant's belly and the cheesemaker's wooden vat.

The Original Cheese Make (in the Belly!)

Warming the milk, renneting the milk, cutting the curd, stirring the curd, forming the cheese, draining the cheese: These fundamental cheesemaking steps are not just human-practiced; for whenever a young mammal drinks their mother's milk, these same parallel steps proceed within their bellies, specifically the fourth stomach, known as the abomasum, in ruminants. We often consider cheesemaking to be a product of our human culture, but everything that happens in the cheesemaking vat has a precedent in nature—in the biological origins of cheese. And understanding this original cheese make can help cheesemakers grasp why and how they handle their milk and curds the particular way that they do.

So when a young ruminant makes cheese, they raise their neck up in the air to drink their mother's milk. And that milk travels directly from the udder into their fourth stomach, still warm from the animal.

The warm milk comes into contact with enzymes and acids produced by the stomach, which cause it to take a new form: The liquid milk is transformed by the action of the stomach's enzymes released into the gastric juices and becomes a semi-firm gel. This is the action of the chymosin enzyme breaking the outer envelope of the casein micelles, allowing them to escape from their water-soluble bundles and link up with neighboring micelles. To aid the delicate process of curdling, the infant animal takes a nap. (In Central American Spanish, when infants doze off after breastfeeding, they say they are "curdling." *Echar un cuaje* means "to take a nap"!)

Once the curd has set, the stomach then cuts it. This encourages whey to evolve and cause the curd to form. Smooth muscles around the stomach cut the curd within to smaller pieces to help this happen. On inflated calf stomachs, these muscles can be clearly seen running the length of the organ, several centimeters apart. The curd is cut cleanly by the contractions of the stomach muscles, cutting the curd into slices several centimeters in size!

After cutting the curd, the stomach churns it, reducing it into consistently smaller pieces, helping the curd to evolve more whey, whose proteins and lactose sugars are digested in the intestine. The churning causes the curd to be stirred around in the stomach, disturbing its structure, and causing more whey to be released through syneresis. The curds become firmer and firmer as time progresses, which ultimately aids their digestibility.

As the curds are stirred in the acidic juices of the stomach, they also become progressively more acidic. The curds, as they achieve a pH of 4.5, soften in texture, gaining a creamier consistency and becoming water-soluble. This creamier curd more easily passes along through the sphincter into the small intestine, where these components are absorbed into the bloodstream wherefrom they nourish the young mammal.

After this biochemical breakdown of milk into cheese in the gut, a microbiological fermentation of the lactose in the whey and cheese breaks it down into more simplified sugars and lactic acid that are also more easily assimilated into the bloodstream. This fermentation is enabled by the presence of an active culture in the intestine derived from the community of microorganisms in mother's milk that ensures its proper digestion. The regular drinking of milk (every 4 hours) ensures that the intestinal culture is well fed and ready to ferment the next milk meal that comes its way. Mainstream science identifies the enzyme lactase supposedly produced in the small intestine as being responsible for lactose's ultimate breakdown. My experiments and research tell me that the breakdown of lactose into its component parts is a microbiological process. The masses of lactose-feeding microorganisms in the infant gut seem to support this idea.

There are many parallel processes at play in the stomach's transformation of milk into cheese. They can be summarized into two categories: the microbiological and the biological. And though they're not exactly in line with the transformation of milk into cheese in the vat, the biological and the microbiological transformations of milk into cheese are mirrored in milk's proper digestion.

In cheesemaking in a vat, we refer to these parallel microbiological and biochemical pathways as fermentation and curd development. To make a certain style of cheese, we must balance the two and assure that the cheese evolves the right manner. When done right, the two evolve in perfect harmony, a pas de deux of fermentation and coagulation, and the cheese becomes the best version of itself.

The Original Cheese Make (in the Barrel)

To further understand the stages of milk's transformation into cheese in the cheesemaking vat, how the process shifts from one that is innate and animal to one controlled by the human hand, it helps to consider the wooden vessels (the *gerle* or the Tina) in which the vast majority of rennet cheeses were originally made.

With the regular making of cheese in a wooden vat, the material informs the make. Milk is poured directly into the barrel, still warm from the animal, without either heating or cooling. The milk is cultured and slightly acidified immediately by the fermented whey from the previous day's batch, which remains at the bottom of the barrel and is absorbed into its slats.

Rennet is added to the still-warm milk (kept warm by the insulating wood) as soon as the vat is filled, to be sure that the milk does not cool excessively before the rennet is added; milk's most effective curdling occurs when the milk is at or near body temperature. The dosage is determined by a familiarity with the coagulant, historically, a young animal's salted and dried fourth stomach steeped in sour whey.

Once the milk has reached its strongest curdle, but before it begins to give off too much whey to be easily cut, the cheesemaker cuts the curd to smaller pieces to help expel more whey. Too long a wait will cause the curd to firm excessively but also cool, losing its responsiveness to the cheesemaking process; too little waiting will cause the curd to be too soft, and can result in a curd that shatters when handled and stirred, making a cheese too firm. There's a sweet spot that is achieved, typically, with the right rennet dose, at 45 minutes at 35°C (95°F), before the milk has soured or cooled excessively and before any visible cream has separated.

The curd is then cut with the blade of a knife to a certain size, depending on the style of cheese to be made: typically around the size of a walnut for a soft cheese, a hazelnut for a semi-firm cheese; and a grain of corn or rice for a very firm wheel. The extra surface area of the smaller curds encourages more of their whey to evolve, helping to firm the curds and make them into their cheese.

After cutting, the curd is stirred to remove more whey. The gentle stirring is performed for certain amount of time depending, again, on the style of cheese being made and its desired firmness. The cheesemaker typically tests the curd for its firmness with the help of one of several traditional tests.

However, only a medium firmness and preservability can be achieved simply through cutting and stirring. Cooking the curd, only possible in a copper kettle traditionally, or milling the curd, a popular but ancient technique that predates cooking in copper, allowed cheesemakers to make a firmer, longer-lasting cheese. Other variations, like melting fermented curd with hot water, are used to make longer-preserving and more delicious pasta filata cheeses. All of these and more evolved originally from the traditional use of the wooden vat, which could only make the curd so firm.

Once the curd is determined to be ready, the cheesemaker forms their cheeses by ladling softer curds out of the vat into their cheese forms, or pressing the curd under the weight of the whey into a firm and solid cheese before draining (à la semi-firm Saint-Nectaire).

The curd is then left to drain on a draining table. Flip it regularly and press softly if the cheese is firmer. Leave the cheeses to drain for several hours up to several days, depending on the style and the desired amount of fermentation, before salting. Now that the curd is resting and slowly becoming its cheese, it acidifies from its fermentation, and the amount of fermentation allowed to pass before salting is a defining character for many styles (the minor amount of acidification in the vat is not an important factor in most cheeses development, aside from lactic styles). The wheels are typically kept covered and safe from flies or drying breezes, both of which can ruin cheeses.

With the addition of the salt and subsequent draining, the make is considered complete. With fresh cheeses, the wheels are consumed after salting and draining, or kept cool to preserve them. With aged cheeses, the make shifts to the affinage once the salt has pulled its moisture out, and the cheeses shift from a primary lactic fermentation within their flesh to a secondary fermentation that dominates their development from the rind.

We'll now explore those various stages of the cheese make and how to best perform them. For most styles of rennet cheese, the stages are remarkably similar. However, for lactic cheeses, the approach to the make is changed considerably. These different cheeses are often singled out for their unique handling methods and deserve a special mention in this chapter.

Warming the Milk

The first step in the cheesemaking process may not be a step at all, because milk comes out of the cow, goat, or sheep at precisely the right temperature for cheesemaking, about 35°C (95°F). At body temperature the enzymes that curdle the proteins of the milk curdle most effectively . . . and the microorganisms that ferment the sugars of the milk ferment most actively. Taking the milk right from the udder and submitting it simply to a filtration before putting it into the cheesemaking vat is the best practice for many reasons including energy efficiency, simplification of cheesemaking systems, food safety, cheese quality, and more.

If the milk has been cooled before cheesemaking, it must be warmed back to just below body temperature, about 35°C (95°F). Too quick a warming can cause the milk to burn or lose calcium as milkstone. Too slow a warming can cause unwanted fermentation before the official cheese make begins.

The cheesemaking process always evolves better if the milk is fresh and still warm from the udder. And though this means that as a farmer-cheesemaker you must make cheese more often, the use of warm milk and natural methods simplifies the cheesemaking to the point where it can become as routine as milking the animals. (Some cheese makes, like the lactic cheese method, barrel butter, and skyr, are simpler to execute after milking than putting the milk into the refrigerator, and ultimately the result is much safer, much better preserved, and much more delicious!)

Culturing Milk

The addition of the culture, be it clabber, kefir, or whey, is the initiation of the milk's microbial transformation into cheese. The starter culture added should always be well fed and showing signs of good fermentation quality—it should curdle and acidify within 12 hours at 20°C (68°F) (6 hours at 40°C [104°F] if thermophilic), and preferably also have a healthy growth of *Geotrichum candidum* atop, visibly obscuring the reflection of light on the surface. Feeding a mesophilic culture daily is the best way to assure this.

The simple use of a wooden vat is an example of the most effective protocol with regard to adding culture. The wooden vat becomes the culture when treated right: a regular daily use—if not twice-daily—and left full of whey between uses to assure the best cyclic fermentation (but also to strain off any whey cream that rises to the top!). The vat is emptied just before the make, and fresh warm milk is poured in overtop any remaining whey. Thus the culture is added to the milk at the soonest possible moment after milking.

Culture should be added to milk in a ratio of around 1:50 to 1:200; for an easy calculation, use 1:100. I personally do not measure my starter, preferring to simply add, by feel, a small amount of culture to a large amount of milk. Surprisingly, the proportion of starter added does not have a significant impact on the speed of the make. For the number of microbes in warm milk double in number about every 30 minutes, so adding double the culture will only speed up the cheesemaking process by 30 minutes. However, adding an exponentially higher amount of starter, say 1:10, will cause an excess in the development of yeasts, which are normally slower to grow than bacteria. Regularly adding a higher ratio of starter adds a larger population of already established yeasts, transforming the ferment. This is the ratio of starter culture used in sourdough, specifically to encourage its yeasts to grow, as well as in the making of alcoholic milkbeer.

What matters much more than culture quantity is culture quality. Adding a tiny, homeopathic dose of a well-tended starter will initiate a strong fermentation, for the healthy culture will grow quickly. However, with a culture that is long-neglected, even adding a much vaster quantity will not initiate a good fermentation.

If you are making cheese on a larger scale, you can prepare less starter—1:200—but you may decide to let the milk ripen for half an hour or an hour before you add the rennet, for it firms more quickly with more developed acidity. Personally, I prefer to add the culture and rennet at the same time, for I feel this is more in line with traditional cheesemaking methodologies, simplifying the process and allowing you to make cheese most elegantly, ecologically, and economically with the warmth of the animal.

If you are adding clabber culture as a starter, shake or stir it well in its container to break up its curd and help it mix into the milk. If you forget to mix, the curd will remain in large pieces in the vat. With kefir, be sure to strain out the kefir grains, and add the strained kefir liquid as the starter. Whey is the simplest of all: Just pour in the whey—just be sure to save some whey at the end of the make for the next day's cheese! Once the culture is added, the milk can be briefly stirred to incorporate the starter's microbial ecology throughout.

With the addition of a natural starter in the ratio of 1:50 or 1:100, you can be assured that the culture will effectively ferment the milk, and that there is sufficient acidity to curdle the milk; you can add the rennet immediately thereafter. Only with the use of freeze-dried starters must you wait between the addition of the starter and the rennet. Industry protocol calls for adding the starter, testing the pH, then waiting 30 to 60 minutes and testing the pH again; only when a pH drop is observed do you know your culture is working, and only then can you add the rennet. This protocol can be ignored with natural starters, because you can have faith in their fermentation.

Renneting Milk

Rennet gives milk its solid form, transforming its proteins into a semi-solid gel, which can be cut and stirred, and sometimes cooked to help remove the whey. The addition of rennet allows a cheese to firm without boiling or souring it, and makes it more manageable, preservable, and digestible. It is added to the vast majority of cheeses and is a most essential ingredient for their realization.

Rennet should be added to the still-warm milk immediately after the culture, so that the milk doesn't lose its warmth. Using a wooden vat exemplifies this. Cheesemakers are forced to work with the warmth of the milk to make their cheese, bringing the milk direct from the milking bucket to the vat, pouring the milk over the whey remaining at the bottom of the barrel over the course of the milking, and immediately adding the rennet once the milking is done and the vat is full. Any delay in time will cause the milk to cool, which significantly slows the activity of the rennet and causes the whole process to lag. This can result in a failed cheese. Without a means of warming the milk (aside from pouring in hot water or tossing in hot rocks!), the cheesemaker would fail in the making of most rennet cheeses.

Most cheeses call for a standard rennet dose, one that causes milk to form a full curdle within 45 minutes at 35°C (95°F). The only exception is lactic goat and sheep cheeses, which only need a quarter dose, and coagulate slowly over many hours at 20°C (68°F). Too little rennet will cause a curd to form more slowly, often resulting in a curd that is overfermented and lacking in fat, which rises as cream to the top of a slowly curdling vat—you should not see free cream on your hand when you test the curd set. Too much rennet will cause the milk to curdle too quickly, often making a cheese too firm too quickly and wasting precious rennet. A similar effect holds true for temperature: Too low a temperature will cause a curd to slowly come together, resulting in lost cream. Too high a temperature, however, can cause the curd to come together too quickly. But temperatures above 45°C (113°F) can inhibit the activity of rennet. The best temperature for a controlled coagulation seems to be slightly lower than that of a ruminating animal, around 35°C (95°F).

When you're using commercial rennet, always consult the manufacturer's advice for the correct dosage rate, adjusting as necessary to get a full curdle in 45 minutes to 1 hour for a rennet cheese. All commercial rennets should be diluted in cool water—or better yet active whey, used also as a starter—before being added. Stomachs left in whey are not too concentrated and need not be diluted. Commercial enzymes can be very potent in their concentrated form and can cause milk to curdle upon contact.

Pour the rennet into the milk while stirring slowly over 5 to 10 seconds, for an even distributing of the coagulating enzymes. The milk should then be gently and thoroughly stirred with an up and down motion to ensure a complete mixing. However, it should be stirred for no more than 30 seconds—just enough to be sure of the rennet's full mixing but not so much as to disturb the coagulation of the milk, which begins very soon after the rennet is added.

With the rennet added, you can take your first break of your morning or evening routine; however, it's best not to leave the curd too long without being watched, for it can overcurdle and become challenging to work with if forgotten. Test the curd occasionally during this break, observing its evolution, as described toward the end of this chapter, to determine its readiness for cutting.

Cutting the Curd

Mimicking the handling of the curd within a young animal's stomach, cutting the curd is an essential step in the making of most cheeses (again, lactic cheeses are the exception). Cutting the curd releases whey from within it and causes fat to escape where the curd structure is broken. The more the curd is cut, the more whey is released, but also the more fat is lost to the whey.

Curd is cut to a variety of sizes, which is a defining characteristic of many styles; in general, the more finely a curd is cut, the firmer the cheese will become, and the less fat it will contain.

Lactic cheeses, those made from long-fermented and coagulated curd, are not cut at all, but simply ladled from the vat into form or cloth. The softest rennet cheeses are typically cut to 2.5 centimeters in size, around that of a walnut, and only very lightly stirred to evolve just a small amount of whey; firmer stirred-curd rennet cheeses are cut to a hazelnut in size, around 1.5 centimeters, to achieve a longer keeping cheese. Semi-alpine cheeses prefer a curd size around that of a kernel of corn—7.5 millimeters—while the firmest alpine cheeses are all cut to the size of a lentil, around 3 millimeters, to drive out as much moisture as possible.

Along with the moisture that is lost, the fat also is lost from the curd. This is because the fat doesn't form the structure of the curd; rather its globules are trapped in the casein curd during coagulation and released from the curd structure when cut. Thus, the more a curd is cut, the more this fat is released, and there is no way around this—softer cheeses, cut larger and stirred less, always preserve more of the milk's fat, while firmer cheeses, cut small to drive out more moisture, always contain less of the milk's cream. The driest cheeses, the alpine styles, are therefore the least creamy, but that's the way they're meant to be. The lost fat can be separated from the whey by a number of traditional means, including making ricotta and skimming whey cream, explored at the end of the chapter.

The curd can be cut with a variety of tools, depending on the sought-after curd size. To achieve the largest walnut-sized curds, draw a long-bladed knife quickly but carefully, machete-like, through the curd vertically in two directions to make the vertical cut. Make the horizontal cut on a diagonal, or cut it later when you stir the curd. Smaller hazelnut-sized curds are more quickly cut with cheese harps or lyras, wires or blades on a frame, which can cut a wide swath of curd into a consistent size and come in two sets, one with horizontal

Curd is cut to drive out its moisture, helping it firm and making it more manageable and preservable. The size of the cut to a great extent determines the direction a cheese will take. This walnut-sized curd will develop into a high-fat, high-moisture cheese that ripens relatively quickly.

blades and the other, vertical. The barley-sized alpine curds are cut from the curd mass by slowly and steadily cutting smaller and smaller with a forceful application of a whisk-like spino. Corn-sized grains for tommes and Saint-Nectaire require less cutting with the spino to achieve their larger size.

The curd is usually left to tighten and firm for several minutes under the whey before stirring commences to keep the still-delicate curd from shattering, except in the case of alpine and semi-alpine cheeses, which quickly knit together into a mass when left unstirred.

Stirring the Curds

After cutting, a cheesemaker typically begins to stir the curd, slowly at first then picking up tempo as the curd firms. The stirring disturbs the curd, encouraging the evolution of its whey; prevents settling and matting at the bottom of the vat; and keeps each curd distinct, encouraging yet more whey to escape.

Rennet curd can be quite delicate after cutting. The first stirs must be very gently executed to assure that no curd is broken; breaking it unnecessarily would give it more surface area, causing it to expel more whey and firm excessively, and could result in losses to the yield and too hard a cheese. But as some whey evolves from the stirred curd, the curds will firm, strengthen, and become denser, and you must increase the pace of stirring so that the curds don't settle and knit.

With softer rennet cheeses, the stirring must be performed with an equivalent softness to preserve their moisture and cream. Fittingly, firmer alpine curds must be more firmly and vigorously stirred to prevent matting. Especially when you are cooking curd, the stirring must not be interrupted so that the curd doesn't have a chance to settle in the hot pot. If alpine curd is left unstirred for even a few seconds, it will settle to the bottom of the pot, knit together, and overcook. With some styles, such as lactic and half-lactic cheeses, the curd isn't stirred even once, preserving the maximum moisture and fat within.

Stir the curd not in circles but in figure eights to ensure that it doesn't simply move together, and settle together, in the vat. The stirring needs to be somewhat chaotic and unpredictable to prevent the matting that could occur

Curd is stirred to further firm it, prevent it from knitting together at the bottom of the vat, and encourage more whey evolution.

with too circular a stirring. Bring the curds from the bottom up to the top so that all the curds are given an even treatment, paying extra attention to the bottom corners of the vats to be sure curds aren't matting there.

As you stir, you'll see some larger curds. Cut these to size, either with a knife or by lifting the curd slightly out of the whey with your fingers and applying light pressure from your thumb. The curd should break in two cleanly along its grain, without shattering.

Stirring is carried through for varying amounts of time and intensity, or to various temperatures, to achieve a desired firmness or malleability of the curd. For a modern camembert, full of moisture and fat, the curd is stirred for about 15 minutes to achieve its best texture (for traditional camembert, even fuller of

After stirring, the firmed curd is drained, often by hand, into forms of the appropriate size to make its cheese.

moisture and fat, the curd isn't stirred at all!); for cheddar and blue cheeses like fourme d'Ambert, the curd is stirred for between 1 and 2 hours. Various methods of testing the curd, described toward the end of the chapter, help you determine when the stirring stage is complete and the curd ready to form.

Forming the Cheeses

Once you judge that the curd is ready to form, cease stirring. With your forms lined up on the draining table, pitch the curd, whey off the whey, and finally ladle the curd into its forms to take its shape.

The curd is often left to settle for several minutes before forming, allowing the removal of some whey from the vat and facilitating the forming. Pitching, as it is known, is this practice of patiently waiting for the curd to descend once it has reached its desired firmness. It is practiced with most styles of rennet cheese to facilitate curd handling and forming.

Wheying off is next. Take a portion of the whey (typically a third of the vat's volume) off the top of the vat, pouring or ladling it through a strainer to catch any loose curds (this is a good time to save whey as a starter for the following day's make). Opening a spigot and letting the whey flow out the bottom of the vat, though most convenient, can damage curds and cause a loss of yield due to shattering.

For rennet cheeses, the curds are ladled from the vat right into their appropriately sized forms, each different-shaped and -sized form making a distinctly different cheese. To avoid cheeses flat as a pancake (unless you're making the pancake-thin tomme Vaudoise), fill the forms substantially more than the final cheese will be tall: roughly two to three times the desired height of the cheese.

Though most rennet cheeses are made by draining the free curd from the vat to the forms, alpine cheeses are best shaped underneath their warm whey to give them the most closed texture, free from mechanical holes. The warmth of the whey keeps the cheeses malleable and pressable and allows them to take their finest form.

Some other cheeses, mostly the blues, are formed as much out of the whey as possible to give their curds the most open structure possible. Stirred-curd blue cheeses are drained using special techniques that leave gaps between their draining curds so that there is room within the cheese for *Penicillium roqueforti* to grow when the wheels are pierced. For stirred-curd blues like fourme d'Ambert, the curd is drained as much as possible through a very slow and patient ladling.

Draining and Fermenting the Cheeses

On the draining table, the cheeses now sit in their forms, draining their whey. There they will remain, for several hours to several days, fermenting until their desired acidity is developed, and being flipped in their forms to assure an even shaping.

The cheeses are flipped frequently, more often for harder styles, less often for softer wheels. Alpine cheeses, very firm at first, and most impressionable when warm, can and should be flipped soon after draining. With most rennet cheeses, the first flip should be attempted within 1 hour of draining. And with lactic cheeses, as well as half-lactic styles, the high-moisture curd should be left to drain for hours before you even think about flipping them. This very soft curd need only be flipped once or twice to achieve its perfect shape.

To be discussed further in the chapter, several types of cheese are pressed either by hand during draining, or with the aid of a mechanical press. Milled-curd cheeses (aside from Stilton) must be heavily pressed to assure a knitting of the firm and acid curds; they should be left in their presses for 12 hours before the first flip is executed.

The cheeses should be left out in the make room, kept warm—20°C (68°F)—and under cover, to protect from flies and drying air, while they drain and ferment, until they are ready to salt. Keep them coddled and protected like newborn babies at all times for them to evolve their best.

Most rennet cheeses are salted once the pH has dropped to around 5.3, typically 8 to 12 hours after the culture is added to the warm milk. Blue cheeses, as well as other rare cheeses that benefit from a full acidification, like tomme crayeuse, are left at room temperature much longer, up to several days, to receive further fermentation. You can test for your goal acidity using the teabag test or other observations, explained below.

If cheeses are salted too soon (the only ones in the book that I recommend salting early are queso fresco, torta, Halloumi, crescenza, and Saint-Nectaire), there will not be enough lactic acid in the cheese to fuel its aging, and it will be prone to unwanted microbiological developments due to the lack of protective acids. Cheeses salted too soon ripen too soon, and typically are mild in taste. If you wait too long before salting, on the other hand, cheeses will overferment, become too acidic and very flavorful, and develop soft, chalky textures and often unwanted *Geotrichum candidum* growth at the rind. Such cheeses invariably end up ripening and tasting like lactic cheeses.

Salting the Cheeses

Salting pulls out moisture from the cheese and slows down all microbiological developments within and upon the rind of a cheese. This very important stage of the make, the final stage for most styles, is addressed in much greater detail in chapter 7, but for the purposes of understanding the various stages of the make, a brief introduction to salting is included here.

Salting ultimately preserves cheese, delaying its degradation through microbiological inhibition, depriving microbes of the moisture that they need to grow and develop. Salting as such slows down the acidification of its paste, and, in the making of aged cheeses, shifts the dominant form of fermentation from a bacterially dominated interior ripening to a fungally dominated rind ripening, as fungi can get their moisture from the humid air in which the cheeses age. There are no aged cheeses (that I know, at least) that are left unsalted, and most fresh cheeses taste better and preserve longer with the right dose of salt. Only some dairy products, mostly fermented milks, are made without salt.

Salt is applied to achieve a salt content of about 2 percent. It can be applied to the rind by hand, through brining in a saturated salt brine for a certain amount of time, or mixing salt into the curd itself, as is done with milled-curd cheeses like cheddar, and soft lactic cheeses such as chèvre.

After salting, the cheeses are left on their draining tables to drain, either in their forms or out, depending on their preferred shape. Leaving the cheeses out of their forms after salting results in barreling, while leaving them in their forms keeps their sides straight. Brining cheeses invariably results in a slight barreling, desirable in some styles, undesirable in others.

Salting is followed by a second period of draining, with the cheeses being left out at room temperature or in their caves for 24 hours. The cheeses should be covered during this draining, to prevent excessive drying and fly strike.

Salt continues to pull moisture out from within each wheel even after the visible salt has disappeared from the surface, or the cheese has been removed from the brine. This period of post-salting draining assures that the cheeses don't bring too much moisture (and salt) with them into the cave or the hastening space.

Additional Cheese Treatments

Cheesemakers may employ some additional treatments, typically to get extra moisture out and make a longer-preserving cheese. These include: cooking and curd washing, two methods of bringing up the temperature of the curds to drive out moisture; the milled-curd method of making and fermenting a cheese before grinding the cheese into bits, salting those bits, and bringing those bits back together into a second, firmer cheese; the pasta filata method of melting a fermented cheese in hot water to make it firmer, more flavorful, and moldable; and the lactic method, which subjects rennet curd to a long fermentation in the vat before forming. Some of these methods then involve pressing cheeses to help the firmer curds become a cohesive wheel.

These addenda to the make, though ancient in their origins, are still to be considered more advanced techniques in cheesemaking, requiring additional work and necessitating certain technological advancements for their best executions.

Cooking the Curd

Cooking is a treatment used to drive moisture out of very small-cut alpine curds. This firms them significantly and causes the curd to become cohesive and malleable, facilitating the formation of large, mechanical-hole-free cheeses. Heating above 40°C (104°F) causes

The lentil-sized curd of alpine cheeses drives much more moisture out. In combination with a cooking and a vigorous stirring, the method makes a much firmer, more malleable, longer-lasting cheese. PHOTO BY CHLOE GIRE

physiological changes in the curd that allow it to adhere and press together, and the more the curd is cooked, the greater the effect. Cooking curds is a technique that defines the making of most alpine-style cheeses.

While milk is damaged somewhat by cooking it, once the milk is turned to curd, a slight cooking can have a beneficial impact by inducing changes that enable the production of very firm, very large, and very long-lasting wheels of cheese.

Cooking is best executed in a copper kettle, the original implement employed for the making of alpine cheeses. Copper kettles transfer heat best from the flame to the curd, which helps to assure an even cooking. And the traditional rounded shape of the copper vat facilitates the forming of the cheese under its whey. Other metal vats will do, but none work as elegantly as this semi-precious metal. The copper kettle is discussed more in chapter 6.

Cooking should be performed slowly and carefully, bringing the temperature of the vat up to its goal in 30 minutes or so. If the cooking is too slow, too much fermentation will happen during the make, changing the cheese; too quick a cook can result in not enough moisture being driven out, and require that a cheese be cooked to a higher a temperature to achieve the right firmness.

The curds must be stirred nonstop during their cooking, without a moment's break. The stirring assures that the heat from the cooking is fully mixed throughout the pot, and helps prevent the curd from burning and sticking at the bottom. It also helps fully firm the curds.

The curd should be cooked to a certain firmness that typically corresponds to a certain temperature. Firmer, longer-aged wheels like Comté are cooked to a higher temperature for a longer period than cheeses like tomme. They take on certain textures at their point of readiness that you can readily judge by hand without a thermometer.

Though you can use time and temperature to determine an alpine cheese's readiness, it's best to use the alpine peaks test, described below. I highly suggest relying on temperature and time only as guidelines and

THE MAKE • 105

getting a feel for the make; cheese was prepared traditionally without clocks and thermometers, and the best makers today still use natural techniques to judge the readiness of their cheeses.

For small batches of semi-alpine cheeses, those that aren't cooked too hot, I prefer to stir with my hand, which also serves as my thermometer, helping me gauge when the heat (usually around 45°C [113°F]) is just right for the cheese. For larger batches, and for higher-temperature-cooked cheeses like Comté and Emmentaler (cooked to 52°C [126°F]), I recommend a long-handled paddle, preferably wooden, to keep a distance from the heat.

Cooking the curd changes its behavior and causes it to become malleable and stick together into a solid and firm mass. The goal to achieve, to different degrees for different alpine styles, is for the cheese to be cooked until it reaches the degree of malleability that, when pressed together, the curds readily adhere to one another. With goat's and sheep's milk, this malleability is achieved simply by cutting the curd small and stirring at 35°C (95°F) for a few minutes. No cooking is required for sheep's and goat's milk alpine cheeses, unless you wish to make the firmest grana or full alpine-styles.

Once the curd reaches your desired texture, take the vat off the heat to prevent overcooking, and leave the curds to form into a wheel, preferably under the whey, which helps press the cheese together. The traditional copper kettle is often suspended by a swinging arm over a wood fire and moved from atop the hot coals once the goal firmness is reached, immediately ending the cook.

Washing the Curd

One or two unique styles of cheese, including Gouda and sometimes raclette, have their curds cooked with hot water to firm them. Adding hot water to the curd, known as washing the curd, firms the curd without the need to cook the vat over fire. This helps create a hard, low-moisture, long-lasting wheel within a wooden cheesemaking vat that can't otherwise be heated.

It is generally understood that this technique was an adaptation to a wooden cheesemaking vat of the alpine cheesemaking techniques involving cooking curds in copper kettles. Washing was supposedly created by Dutch cheesemakers upon invention of their most famous cheese, but the existence of other washed-curd styles like raclette—also traditionally made in a wooden vat that can't be cooked—paints a picture that perhaps this method was once more widely used historically, when cheese was most often prepared in a wooden vat that couldn't be heated, and that could only make a cheese that could age 3 to 4 months, tops. With the washing of the curd in hot water, a firmer, drier, longer-aging cheese could be realized in that same wooden vat.

To successfully wash the curds, have on hand a quantity of very clean water equal to about half the volume of milk, warmed to 50°C (122°F), just too hot to touch. Prepare curd like a normal alpine cheese by cutting it small and stirring for several minutes. Once the curd has firmed slightly, just from stirring, leave it to settle, removing a portion of the whey (about one quarter of volume of the vat). Then add an equal volume of the hot water slowly back to the curds, stirring all the while, to bring the temperature up to 40°C (104°F). Stir the curd for some time to firm it, then repeat the process one or two more times until the cheese reaches a desired firmness and temperature (about 45°C [113°F]), where the curd presses together well within a squeezed fist (the alpine peaks test described below). At that point, leave the curd to settle, then gather it together to form a firm wheel under the whey.

Washing the curd may not remove lactose from this style of cheese, as many believe. My current understanding is that Gouda ferments much like other styles due to its high lactose content; it is salted at a similar acidity to other styles and develops a very low pH as it ripens. Gouda often refuses to melt when well aged due to its low pH, like a Parmigiano-Reggiano. Lactose itself is not a very mobile sugar; its ability to flow isn't the same as other sugars. So the lactose likely remains in the curd even when washed. However, when the lactose is transformed into lactic acid during fermentation in the vat, that lactic acid is much more mobile, and may reduce the acidity of the cheese.

Milling the Curd

Milling the curd is another traditional technique employed to make a cheese as firm as possible without cooking or washing its curds. Also evolving originally from the use of wooden cheesemaking vats that cannot be heated, milling curd involves first making a cheese as firm as possible without cooking, then fermenting that cheese to a certain degree before breaking it into pieces, salting the small pieces of broken curd, then reassembling them, often pressing the salted and drained curd into a second cheese.

Cheeses like cheddar, Lancashire, and Stilton, and even some pasta filata cheeses like caciocavallo, are considered members of the milled-curd family. Some fresh cheeses like queso fresco and the *roqueforti*-ripened Gorgonzola might also be considered milled curd, but these cheeses aren't fermented before milling and salting, which changes their evolution considerably.

Fermenting a first cheese, then milling it, salting it, and reassembling it into a second one enables a cheesemaker to pull moisture out from within a soft, uncooked cheese, permitting it to be more effectively preserved. Without milling, an uncooked cheese can only be salted from the exterior; doing so limits the size to which these cheeses can be made, as the salt is slow to penetrate and pull moisture out, allowing fermentation to continue unabated. A large uncooked wheel will invariably over-acidify and develop soft textures or unwanted gas due to a continued fermentation within the cheese that isn't slowed by the salting from the rind.

Making a rennet cheese, fermenting it, then milling that fermented curd and salting it permits the cheese to be much more effectively preserved, and lets you make as massive a wheel of cheese as possible without cooking it. Milling the curd means you can make a wheel much, much larger than even an alpine wheel without flaws like blowing resulting from slow salt penetration; the current record for the most massive wheel of cheese is a cheddar, an eminently preservable milled-curd cheese, clocking in at 25,000 kg! Even alpine wheels made with cooking can hardly be made over 50 kg without developing flaws like eyes. The maximum size for an uncooked and unmilled curd is about 2 kg before problems from excessive fermentation arise.

Milled-curd cheeses can be quite diverse within the category because of the many different ways they and their curd can be handled in the make. In the making of more American cheddars, the firmest of the milled cheeses, the rennet curd is often cooked to a higher temperature (this is often referred to as the scald) then drained into massive slabs that are stacked atop one another to press out and drain out as much moisture as possible before milling. With traditional British cheddars, however, the curd is not cooked to as high a temperature (or at all), resulting in a moister and more crumbly paste.

Differences in the milling technique can also create significant differences in a cheese. Milling can be performed with a number of different tools: American cheddars tend to be cut with the help of sharp blades into larger finger shapes (à la cheese curds)—french-fry cutters are cheap and readily available for the purpose. Peg mills give milled-curd cheese a more British crumble. Cheeses such as castelmagno and bleu de Termignon, which are very rare and very closely related cheeses of the Alps (one from the Italian side, the other from the French), are milled super-fine in the blades of a meat grinder.

A rennet cheese can be milled to pull out moisture from within a cheese, making it firmer and longer lasting. The milled-curd method is the defining technique for making many hard but uncooked cheeses, including cheddar.
PHOTO BY CHLOE GIRE

Normally, milled and salted curds must be pressed back together using significant pressure to assure that the cheeses are fully knit together without allowing the passage of air to the interior, which can cause blueing and premature decomposition. Cheddars and cantals are pressed with mechanical presses, for example, ensuring that as the firm and acidic curds knit together, too little pressure and air pass to the interior to permit the growth of air-dependent molds like *Penicillium roqueforti* and eventually the degradations of cheese mites, which can tunnel through the open structure of a cheese and open it further to the effects of air. Some milled-curd cheeses, however, are intentionally left with an open structure to encourage such decay. Stilton is a milled-curd cheese left unpressed, which helps assure the development of its distinctive blue veins.

Spinning the Curd

Spinning the curd is a process employed in the production of pasta filata, or spun-curd cheeses, like mozzarella, Oaxaca, and caciocavallo. These styles employ an elaborate evolution of the milled-curd method where curd is made and then fermented for several hours, to the point where it stretches strongly when submerged in hot water. The curd is then milled and sometimes salted to remove moisture and help preserve it, much like milled-curd cheeses, then melted with nearly boiling water to turn it into a hot and very plastic dough.

This melting is caused by another remarkable transformation of the miraculous casein protein. At very precise acidity (pHs between 5.3 and 5.0) and very high temperatures (above 55°C [131°F]), curd melts most magnificently, due to a loss of calcium from the casein proteins that turn the curd to liquid. Though we may understand how it happens, no one can explain why exactly the curd behaves this way, though I have my own theories. The melted curd can be handled and manipulated into various forms while hot, then cooled in water or brine. Once cool, the calcium is reabsorbed into the casein and the casein turns solid again, giving permanence to the shape of the cheese.

This plastic dough can be stretched and shaped by the cheesemaker into the most elaborate and beautiful forms in the world of cheese. Akin to the spinning of fiber, the curd's proteins are aligned by the handling, causing the cheeses to develop strength and texture not found in any other cheese.

The cooking also changes the flavor of the cheeses, creating more aromatic compounds, breaking down proteins, and making the pasta filata style the most savory and delicious (and possibly nutritious) of the fresh cheeses (there's something special about the combination of fermenting and cooking milk). When these pasta filata styles are aged, as in the case of caciocavallo, they can take on the most complex flavors possible in cheese.

The melting ultimately helps preserve cheese by removing moisture and firming the curd, even pressing it together. With the cooking, beyond the limit of making grana cheeses, these pasta filatas can achieve as great a preservability and longevity as some of the most renowned wheels. In Italy they are often made as enormous as possible expressly for this purpose!

Pressing the Cheese

Pressing is an essential step in the making of some firmer cheeses, like alpine or milled-curd styles. Pressing, however, does not make a cheese firmer; it primarily causes firm curds to knit together better. (What does make a cheese firmer is cutting curd smaller, stirring curd longer, and cooking curd to a high temperature.) But pressing isn't just an aesthetic treatment: It removes open pockets and cracks from a cheese's paste, creating a more monolithic wheel that has no pathways of air to the interior, that preserves longer as it is less prone to fungus or cheese mites than a more open wheel.

Pressing can be achieved through passive pressure or mechanical pressure, depending on the style. Typically, cooked- or washed-curd cheeses, like alpine styles or Goudas, can be pressed effectively with only a passive pressing, especially when pressed underneath their still-warm whey. Milled-curd cheeses need more active pressure, and generally require a mechanical press to get their cooler and more acidic curds to knit together well.

Passive pressing works with the cooked cheese's own plasticity to press together the curd. Alpine styles retain their malleability and therefore their pressability only when warm, so the guiding principle in pressing alpine cheeses is to press them while warm. Forming and pressing the wheels together under their whey helps achieve their best pressing, as it preserves as much warmth as possible, and uses the weight of the whey to press them firm. Stacking cheeses atop one another on a draining table to passively press also keeps them warmer, improving the pressing, as does pouring warm whey (or water) atop the wheels while they're in their forms.

Active pressing is required only for pressing together milled-curd cheeses. Cheddar and other milled cheeses are much more brittle at their time of pressing, as the acidity resulting from their longer fermentation pre-pressing strips calcium that normally makes curd malleable and pressable. Milled cheddar curds need significant pressure, and a longer period of it, to knit the rigid curds together, and this is no small task. A cheddar left without enough pressure will have gaps between its curds, which will leave the cheese susceptible to blueing, mites, and premature aging. Stilton, a unique milled-curd cheese pressed together only under its own passive weight, is left with an open structure intentionally, which leads to better blueing; but the cheese must be made to a certain height, for a shorter wheel won't press sufficiently to make a ripenable wheel.

Milled cheeses will need a mechanical press such as a lever, a screw-type, or even a pneumatic press for perfect knitting. A small Lancashire likes to have at least 25 kg

Pressing cheeses by hand into their forms suffices for alpine cheeses if they're brought together under their warm whey and kept warm in their forms. Only milled cheeses like cheddar need a mechanical press to knit their firm and acidic curds together. PHOTO BY CHLOE GIRE

THE MAKE • 109

of pressure, while a large cheddar needs over 100 kg; the amount of pressure needed depends on the firmness of the milled curd and (more importantly) the upper surface area; these cheeses are often made taller than they are wide, to make the most of a press's pressure. The typical amount needed to press a cheese is often measured accordingly in pressure, like kPa or PSI. Milled-curd cheeses are best pressed in open-ended hoops on followers, so that they can be flipped easily (a simple flip suffices, but the cheeses should have their cheesecloths rearranged each time). They are typically packed into their forms with finer cheesecloth than alpine cheeses, and mostly to help with whey drainage.

Alpine cheeses are typically wrapped in strong cheesecloth before being pressed; the cheesecloth helps you handle the cheeses, to get them out of their whey and to flip them in and out of their forms. The cloths also help assure that whey is well drained from the pressed cheese and that the cheeses don't develop pockets from moisture trapped directly against their usually impermeable forms. And strong cheesecloth also gives the most beautiful surface impressions on cheeses if they have their rinds washed during their affinage (you can see the weave of the cloth on the rind of the aged cheese).

Fermenting the Curd (Lactic Cheeses)

Normally rennet cheeses are transformed from milk still warm from the udder, with the milk in its unfermented, sweet state; the cheeses are fermented to a certain degree after the cheese make. Lactic cheeses turn this standard make upside down by fermenting the curd fully before the now acidic curd is transformed into its cheese. Though they are made with rennet, the term *lactic* refers to the long lactic acid fermentation that gives this family of cheese its distinct qualities, including a soft creamy texture, a mildly sour tang, and a full flavor, fast!

Lactic cheeses, always the outlier, are made substantially differently from the diverse styles of rennet cheese discussed up until now. The long, slow fermentation that precedes the ladling of the curd results in a substantial development of acidity, dropping the pH to 4.5, and a substantial development of flavor. No other style

Many styles of fresh and aged lactic cheeses—including faisselle, crottin, and Valençay—are carefully ladled from the vat into forms, cutting the curd as little as possible. This practice is known in French as *moulé à la louche*.

of cheese develops as much flavor within 24 hours. The lactic method also softens the curd and makes it spreadable, due to the effects of the advanced acidity upon the renneted curd (which is explained in chapter 4).

With this and only this style of rennet cheese, milk should be cooled from animal-warm to ambient temperature before culturing and renneting. Lactic cheeses evolve best when coagulation and fermentation are slow; cooling the milk to 20°C (68°F) before commencing the make creates a softer, more consistent-textured curd (i.e., with fewer cracks), and a more flavorful cheese. Cooler temperatures lead to a more diverse heterofermentation, and deactivate the biological effect of rennet to a certain extent. As such, these cheeses are best prepared in smaller batches and in cooler weather to facilitate a faster cooling. Larger quantities of milk hold on to their warmth much

Lactic curd can also be drained in natural cloth to make a free-form soft cheese for fresh eating or aging.

longer and should be actively cooled with cold water to prevent excessive fermentation before the lactic make.

In the lactic make, rennet is added to the milk just after culturing, as in all other styles. In the making of lactic goat and sheep cheeses, however, it is best practice to use a reduced rennet dose. Adding a quarter the normal dose of rennet causes goat's or sheep's milk to curdle more slowly and delicately, resulting in a more uniform texture, without the cracks that can often develop if the milk is curdled at too warm a temperature or with a higher rennet dose. It's because goat and sheep cream is slow to rise that cheesemakers can be more frugal with their renneting. Lactic cow cheesemakers must use the full rennet dose to capture all the fast-rising cream in the curd.

Lactic cheeses are left to curdle and ferment for at least 12 hours, but typically overnight, about 24 hours, until the pH drops to 4.5 and the curd drops below the whey, softens in texture (it also loses its jiggle), and develops a mild sourness and intense flavors, all of which make it instantly recognizable as lactic. After 24 hours, a velvety veil of *Geotrichum* should be seen growing atop the whey when the light is just right, indicating that primary fermentation is complete, and that this important ripening microorganism is established.

Lactic curd can be carefully ladled from the vat into cheese forms to give the cheese a shape (cylinder, pyramid, what have you), or scooped into cloths and then hung to drain their whey free of form; either can be eaten either fresh or aged.

If you're ladling, the process is best done with a well-designed lactic ladle (with the handle and scoop in line with your wrist), which cuts the curd on its sharp edge into large ladles-ful without smashing it into smaller bits that don't make as delicately textured or flavorful a cheese (the cheese is also best when it's exposed to as little air as possible). To ladle lactic curd, ladles-ful of curd are carefully cut from the mass, one layer of ladles at a time, from one side of the vat to the other (in a square vat) or in concentric circles from the outside toward the center (in a round vat). The whey is then ladled out of the depressions before the next layer down is ladled out. If the ladling is practiced right, only large scoops of curd are transferred into the forms, without any curd breaking.

If you're draining lactic curd in cloth, you can transfer the curd into the cloth with a large scoop, and then hang the curd in a way that allows its whey to freely flow. Tie the cotton cloth together by its corners around the curd, and then suspend from a spoon or a hook over a basin below to catch its whey. The pressure from the hanging bag helps extract more moisture than leaving the curd to drain on a surface or in a colander. Typically, the salt is added only once the curd is fully drained, usually after 12 to 24 hours (sooner if the cheese is to be eaten fresh; later if it will be aged).

Cheese Testing

Some monitoring of the cheesemaking process is essential to assure that cheeses evolve their best. For consistency and safety's sake, cheesemakers can employ numerous tests to judge the evolution of their cheeses over time; historically, cheesemakers likely used simple natural tests like the following to determine when their cheeses were ready for cutting, draining, salting, and so on.

The success of a cheese's make and affinage often rests on these tests. Testing for temperature helps assure that the milk ferments and curdles right. Testing the readiness of the curd for cutting helps assure that the curd is evenly cut without losses. Testing for firmness of the curd helps assure enough moisture has been driven out of it to properly conserve it. And testing for acidity before salting (or melting) assures that the cheese has fermented to a degree that permits its proper evolution and ripening.

Various aspects of the cheese's production, like temperature of the milk, the temperature of the make room, and the solids and relative protein content of the milk, change from day to day. Testing helps achieve a certain consistency in the cheese make from day to day despite these discrepancies.

Monitoring Temperature

Although I do give specific temperatures for the making of particular styles of cheese, I do not measure temperatures in my cheesemaking with a thermometer, because traditional cheesemakers didn't use them and I prefer to rely, as they did, on careful practice and my senses to achieve the desired results. You can feel comfortable using your hand, or other senses, to judge the temperature. And the cheesemaking process becomes more fluid and simplified if you do—your hand knows in an instant if the temperature is right, whereas a thermometer requires time to take a reading, and doesn't always read as reliably.

Because cheesemaking is very much an animal process, we, as animals, are very good at getting a sense of the temperatures needed for cheese's best evolution. Judging the temperature for beginning most cheeses, 35°C (95°F), is as simple as testing for a slight warmth on your wrist. This judges temperature more accurately than using a finger, which can be colder than the body. That sought-after temperature is slightly less than cow-warm, as the milk traditionally would cool slightly as it was milked and collected, before the cheesemaking process was begun. For lactic cheeses, which ferment and curdle best at 20°C (68°F), I wait until the curd has fully lost its warmth before beginning the process.

For the cooking of full and semi-alpine cheeses, I do give temperature guidelines (45°C [113°F] for semi-alpine, 52°C [126°F] for full alpine), but I prefer to rely on the texture of the curds to get a sense of a cheese's ultimate firmness—for it's not just the temperature that matters, but the malleability of your curd, which changes with temperature. Getting a sense of the sought-after texture provides a better sense of how much cooking is required. I elaborate on this idea in "Judging the Curd for Forming," below.

For mozzarella making, instead of aiming for a certain temperature to melt the curd (60 to 70°C [140–158°F]), I bring water to a boil, then let it cool slightly (the ideal temperature for the melting water is 90°C [194°F], slightly less than boiling) and add the water stepwise to the fermented curd, a couple of ladles at a time, being careful about my practice to slowly bring the temperature of the curd up to where it melts perfectly. This slow practice helps to achieve the perfect temperature for stretching without measuring.

For yogurt making (and paneer making), their high temperatures for cooking (80°C [176°F] is often a commercial target) can be naturally judged. When cooking the milk for yogurt, I often bring it to a boil (but stop before it starts rising up in the pot and spills out over the stove—you should never neglect a potful of cooking milk!). The longer cooking time to reach a full boil makes a much better yogurt.

As for the yogurt's fermentation, after boiling I cool the milk to 40°C (104°F) and keep it there for many hours; this key thermophilic temperature can be judged by sticking a finger in the milk. If you can tolerate the heat for 10 seconds without pain, then the thermophilic microbes that make the yogurt will be happy!

Testing the Curd Set

For most cheeses, the curd is considered ready to cut once the milk has transformed to as firm a gel as possible before it gives off whey. I look for the same standard of firmness for every style of cheese that I make, with any milk, from any breed or species, at any season, with the exception of lactic cheeses.

There are several ways to judge the readiness of the curd for cutting, including the clean break test and the whey test. Once the curd begins to give off its whey, that's the clearest sign that it has reached its utmost firmness before it firms excessively. When the curd has reached the stage where there is a very slight puddling of whey atop, I consider it ready for cutting. If you wait too long, the curd will give off more whey and begin to contract from the sides of the vat; once too much whey has evolved, the curd will be harder to cut. It may unmoor from the vessel and spin in the pot when cut, making it challenging to cut it straight.

The clean break test is another way of determining the strength of the curdle. To do this test, poke your index finger into the curd at an angle (you should feel a pop as it breaks the surface tension of the curd) then lift it straight up. If the curd rises above your finger, then breaks cleanly in two, it can be considered ready. But if the curd barely lifts, then breaks into smaller pieces, the perfect set is still a few minutes away.

If you wait too long before you cut, the curd can be too firm and difficult to cut, which can result in a cheese with more moisture in it; whereas cutting the curd too soon results in a curd that's too soft and shatters too readily when handled, and actually makes a firmer cheese.

Judging the Curd for Forming

When stirring curd, I prefer to stir not for a specific amount of time but to achieve a certain desired firmness in the curd before forming. This firmness can be precisely measured through various hand tests.

Many of these tests relate to the practicality of their cheese makes, so measuring curd firmness can be realized quite objectively. For example, for many soft rennet cheeses, the curd needs to be ladled out of the pot and into a form; thus the curd needs to be stirred for several minutes until it is stronger and easy to handle. Also, alpine cheeses are cooked until their curd readily adheres to itself and can be easily pressed into a cheese—so the curd's adherence is tested to determine if the cooking is complete.

Relative firmness can be judged as many cheesemakers do by taking a small handful of curd and examining it, looking for familiar textures and strengths and even chewiness. But if you're inexperienced or trying out new techniques, more thorough and objective methods are in order. And though these methods might seem at first to be crude, they have been tested by time, and have likely been enabling cheesemakers to accurately measure their cheeses' readiness for draining for millennia.

For softer rennet cheeses, as well as the stirred-curd varieties, I deploy the splat test: Drop a single average-sized curd from measured heights to determine its strength, by observing its behavior on impact. As the curd firms, it will bounce, not splat, from higher and higher heights. For example, upon cutting, a hazelnut-sized cow's milk curd will splat on impact from 30 cm; after 15 minutes of stirring at 35°C (95°F), the curd will bounce slightly from 30 cm, but splat from 60 cm; after 30 minutes, it will likely bounce slightly from 60 cm; and after 1 hour of stirring the curd will bounce from 90 cm.

When you're testing readiness for certain varieties, you can determine a goal firmness by regularly splat-testing and stopping the stir once the desired drop-without-splat is achieved. For example, if you're making a modern Camembert, you stir the curd until it bounces when dropped from 30 cm; for cheddar and fourme d'Ambert, the long-stirred curd should withstand an impact from 90 cm! But be sure, for consistency's sake, to test a standard-sized curd in your vat; larger curds will firm most slowly, while smaller curds will firm much faster, and bounce from higher heights (a good reason to cut your curd as consistently as possible).

For firmer, cooked-curd cheeses, the alpine peaks test is the test of choice. This test measures the malleability of curd, which increases as the curd is cooked, and can be used to determine when a wheel is ready to be formed. With alpine styles, the curd is cooked and stirred until this desired texture is achieved, testing the curd every 5 minutes to determine its readiness.

To perform the alpine peaks test, retrieve a small handful of curd from the vat and let it drain flat for a moment in an outstretched hand. Then squeeze the curd between your fingers, lightly at first but with increasing firmness, to observe its behavior. During the early stages of stirring, the curd will not adhere, instead squeezing out loosely from your enclosed hand. After a small amount of cooking and stirring, typically up to 45°C (113°F) (but at 35°C [95°F] for goats and sheep), the curd will form jagged peaks when squeezed between the fingers, indicating an increased malleability and a strong tendency to knit together. After a greater amount of cooking and stirring, typically up to 52°C (126°F), the curd will achieve an even firmer pressing, without gaps or cracks, and the pressed curd will squeak loudly on your teeth when chewed. This incredible malleability allows these high-temperature-cooked curds to be fully pressed into a perfectly cohesive wheel without mechanical holes and indicates that the cooking is complete.

These tests are consistent across seasons and species; they can be used in the same way for the same style of cheese made with sheep's, goat's, and cow's milk, and of all their breeds. Most fittingly, the test relates the relative firmness and different evolutions of these different milks' curds and allows an easy comparison and adjustment of technique. Sheep's milk curd will achieve its goal firmness much faster than cow or goat; Jersey will typically reach its goal firmness faster than Holstein or other high-producing but low-solids milks. Seasonally, the different fat and protein contents do cause curds to evolve differently, but the splat test and alpine peaks test remain faithful even with the curd's changing behavior and allow you to make cheese most consistently over time.

Many natural tests can be used to keep track of the evolution of a cheese. The alpine peaks test judges the malleability of the curd, which increases with temperature during its cooking. You can accurately determine the degree of cooking needed to make an alpine cheese with this simple test: Drain a handful of curd, then squeeze it in your hand, lightly at first, then more firmly; if you see alpine peaks between your fingers, the curd is sufficiently cooked!
PHOTO BY CHLOE GIRE

114 • MAKING AND AGING

Monitoring Acidity

Acidity monitoring is considered essential in standard cheesemaking practice for quality assurance, food safety, and consistency between batches. In natural cheesemaking, there are effective ways of monitoring acidity to assure that the process evolves safely and effectively. I didn't much recommend acidity testing in my previous book, and still do not encourage the use of pH meters, but now I recommend a number of effective natural tests you can use to accurately judge the acidity levels in your dairy fermentations.

But before I describe the natural methods of measuring acidity, a quick acidity lesson is in order. Scientifically speaking, acidity is a measure of the hydrogen ions present in a liquid (mathematically, it is the inverse of the concentration of the hydrogen ions). The most common method of measuring acidity is the pH scale (though many cheesemakers in Europe and the U.K. use Dornic degrees, which is a bit more DIY than pH). The pH scale is logarithmic, with each measure on the scale 10 times more than the number that follows it. The 0 on the pH scale is the highest possible concentration of hydrogen ions, and thus the most acidic; 14 is the lowest possible concentration (and the highest in hydroxide), and therefore the most alkaline. In the middle is 7, a neutral condition, found in pure water, with an equal concentration of hydrogen and hydroxide.

Most cheesemaking happens in the realm of the neutral and acidic. Milk comes out of the udder at a pH of just below 7, and curdles into fermented milk at a pH of 4.5 due to the creation of lactic acid (which is the same pH as lactic cheeses). Mozzarella stretches at a pH of 5.3; coincidentally (or not), most cheeses are salted at this same acidity. Aged cheeses like Camembert and blue cheese change in texture and deliquesce (liquify) as their pH's climb back up to 7, largely due to the rind ecology's consumption of lactic acid.

The pH scale was invented in Denmark (along with freeze-dried starters and commercial rennets) by scientists looking to quantify the conditions of their fermentations (in their case, though, it was with industrial beer production). It is supposedly an objective test, and relies on laboratory tools and ingredients, and the response of electrodes to the hydrogen ion concentration in liquids (thus it does not work on solids, like cheese!).

But by no means is pH testing with pH meters the only objective way of measuring acidity. Arguably these tests of industry are *subject* to all sorts of difficulties in their use that make their readings less reliable than natural tests. And natural tests actually provide a better grasp and measurement of evolving acidities in a cheese.

Natural acidity monitoring can be used to ensure best practices in natural cheesemaking. From judging the fermentation quality of a starter to assuring the right amount of lactic acid in a cheese for its preservation, testing can help a cheesemaker gain effective control of a cheese's evolutions and foster the development of milk's best microbes. And natural acidity monitoring can be quite simple: For example, milk fresh from the udder, still warm, will always have a pH of 6.8. Seeing clabber or kefir culture curdle with its regular feeding is an objective reading of a pH of 4.5; and watching cow's milk curd stretch in hot water is a clearer indication of a pH of 5.3 than any pH meter can offer.

With natural cheesemaking, the daily evolution and acidification of the starter culture offers substantial assurances that our cultures are active and best able to ferment our milk into cheese. Seeing a regularly fed clabber culture curdle (a sign that the pH is at or below 4.5) within 12 hours (and tasting its acidity) is a reliable indication of the culture's activity and should give you the certainty that the culture will cause the milk in the vat to acidify in its most timely manner. A freeze-dried starter's activity cannot be observed in a similar manner. Acidity meters are therefore the only way to assure makers that the freeze-dried cultures they use are acidifying the milk—that they are not inactive and have not been killed by phages or competing microbes.

Cheesemakers can also use various natural methods to test the evolution of acidity during the cheese make. A cheesemaking process, however well controlled, will evolve differently depending on the amount of starter added, the temperature of the milk and the make room, and the firmness or moisture content of a cheese. Each of these factors can affect the development of acidity, which can have a great effect on the taste and texture of a cheese.

When you're using natural methods—the freshest milk and natural starters—you only need to measure the acidity during the late stages of a cheese make; measuring in the early stages is only necessary with more industrial processes and ingredients, whose fermentation is irregular and prone to failure. With a more natural approach, the fermentation of a cheese can be understood without the need for constant testing.

The acidity of a cheese can be discerned through a number of means, including aroma, taste, texture, and a test I call the teabag test. Some useful milestones along different cheeses' journeys can be measured as such.

For example, when a lactic cheese develops its goal acidity, its curd will soften and sink beneath its whey. An acidity meter isn't needed to determine this point; simply taking a spoonful of curd and observing it and tasting it can give you results with a great degree of accuracy. At a lactic cheese's goal pH of 4.5 or less, the curd will be firm but brittle, and soft and creamy when smeared between thumb and forefinger; it will not jiggle the way it does at higher pHs. Lactic curd will break into moist flakes, revealing the curd's internal texture; when you taste it, the mouthfeel will be slightly grainy or chalky, and the flavor will be tart. Simply observing the curd sunken beneath the whey, and smelling the notes of fermentation wafting from the pot, are assurances that the curd is ready to drain; while witnessing the slight growth of *Geotrichum candidum* atop the whey obscuring the reflection of light from a window or a lightbulb altogether assures you that your curd has gone lactic and has a pH of 4.5.

With rennet cheeses, the only time I test their acidity, is to determine the best time to salt or stretch the cheese. The acidity of the curd can be very accurately measured by examining its behavior when melted in hot water, as performed in pasta filata cheesemaking (but the test also applies to many other styles to determine the best time to salt).

When a small piece of curd is submerged in hot water, its behavior is a very accurate indication of the cheese's acidity. At pHs between 5.3 and 5.0, cow's milk curd will melt and stretch to varying degrees when steeped in hot water. At pHs above 5.3, the curd will not stretch fully; and at pHs more acidic than 5.0 the curd will crumble. As such, you can observe the evolving acidity of the cheese simply by regularly performing what I call the teabag test.

To perform the teabag test, prepare a mug of boiling water (it's helpful to have a hot-water kettle in the dairy for this purpose). Cut a small piece (one teaspoon) of curd to be tested and submerge it in the water for a minute or two, as if you were steeping a cup of tea. Then, with the help of a spoon, retrieve the curd and hold it between your thumb and forefinger. Dip it again into the hot water and lift it up and down, in and out of the water, like a teabag (without pulling on it). Depending on how the curd behaves once it gets hot, you can get an accurate measure of the curd's acidity:

If the curd:

- Does not stretch, its acidity has not yet developed and its pH is above 6
- Stretches slightly, but breaks, its pH is 5.7
- Stretches but breaks once thin, its acidity is 5.4
- Stretches and forms a *strong string*, its pH is 5.3
- Stretches but forms *loose strings*, its pH is 5.1
- Melts but does not stretch, its pH is 5.0
- Crumbles and does not melt or stretch, its pH is at or below 4.8

Though I learned to love this scale for making mozzarella, the most important application of this acidity test is to determine the best time to salt a cheese. When you're making a certain style of cheese, the acidity at which you salt the cheese can determine the way in which it ages and ripens, its appearance, its texture, and its taste.

For mozzarella, fermentation to 5.3 brings about the proper acidity evolution for the cheese to stretch: at just the right acidity, the cheese's texture will be soft and giving, and the flavor well developed but not sour. Because we're looking for a textural and flavor quality for our mozzarella when melted, judging the texture and flavor of the melted curd through the teabag test is a much more accurate way of determining the goal acidity than using a pH meter. With a teabag test that shows melting curd, the mozzarella will have the perfect plasticity. Mozzarella makers in Italy always trust this practical test over more scientific instrumentation.

But this test can also be used on non-pasta-filata styles, whose best salting time can be determined based on the cheese's tendency to melt (or crumble for certain acidic styles). To me it's no coincidence that most styles of cheese are salted at a pH of 5.3, when curd melts in

The teabag test is a reliable method of testing the acidity of the curd to determine the best time to salt (or stretch) a cheese. The curd melts to different defined degrees depending on its level of acidity. Cow and buffalo curd will stretch infinite strings at a pH of 5.3.

hot water. (Cheesemakers in the nineteenth century relied on a similarly performed hot-iron test to determine the fermentation of the curd!) This is an observation that cheesemakers must have understood at least since spun-curd cheeses that rely on this phenomenon were invented, likely over 2 millennia ago! And many traditional cheeses are salted at specific acidities that could have easily been assured using this simple test.

For example, when making a Camembert we can use the teabag test to determine the cheese's acidity, and therefore the best time to salt, as explored in chapter 7. Once the curd has been draining in its form, it will begin to knit together and firm, and its acidity can be tested. Under normal circumstances, the curd will develop a strong stretch—a pH of 5.3—after around 8 to 12 hours from the addition of the starter to milk, and at that point the cheese can be salted, preventing further acidification, creating a cheese will eventually age into what we know as a Camembert.

If, however, the Camembert is left to acidify for 24 hours from the addition of the starter, the curd will give off loose ribbons indicating a pH of around 5.1. If salted then, the more acidic curd will yield a more acidic Camembert. If the cheese is left to ferment for more than 24 hours, however, the curd will crumble in the teabag test (yet stretch at the rind!), indicating a pH below 4.8 (but with a pH about 5.0 at the rind due to the growth of lactic-acid-consuming yeasts). A Camembert salted at this point will evolve as a lactic cheese, with a very soft and brittle interior, and proteolysis that's restricted to just beneath the rind.

Testing a sample of the cheese from the interior will give a more accurate portrayal of the developing acidity. The acidity on the surface of the cheese can be different from that of the interior, due to the development of yeasts on the rind that eat lactic acid and neutralize acidity. (This explains why the Camembert in the above example had a different stretch test result and therefore a different pH at the interior than at the rind. A pH meter could never provide such nuance!) To get the best representation of the acidity of a cheese, a small part of the cheese must always be destroyed; cheesemakers often set aside one cheese out of the batch to serve as the sacrificial source of samples, to test all the way up to its eating!

Regrettably, the teabag test is not as effective at gauging the acidity of goat curd and sheep curd. Though the curds of these milks do behave differently when melted at different pHs, they do not stretch to the same degree as cow curd or buffalo curd. The test can still be used to observe the changing acidity of a fermenting cheese like a Camembert, but at a pH of 5.3 the same endless stretch will not be observed in the goat curd as in a cow curd. We must simply change our expectations when working with these milks and consider a partial stretch to be equivalent to a full cow stretch. Similar to cow curd, over-acidified (pH below 5.0) goat curd and sheep curd will crumble.

Nevertheless, despite this slight shortcoming with natural techniques, supposedly more advanced pH meters only record the acidity of the *whey*, and not the cheese being tested, which often differ. Observing how the cheese melts (or doesn't) offers a more accurate measure, as it shows the actual acidity of the *cheese*.

Boiling Milk

Likely half the world boils their milk before they make their cheeses. Across much of Eastern Europe, Asia, and Africa, cooking is the first thing done to milk when it leaves the udder. Indeed, boiling is a significant and obligatory step in the making of a large number of dairy products.

Though I am generally recognized for my advocacy for raw milk cheesemaking, there are times when I believe boiling milk is in order. In the preparation of many traditional fermented milks, cooking is employed to improve the flavor, texture, nutrition, and yield of the final dairy product. I call for cooking milk to very high temperatures in several techniques in this book, and would not recommend making the dairy products otherwise. These include yogurt, yogurt cheese, shankleesh, gros lait (mesophilic yogurt), milkbeer, clotted cream, and cream cheese.

Cooking milk normally damages it, reducing its responsiveness to the cheesemaking process, specifically the way rennet works its magic on milk proteins. In my technique for cream cheese, cooking milk softens the effect of rennet coagulation, making the curd creamier in texture and higher in moisture, but in no other rennet cheeses does cooking milk help—it severely damages milk's ability to coagulate with rennet. However, cooking is required for preparing fermented milks like yogurt if they are to evolve the right way and taste their best. A proper cooking causes a denaturing of proteins, which thickens the curd while it softens it; concentrates the flavor and texture of the curd through evaporation; and caramelizes the sugars, adding sweetness and flavors that counter the acidity of the final product.

The effect of pre-cooking the milk in a fermented milk like yogurt or milkbeer should be seen as equivalent to cooking any other food, generally rendering the food more digestible and therefore more flavorful. Our bodies have evolved alongside our technological ability to cook foods—as well as ferment them—so we find their flavors more attractive, especially those that combine these two transformative processes. We all will find the flavors of yogurts more attractive if the milk is cooked first (try making parallel batches of yogurt with both raw and boiled milk to see for yourself!).

Cooking milk should not be equated with pasteurization, a microbiological control that is most often used to make milk that's unsuitable for use "safe" to consume or make cheese with. Traditionally, milk was cooked to make its products more flavorful and digestible. Cooking also improves yields due to the denaturing of the albumin protein, which makes a thicker yogurt, and results in increased yields when making yogurt cheeses. The "pasteurization" that occurs when the milk is cooked is beside the point, as the boiled milk is well fermented with backslop once cooled.

When boiling milk is called for, a gentle cooking is always in order. Milk is damaged significantly when heated. And to assure the best preservation of the milk and prevent losses from scalding, milk should be cooked as carefully as possible. Some of the damage of cooking is undone when that cooked milk is re-fermented, and yogurt should not be seen as lacking nutritionally compared with raw milk. It is, in fact, likely much more nutritious because of its improvements in digestibility rendered by a combination of cooking and fermentation.

Typically, when milk is to be cooked, it is brought to 80°C (176°F) then cooled. But my preference is to bring it up to a full boil at 100°C (212°F). Bringing the temperature to 86°C (187°F) is the bare minimum, as at that temperature the milk's albumin (and casein) proteins are denatured, rendering the resulting dairy products softer-textured and thicker. But to make the best yogurt, a longer, higher-temperature cooking results in more evaporation and resulting concentration of milk, making a thicker, more enjoyable dairy product. I also prefer to bring the milk to a full boil as this is an easier point to judge without a thermometer than is 80°C (176°F).

When milk was cooked traditionally, it was typically done in copper kettles or clay pots. Copper distributes heat from a flame best and assures an even cooking without burning the milk at the bottom of the pot; clay slowly diffuses the heat and keeps milk from burning. Copper kettles should be called for when cooking

Milk is often boiled before its fermentation. Though it destroys a milk's capacity for making a rennet cheese, cooking to a high temperature improves the flavor, texture, nutrition, and yield of fermented milks and many rennet-free cheeses. Many traditional dairy products, like yogurt, gros lait, and milkbeer, miss their mark if the milk is used raw.

THE MAKE • 119

milk for yogurt and paneer, but by no means are they necessary. Lacking a copper pot or vat, a hot-water bath is helpful for reducing damage to milk. But my preference for all pots is to cook milk directly over a medium fire, stirring constantly.

While cooking milk, be sure to stir, stir, stir! Certain milk proteins (lactoglobulins) denature at the surface of milk when cooked above 45°C (113°F) (35°C [95°F] for sheep), causing a milk skin to form. This cannot be undone and leaves inconsistent textures in the final product. Stirring also helps ensure the best heat transfer between the milk and its cooking vessel, distributing heat and preventing burning at the bottom of the pot. A pot of milk that's stirred consistently during its cooking should not burn at its base and will preserve the milk best without waste or off flavors. Also, constant stirring keeps your attention on the pot, precisely where it needs to be while cooking to prevent that precious milk from boiling over when it reaches its boiling point.

After cooking, the milk should be quickly cooled to fermentation temperature, be it 40°C (104°F) for yogurt, or 20°C (68°F) for cream cheese, gros lait, or milkbeer. It is best to cool milk in a cold-water bath (changing the water as it warms), stirring as it cools to keep that milk skin from forming, but also to improve heat transfer, and to cool the milk quickly. Quick cooling is especially important with large batches of cooked milk, which may begin fermenting spontaneously during a slow cooling.

I enjoy the process of cooking milk, recognizing the beneficial impact it will have on certain dairy products and meditating on its practice. But no such benefits are realized through pasteurization. It's a complete waste of a farmer-cheesemaker's precious time, utterly unnecessary, and ultimately damaging for small-scale natural cheese producers and the nourishing foods they make.

Whey Use and Reuse

With nearly every cheese comes a considerable quantity of whey, often up to ten times the volume of the cheese produced from milk! A responsible cheesemaker should consider the whey as a product of their cheese make, and not simply as waste. And there are many thoughtful ways that whey can be used in the natural dairy.

Whey can find many uses in the dairy on the day of its make. For example, it can serve as a source of heat for keeping alpine and semi-alpine cheeses warm to preserve their pressability. I prefer to press my cheeses under whey in the vat, using the warmth and weight of the whey to press the curd firm before I place the cheeses into their forms. Whey can also be poured into buckets and used as a weight to press milled-curd cheeses.

Whey can be transformed into a number of whey cheeses, including ricotta and brunost. When making ricotta, leave the whey to ferment for an hour after the make, then bring it to a boil to separate the remaining protein and fat. After you strain out its curds, whey still remains, but it has little protein fat or other nutrition left. Only with the making of brunost, a long-evaporated whey cheese, is there no whey remaining after the make.

Even if you're not using the whey for any other purpose, it can be left out to sit for a day in barrels to separate its fermented whey cream, often a significant loss from standard cheesemaking practices; it's easily recovered by skimming the whey. (You can also put the whey through a centrifuge to separate its fat.) This is especially warranted with alpine and semi-alpine styles, whose whey contains a large portion of the milk's cream, as well as the hot water used from stretching mozzarella, which contains as much as half the fat from the milk! Instead of skimming excess fat from the milk before the make (something commonly done with modern alpine styles)—a damaging process that affects milk's capacity to turn to good cheese—consider skimming this fat once the cheese's whey has sat. That whey cream can then be churned into a delicious whey butter.

Whey can even be fermented and drunk as is (I prefer its taste and texture when fermented) or even aerated as it is fermented (in a gourd or through regular stirring) to turn it into the ancient alcoholic drink beloved by Vikings known as blaand (apparently the old Norse word for its stirring spoon!).

Whey can be used to prepare saturated salt brines (more on them in chapter 7). The whey can be left to ferment to its ideal acidity (the same as that of the cheese to be salted, usually around 5.3, which it typically achieves within the same time frame as its cheese). Then salt can be added to make it saturated. (Some producers boil the whey first, to remove unwanted whey cream and proteins.)

Fermented and lightly salted whey makes the most perfect morge (see chapter 9), containing diverse populations of surface-ripening yeasts that can help washed-rind cheeses establish their washed-rind ecologies hastily and more completely. Whey brines maintain nutrients and flavor within the cheese, rather than stripping them away as washing with water does, and likely promote a better microbial ecology than water brines. Cheeses washed with whey are more savory and complex, and seem to preserve better and age longer than those washed with water.

Fermented whey can also be used as a natural cleanser, produced in-house, that is most effective at controlling unwanted microorganisms in the dairy environment. Fermented overnight (then skimmed of its cream), the whey develops a mild but cleansing acidity and a protective microbiology. It can be used to wash hands, or to scrub out cheese vats or wooden or steel draining tables. It is effective for washing out cheesecloths and natural cheese forms, which should never be washed with chemicals or even water, to prevent yeast overgrowth and unwanted gas production in cheeses. (Unwashed natural materials will inevitably grow yeasty biofilms that will cause gas development where cheeses come in contact with them.) Dairies can operate entirely without running water (or a drain) if cleansed with whey. And natural cheesemaking, when practiced as a biological system, shouldn't have water added—much as young cows, goats, sheep, and human infants shouldn't drink water.

For an extra cleansing effect, the whey can be left at room temperature and stirred several times a day over a week to become whey vinegar, as described in the technique for milk vinegar (see chapter 15).

I might even prescribe cleansing udders with 1-day-fermented whey, inoculating the milk at its source with protective populations of healthful microorganisms, and eliminating the need for industrial cleaners, antimicrobial teat dips, and moisturizers. The slightly acid whey also helps remove soil, manure, and other debris, and its butterfat and lactic acid (the original alpha hydroxy acid) moisturize and strengthen the mother's sensitive udder. Such probiotic and protective practices may reduce the incidence of mastitis. The only thing better is a calf's saliva!

Fermented whey can be used to clean out stalls and outbuildings in wintertime, imbuing the barn with beneficial bugs, improving the microbial ecology of the dairy and its cows at its foundation.

A growing pig will devour quantities of whey every day, gorging on its lactose and lactic acid, as well as its proteins and fats. There's no more economical way of makin' bacon! It's also perfect for plumping poultry and excellent for egg production; whey contains abundant albumin, the primary protein in eggs. Warm sweet whey can be fed to calves, though it cannot compare to the nutrition of their mother's milk.

Whey can be used as an excellent, well-balanced, non-burning (it must be broken down by microbes) soil amendment in vegetable garden. The many macro- and micronutrients it contains are well balanced for plant growth. It is rich in nitrogen, phosphorus, and potassium (from its proteins), and it nourishes soil microbiologies as much as the plants that grow in them, due to its unique ability to foster fungal development with its abundant lactic acid.

Fermented whey can be used as a foliar feeding spray in gardens, orchards, and vineyards. Several fungal diseases, including powdery mildew, can even be kept in check with its timely application. Some natural winemakers spray whey to reduce sulfur applications to their vines.

In the kitchen, whey can be used in place of water for cooking almost anything, vastly improving its flavor and nutrients. Meats can be slow-roasted in sour whey, its lactic acid caramelizing beautifully on the skin. Soups and broths can be prepared with whey remaining after ricotta, and whey can be used to boil pasta or vegetables (as with the method for making Halloumi). I even mix mesophilic whey in place of water into bread doughs to add much more substance to a loaf; the whey, with its yeast and lactic acid, will cause a loaf to rise in 12 or so hours all on its own, without even adding a starter!

And at home, whey can be used as a moisturizing and cleansing face wash, body wash, or even shampoo, with its gently exfoliating lactic acid. Use fermented milk or whey daily as a sort of facial peel (simply scrub with water when you're done) and your skin will soon be as soft as a baby's bottom. (In Indonesia, kefir's popularity has surged expressly for this purpose.) If Cleopatra was as beautiful as the legends say, it's likely she bathed herself, not in milk, but whey!

However, whey finds its most exalted purpose when returned to pastures and hayfields, nourishing the soil that grows the grass that feeds the animals that provide the milk that makes the cheese that produced the whey. Not as acidic as is often perceived, whey has lactic acid that feeds many microbiological developments in the soil, and its use does not acidify soil—it may indeed be a good ingredient for raising soil pH. Among other capacities, it nourishes soil food webs, builds fertility, grows healthy plants, and helps sequester carbon! The lactic acid is broken down into less acidic compounds as it is consumed by fungi aerobically in healthy uncompacted and unsaturated soil, much as it fuels the microbiological developments on the rind of a cheese. Ultimately the practice of returning whey to the soil is one of the most sustainable practices a dairy can implement.

Skimming Cream

Centrifuges are an unnecessary addition to the milk processing chain. They are expensive to purchase, often imported from overseas, easily damaged, tricky to operate, and difficult to clean. They likely damage milk, its butter, and its cheeses. I highly recommend against their use.

The technology, invented in the late 1800s by the deLaval company (which has directed the industrialization of dairying perhaps more than any other) separates lighter cream from heavier skim. Its adoption was the first stage of the industrialization and specialization of dairy farms that formerly transformed milk into cheese and/or butter in their dairies. With the advent of the separator came the commodification and sale of sweet cream, a previously nearly unknown product, impossible to produce without industrial technologies. This was before the more voluminous milk could be easily transported by truck and when dairy farmers started selling themselves short.

Many small dairy farmers instead separate their cream by hand after leaving it to sit for a day in the refrigerator. Refrigerating, however, is a costly technology that also damages the milk, the cream, and especially the butter, leading to rancidification and the growth of cold-tolerant microbes.

Instead, consider leaving the cream to separate itself at room temperature, fermenting is as it separates, as in the traditional technique for barrel butter (see chapter 14). Cream can be fully separated from cow's milk and buffalo milk (sheep's and goat's milk need different treatments) by leaving the milk to sit and ferment undisturbed until it's thick and lightly soured. Boiling milk before fermenting it, as in the case of yogurt and clotted cream, enables a more substantial separation of cream, and improves the yield and efficiency of buttermaking.

The thickened cream will be much more simply and effectively separated from the now semi-solid curdled milk below, further simplifying the skimming. Adding a very small amount of starter (I recommend 1:1,000) initiates a slow fermentation, increasing the cream yield.

The skimmed milk curd below can be hung to make skyr (a traditional by-product of buttermaking) or other fermented cheeses like graukäse, tvorog, handkäse, and cottage cheese, which, though perfectly good food, do not compare in quality to rennet cheeses made with their full dose of milk fat.

Cheesemakers have a choice: make cheese or make butter, prioritizing one of the dairy products at the expense of the other. When making cheese, butter suffers; when making butter, cheese suffers. However, a portion of milk's cream can often be separated by gravity from the *whey* after a rennet cheese make, especially when making alpine styles, which lose much cream to their whey because of their small-cut curds.

Under no circumstances does milk need to be skimmed before the cheese make—something that makes every cheese suffer. It's always a better bet to leave your fat in your milk when making cheese, for that's where half of the flavor, nutrition, and texture of your cheese reside.

CHAPTER 6

The Tools

The make of a cheese is defined by the tools that make it. And these tools in turn are defined by the materials used to make them.

The nature of the tools used by a cheesemaker dictates the way their cheese evolves, be it the influences of the metal on the way a cheese vat cooks, the qualities of the cheese forms and how they drain a cheese, or the weave of cheesecloth and how it holds its cheese.

The cheesemaking tools that make all styles of cheese are all natural in their origin. These include wooden vats and copper kettles, whisks, wooden ladles, linen cheesecloth, beechwood hoops, clay forms, reed baskets, and the wooden draining tables they rest upon. These are the original tools of cheese, and they are essential for making a cheese right. For the traditional materials used by cheesemakers define even the contemporary industrial practice of cheesemaking.

A second significant philosophical difference between North American and European cheesemakers is that European cheesemakers make cheese as their traditional tools dictate, while American makers make their cheeses despite their tools. European makers are equipped with the specific tools that make their particular cheese what it is, while their North American counterparts often make a dozen different cheeses with the same set of tools, none of which are specific to any of their makes. Forcing the different cheeses into these inflexible tools makes for clumsy production methods that often result in poorly executed cheese makes and flawed cheeses.

This section will explore the traditional cheesemaking tools and how the nature of their original materials has influenced their function. We'll question why some vats are square and others round; why cheese forms have the forms they do; and why curd cutters and ladles are shaped the way they are. Almost always, the reason goes back to

The materials used by a cheesemaker inform the evolution of their cheeses. The impression of the linen cheesecloth used to drain an alpine cheese remains visible on the rind through its entire affinage.

the relationship of cheesemakers to their land and the influence of their material culture on the cheeses they made. And if we wish to bring these natural tools back into the dairy (we should for many reasons!), we must understand the ways they help us safely make our cheese.

The Cheese Vat

The cheese vat is the vessel in which the milk is curdled and the cheese is made. Cheesemakers historically used either copper kettles or wooden vats for this purpose. Modern cheesemakers' stainless steel vats largely mimic their functions.

The wooden vat was the original vessel used to hold milk for cheesemaking (after animal skins). The wood itself would have created the ideal environment to make cheese. Rarely in use today, due to restrictive regulations on the use of wood in industrial dairies, most styles of cheese owe their origins to these historical vessels.

Wooden cheese vats are coopered together from staves of chestnut, oak, or other hardwoods, held in place with hoops made of branches, copper, or iron. They have mainly circular shapes because of the nature of their construction. Wooden gerles still used in the Auvergne region of France today (as well as Tinas in Sicily) are constructed with straight, round walls and flat bottoms, and are made to be about as tall as they are wide. Though typically around 200 to 400 L in size, they can be made quite large, up to 1,000 L, and of course much smaller. The chestnut wood used to make the gerles (and the Tinas) is lightweight and strong. The vats were made to be moved from pasture to pasture with the animals, because the cheese was often made out in the fields with the animals and their still-warm milk.

The milk inside the wood could not be heated, so cheesemakers were limited in what they could make. Thus styles of cheese made in wooden vats are, with a few important exceptions, relatively quick-ripening, soft, uncooked cheeses. Cheesemakers were forced by the nature of the vats to use their milk while still warm from the udder and had to make cheese at every milking. Additionally, the wood insulated the milk, keeping it warm through the entire cheese make, enabling the cheese to progress in its best biological way.

Two workarounds evolved for making firmer, longer-lasting cheeses in a vat without cooking: adding hot water to the vat to cook the curd without fire, making a washed-curd cheese like Gouda; and the milled-curd method used in cheddar, which involves making the curd as firm as possible without cooking, then draining the curd of its whey and fermenting it for several hours before grinding it into smaller pieces, salting the curds, and pressing the curds together into a now much firmer wheel. Both Gouda and cheddar are now made almost exclusively in steam or hot-water-jacketed stainless-steel vats that permit the curd to be cooked if the cheesemaker chooses. But cheesemakers still produce the cheeses according to methods that were originally dictated by the use of wooden vats, for

Wooden vats inform the make of many classes of cheese, including milled-curd and washed-curd cheeses. The wood, which could not be heated, could only make a cheese so firm. The cheese being made in this vat, Saint-Nectaire, can only be aged so long because of its high moisture content, but if milled and salted, it becomes a cantal, a firm and preservable cheese, like cheddar.
PHOTO BY CHLOE GIRE

it is these particular adaptations to making cheese in wood that make the cheeses taste the way they should.

Wooden vats were likely once used by farmer-cheesemakers all across Europe. Most classes of cheese, including lactics, half-lactics, rennet cheeses, stirred-curd cheeses, washed-curd cheeses, and milled-curd cheeses, likely evolved because of their use. Wooden vats and the many methods married to them predate Roman influence in Europe, and represent some of the earliest styles of cheese known.

As explored in chapter 3, the regular use of wooden vats also created the natural circumstances for carrying culture forward from one batch of cheese to the next, enabling the natural fermentation that is responsible for cheese to evolve well. More information on using a wooden vat as a starter culture is provided in appendix D.

The copper kettle is a relative newcomer on the cheese scene, though it's still of course quite ancient—it only made its way to Switzerland, for example, in the 1500s. Its arrival there heralded the beginning of Swiss alpine cheeses as we know them. And though no one knows what culture began the tradition of cooking cheeses, with the advent of the copper kettle (the metal is believed to have originated in the Middle East) we gained the capacity to make cheeses firmer and longer lasting through cooking over fire.

Copper has unique qualities that enable curd to be cooked in its best way. For copper conducts heat perfectly throughout the pot, assuring an even cooking of the curds within their whey. The metal transfers heat most fluidly and helps cook milk (or whey) evenly without hot spots or burning, unlike stainless steel.

The copper kettle, with its circular shape and rounded bottom, is fundamental to the making of cooked-curd alpine cheeses. Its rounded bottom allows the cheese to be suspended over a fire to be more easily and evenly cooked. If it's held by a swinging arm, the kettle can be moved on or off the fire as it is cooked to temperature. The shape also permits cheesemakers to collect the curd that knits together into one big lump of cheese after cooking using a flexible rod with the end of a large alpine cheesecloth rolled around it; this permits the easy removal of the often massive cheese from the very hot whey (you can't do that in a flat-bottomed vat!).

Though it has some extraordinary qualities for cheesemaking, a copper vat needs some special care and attention to preserve its best functioning. Copper tarnishes on exposure to air and this oxidative process is intensified if the material is in contact with an acidic liquid. It's best never to let milk or whey ferment excessively in the copper vat and never to make lactic cheeses. And be sure to dry out the vat over light heat after its use, without even washing with water, to keep the copper from being exposed to moisture, acidity, or air (the dried sweet whey or milk actually prevents the copper's exposure to air). And then, to prepare for its use, scrub the vat with fermented whey to strip away any oxidation that occurred before the milk is next added. If unused for many weeks or months, or exposed to moisture, the vat must be scrubbed with vinegar (or milk vinegar) then rinsed with water or whey to remove the toxic tarnish.

Copper also contributes to the incredible flavor of traditional alpine cheeses and is still used today in many regions of Europe. Often its use is even required for a cheese to qualify for its PDO, as is the case with Parmigiano-Reggiano and Comté, among others. Conversely, the use of copper kettles may be restricted in North America because of the unwarranted concern that the porous copper is challenging to sterilize and can carry unwanted microbes from batch to batch.

Copper is an element that is present in trace quantities in all living beings. Our bodies need small amounts of it in order to function properly—our enzymes, oxygen transfer, and many metabolic processes depend on its presence. That's why it adds desirable flavor to a cheese. But too much copper could be problematic: The verdigris that forms on long-tarnished copper can be quite toxic, and seeing such a green color in a cheese is a bad sign. But with proper cleaning and care, a copper kettle should not be a cause for concern.

Square vats are a relatively recent addition to cheesemakers' tool kits, and only came to be when tin and steel made their way into modern dairies in the early industrial era, replacing the traditional wooden vats that could only be made round (they have to be made round because of the way wood swells when wet). These square vats (and the cheeses made within them) could be scaled up in ways that wooden vats could not: They could be constructed as long as a room, and just wide enough that a cheesemaker could reach the middle from each side. Their large size and shape heralded an advancement in technology that enabled cheesemakers to make cheese efficiently in a factory setting, with consistent curd size and consistent curd handling, and led to the evolution of mass-produced cheddar in dairies around the world, but also largely led to a degradation in cheese quality.

When dairying evolved into an industrial craft, and cheesemakers abandoned their wooden vats for metal ones, an era of cleanliness was ushered in, and sterility is still the creed in most cheesemaking circles today. But cheesemakers at the time neglected to acknowledge the role that the unwashable wood would have played in carrying culture forward; cheese fermentation often failed from the cleanliness. These are the circumstances in which laboratory-raised starters first arrived—to improve the quality of fermentation that was suddenly, for some unknown reason, failing, a problem that likely was not an issue before the switch to steel (this is the time when Pasteur invented his technology to control unwanted fermentations in winemaking). The problem was compounded by the abandoning of traditional rennets, prepared by steeping stomachs in fermented whey, in favor of commercial rennet prepared in a sterile fashion.

The square shape, however, enables a more consistent curd size, cut quickly with the help of cheese harps (see "Curd Knives," below). Such a vat is ideally suited to the making of stirred-curd blue cheeses, cut to a consistent size, or milled-curd cheeses. The square shape also permits the vats to be easily drained (as opposed to wheying off from above): A sloping bottom (often with a V shape) permits the whey to flow out to a spigot. Still, even though they may make wheying off easier, spigots create two problems for cheesemakers in their vats. One is that the strong flow of whey from the curd can disturb the curds and break them, releasing unrecoverable fines and fats into the whey stream; the other is that the spigots are a hassle to clean! Though it may be more work, cheddars and other firm milled-curd cheeses evolve more effectively if they are made in wood and wheyed off from above, as with the French equivalent, Salers.

Stainless-steel pots or vats are commonly used by home cheesemakers and commercial cheesemakers today, so I'll discuss their particularities and shortcomings here (and again later in this chapter). Stainless steel is nonreactive and permits the heating of milk and curd, allowing you to practically approach almost any style of cheese. It is best, however, to avoid heating the milk directly in the stainless steel: Most cheesemakers warm the milk in steam or hot-water-jacketed vats, which give a more even heat, and a home cheesemaker may not be able to do the same to avoid burning the milk. Nonetheless, any warming of milk causes it to lose its calcium, resulting in milkstone deposits on the steel that must be cleaned off with mild acids like vinegar. Ideally, milk should not be warmed in the pots, but rather used from the animal still warm, and the pot should only be heated once the curd is cut for the making of alpine cheeses. For other styles, a small stainless-steel pot may lose its warmth over time, changing the way a cheese evolves, typically for the worse. To assure the cheese stays warm for its proper evolution, it is still best not to heat the pot, which can easily overcook the curds; instead, make the cheese more quickly (adding culture and rennet at the same time helps) and keep the vat as warm as possible with either insulation or a warm-water bath.

Curd Knives

There are a variety of cutters you can use to cut your curds, depending on the style of cheese you're making and the type of vat you're using.

The original curd cutter is the hand, a practical tool for cutting the curd still used by many makers across Latin America. With its fingers opened wide, the hand can be moved around the pot to make the horizontal cut, then up and down in the pot to make the vertical cut, achieving a curd size that's relatively consistent. The hand is better for attaining a relatively small curd size, like the corn size found in semi-alpine cheeses such as tommes. The hand cuts well along the curd's natural lines of cleavage—the grain of the milk.

For small cheesemakers, cutting large-sized curds in a round pot, a simple curd knife is the best option. Curd knives are long-bladed, straight-edged knives, often with a rounded tip (so as not to scratch the pot), but any long-bladed knife will do. Curd knives are ideal for cheeses like Camembert and Brie whose curds need only vertical cuts (the horizontal cuts are made during ladling).

A perfect horizontal cut cannot be achieved with a knife in a round pot, but cheesemakers can make a cut with a knife on a diagonal to achieve an almost horizontal cut (of course, the bottom of the pot will not be well cut, but upon stirring the curd later, the large curds can be cut to size with a knife). When making medium-sized curd cheeses—milled-curd and stirred-curd cheeses—the curd is best cut more uniformly in size, and most commercial cheesemakers use harps for the horizontal cut.

Cheese harps are a more modern curd cutter, used especially in the making of medium-sized curd cheeses in square vats. Harps, as you'd expect, are collections of wires, or blades, that are pulled through the curd to precisely cut it. The knives come in two sets: with horizontal wires/blades and with vertical. The vertical harp is pulled through the curd in two directions to cut the curd vertically, then the horizontal harp is used to achieve the horizontal cut. They work especially well in square vats, where the curd is to be cut as uniformly as possible.

The spino is a natural curd cutter that is a slight improvement on the human hand. It is essentially a wooden stick with its branches protruding from all sides at one end, which can be made with the branch of any tree; but is especially effective when made with buckthorn (*biancospino* in Italian), whose strong branches with long spines are perfect for cutting curd on the horizontal.

The spino mimics the form and function of a hand, with the spines protruding from the branch working like fingers cutting through the curd; however, its longer length allows a cheesemaker to make their tommes in a larger pot. It also serves as a natural source of starter and is often left unwashed!

Another option for cutting small curd for alpine styles is a whisk. Originally whisks were fashioned from sapling spruce or pine trees, cut in the late spring, at the beginning of the *alpage*, when the trees were rapidly growing and the bark was easily stripped and the branches flexible. The naked branches of the greenwood were trimmed then pulled back to make the whisk and tied together with string or woven ring. The whisk enabled cheesemakers to achieve a very small curd size, essential to the making of their very firm, high-temperature-cooked wheels. The steel whisks used today by cheesemakers in the Alps to cut their curds originated from these first whisks.

Curd Stirrers

Once the curd is cut, cheesemakers use various tools to stir the curds in their vats. Having a good paddle helps cheesemakers get a handle on their curds, encourages the curds to firm, and prevents them from matting together.

The hand, with its fingers together, makes a perfect paddle that allows a cheesemaker to feel what's

Wood is good not just for cheese-ripening boards but for nearly all cheesemaking equipment. Wooden vats, curd cutters, cheese forms, cheese sabers, and butter paddles all make cheese better than their more modern counterparts.

happening in the pot better than any other tool. If the curd is breaking or matting at the bottom, or if the temperature is rising too rapidly, your hand will quickly tell you.

A *Käsesäbel* (literally "cheese saber" in German) is a strong wooden paddle used to stir the curd for alpine cheesemaking. One of my favorite cheese tools, the cheese saber is especially useful for higher-temperature curd stirring, and for preparing spun-curd cheeses that are melted in boiling-hot water. Steel stirrers don't work for this, as they transfer too much heat and burn the hand, while mozzarella sticks to the hot metal! And obviously you should not use a plastic paddle in boiling water. This is one of the most practical wooden tools a cheesemaker can work with. Smaller versions of these wooden paddles work well for home makers.

If you are using wooden stirrers, be sure to dedicate these to cheese, and never wash them. The unwashed paddle holds on to the milk and its whey, day after day, encouraging its fermentation and contributing to the successful evolution of the cheese.

The Hand

The hand is a cheesemaker's most personal and helpful tool for making cheese. Having a hand in the cheese also enables you to better gauge the process's progression and leaves a lasting mark on the quality of a cheese. You shouldn't be afraid to have your hands in the vat; both we and our milk are biological in our origins.

The hand is the most versatile tool in the cheesemaker's tool kit. It is at the same time many different tools and myriad gauges. It is a curd knife and a curd stirrer, a thermometer, a scoop, and a cheese press; it is a curd tester and an acidity meter if well trained. Good cheesemakers have their hands in their vats through nearly the whole cheesemaking process; even if mechanical aids are used, a cheesemaker's hand makes all the calls!

With the fingers apart, the hand can break the curd into smaller, though somewhat inconsistent-sized, curds. And with the fingers together, the hand becomes a paddle that perfectly stirs the curd.

The hand can be used as a remarkably precise and reliable thermometer (you can never fully trust a digital readout) in certain temperature ranges, and can be relied on for testing the warmth of milk for culturing and rennetting as well as for making yogurt. Your finger can comfortably stay 10 seconds in milk at 40°C (104°F) or even during the making of a some cooked-curd cheeses. Tomme de montagne, for example, is cooked to the point that anyone's hand can just about stand—45°C (113°F). When trained with the help of a thermometer, you can easily come to judge the temperature of milk to within one degree!

I use my hand for judging curd readiness for each step of the make, testing the strength and firmness as it evolves. In most cheeses the readiness of the curd for cutting can be judged by feeling for a slight puddling of whey atop the pot. The curd's readiness for forming can be tested by hand using various techniques, depending on the cheese, such as the splat test and the alpine peaks test. Even the readiness of a cheese for salting can be gauged by handling the cheese and judging its changing elasticity, an indication of its changing acidity.

Hands are useful for bringing alpine and semi-alpine cheeses together beneath their whey and forming perfect wheels before removing them from the vat. They can even apply just the right amount of pressure to press these firm wheels together when the cheeses are kept warm with whey poured over the top.

To cleanse your hands before the make, consider scrubbing them in fermented whey. I always keep a reservoir of the previous day's whey and use it to clean all my natural tools, including my hands. The whey is moisturizing, unlike commercial cleaners, and does not harm the skin. It is also especially effective at removing unwanted microbes and replacing them with beneficial ones that can only improve the fermentation of your cheeses.

Ladles and Scoops

The ladles and scoops used by cheesemakers are often specialized and specific to the cheeses being made. Having the right scoop or ladle on hand can improve your curd handling and often makes a particular cheese make better, quicker, and less laborious. It's also helpful to have bowls for removing the whey from atop the curd before forming.

Lactic cheeses are often ladled from the vat with a straight ladle, so as to have the most ergonomic action when filling many forms, with the scoop being in line with handle (I find myself bending soup ladles straight all too often for this purpose). The sharp-edged scoop of the ladle, originally made of copper, cuts a perfect hemisphere of curd cleanly without breaking, minimizing losses and preserving the beautiful grain of the milk in the cheese. Lactic and half-lactic cheesemakers spend much time hand-ladling, assuring the curd isn't broken by rough handling. If this is done right, with the right ladle, the results shine through, especially after affinage, so much so, that cheesemakers in France often label their cheeses "moulé à la louche." Traditional Camembert makers also ladle curds into their hoops a specific number of times (exactly seven) to ensure a consistent height of their cheeses. More industrially produced lactic cheeses made in France are first drained in cloth and then packed into their forms by hand, an approach that saves time and therefore money but makes a poorer-quality cheese, without the grain of the milk. Increasingly, producers of traditional lactic cheeses are labeling their cheeses as moulé à la louche à la main— molded by ladle, by hand—for some producers use robots for the ladling!

For making Brie, cheesemakers use a *pêle à Brie*, a Brie shovel. These perforated scoops are made today of steel (though historically, like the ladle, they were made from copper) cut round, with a slight concavity, with their handle positioned directly above the scoop to make easy work of ladling. The pêles have a sharp edge that slices cleanly through the curd, making horizontal cuts (Brie is only cut vertically) as they scoop. And the shovel has the precise diameter of the cheese form, so that the curds can be placed into their hoops one full layer at a time, to assure the most efficient and consistent work, and the best curd texture. The exact same tool is used in the ladling of curds for the making of Stilton.

Cheesemakers making other milled-curd cheeses, like cheddar, traditionally used curd scoops or shovels carved from wood for salting their milled curds and packing them into forms. A similar tool is often seen in alpine cheesemaking operations, used to handle the curd in the copper vat. For making alpine cheeses, though, the curd is removed from the whey, most often in one massive lump, with the help of very strong cheesecloth.

Cheesecloths

Fine cheesecloths can be used for draining lactic cheeses, protecting young cheeses from flies, or assisting with the of pressing milled-curd cheeses like cheddar. Coarser-textured cloth is more suited to handling and pressing alpine cheeses. And very fine cheesecloth (the store-bought stuff with a very open weave) is really only useful for bandaging the rinds of cheeses like cheddar; the technique for doing so is explored in chapter 9.

Finer cloth is appreciated for the draining of high-moisture lactic cheeses, which can slip through the open weave of a coarser cloth. Usually made of cotton, the fine weave also absorbs just the right amount of moisture, helping the curd drain well. In commercial dairies, this fine cloth is often sewn into bags to help with the draining of larger quantities of cheese, which are hung from hooks over receptacles that catch the dripping whey. Industrial cheesemakers tend to use disposable plastic cloth for this purpose. The plastic cloth doesn't absorb moisture, which slows drainage and also, conversely, makes the cheeses within more prone to drying along their edges, a phenomenon that doesn't happen with natural cloth.

Fine cheesecloth is also used to wrap semi-firm cheeses like tomme or Saint-Nectaire in their forms to aid their draining, and can help keep milled curds in their presses as they're pressed into cheeses like cheddar. Fine cloth can also be draped over draining cheeses for fly control. Cloth coverings also help keep cheeses from drying out and losing the integral moisture that helps them ripen right (cheeses are best drained with only gravity and salt, and not with drying air).

Alpine cheesecloth is worth seeking out if you want to make traditional cooked-curd cheeses. Woven from thicker strands of hemp or linen, with a very open texture, these cloths are much stronger and more durable than typical cotton cheesecloth, permitting the handling of very heavy cheeses without tearing. The large weave also leaves a beautiful surface texture when wheels are wrapped in it, giving alpine cheeses an intricate rind

Many different grades of cheesecloth are available for many different purposes. Fine cotton cheesecloth is perfect for draining lactic cheeses, while stronger-fibered hemp or linen cloths are necessary for handling massive alpine cheeses.

patterning as they age. You can judge an alpine cheese by its cover: Industrially made alpine cheeses have an imprint of their plastic forms on their rinds, but the best ones, those made more naturally, still show the weave of the cloth on their crust even after a year of aging.

Alpine cheesecloths can be quite sizable: almost 2 meters in length and 1.5 meters wide for making the most massive wheels. Their uneven size enables you to wrap the long end of the cloth around a flexible rod or branch, which you can use to pull the remaining cloth underneath cheeses like Comté in their round-bottomed vats; you can then pull the cloth up around the coalesced cooked-curd cheese to remove it from its exceedingly hot whey. The cloth is then tied onto a strong rod with the corners of each side, securing the cheese within a bundle to the rod, allowing a cheesemaker to lift the new cheese out of the vat within its bundle. (I cannot help but imagine a stork carrying a baby!)

It's probably best to change cheesecloths daily, if cheeses are spending extended time in them (as in the case of blue cheeses, big alpine cheeses, and some aged lactics), to avoid yeastiness. Saturated in lactic acid, they tend to turn orange-pink like washed-rind cheeses. Cloths can be gently washed with scent-free soap to refresh them between uses, or, better yet, scrubbed with fermented whey before being wrung and hung to dry.

Cheese Forms

Cheese forms are the vessels that give cheeses their shapes; they're typically rounded, with or without a bottom, often with small holes all around to facilitate drainage. Cheese forms are made in myriad different materials, shapes, and sizes, representing different cheesemaking traditions from all around the world, but in general they come in several different formats governed by the natural materials originally used to make them, and the needs of the cheeses made in them.

Cheese forms can be woven baskets, for rennet cheeses. Clay pots with drainage holes are perfect for lactics. Alpine cheeses are often traditionally pressed and formed in wooden hoops while cheese forms for cheddar and other milled-curd cheeses that have to be heavily pressed are very strong to resist the pressure, and were historically made, like barrels, of coopered wood. All of the many different modern forms made of plastic or stainless steel have their origins in these traditional forms.

Open-ended hoops, like those used for Camembert and Brie, give these ladled cheeses their sought-after shapes. The straight sides of the hoops (originally made from lengths of clay pipe) make it impossible to flip these soft cheeses by removing them from their forms . . . so instead the cheeses are flipped upside down within them. To flip one, wait until it's fully drained and salted on one side, then after an hour give it a first flip by carefully sliding your hand under the cheese in its hoop. Flip the hoop and its cheese quickly together back onto the draining mat.

Alpine cheese hoops were usually made from large beechwood bands, known in German as *Järbe*. These hoops are held in shape with a wooden slider on strings that can be adjusted according to the milk supply to make a cheese of a certain height—essential for the making of these cheeses, which must be a consistent height to ripen consistently. A variation on alpine hoops from more Mediterranean climates is an adjustable band woven from esparto grass. These forms gave the beautiful herringbone pattern to the sides of the region's alpine-style cheeses, like manchego.

Cheddars, cantals, and other cheeses that need a heavy pressing to knit together were pressed in much stronger, often reinforced forms. These cheese forms, traditionally coopered wooden hoops with bottoms, made the pressing, flipping, and removal of the cheese simpler. Gouda forms, still made from coopered wood today, were historically carved from willow, a flexible and strong wood that doesn't crack or expand when wet, making it perfect for pressing a firm cheese. (Dutch klompen are carved from the same wood for similar purposes and make great waterproof shoes for the dairy.)

Baskets woven from willow or reeds (often *Juncus* species of grass, from which the term *junket* may originally derive) were the original cheese form for many styles of cheese across Europe. Willow baskets would have made stronger forms for firmer cheeses; reed baskets would be used for higher-moisture and more delicate softer cheeses. Both types typically had twined sides that left vertical slits, permitting the easy removal of the cheeses formed within. Though largely neglected in dairying today, basketry was considered an essential art for the milkmaid/milkman of yore and would have likely held a very strong significance. The starting point for woven cheese baskets is a Saint Brigid's cross, a symbol made of reeds that's used in the Celtic-Pagan festival of Imbolc, which celebrates the first milk of the spring and the butter and cheese made from it.

Such baskets should be held in such significance again, for they truly make a better cheese: They drain a cheese in its best possible way while also keeping the cheese humid and preventing its drying; they give a more intricate patterning to the rind; and the natural material holds on to beneficial microorganisms that repel unwanted ones. Many plastic cheese forms are fashioned to resemble handwoven baskets for their appearances only. Ironically, identical patterns can be seen on artisanal and industrial cheeses made all around the world! And they don't drain as well as naturally woven baskets, nor do they keep cheeses humid or hold on to protective microorganisms. It might seem nostalgic to some, but the traditional form of the handwoven basket makes for a much more ecological and beautiful cheese, with its own particular character, that is more reflective of the lands and hands that made it. A handwoven basket is truly a complement to a handmade cheese.

Clay may also find its most perfect purpose in the forming of cheeses; it might well have been used to make cheese for over 8,000 years. Archaeologists scratched their heads for years over the purpose of ancient shards of perforated pottery found at excavations across Europe. It took a protein scientist who tested the residues on the perforations of such a shard found in Poland dated to 6000 BCE to discover dairy proteins, a sign that a cheese was made in the vessel millennia ago. But any cheesemaker could have told them

Cheese forms were originally made of many different materials, including woven reeds and willow baskets, clay pots, esparto bands, and beechwood Järbe.

that, for perforated potted forms were used for cheesemaking up until the 1930s when tin and later steel and plastic came to replace them. There may be no other object as satisfying for a potter to pot, for every form made will replicate itself hundreds, indeed thousands of times over with every cheese formed within it!

Lower-temperature-cooked clay works best for cheese. More porous than high-temperature-fired porcelain, earthenware and stoneware help pull out moisture from the cheeses and cool them as they drain. And what better way to form a product of the earth than in earth itself? Indeed, the nuances of this earthen material birthed the many diverse shapes of traditional lactic cheeses, which were mostly drained in clay. You can read one such origin story in the Valençay technique in chapter 12.

Clay is likely best unglazed; to help the material pull moisture away from the cheese; and giving cheese rinds a smooth finish. The glazing of pots is in essence, glass, which poses its own hazards if the pot is broken and is best avoided. Though the glazing may make the pots easier to clean, the glass will slowly crack over time. By leaving the pots unglazed, the pots absorb whey and come to life, much like the cheeses formed within them. Unglazed clay forms (as well as woven cheese forms) used over and over and over will come to develop biofilms of *Geotrichum candidum*. I scrub these off with fermented whey and a soft-bristled brush, then use them again or leave them to dry.

Draining Tables and Mats

Draining tables are the surfaces upon which cheeses are placed to drain after their make. A good table assures that whey flows well from the curd, without the cheeses sitting in a puddle.

In their most original form, draining tables were carved from large trees sawed in half or chiseled from wide, flat stones; they are inclined so that the whey flows from the many cheeses placed upon them, and their end is often fashioned to a point, so that the whey flows off in a stream and can be collected in a bucket for its various uses.

Commercial draining tables used today are typically made of stainless steel and are inclined toward a spout that drains the whey into buckets below; having a squeegee on hand helps keep the table clean and the whey flowing.

A micro-scale cheesemaker can use stainless-steel hotel pans (commonly available from restaurant supply shops or industry auctions): a shallow perforated one for the draining table, and a deep one below to catch the whey that drains from the cheeses. You can fit up to a dozen small cheeses or two large wheels on them, easing cheese handling and tidying the make. Home cheesemakers can also improvise a draining table by placing a baking rack atop a deep casserole dish to keep their cheeses clear of their whey (though the rack may rust if it's not stainless steel).

Often draining tables do not do a perfect job of draining due to their impermeability, so cheesemakers typically place cheeses in their forms atop draining mats to keep the whey flowing beneath the cheeses. Cheeses placed against an impermeable surface do not drain as well and can develop pockets of whey, resulting in holes on their rinds.

Draining mats can be woven from straw, bamboo, or finely split wood or can even simply be straw loosely laid upon an inclined table. Many traditional cheeses from Europe were historically laid to drain on straw, and recognizable aspects of their appearance sometimes derive from this: Taleggio, which was once drained on straw, doesn't look quite right without straw-shaped impressions on the cheese's top and bottom. Today the standard plastic cheese forms for taleggio are fabricated to leave straw-like impressions on the cheese's rind! Real straw, however, does a much better job leaving straw impressions, and also improves a cheese's drainage over plastic.

Draining tables in continuous use, either wooden, stone, or steel, can be scrubbed down daily with the previous day's fermented whey to clean off stuck-on curd and keep yeast growth controlled.

Modern Cheesemaking Materials

Typically, only two materials are permitted for cheesemaking in a certified modern dairy, at least in the United States and Canada: stainless steel and plastic. Though both will work for a natural cheesemaking, neither material is ideal for the purposes of making cheese. And for cheesemaking to be considered sustainably practiced, we must fight for the approval of natural materials like wood, clay, straw, and reed in our dairies.

The only natural material commonly permitted in the US and Canada is wood for aging cheese (though not for making it). Even copper is questioned by inspectors for its permeability, though its use is not expressly illegal. In Europe, the use of natural materials is more permissible, due to the recognition of the importance of these materials to the continuing traditions of cheesemaking, and the contributions they make to cheese quality.

These days, steel and plastic are the only materials in which cheesemaking equipment is generally manufactured, though plastic is increasingly replacing steel. Many cheesemaking operations today have become completely plasticized, apart from the cows and their milk!

When I started out my cheesemaking journey many moons ago, I apprenticed at a commercial cheesemaker that was in the midst of a transition from stainless steel to plastic and was still using some stainless-steel cheese forms. These days, most cheese forms are only available in plastic, and a cheesemaker can be hard-pressed to find molds made of anything but in a standard cheesemaking supply catalog.

But plastic isn't as fantastic as it's often played out to be. The material absorbs fats and can be difficult to clean. And despite the industry's claims, it's not easily sterilized—we all know about the problems with plastic cutting boards!

Plastic derives from the petrochemical industry and should be seen as an objectionable material. It is most likely toxic, and is not very durable. It will only last a year with heavy use and will degrade its chemicals and its microplastics into our cheeses and our bodies. Microplastics are now commonly found in our blood and in mothers' milk, being especially attracted to fatty tissues. And this invasion is not benign: A study published in the *New England Journal of Medicine* in March 2024 showed a very strong correlation between microplastics in fat deposits in our arteries and the incidence of heart attacks, strokes, and death.

Plastic's low cost up front is misleading, for it must be regularly repurchased and is constantly adding to the operating costs of dairies. The material is not recycled in most dairies; even if it were, it is not truly recyclable.

A cheese today will spend almost its entire existence in contact with plastic. From the wrap of the animals' silage to the milking machines, the cheesemaking vat, the cheesecloths, the cheese forms, the draining mats, the aging shelves, and ultimately its packaging, cheese is exposed to plastic at every stage of its life. Our relationship with this material will need to be reckoned with as we recognize the challenges that it poses to our planet and ourselves.

Stainless steel also presents problems for a cheesemaker, not the least of which is its high cost, the main reason it's being abandoned today for plastic. Stainless steel is extremely energy-intensive in its production, and steel equipment is typically manufactured overseas. The material is also not as durable as commonly believed; surprisingly, a well-maintained wooden vat can last just as long as a stainless-steel one, which can be ten times as expensive to purchase new! Certainly, a copper vat will long outlast a stainless one—I've seen some that have been in continuous use for well over 100 years.

CHAPTER 7

The Salt

What would cheese be without salt? Excessively moist, overfermented, fast ripening, and bland! Salt is a preservation agent that controls moisture, and therefore restricts microbiological developments so that a cheese's aging can be better controlled by the cheesemaker. It is an exceedingly important part of what makes every cheese what it is meant to be. Most cheeses simply couldn't be made or ripened without it.

Salt is integral to the development and preservation of nearly every style of cheese, the only exceptions being fresh dairy ferments, or those cheeses meant to be eaten in their first days. Cheese simply does not last without this aid to conservation.

There are three ways of giving cheese its salt: surface salting, saturated salt brining, and mixing salt into the curd. This chapter explores these different ways to add salt to a cheese, along with how salt works upon a cheese, how much to apply, and why. And for the overly ambitious, we explore how to make salt from seawater.

Salt must be understood as a tool used by cheesemakers to get the moisture out of, and therefore preserve, a cheese. While we normally apply salt to flavor our food, the application of salt to cheese is much more fundamental, and the integral ingredient must be applied in the proper amount—no skimping on salt is allowed! It is essential for controlling the fermentation of most preserved or even fresh foods like bread, which is rarely baked without salt, in part for its moderating effects on the sourdough fermentation and its inhibition of mold, much like in cheese.

Salting does, of course, also improve the flavor of a cheese (this is because our body needs this important mineral), and soft lactic cheeses or milled-curd cheeses that have salt mixed into them by hand can be salted quite accurately to taste.

Salt was once a much more valuable commodity, required for the seasonal preservation and fermentation of almost all of our food before the days of refrigeration, long-distance transportation, and agriculture's industrialization. We now eat produce delivered to us fresh at all times of the year, from wherever in the world that food may be in season, but historically, when our diets were more regional and more seasonal, salt permitted the preservation of much of our food from its harvest through the year. Fermented vegetables, cured meats, and cheeses, all put away seasonally with the help of salt, required a steady supply of the once-valuable mineral. Regions that mined salt, like Salzburg ("salt city") in the Austrian Alps, flourished with the wealth generated from its salt sales. The word *salary* is said to originate with the Roman *salarium*—the money given to soldiers for their essential purchases of salt.

As we once again realize the value of salt for the preservation of more sustainably raised foods, its local production is undergoing a renaissance of sorts. Cheesemakers and fermenters are beginning again to use local salts, helping to define their taste of place.

Salt Loves Water

Salt pulls water toward it through a process known as osmosis, an effect that governs the interplay of the two compounds NaCl and H_2O. If salt is placed in contact with an object that is less salty and full of water, the salt will pull moisture from the object toward itself, and the salt will migrate through that water toward the object, until a new water-and-salt balance is achieved.

When dry salt is applied to the still-moist rind of fresh cheese, that salt will adhere to the cheese and from there work its osmotic magic. The salt will pull moisture out from the cheese, dissolving its crystals into a liquid brine on the rind. With the salt now transformed into a brine, salt itself begins to flow backward into the cheese. It continues to do so, until the levels of salt and moisture at the surface of the cheese are in balance with the brine on the rind. The now slightly salty moisture that was pulled to the surface then drips from the cheese. This effect can be very clearly visualized by placing a few grains of salt on a ripened Camembert. As the salt sits in contact with the rind of the ripe cheese, it pulls moisture out from within. That moisture swells the salt crystals on the rind until they turn to liquid, which forms perfect beads on the surface due to the moisture-repelling qualities of the fungal rind.

But that's not the end of the effects of salting. The salt continues to pull out moisture from within the cheese toward the rind as the mineral migrates slowly through the wheel with the help of its osmotic influence. Over time, salt will slowly penetrate the cheese from the surface to the core as moisture is pulled outward. It may take days or even weeks for the cheese to come to a salt-and-moisture equilibrium, depending on its size, firmness, and moisture content.

Very hard full-alpine cheeses have very slow movement of salt within them. This is a significant part of the reason why the large cheeses are made short in stature: to assure good salt penetration in an effort to control unwanted microbiological developments. A poorly salted alpine cheese can swell as the result of continued fermentation and the resultant gas formation by milk's own propionibacteria. Though this is a desired effect in the making of Emmentaler, it is unappreciated in most alpine styles.

Salt may have a controlling effect on pathogenic microbes, but it really has an inhibitive effect on all cheese microorganisms, beneficial or not, slowing the development and fermentation, both good and bad, that exists in all cheese styles. A proper fermentation does a much better job of controlling unwanted microorganisms than a proper salting. Milk will ferment well without interlopers and ripen well without "contamination," even without salt, so long as an active and microbially rich starter is added to it.

Salting gives control over the complex interrelationship of moisture and fermentation within the diverse makes of cheese, allowing certain microbiological and textural developments to occur instead of others. For instance, the time at which a cheesemaker salts a Camembert can influence the amount of lactose that is fermented into lactic acid. This controls whether the cheese develops into a more modern enzymatic style Camembert or a half-lactic traditional Camembert. Camembert can even become fully lactic in taste and texture if salting is further delayed. Salting a cheese tends to shift its fermentation from anaerobic bacterial fermenters to aerobic fungal fermenters. The timing of its application therefore controls the transition from primary fermentation to secondary fermentation, significantly affecting the evolution of a cheese.

Fermentation is dependent on moisture, because bacteria and fungi responsible for cheese's many transformations need water to thrive. Salt, however, binds moisture to it within the flesh of a cheese, making the water less available to all microorganisms that might ferment it. Primary fermenters like lactofermenting bacteria need more moisture to thrive, while ripening organisms (mainly fungi), like *Penicillium roqueforti* or *Geotrichum candidum*, seem to need less moisture to do their work. These rind-ripening organisms may get their necessary moisture from the humid cave air, which is not locked up by salt. Therefore, adding the right amount of salt, and at the right time, is a significant control over the degree of primary fermentation that a cheese receives before secondary fermentation begins. The effect of the timing of salting will be explored later in this chapter.

Salt has other effects on cheeses, more gastronomical than microbiological. For salt also helps develop cheese's best flavor.

The Two Percent

Two percent is the recommended salt application by weight for most styles of cheese, preserving a cheese against rapid microbiological degradations without making it overly salty to taste. Two percent is the amount of salt in cheese that shifts fermentation from primary to secondary by slowing the development of anaerobic lactic acid microorganisms and encouraging aerobic rind-ripening ones. A 2 percent salting locks up the right amount of moisture within the cheese, limiting its use by lactose-fermenting bacteria but still permitting fungal or other secondary fermentations on the rind.

But a 2 percent salting is not just specific to cheesemaking; any fermenter will recognize the number, for 2 percent is the amount added to cabbage to control for the best quality and flavor of sauerkraut; 2 percent is the concentration of salt that makes for a perfect pickle; and 2 percent salt (dry weight) is about the quantity of salt that regulates fermentation in sourdough breads, but also makes the crumb most flavorful without tasting overly salty.

Two percent salt, significantly, also makes most foods taste their best. Two percent salt brings out the best in a broth. Adding 2 percent dry salt to a chicken makes for crispy skin and tender, juicy, and flavorful meat. And 2 percent gives cheese curds just the right savoriness and squeak! An easy way, therefore, to judge for 2 percent is with your tongue. Two percent salt tastes satisfyingly salty, without an offensive saltiness. And this, it's important to note, is not a personal taste. Though some might prefer a butter salted or unsalted, salt tolerance, like temperature tolerance, is universally affected by our bodies' chemistry, and its intolerance to the destructive effects of high concentrations of salt on our cells. And though often vilified in a modern diet, our bodies crave salt, as they need it for their proper balance and function. And this intensive, attractive flavor must be sought out when making cheese.

As explained earlier, not all cheeses are best salted to 2 percent, which is actually around 3 percent when applied by hand, because much salt is lost in the process. There are some exceptions to the 2 percent salt rule. Butter, with very little moisture, needs less salt to be preserved—even 1 percent salt in butter is salty enough for most tastes. And some fresh lactic cheeses, like faisselle, are best without any salt but must be eaten within a day of their make (1 week max with refrigeration) so that they don't become moldy.

When salting cheese, cheesemakers invariably rely on feel. Precisely measuring the salt applied to a cheese is a challenge. There is no convenient tool that can tell you when a cheese has received enough salt, other than observation, touch, and taste. And understanding the appropriate amount of salt that a cheese needs can only truly come with experience. While you're gaining experience, however, a scale or a measuring spoon can help.

It should be recognized that the most sought-after rind-ripening microorganisms and their supportive ecologies, *Geotrichum candidum*, *Brevibacterium linens*, and both *Penicillium roqueforti* and *P. candidum*, all have similar salt tolerances to ourselves. Thus we can trust our tongue when it comes to a test of salinity. Furthermore, we are going to be eating these cheeses after all, and we want them to taste as balanced in salt as they should be, without want or excess. Salting to 2 percent preserves a cheese to its best degree, without making it overly salty to taste.

Fortunately, cheeses can tolerate slightly more or slightly less than the prescribed quantity of salt, although some changes will certainly occur as a result of the difference. Precision in salting is important, but it isn't entirely necessary. If some variability is acceptable in your batch of cheese, you can salt with less precision, but you should expect that cheeses that receive less salt will retain more moisture and continue to ferment, taking on a more lactic texture and flavor, and will ultimately ripen faster; those that receive more salt will age more slowly, and may not develop their best rind ecologies, textures, or flavors as quickly because of excess moisture losses. A consistent salting results in cheeses with a much greater consistency within the batch, and also between batches.

Two percent is a considerable amount of salt, and most new cheesemakers may have trouble applying the right amount. Keep in mind, however, that it is better to undersalt than oversalt, because more salt can always be added if the cheese is found to develop its rind ecologies too quickly. Cheese will continue to primary ferment if it isn't sufficiently salted, which can change it; however, if a cheesemaker oversalts their cheeses, there is nothing that can be done to remove the excess.

The Timing of the Salting

The timing of the salting has a significant impact on the evolution of a cheese—so much so that the salting schedule is a defining element of most of the styles of cheese in this book. This schedule can have a great impact on the way a cheese ripens, the cheese's texture, and its final taste. Salting at the wrong time can also lead to significant flaws and the growth of unwanted (but not necessarily dangerous) microbes in many cheeses. So, it's important to know the best time to salt for each style and why.

Because salting controls moisture, and because bacterial fermentation needs moisture to progress, salting controls this primary fermentation and acidification in a cheese. In essence, the timing of the salting determines how far the primary fermentation of a cheese is allowed to proceed before the secondary fermentation, or ripening, begins. This can have dramatic effects on the way a cheese ripens. A tomme, for example, must be salted at a pH of 5.3 (using the teabag test, explored in chapter 5, a small piece of the cheese will stretch when submerged in very hot water); if it is salted at a pH of 4.9 it evolves into a different cheese, tomme crayeuse, literally "chalky tomme." The different pHs at salting of these two cheeses changes the interior texture. The tomme is semi-firm, while the tomme crayeuse is chalky, due to the advanced acidification of the curd. The tomme grows *Penicillium roqueforti*, while the crayeuse grows *Geotrichum candidum*, due to the longer fermentation time at room temperature, and this determines how the cheese ripens. The tomme stands tall and breaks down evenly throughout its paste, while the crayeuse breaks down and flows close to the rind, due to the buffering effect of the acidic paste on the raising of the pH of the cheese, restricting deliquescence to just beneath the rind. Yet despite all these seemingly significant differences in cheese qualities, the only real difference between the two very different cheeses is how much their primary fermentation is allowed to proceed before the cheeses are salted.

Because the primary fermentation is more controlled by salting than the secondary fermentation, salting at a certain point in a cheese make preserves the cheese at that certain acidity and turns the tide of microorganisms to those secondary fermenters, usually fungal, that grow upon the cheese's rind. There are several consequences to the degree of primary fermentation (or pH level) that proceeds before salting, the most important being the microbes that develop the rind, the flavor of a cheese, its texture, and its curd breakdown pattern. And we'll use three other cheeses—a crescenza (salted at a pH of 6.5), a Camembert (salted at 5.3), and a Saint-Marcellin (salted at 4.5)—to examine the effect of changing the salting schedule on the cheese.

When a cheese receives more primary fermentation, its paste develops more lactic acid, which encourages the growth of certain microorganisms over others. *Geotrichum candidum*, especially, develops much more readily on a cheese that's longer fermented before salting, and this shows with the Saint-Marcellin, which is left for 2 days at room temperature before salting. A Camembert has more of a tendency to grow *Penicillium candidum*, which prefers to grow in conditions that are less acidic, while a crescenza will tend not to develop a rind ecology because it more or less lacks lactic acid due to its very early salting.

With respect to flavor in a cheese, the degree of primary fermentation (its lactic acid fermentation) has a great impact on the flavor development of a cheese. In general, the more primary fermentation is permitted before salting, the more flavorful the cheese will be in its early stages. A cheese salted with less primary fermentation needs more secondary fermentation to develop its flavor. Accordingly, the lactic Saint-Marcellin is delicious to eat right after salting as what is essentially a cream cheese, while the less primary fermented modern Camembert has very little flavor at the same young age.

Furthering the flavor factor, the more primary fermentation is permitted before salting, the more secondary fermentation will take place before a cheese is ready to eat, resulting in even more complex flavors in a fully ripened cheese. If the secondary fermentation follows a more complete primary fermentation (that is, if the cheese is salted later), the final products of the primary fermentation will be more numerous, and the final flavor of the cheese will be more substantial. The aged Saint-Marcellin is therefore much more complex than a Camembert in part for this reason.

All three of these cheeses are considered to be fully ripened when their pHs rise to around 7.0, the point at which cheese deliquesces (more on this term in chapter 10). If the cheeses are salted at a lower pH, it takes more secondary fermentation to raise the pH to its final point, which results in more significant flavor development. For this reason, as well, the full ripened flavor of a Saint-Marcellin at 2 months is the most complex of the three cheeses salted at different pHs. The Camembert sits nicely in the middle with a strong but pleasant animal flavor, almost like cured salami, while the crescenza reaches its readiness and begins to flow within a week or so; but it has a much simpler flavor profile, tasting mostly of milk because its primary and secondary fermentations were limited by its early salting.

With respect to texture, the more fermentation is allowed before salting, the softer and more brittle the cheese's texture becomes in its early stages. This initial curd texture carries through to the cheese's final days by influencing the extent of curd breakdown.

A cheese salted at a very high pH, like crescenza, almost before any primary fermentation (it's salted in the milk, at a pH above 6.5), will remain almost fluid, and will not retain its shape when removed from the form. A modern Camembert (salted at a pH of 5.3), however, will be strong, elastic, and very flexible, and a lactic Saint-Marcellin (salted at 4.5) will be firm but brittle so that it stands very tall but crumbles if not carefully handled. The differing textures of these cheeses relates to the calcium content of their curd, which preserves the strength of the bonds between the casein proteins in their micelles. More calcium is retained in cheeses salted at a high pH, which keeps the curd flexible, elastic, and strong. When a cheese is salted at a lower pH, more calcium is lost from the curd, which makes it softer and more brittle. The texture achieved when the cheese is salted remains throughout the cheese's ripening but is changed ultimately by the effects of rind ecology and the resulting curd breakdown.

With respect to curd breakdown, the degree of primary fermentation permitted before salting has a strong influence on the way a cheese ripens. With a lactic Saint-Marcellin, fermented to a pH of 4.5, the lactic acid content is very high, which restricts, or buffers, against the raising of the pH from the fungal metabolism at the rind. This acidity limits the liquification of the curd, which is, in essence, the effects of the rising of the pH that causes cheese's liquification, restricting that liquification to the rind and slowing its progression toward the center. A Camembert salted at pH 5.3 has less lactic acid, which permits the fungus at the rind to raise the pH at the interior evenly with the rind, which results in a more even breakdown throughout the cheese. A half-lactic traditional Camembert, salted at 5.3—but with more moisture in its curd, which allows for more fermentation after salting—will ripen between the modern Camembert and the Saint-Marcellin, with curd breakdown somewhat restricted to the rind by its acidity, and a more brittle, almost chalky core. Many consider this texture to be the symbol of absolute perfection for a Camembert, but those used to a more modern version will see the unripened core as a sign of an unripe cheese.

Every cheese style has its preferred pH at salting, for the texture and chemistry of a cheese at salting is a strong determining factor for that particular style's evolution. Typically, though, higher-moisture, higher-fat cheeses are at their best and most flavorful when allowed to ferment longer before salting.

Accordingly, stirred-curd blue cheeses like Roquefort or fourme d'Ambert develop their ideal flavors and textures when they are allowed to ferment longer before salting—often up to 3 days at room temperature! These creamy blues are typically salted when the pH descends below 5.0 and the cheese crumbles when steeped in hot water. And the milled-curd Stilton achieves the creamiest and most flavorful version of itself if its curds are left to more fully ferment before milling and salting. An advanced fermentation in these cheeses results in more lactic acid, which fuels a longer and more substantial affinage before the cheese is ripe, and results in better blue flavor development and color, giving these styles their sought-after, creamiest textures. The greater acidity in the curd also makes these cheeses more brittle and lets them stand taller without barreling—allowing, in the case of Roquefort, the cheeses to be kept in the cave on their sides without losing their round shapes. This high amount of fermentation also enables the cheeses to develop their sought-after creamy textures when they age. If they are salted sooner, they will tend to bulge and develop more of a flowing, Camembert-like texture, as is also the case with Gorgonzola, salted at a pH of 5.3.

Most alpine cheeses, though, are fermented to a pH of 5.3 before salting, as too much acidity development can cause a weakening of the usually elastic alpine curd, giving the cheese unwanted chalky textures (à la tomme crayeuse salted at 4.9). Excess primary fermentation in alpine cheeses also results in curd breakdown concentrated toward the rind like a lactic cheese, giving these normally firm cheeses an oozy, almost Camembert-like texture beneath their crusts when aged. Both of these normally unwanted developments shorten a cheese's life span and its potential for both preservation and flavor development, significant flaws for those cheeses made to be very long preserved and easily transported.

Surface Salting

Surface salting, also known as dry salting, is a simple yet effective method of pulling moisture out from a cheese. To surface-salt, salt is applied liberally to the rind of a young cheese, still wet with its whey, usually within 8 to 12 hours of the beginning of the cheese's make, depending on the goal acidity for the salting.

To dry-salt a cheese, apply salt to achieve roughly 2 percent of the weight of the young, wet cheese in salt; this is actually around 3 percent if you weighed it, as much salt runs off the cheese with the moisture that's extracted. Few cheesemakers, therefore, actually weigh their cheeses, then weigh their salt; most salt their cheeses by eye, building up the consistency of their salting over years of making the exact same cheese. Precisely measuring 2 percent can lead to undersalting, unfortunately, as some salt falls off the cheese during its application, and much of it runs away with the whey it extracts.

Salt can be applied by rubbing the salt over the surfaces, or by sprinkling it on with a large saltshaker. Small round cheeses like crottins can be rolled a certain length on a table sprinkled with salt to speed up the process.

Two percent salt looks slightly different on the rind depending on the shape and size of the cheese being salted. A very thin cheese or one with a square shape, which both result in a high surface-area-to-volume ratio, will need a lighter coverage of salt on its rind to achieve 2 percent than will a cheese of the same weight that is taller and has a lower ratio. Larger wheels of alpine cheese need to be surface-salted two or more times by hand over the course of several days, as the cheeses do not have enough surface area to hold on to the 2 percent salt needed to remove their moisture.

Some cheesemakers apply salt only to the top side of the cheese; then, once the salt has dissolved, they flip their cheeses and salt the other side, as the salt applied to the bottom of a cheese doesn't stick as well. But I find that salting on both sides at the same time helps stop primary fermentation more effectively. If using coarse salt with alpine cheeses, the salt should certainly be applied only on top, as the large grains can leave divots on the underside of the cheese that can be challenging to wash, and that can play host to cheese mites. Traditional Camembert and Brie, similarly, are only salted on their top sides; then flipped after several hours before being salted on the other side, now on top. Their high moisture makes them challenging to flip before they've received their first salt.

Dry salting can also preserve a cheese's best shape, limiting the barreling of the wheel. At the pH at which cheeses are typically salted, 5.3, their paste remains gel-like and retains the ability to flow. If they're hand-salted then left in the form after salting, the cheeses will keep the shape of the form, creating a firm rind that keeps the cheese in shape. But if the cheeses are taken out of their forms after salting, they will barrel slightly; the same effect will happen with cheeses placed into a brine to salt. To keep cheese with straight sides, desirable with many styles, avoid brining, and leave the cheeses in their forms through salting and draining until they are ready for their affinage. For some cheeses that look their best with a barrel shape—think grana styles and Gouda—it is generally considered best to brine.

The softest and moistest styles of cheese, like lactic and half-lactic styles, are surface-salted, as they are too delicate at their time of salting to be placed into a brine.

Salting all the surfaces of all the cheeses by hand does make more work for a cheesemaker, and the results can be less consistent than brining (described below), but there are none of the many complications associated with maintaining a working brine. Ultimately, though, surface salting and brining both pull out moisture from within a cheese the same way.

Natural salt can be applied by hand to the rind in one application for small cheeses. Larger cheeses need several applications of salt for their proper preservation.

Brining

To brine a cheese, a wheel is placed into a saturated salt brine made of fermented whey and a significant amount of salt, for a certain amount of time, to pull out just the right amount of moisture and apply just the right amount of salt.

Brining is slightly more complex than surface salting in its procedures, but typically more efficient and consistent in its effects. And while it may be a bit more work to maintain a salting brine in good working order, brining results in significant savings over time.

Now, it may seem counterintuitive that a cheese would be placed into a liquid brine to have its moisture removed, but brining works because the brine is *so* salty that the salt in the brine pulls moisture toward it from the wheel, while salt slowly migrates in. Osmosis works just as effectively when a cheese is submerged in a brine as when cheeses are dry-salted. In fact, the application of salt to the surface of a cheese is equivalent to the effect of brining, for the crystals of dry salt on the rind of a cheese slowly transform into puddles of saturated brine as they pull moisture from it.

A salting brine must be overwhelmingly salted—actually beyond a saturated brine—with extra salt to have the right effect. It will be so salty that you cannot salt it to taste—the salt will burn your mouth if you try. Instead salt the brine to sight while stirring, adding so much salt that you cannot dissolve any more, and keep on adding it until there is a reservoir of undissolvable salt below. In other words, the brine should be super-saturated. Water's saturation point with salt is around 27 percent at room temperature (this is about the same with whey), but this level of salt need never be measured, because by adding much more salt to the brine than can be dissolved, the brine is kept at an effective 27 percent consistently. Adding salt to just 27 percent using a salinity meter can lead to an overreliance on measuring and an inaccurate interpretation of the salt level of the brine, for as soon as a cheese is placed into the brine, its salt concentration declines.

Having a simply saturated brine is not enough to maintain a balance of salt; the brine must be super-saturated (with more salt than can be dissolved). To maintain its effectiveness, a working brine needs a

A saturated salt can also be used to give cheeses their salt and pull their moisture out. PHOTO BY CHLOE GIRE

reservoir of salt to keep its levels consistent as it is used. But every time a cheese is placed into the brine, some salt migrates from the brine into the cheese. To keep the brine at its saturated levels there must be excess undissolved salt at the bottom.

So to keep the brine at its fullest salt content, a cheesemaker must only ensure that there is always a layer of excess salt at the bottom of the brine, and add more salt when that layer diminishes. A brine can be used indefinitely so long as it is consistently salted.

In a properly made salting brine, only salt and moisture flow. But if a brine is deficient in calcium or potassium or other minerals, those minerals will also flow from the cheese to the brine, resulting in a loss of flavor and a degradation in cheese texture. To maintain a mineralogical balance between the cheese and the brine, make the salting brine from whey at a pH around the cheese's acidity at salting, which is in better balance with a cheese than water.

Water and cheese should probably never mix. If a cheese is placed in a brine made of water, many of its minerals and its acidity can leach into the water. A water brine must have its acidity levels adjusted, and calcium, magnesium, and other minerals added, in order to maintain a balance with the cheese. But all of these elements—minerality and acidity—are found in a well-prepared whey brine. It is generally understood that a water brine gets better over time, for as its brine slowly pulls whey out of cheeses, it becomes a whey brine. So why not start the brine off right by preparing it with whey?

The whey should be at about the same acidity as the cheese to be salted. Most cheeses that are brined are salted at a pH of 5.3. Cheeses that are salted at a lower pH should probably be surface-salted because of their brittleness. To start a brine, some recommend boiling the whey when it reaches the right acidity (about the same time it takes for a cheese to reach the pH of 5.3) before using it, to remove unwanted fat and protein from the whey in the making of ricotta; this can help, but by no means is it necessary. Some salty cream does rise to the top of an unboiled whey brine when salt is added to it, however, which should be skimmed from the top.

Cheeses should be placed into a brine for a certain amount of time based on their weight and firmness. A softer, higher-moisture cheese is salted faster and needs less time in a brine than a firmer, drier cheese. Uncooked rennet cheeses like Camemberts should spend about 4 hours in the brine per kilogram, while cooked curd cheeses like alpine styles should spend 8 hours in the brine per kg. A 0.25 kg Camembert should therefore spend 1 hour in the brine (being flipped halfway through), while a 3 kg tomme would spend 24 hours in the brine (being flipped after 12 or so hours). Grana cheeses like Parmigiano-Reggiano, weighing in at 50 kg apiece, spend over 20 days in their brines, being flipped daily. Surface salting has the same effect as brining and can be practiced with the same timing. Cheeses can have salt regularly applied to the surfaces until they have had the same amount of time in contact with salt at their surfaces as they would spend in the brine. So long as there is undissolved salt sitting on the surface of a wheel, it affects the wheel the same way it would in a saturated brine.

Cheeses placed into a brine should not be overcrowded, as pressure from surrounding cheeses can distort their shape. As well, when cheeses are first placed into a brine, their top surfaces, floating above the level of liquid, should be generously salted (as in surface salting) to assure an even removal of moisture around the cheese. If the cheeses aren't salted from both sides in the first hours, they can become misshapen (they muffin-top) by an unbalanced moisture migration. Flipping the cheeses at their halfway point in the brine, and at least every day if they're in for a long period, will also help assure an even salting. Any imperfections in the cheese shape resulting from improper brining will be carried through the rest of the cheeses' life.

The liquid level in the brine may rise over time, because every time a cheese is placed inside, the brine pulls out whey. You may have to remove and discard some brine occasionally to keep the liquid level. The brine should never go moldy, or grow films of microorganisms, as the saturation of salt severely restricts bacterial or fungal growth.

The brines themselves should be kept in noncorrodible materials, like ceramic-lined tubs, wooden barrels, or even, though hopefully not, plastic containers. The containers or basins of brine should be covered to keep out unwanted debris, flies, mice, and so on. Brines can be kept at room temperature in the dairy, or in the cooler cave.

Salt can be applied to lactic cheeses after hanging by kneading it into the curd within the cloth. The cheese is then hung again to allow its whey to flow. PHOTO COURTESY OF MAX JONES

The temperature of the brine will, however, affect the rate of salt and moisture movement, as well as impacting the firmness of the cheese and its ability to uptake salt. The process should be slightly slower at colder temperatures due to the slowed movement of salt and water, and the increased firmness of a colder cheese. I prefer to brine my cheeses at room temperature rather than in a cave simply for reasons of attentiveness and presence of mind—if the cheeses remain in sight, they are more likely to be tended to. And cheeses being salted in their brines must not be forgotten!

I believe that many different styles of cheese can all be aged in the same brine; "contamination" doesn't carry forward through a properly prepared brine, or in a well-made cheese. It is typically only after a cheese is brined that its rind ecologies begin to develop and a cheesemaker has complete control over what a cheese becomes regardless of what microbes it may be exposed to in a brine. For example, even if some *Penicillium roqueforti* spores made their way from a blue cheese to the rind of a Camembert through the brine, so long as the Camembert is aged in the proper manner, with a regular rind washing during the first week of aging, the rogue *roqueforti* will never establish itself, and the Camembert will grow a perfect downy coat of *Penicillium candidum*.

Salting in the Curd

Several styles of cheese are salted in the curd—that is, they are salted from within, rather than from the rind. This technique is only practical for a very few select styles, those whose curds are drained then broken up before being put back together again. This is the case with the very softest fresh lactic cheeses like chèvre and cream cheese, which have their salt kneaded into the curd, as well as soft aged lactic cheeses like le sein, which are salted and drained before the cheeses are given their final shape. But this is also the technique used with some of the very hardest milled-curd cheeses like cheddar, which have salt mixed into their milled curds before the curds are pressed into a wheel; as well as some pasta filata cheeses, like caciocavallo, whose fermented curds are milled and salted before they're melted with hot water and stretched; and queso fresco, which is salted in the curd to limit fermentation.

With fresh chèvre, brebis, vache, boufflonne, and cream cheese, the lactic curd should be left to drain as fully as possible in cheesecloth before being salted. Salting at the right moment assures the best preservation of these cheeses, and the slowing of any secondary fermentations that would aid their transformations into moldy or yeasty cheeses. Typically, they are salted within 12 hours of draining, once the curd has stopped dripping whey, but before any signs or smells of *Geotrichum candidum* can be detected on the cheese or on the cheesecloth. If the fungus can be seen or smelled (it smells yeasty like sourdough bread rising, or wine fermenting, and feels slippery on the cheesecloth), the salting will be too late!

At the right point of firmness, the lactic curd should be salted to about 2 percent by weight, which can be determined through weighing, or through tasting—a 2 percent salted curd should taste intensely good. It's easy to judge once you get a feel for it: 3 percent salt will taste too salty, like the sea, and a 1 percent salting will leave a taste that's sadly lacking in salt. I typically add a small amount of salt, then knead it in thoroughly by hand with the help of the cheesecloth. I keep the cheese within the fine cheesecloth and knead with the cloth between my hands and the cheese. I use the ends of the cloth to fold the cheese over, and press with the butts of my palms again and again until the curd is uniform and the salt mixed thoroughly within; I add more if needed to taste, and knead again, repeating the process with ever-smaller quantities of salt until the taste is as it should be. It may seem challenging to add so much salt to your cheese, but it is essential, for a more substantial salting does a better job keeping these cheeses tasting fresh.

After salting lactic cheeses, the curd should be left to drain again, preferably by hanging, until the whey stops dripping. The addition of salt pulls additional moisture out, and the salting can be considered complete once the whey stops dripping, which can take between 1 and 2 hours. You can leave the curd to drain longer, but the longer it's left the more likely it is that yeasts will begin to develop on the curd. So if a fresh lactic cheese is your goal, pack the cheese away (you might knead it again first to mix the curd and make it smoother) and refrigerate it before leaving it to drain too long.

With aged lactic cheeses that are pre-drained before shaping by hand, like le sein, the process is similar, if a bit more relaxed, time-wise. For in the case of these aged lactic cheeses, the growth of *Geotrichum candidum* is to be encouraged! To salt these unique lactic cheeses, the curd should be hung for about 24 hours in cloth at 20°C (68°F) to evolve as much whey as possible before salting, when the growth of *Geo* should be just barely visible on the draining cloth. The cloth bag should then be opened, and 2 percent salt added to taste by kneading the salt into the curd. The curd can then be drained for 12 to 24 hours before it is then given its shape. At this point, the curd is very dense, having had much of its moisture removed; it will have a texture like modeling clay. You can then model it into various shapes—little cows, goats, iconic landmarks, hearts, or other significant symbols—and age them into moldy versions of themselves.

Firmer milled-curd cheeses are salted from within in a similar fashion. This internal salting effectively preserves the acidity and improves the firmness of the cheese, much more than the surface salting or brining would. Salting the milled curd has an immediate effect of halting primary fermentation, while surface salt can take days to penetrate to the core of a cheese, where fermentation can continue onward, especially when a cheese is not cooked and high in moisture (as most milled-curd cheeses are). This milling and salting of the curd from within was historically part of a very effective way of making a cheese as firm as possible, and therefore as long lasting as possible, without cooking it.

Salting is performed once the curd has reached a goal firmness and acidity, preferably both at about the same time, according to the cheese's best make. For cheddar the curd will be very firm, like a raw chicken breast, and with an acidity around pH 5.3 (a very strong stretch when melted) before it is milled then salted. For Stilton the curd will be moister, and with a more advanced acidity of pH 5.1 (a loose stretch when melted). Before salting, the curd must be milled to a small but consistent size to assure effective and consistent salt penetration, as well as a thorough moisture drainage before the curds are pressed together into the cheese.

The milled curd is then salted to 2 percent, much like lactic cheeses, and can be done by weight (apply to 2.5 percent) or by taste. The salt should be liberally mixed into the curd by hand to assure an even coating and a consistent salting. Many cheddar makers come up with a formula for salting based on the milk volume in the vat and its typical solids content. From its expected curd yield, they calculate the weight of salt needed, then measure and apply it. I prefer, however, to look for a salted curd with an assertive saltiness and squeak, and this, I believe, can be just as accurate a measurement of salt.

The salted milled curd should be left for 30 minutes to an hour after salting to assure the best whey drainage before pressing; and the curd is best left on an inclined draining table or vat to facilitate the flow of whey. An occasional mixing of the curds (every 5 minutes) ensures they do not clump together and permits continued drainage. The now firmer curds can be packed into their cheesecloth-lined forms and pressed together to make the final wheel of cheese or left to press by gravity alone, as in the case of Stilton.

Post-Salting

The salt application is typically followed by a period of draining, before the cheeses are ready for their ripening. This second step of the salting process is not to be neglected, for it allows the salt to complete its job of draining the moisture away from the cheese.

Salt applied at the rind slowly pulls moisture out from within a cheese, and continues to do so for days after salting. As the salt migrates toward the center of the cheese, moisture is consequently pulled out to the rind. This movement continues even after the last grains of salt are visible.

To allow for the continued migration of its moisture, the cheeses should be left to drain on a draining table for about 1 to 2 days after salting to help their whey to drain. Cheeses salted in a brine also need to be drained for a day after being removed from the brine before being placed into a cave; otherwise, they will spill salty brine all over their boards.

During this draining period, the cheeses should be flipped once or twice daily to allow moisture to flow evenly from all sides. They should rest atop a surface that permits their best draining. Traditionally, draining tables were often covered with dried straw to encourage

Milled-curd cheeses are salted from within by mixing salt into the milled curds and allowing them to drain before pressing the cheese together. PHOTO BY CHLOE GIRE

drainage and prevent cheeses from sitting in puddles of their salty whey.

Cheeses should always be covered during draining to prevent excess drying (except for a few rare exceptions), and to keep flies at bay. If cheeses are exposed to too much drying air during the draining, too much moisture can be pulled out, and the cheeses can fail to ripen right. If a cheese yellows and shows dry spots on its rind after salting, that's a sign that the cheeses haven't been sufficiently protected.

Even when salted, cheeses remain susceptible to fly strike. Never let your guard down with regard to flies, not even when you're working in a space that is supposedly fly-free. All it takes is one to lay its eggs.

Types of Salt

Many different types of salt can be used for salting cheeses, with many different effects, but one important consideration needs to be taken with all of them: The salt, just as is the case with milk, should be as natural as possible.

Importantly, salt used in cheesemaking should not have iodine, an antimicrobial agent that may affect the microbiological evolution of a cheese and may impact the beneficial species more than the pathogenic ones, which are more opportunistic and thrive in the absence of other microbes. And it should be free of anti-caking agents, like yellow prussiate of soda, that keep the salt free flowing as its minerals slowly absorb moisture back from the air. On the package, the ingredient list should simply read: s-a-l-t, salt!

Salt can come from the earth or it can come from the sea, though salt deposits on earth originally derive from the sea. Salt is mined directly from salt deposits under the earth or at its surface (think blocks of Himalayan salt); or brought up from deep salt caverns by first pumping in water, dissolving the salt from its geological formations, then pumping out the saltwater brine. That brine is then evaporated with the help of boilers to create salt of different grain sizes. Sea salts are generally evaporated with the help of solar evaporation, though many are cooked.

Salt can be refined, meaning that its impurities (anything other than NaCl) are removed by filtering and chemical processes. Unrefined salt can have a tendency to absorb moisture from the air (the magnesium in unrefined salt also attracts water) and become damp. This can change the density and the weight of salt and affect its ability to pull moisture from a cheese. Unrefined salt, when used for surface salting, will also leave stains on the rinds of cheeses (these are just minerals and microbes) as well as deposits of undissolvable mud from the bottoms of saltpools. Neither is a great cause for concern, but rather a sign of quality. I personally find that unrefined salts give cheeses a slightly more interesting flavor due to the bitter tastes of the other minerals.

Salt can be large-grained or fine-grained, and crystalline or not. Salt that is too fine (in other words, powdery) can be difficult to apply to a cheese by hand, whereas salt that is too coarse will not stick to a cheese's surface and will pull out moisture slowly, often leaving lasting divots on the rind of a cheese. Generally, cheesemakers use fine-grained salts when surface salting (and salting in the curd); large-grained salt is only suited for a brine.

Crystalline salt, like kosher salt or flaked sea salt, is evaporated from brine in a way (see below) that causes the salt crystals to grow slowly over time into pyramid-shaped forms. The crystals have a natural tendency to grow this way when undisturbed in open pans because of the molecular idiosyncrasies of the sodium (Na+) and chloride (Cl-) ions.

Such pyramid salt sticks well to flesh and is useful in the process of making meat kosher, wherein salt is used to pull the un-kosher blood from animal carcasses after slaughter. Presumably this would mean that crystal salt will stick well to cheese, though few cheesemakers use it given its greater cost and lower density (much of the volume of kosher salt is air—it can have half the density of fine salt), which makes it harder to accurately measure. I, however, approve of its use for cheesemaking, because it's often an accessible and affordable salt for home cheesemakers to purchase, easy to handle and apply to cheese.

Making Salt

If you are fortunate to live close to the sea (or to salt flats), and you have the desire to make salt from your home region instead of importing it from afar, you can prepare your own salt for your cheesemaking. Salt, however, is among the oldest of the traded commodities, so don't overconcern yourself with the need to make your own. Few, if any, cheesemakers did so historically or do so today. Even if you never will make your own salt, though, it may benefit a cheesemaker to understand its origins.

Seawater should be taken from shores where the water is clean, and lightly filtered through fine cloth to remove debris. Buckets of seawater should be collected—each 20 L bucket will only yield 700 g salt (with 3.5 percent salt), enough to salt just over 20 kg cheese (if applying a total of 3 percent).

It takes much time and energy to evaporate moisture from salt; seawater is only about 3.5 percent salt, so separating it demands removing the remaining 96.5 percent, which is mostly water. This can be done by boiling the seawater in nonreactive pots preferably outside over fire or gas.

Once 90 percent of the water is evaporated and small crystals of salt are beginning to form, the concentrated brine should be placed into shallow pans and left to evaporate more slowly to form its finest crystals. The slower and gentler the handling of the salt during this secondary evaporation, the larger the crystal formation will be (much like handling curds). And if the salt is handled just right, the crystal formation will be perfectly pyramidal. Salt can also be fully evaporated without any cooking in shallow pans or basins in the sun or a solar-heated greenhouse.

The final evaporative process will never fully complete itself, for as the crystals of salt form out of the brine, other minerals from the seawater get left behind in the liquid, and these minerals hold on to moisture. The primary ocean minerals left after salt crystallization are magnesium, potassium, and calcium. The final crystallized salt should be carefully scooped out and left to drain to separate the mineral-rich brine and help it to dry. Oddly, this brine, known as bittern or, in Japanese, *nigari*, can be used to coagulate soy milk, and is still used today in traditional tofu production.

CHAPTER 8

The Cave

The cheese cave is the cool and humid environment in which cheeses age best. More specifically, an aging cheese's preferred ripening conditions are temperatures between 8 and 12°C (46–54°F), with a humidity of 90 percent.

These conditions are ideal for the vast majority of styles for the maintenance of their integral humidity, the best development of their rinds, and their ideal ripening. And these perfect aging circumstances were originally found underground, where the earth maintains coolness and humidity year-round.

Few cheesemakers today take advantage of the natural cooling and humidifying effects of the ground beneath their feet, preferring instead to construct aboveground cheese-aging rooms (often called caves) and installing cooling and humidifying and aerating systems, all environmental controls that aim to mimic original underground caves.

But any cheesemaker wishing to adopt more natural cheesemaking practices should take pause and consider that the standard methods of cooling and humidifying aging rooms do not fall in line with a natural cheesemaking philosophy. And many systems used by cheesemakers today do not create ideal circumstances for aging.

Of course, not every cheesemaker can or should build their cheese caves entirely underground. Still, if they consider their constructed aging environment to be akin to the original underground cheese cave, cheesemakers can gain significant improvements to the energy efficiency, climactic suitability, and general functionality of their aging spaces, all of which help contribute to a more successful affinage.

Considering ripening cheeses' needs is exceedingly important to providing for their best care during affinage. Along with the use of good milk, the addition of the appropriate microbial ecologies, and the care and handling specific to your style of cheese (discussed in chapters 9 and 10), creating the right ripening environment creates the ideal circumstances to support a proper affinage.

This chapter will explore the needs of an aging cheese, and how the environmental conditions affect the developing wheel. It investigates the traditional underground cheese cave and how it provided for those needs, along with the ways a more modern cheese cave can mimic a traditional one. In addition, the chapter addresses hastening spaces for the rind development of lactic cheeses, and how natural materials can aid in the creation of cheese-aging spaces.

Cheese Needs

For a cave to function at its best, it must provide for three basic cheese needs: high moisture, cool temperatures, and the right flow of air. These needs are specifically the needs of the fungi (or bacteria) growing on the rinds of a cheese that are largely responsible for their ripening.

The needs of the fungus on a rind are parallel to the needs of the fungi of the soil. *Geotrichum candidum*, a favorite rind ripener, is named "hairy, white earth fungus" because its original habitat is damp earth where it feeds on and decays organic matter. To re-create the conditions this microorganism needs, we need to look to the soil for inspiration. The diverse microorganisms responsible for decomposition of organic matter need the right moisture, relatively low temperatures, and just the right amount of air to do their work. And the same is true of the many microbes responsible for cheese's decomposition, which originally derive from the earth beneath our feet.

High humidity—around 90 percent relative humidity—helps maintain critical moisture within the cheese that allows for continued microbiological developments and a proper ripening. Cool temperatures, around 8 to 12°C (46–54°F), are ideal for most styles, permitting cheeses to be easily handled and preserved for long periods without the need for constant care; they also encourage the microorganisms that evolve most cheeses' best taste. And airflow allows the growth of those rind microorganisms and permits safe working conditions for cheesemakers.

In addition, a cave should be dark—light will encourage the growth of green algae in the cave and on a cheese—and it should be quiet, for the well-being of the cheesemakers working inside of it.

Cool Temperatures

Low temperatures are necessary to permit the long aging of cheeses; to ease handling of the wheels; to assure appropriate rind microbiologies; and to develop cheeses' best flavors.

Cool weather slows microbiological developments in a cheese, which also slows curd breakdown and results in a less frequent need to turn cheeses and treat their rinds. It also slows down the macrobiological cheese mite activity.

Lower temperatures also tend to slow the development of bacteria, while encouraging more cold-tolerant yeast and fungi to develop on the rind. Low temperatures are especially important for the development of *Penicillium roqueforti* and *P. candidum* rinds (their preferred growing conditions are the cold and humid cave), but also assure that cheeses' evolutions are dominated by rind organisms, and not by interior bacteria that can continue developing excessive acidity within cheeses in warmer weather. Cheeses left to ripen too long in a warm environment can develop lots of lactic acid and consequent chalky textures in the curd, which causes them to ripen differently. Such a scenario can cause rennet cheeses to evolve more like lactic cheeses, breaking down excessively at the rind and developing softer textures in their paste.

Most fermented foods are improved by cooler aging conditions, including beers (specifically lagers) and hams, which in the south of Europe are often ripened in the cooler mountainous regions. Typically, warmer-weather ripening leads to stronger flavors and more acidity in all these foods, and lower temperatures make for more nuanced and subtle tastes. Higher temperatures encourage more bacterial acidification in a cheese; yeast and fungi typically dominate at lower temperatures, and provide more nuanced and flavorful fermentations.

Low temperatures also keep cheeses firm: The deliquescent proteins in cheese tend to be less liquid at lower temperatures, and cheese's fat and fatty acids, like butter, are more solid when cold. Together, the texture of the fat and proteins gives the cheese its particular texture, and both hold the cheese together better at lower temperatures, especially in softer, higher-moisture cheeses. Camemberts, blue cheeses, and lactic cheeses especially all need low temperatures as they age, for these help keep the cheeses firm, and make their regular handling simpler and less urgent. The cooler temperatures of a refrigerator (4°C [39°F]) can be helpful for holding together the ripest, softest cheeses as they ripen. Colder weather also slows down the ripening of cheeses, permitting them to be held longer before they lose their best tastes and textures.

Firmer cheeses like full alpines and cheddars can be ripened at temperatures of up to 20°C (68°F). (In some regions these cheeses were historically aged in aboveground storehouses.) However, higher temperatures result in a faster affinage, more acidification, and the need for more frequent cheese care.

High Humidity

The constant humidity found just a few meters underground helps to create the ideal environment for aging cheeses. This is not coincidental, as the vast majority

Cheeses ripen best in underground caves as the microbiologies that ripen them right, *Geotrichum candidum*, *Penicillium roqueforti*, and *P. candidum*, originate from the soil beneath our feet. PHOTO COURTESY OF QUEIJARIA BELAFAZENDA

Humidity in the cave helps most cheeses ripen right by preventing them from drying and allowing the sought-after ripening species to thrive upon the rind. PHOTO BY VIRGINIE GOSSELIN

of fungal decay in temperate forests and fields occurs below the soil line, where the humid conditions are ideal for the microbiological developments largely responsible for cheese's affinage.

A humidity of about 90 percent is essential for preserving the integral moisture of cheese, and to promote the growth and development of the most sought-after rind ecologies. Ripening cultures need a certain amount of humidity both within the cheese and in the cheese-aging environment to continue working their ripening magic on cheeses. If its cave is too dry, a cheese will dry; and if a cheese dries, its fermentation dies, too.

But too much moisture can also be detrimental to cheese's ripening, because fungal cultures don't function right when they're inundated with water. Cheeses kept in too wet a cave will find their rinds dominated by moisture-loving bacteria and yeasts—washed-rind ecologies that will cause the cheese to develop strong flavors and orange/pink colors. Humidity in a cave nearing 100 percent will result in too much water retained on the rind of cheese, from too little natural diffusion of moisture into the air.

Interestingly, cheeses' ripening fungi, including *Geotrichum candidum* and *Penicillium candidum* and *roqueforti*, are water-repellent and will cause excess water to bead on their surfaces, preventing them from being overwhelmed by moisture. This seems to be a protective mechanism that maintains their best growing conditions. Continued exposure to excess water, however, will cause the fungal ecologies to suffer, eventually becoming inundated with moisture and overwhelmed by the washed-rind ecology.

Cheeses kept in too dry a cave will suffer a different fate: They will slowly dry to their core, halting their affinage almost entirely and resulting in a hard paste that does not develop its sought-out flavor or texture. Ideally, moisture should only leave an aging cheese through microbial respiration, and not through dry air. Only the most massive grana cheeses, clothbound cheddars, and certain lactic cheeses benefit from being aged in a drier cave.

But high humidity in a cave isn't enough; a proper cheese cave needs to have some moisture absorption capacity to keep its humidity constant, as aging cheeses give off considerable moisture as they ripen.

A properly ripened naturally rinded cheese should only lose excess moisture through the respiration of its rind ecology. A Camembert, for example, will come to the perfect ripeness so long as it is kept in an environment that allows excess moisture from continued fermentation to move away from it; any additional moisture loss from dry air will cause the cheese to dry excessively and not develop its best texture or flavor. The perfect Camembert curd breakdown comes about naturally through this metabolism, so long as the moisture levels in the cave are balanced and no excess moisture flows into the cheese or is pulled from it. And the same holds true for just about every style of cheese.

A cheese will release up to a third of its weight as moisture and gases as it ages through respiration! It is essential to have some means of absorbing and or releasing this moisture from the cave environment, to be sure that humidity does not build up within, creating too wet a cave for proper affinage. Natural materials like wood, straw, or hay in the cave provide moisture-absorbing capacity that can help regulate humidity levels over time.

A cheese left sitting against an impermeable surface like plastic, stainless steel, or to a certain extent wood, for too long will inundate itself with this excess moisture, altering its affinage ecologies in the direction of *Brevibacterium linens*. This problem is particularly acute with plastic. If you are using plastic containers to ripen cheeses, they must not sit directly on the plastic, for the cheeses become overly wet at the contact point. Leaving a cheese unflipped on that plastic surface for even 2 days can permanently alter the rind ecology. Commercial cheesemakers therefore use plastic mesh mats that allow air to flow beneath the cheeses, but everywhere the cheeses touch the plastic is negatively affected. Flipping a cheese regularly, and using breathable and absorbent bedding surfaces like wood, hay, straw, or other natural materials, helps alleviate this issue, assuring the best rind development.

Cheeses aging in a vacuum seal, in wax, or under a brine do not need humidity to ripen right; however, these treatments restrict the passage of air to their surfaces, which limits aerobic fermentation, and therefore halts rind development. This, in turn stops the evaporation of moisture from the cheeses as well as the breakdown of their components. In addition, cheeses aged these ways preserve their weight without moisture loss, which, I suppose, is an economic advantage to cheesemakers. Still, in the absence of air and its resultant fermentations, they take significantly longer to develop their flavors, and in the end become more acidic through continued primary fermentations, like a sharp cheddar. This is what's sought after with a cheddar (or a gouda) but not with most other styles.

Air

Naturally made cheeses aging in a humid cave breathe in and out as they age. Cheeses exposed to air and its oxygen (as opposed to those ripened in plastic or wax, or even below a brine) will develop natural rinds that respire on the surface, breathing in air and breathing out carbon dioxide, moisture, and other cheese breakdown by-products. The living rinds metabolize the cheese within, slowly breaking down its proteins, fats, and lactic acid as they develop the cheese's character. As they do, they release moisture and other gases into the air of the cave.

The cheeses within the cave need air for their best rind development and for their ideal evolution. Oxygen is essential to the growth and development of aerobic surface ecologies, but the by-products of that aerobic fermentation, toxic gases like alcohols, ammonia, and carbon dioxide, are released into the environment and must also be replaced with fresh air.

Exposing cheeses fully to air is also necessary if they are to develop their rinds all over, so keeping them on breathable surfaces and flipping them regularly are musts. And though they crave air, cheeses prefer to be in environments with little to no air *movement*, like the still air within a cave, to encourage the best conditions for fungi to develop on their rinds.

Air is fundamental to the growth of all the desirable rind regimes. *Geotrichum candidum*, *Penicillium candidum*, *P. roqueforti*, and *Brevibacterium linens* ecologies all need air to develop on the rinds (or in the veins) of a cheese. These aerobic surface ecologies are dependent on oxygen for them to break down the cheeses they ripen in their most exceptional and delicious ways. *G. candidum* even wrinkles the surface of the soft and acidic cheeses it grows upon to maximize their surface area, giving fungi more room to grow and more air to breathe, enabling the cheese to ripen more quickly.

To get blue veins (or eyes) to develop in the interior of blue cheeses like Stilton or Roquefort, the cheeses must be pierced to enable air to flow into their flesh, allowing the *Penicillium roqueforti* to blossom within. Piercing cheeses allows aerobic ecologies to ripen them from the inside, hastening their affinage.

Limiting air exposure to a cheese—by submerging a cheese under a brine, in oil, or under pomace, or covering it with a cloth binding (I suppose I should also include here waxing and vacuum sealing)—stops the development of those surface ripening microorganisms altogether. Anaerobic fermentations dominate those cheeses' developments, slowing their aging considerably and changing the pathway of decay; curd breakdown is halted entirely because of the lack of air, and many of the cheeses' most desirable flavor evolutions simply do not happen. However, all is not lost: Restricting exposure to air restricts deliquescence, preserving the cheeses longer and easing the workload of maintaining rinds by reducing the need for treatments like flipping.

Not flipping a cheese can limit airflow to it, stopping rind development entirely. A cheese left against a surface (especially an impermeable one like plastic or steel) for a week without flipping will become overwhelmed by its own moisture (from respiration) and have its rind inadvertently dominated by a washed-rind ecology. If that cheese is left against its supporting surface even longer, fungal growth will fuse it with its supporting surface, robbing oxygen from its underside and halting rind development there entirely. Flipping cheeses regularly helps assure that this damaging scenario doesn't arrive.

But it's not just exposure to air that matters; air quality is important to a ripening cheese, too. The composition of the air in a cave is changed by its cheeses' decomposition; the exhalation of surface ecologies can change the ripening environment, and not just by releasing moisture. As cheeses ripen and their curd is broken down, various gases are released alongside the moisture that these rind ripeners give off. Cheeses *and* cheesemakers left in a completely closed ripening environment can become overwhelmed by these often noxious gases.

The proteins in cheese, for example, are broken down with microbiological activity into amino acids; with further proteolysis, amino acids are transformed into ammonia, which escapes from the cheese into the cave air. Similarly, as the cheese's lactic acid is fermented by the microorganisms on the rinds, carbon dioxide (and alcohols and other esters) are released. The buildup of ammonia and carbon dioxide can potentially hinder the development of aerobic rind-ripening microorganisms, or even raise the pH of cheeses and ripen them prematurely, as can the loss of oxygen from the cave air that's required for their metabolism.

Air conditions that are harmful to a cheese's ripening are also harmful to the cheesemakers working in the cave. Some air exchange, therefore, is essential for the best ripening of a cheese, as well as for the health and safety of cheesemakers. True cheese caves naturally provide this, but more modern versions must be designed with this essential exchange in mind.

And though air is necessary for a cheese's proper ripening, too much air movement can have an adverse effect on affinage. A cave-like stillness is an important aspect of the final cheese need that must be provided for; for the delicate fungal filaments that develop on the rinds of cheese need to be free from any disturbances to form. Roots of a tree and their mycorrhizal symbionts only function together if their environment is still as soil (and relatively cool and humid). The same is true of *our* fungal symbionts growing on the rind of our cheeses.

The Original Cheese Cave

Humanity has conserved our food in cool caves and other underground places for a long period of our history, and all naturally preserved foods including wine, cured meats, and fermented vegetables benefit from being aged among one another in these same spaces. The earliest example of cave-aged cheeses may be those of the Middle East and the Balkans aged in the skins of goats, like Divle cheese of Turkey or darfiyeh of Lebanon. Today these are still made with remarkably ancient methods, and aged in natural caves for their best preservation.

Original cheese caves all have one thing in common: All take advantage of the cooling, humidifying, and air-exchanging effects of the earth to provide cheeses with the circumstances they need for a successful affinage. And though not all cheeses were historically aged underground, those aged aboveground were often kept in spaces naturally cooled, humidified, and aerated through special design considerations. A natural aboveground cheese cave I visited in the Netherlands keeps its cool with the help of 60 centimeter thick peat walls and a generous thatched roof. The earth provides all that cheeses truly need to ripen right.

Just 3 or 4 meters underground, the earth cools and humidifies considerably. The soil and stones there are not affected by the changing weather aboveground but are kept at a temperature as consistent as the average annual temperature. Even during dry spells, the deep earth holds on to moisture that flows down from rain. This is not just the case in temperate regions; in tropical or desert climates, too, conditions can be cool and humid enough to age some styles of cheese (alpine, washed-curd, and milled-curd cheeses) just a few meters below the surface. Importantly, the traditional cave, with wooden shelves, doesn't simply regulate coolness and humidity naturally, but also provides another affinage essential: air.

Traditional cheese caves breathe in and out as the cheeses aging within them consume air and then give off moisture and gases. The caves then attract air from the surrounding soil and exchange ripening gases produced by the cheeses. As cheeses age in an oxygen-rich environment, their microbial ecologies continue to degrade them, their components broken down by continued fermentation into various gases. These noxious gases leave the air of a traditional cave through the soil, through cracks in stone, or through the entranceway. The natural surfaces of the cave also permit the oxygen from the soil to refresh the air within.

Traditional caves are often rounded in shape, avoiding flat ceilings, which couldn't easily be built with materials available historically. Arched ceilings ensure that accumulated moisture doesn't fall upon the cheeses aging on their racks, which can result in unwanted microbiological developments, the transfer of dirt or other matter, or increases in cheeses' moisture content (an extra level of shelves can be placed atop the ripening cheeses to shield them if need be). We don't appreciate being rained upon, and neither do our cheeses!

The climate within traditional caves is gentle and unchanging, very consistent from season to season and hour to hour. The natural earthen walls of the cave, combined with the presence of wooden boards, regulate the ripening conditions inside, creating a perfect atmosphere for affinage entirely without intervention.

Design Considerations for More Natural Caves

If you wish to adopt more natural cheese-aging practices, look to the original cheese cave for inspiration. Even if you don't build your cave fully underground, you can still take into consideration the nature of the underground cave to build the best possible aging space. For traditional caves, much like traditional cheesemakers, have much to teach us about the proper ways to ripen cheese.

Building a cheese cave submerged into the ground even slightly takes advantage of the cooling and humidifying effects of the earth. Having the floor of the cave just 1 meter under the level of the ground can result in significant natural cooling. Burying the cave slightly underground can also reduce the costs of heating in the winter.

Constructing the cave on a north side of a slope in the Northern Hemisphere and the south side in the Southern Hemisphere can shield the cave from the sun's warming rays. So can planting nearby deciduous trees, which shed their leaves in winter when the light will be appreciated for its warming effect.

If you are building underground, be sure that the cave is in a geological situation that is humid but not susceptible to surface or groundwater flows. A cave that is inundated with water during a rainfall event is not a design goal for a cheese-aging space. You may have to design drainage into the cave if it doesn't drain naturally.

Having floor, walls, and ceiling made of natural stone, brick, or earth can allow the moisture from below to humidify the cave, and gases to exchange. Applying natural plasters or lime to walls allows moisture and air to flow from the surrounding soil while keeping the cheeses clean from dirt. In North America, modern cheese caves are generally required by regulation to be sealed off from their environs; if they are built underground, they need to be outfitted with impermeable surfaces to prevent the inflow of moisture or air, and their accompanying microorganisms. The impenetrability of modern cheese caves also results in a buildup of gases from the microbiological breakdown of cheeses, particularly ammonia.

Having some air exchange with the earth is important for dissipating built-up gases within the cave. In absence of permission to do so naturally through exposed earth, some sort of air exchange must be conceived to assure that oxygen comes into and ammonia is released from the cave. In a small cave, simply opening up the door (or the lid) on a regular basis (two or three times a week) will suffice to refresh the air inside. But a larger cave must have air exchange designed into its system.

In lieu of cooling with forced air, which dries the aging space, consider a more passive, cave-like cooling system that chills the cave air with water. Pumping cold water through a cave like a cold-water radiator creates a gentler cooling effect that preserves the cave's humidity without creating cold or dry microclimates. A slightly more complex cooling system that chills water, then pumps that cold water through the cave, eliminates the need to humidify the aging space and the need for air movement.

Avoiding forced-air cooling from air conditioners is the most significant consideration in a more natural aging environment. Passive cooling and humidifying technologies, like cooling with water, are better able to provide for cheeses' needs, one of which is to be in an environment that's free from excess air movement.

Cooling a reservoir of water to 8°C (46°F), then pumping that water through an array of pipes in the cave and back to the reservoir, can have the effect of keeping the cave at the desired temperature of 8 to 12°C (46–54°F) while also preserving the necessary humidity. The cost of installation may be higher, but the simplification of the systems will save you energy and money in the long term.

Cooling with water helps preserve the integral humidity of the cave by not stripping moisture away. As air of the cave comes in contact with the cold-water pipes, some condensation occurs, but because the temperature difference isn't great, not much moisture is pulled out. Instead all that moisture remains in the cave, helping to keep it at its ideal humidity.

Standard designs for built cheese caves call for humidification to counteract the drying effects of air-conditioning, both of which cause countless problems for aging cheese. A standard aging room today is outfitted with air conditioners to cool the air and humidifiers to humidify it; but these two systems work against each other, resulting in energy inefficiencies and climate inconsistencies that can wreak havoc with the cheese within. For example, as air conditioners cool the air, they dry it; and as humidifiers humidify it, they warm it. This means cooling and humidifying systems work overtime to counteract each other, resulting in excess energy use. In addition, some zones within the cave will be too cool and dry, while others are too warm and wet, causing variabilities in cheeses aged in different corners. Cheesemakers often install fans within the cave to move the air so that the microclimates created by environmental controls do not persist, but this results in excessive air movement that prevents cheeses from ripening well, and even more cheese drying. The working environment in such spaces can also be challenging because of noise and wind from all the mechanical systems.

Traditional caves have no such needs: Without air-conditioning, cheeses produce all the humidity they need to keep their environment humid through their own respiration. The concern in a natural cave is not adding humidity but removing it—and wooden boards provide for exactly this function! Wooden boards do not simply support cheeses; they support a proper cheese-ripening environment.

Wooden Boards

Wood is good for the ripening of cheese. Wooden shelves perform stacked functions as they support the weight of the cheeses they bear. Wooden ripening boards absorb excess whey and moisture from the cheeses they hold; they slowly release that moisture into the aging environment, preserving its humidity; and they serve as a reservoir of beneficial microorganisms, protecting cheeses from exposure to unwanted bacteria and fungi. Modern materials such as plastic and stainless steel do not offer these advantages. Plastic and stainless steel do not interact with cheeses in their caves, neither absorbing moisture or whey, nor providing shelter for protective microorganisms. They do not help to regulate humidity, as they do not serve as moisture reservoirs. Thus modern cheese caves are especially sensitive to shifts in humidity.

Cheeses are best ripened on long wooden boards, with multiple cheeses on each. Those boards should be supported by a strong frame that holds many levels from nearly the floor to the ceiling. Alternatively, removable cantilevered cheese boards, supported by a special frame mounted upon a wall, can make even more efficient use of a small aging space. Cheese supports should be constructed in a way that permits the easy movement and cleaning of ripening boards. If cheese mites are to be controlled, the cave must be thoroughly cleanable.

Any untreated wood can work. Preferably, though, the seasoned boards should be made of a strong wood from your locale, which can contribute to the terroir of your cheeses. You may choose to avoid overly aromatic woods like cedar, though such a choice can contribute flavorful qualities to a cheese. Boards can be rough-sawn (the uneven surface allows for better aeration of a cheese) or planed, but should be cut thick (with a 2 cm minimum for strength and to avoid warping), and wide (as wide as the cheese to be aged upon them) to support the weight of the cheese. The boards can be used for decades if they're made and maintained right.

Cheeses should be first placed in the cave upon boards that have been fully cleaned and dried. A cheese will spend its entire affinage upon that same wood board, being flipped and tended to regularly, until it's ready. When cheeses are removed from the cave, so too are the boards, which are then scrubbed simply with water—or better yet fermented whey, to improve their microbiology—left to dry, and then returned to duty. When you place the cheese back on a dry board, the wood works as a humidity regulator, absorbing excess moisture from the cheese in its early, wet stages and releasing that moisture into the cave air over time. Unlike modern cheese caves, which need to have humidity added to counter the drying effects of air-conditioning, more traditional caves have their humidity more naturally regulated with the help of wood.

Some styles of cheese, like Camemberts, lactic cheeses, and soft washed-rind cheeses, age better on straw mats placed atop supportive wooden boards. These high-moisture cheeses benefit from having extra aeration to assure good rind development, and the straw mats help keep the wooden boards clean.

Hay and Straw

Hay and straw are two common agricultural products that can be used as a functional bedding material during the affinage of softer cheeses. Like wood, they absorb excess moisture released from cheeses in the early days of affinage, preventing excess humidification; and then they slowly release that moisture over time, helping to keep the humidity levels consistent in a cave. However, hay or straw allow greater aeration of cheese surfaces than wooden boards, by reducing the contact surface. They also result in beautiful surface impressions upon cheeses. These two considerations make hay or straw well suited to ripening high-moisture, surface-ripened cheeses.

Hay, consisting of grasses and flowering plants cut when green and dried in the sun, is primarily used as an animal feed. But good-quality hay can also be used as a bedding for aging cheeses, improving the ripening of Camemberts, crottins, and their close cousins. Though a valuable agricultural product for milk production, hay also provides great benefits to an aging cheese. The aromatics of hay contribute meadowy essences to a cheese. And the pigments (including carotene) from some species of grasses leave behind colorful red/orange/pink imprints on the rinds of cheeses they touch.

Hay or straw can be loosely laid in thick (5 cm) piles atop wooden boards in a cave (or in a plastic home cave; see "Home-Scale Affinage," below), with the young cheeses positioned atop. Loosely packed, thick-laid hay provides extra moisture absorbing capacity. Bits of hay will adhere to a cheese. Some cheesemakers may choose to remove these, but the dried wildflowers and leaves of grass make a beautiful addition to a cheese's rind. If my cheeses are ripening on hay, I often won't flip them, but instead only lift them regularly so that just the undersides touch the dried grass.

Straw is the dried stalks and leaves of grain crops like wheat, barley, or rye that remain after the seedheads are harvested. Considered today a worthless by-product of grain growing, straw was once a valuable cultural and agricultural resource, used for animal bedding, mattresses, decorations, natural building material (straw bale construction, of course, as well as cob), thatching, and baskets. Historically (and still to this day in France) it was used as an absorbent bedding material for cheesemaking.

Straw, more utilitarian in its relations with cheese than hay, can be woven, but only if it's been cut and harvested by hand into supportive mats that you can place under a ripening (or draining) cheese, allowing a simplified handling and giving a cleaner rind than loosely laid straw or hay. Hand-cutting and drying these materials makes them easier to work with, by keeping the stalks long and in line. This extra work helps keep cheese caves neat, and cheeses even neater. Straw mats can be placed atop wooden boards to provide full support and moisture absorption for ripening cheeses. The open structure of the mats assures the movement of air to the surface of the cheeses while absorbing some moisture, improving rind development but also making the cheeses much more beautiful with impressions from the straw. The mats should lie flat to assure a well-supported, nondeformed cheese. They can be washed with water or whey between batches then hung to dry, or they can be sacrificial—just compost them after every batch of cheese to save cleanup.

Hay and straw will go moldy as they decay with the cheeses in their caves, but this should not be a concern. The absorption of moisture from the cheeses, and with the material's own community of beneficial microorganisms, keeps unwanted microbes in check by a fundamental forage fermentation. Forage itself is the origin of the beneficial microbial species that cause forage to ferment in the first stomachs of ruminants. The microbial species in mammals' milk and cheese may well derive from this same source. Furthermore, the decomposition of hay or straw does not affect the microbial ecology of cheeses ripened on them. The cheeses' own strong microbial ecosystem/immune system keeps the microorganisms from the straw or hay from contaminating them (they may even be the same microorganisms), and allows them to develop perfect rinds even when in contact with different-colored fungi on decaying grasses.

Unlike wood, these natural materials that can aid the ripening and development of soft cheeses are unfortunately restricted in commercial cheese production because of contamination concerns. But the same circumstances that make wood safe to use for cheesemaking also make hay or straw safe for affinage.

Hay or straw can, however, contain foreign matter (soil, snakes, mice, and so on), especially when bound together in a baler. Beware of hay and straw sprayed or otherwise contaminated with herbicides and pesticides. Certified organic straw or hay should certainly be sought out for this purpose. Fermented hay will not work, as the material does not absorb moisture. Hand-harvesting these materials makes for a cleaner and more biosecure cheesemaking material.

Hay or straw can also be used as a packing material, to prevent shifting of cheeses in a shipment, as well as cushioning their impacts against one another. And the hay or straw, especially when woven, also makes for an impactful visual display at a cheese shop! Wool is suitable for these purposes as well, and has a history of being used to cushion and protect soft sheep's milk cheeses while also contributing delicate lanolin flavors.

Home-Scale Affinage

A small-scale or home-scale cheesemaker can have a successful natural affinage by taking the usual cave considerations to heart, using materials and equipment you already have at home or on the farm. A refrigerator or a cellar can provide space for cheeses to age. However, on their own, these common household areas typically do not provide all that cheeses need. To have success aging cheeses in the cellar or refrigerator, a home cheesemaker must consider the basic purposes of the cheese cave: to keep cheeses cool; to regulate humidity; and to have air to breathe.

The first and most important home cave consideration is keeping cheeses cool. The ideal temperature for aging cheeses is again 8 to 12°C (46–54°F), but fortunately they can also age at lower temperatures, only more slowly. For the home cheesemaker the easiest and most accessible cool space in which to age cheeses is the household refrigerator, and these typically operate at 4°C (39°F). You can also adjust their climates to operate at a slightly warmer 8°C (46°F) using thermostat override kits available from home-brewer websites. Wine fridges with adjustable temperatures can also be used, as can a cool cellar, but all these refrigerated spaces tend to be on the overly dry side (aside from older refrigerators with an icebox, which tend to run on the humid side). Cheeses left within them exposed to their air will not ripen well.

To improve the humidity of the home ripening space, a start-up cheesemaker might consider ripening cheeses inside wooden boxes or plastic or glass containers inside a refrigerator. Refrigerators operate on the dry side, so keeping cheeses in containers preserves their humidity without exposing them to the drying air. Glass or plastic containers lined with a natural material can serve as miniature cheese caves that create the perfect affinage space within a cool but dry environment. Using a plug-in humidifier or leaving out pans of water to passively humidify a refrigerator generally is not enough to counter its dryness.

Putting a natural material inside plastic or glass containers can absorb excess humidity from cheeses and then release it into the environment over the course of ripening. Cheeses produce enough humidity during their affinage that you do not need to add any moisture to the aging space. The natural material also keeps the cheeses from being trapped against plastic, and from sitting in puddles. Hay, straw, dairy-dedicated cloths, bamboo mats, or wooden boards help absorb moisture from the cheeses while elevating them above possible puddles on the bottom of the plastic container.

Cheeses will continue giving off moisture through their long affinage. In an impermeable plastic container, you must remove excess moisture regularly by wiping the sides with a clean, dry cloth (preferably every time you tend to the cheese, every other day or so); placing a cloth under the lid can also help pull excess moisture out of the cave without the need for a regular wiping. If the cheeses are in too humid an environment, excess moisture from their ripening will not evaporate from the rinds and the surfaces will get too wet, resulting in unwanted washed-rind ecologies or excessive fermentation. Aging cheeses in a wooden box will naturally regulate their moisture without the need for wiping extra moisture away.

As for the final affinage need, air: In a small aging space, simply opening the door or lid to the aging environment on a regular basis exchanges air and built-up gases with no added effort or adaptations. Your regular schedule of handling and flipping the cheeses introduces enough air for their continued affinage. No extra considerations need be taken.

Cellars beneath a home can also work well as a cave. Cellars, however, are not generally deep enough in the earth, and even when exposed to soil, they do not provide enough humidity to age cheeses well. I usually recommend that cheesemakers keep their cheeses in cellars inside wooden boxes or plastic containers to preserve the humidity in their immediate environs; or to artificially raise the humidity in the cellar itself with a humidifier, or with the regular pouring of water on the cellar floor.

Also, because they are not typically deep enough in the earth, cellars' climates can change seasonally: They generally are slightly cooler and more humid in the wintertime, and warmer and drier in the summer, so cheeses may ripen differently within from season to season. In the summer, soft, high-moisture cheeses like Camembert may dry or conversely overliquify in a humid summer cellar; the winter climate may be more agreeable to their aging. Furnaces or freezers can also raise the temperature and lower the humidity in a cellar, making it unsuitable for aging cheeses.

And just as home-scale cheesemakers can easily create a cool and humid aging space, you can also easily create a warm and humid hastening space (more on this below) for *Geotrichum*-ripened lactic cheeses. The same wooden boxes or plastic containers lined with a natural material that work as an aging space when placed in the refrigerator or cellar will function as a hastening space to grow *Geo* when left at room temperature (20°C [68°F]). After the 2- to 3-day hastening period, and with their white rinds established, the cheeses can be left in the container and the container placed in a refrigerator. Now the hastening space becomes the aging space!

Hastening Spaces

Many cheeses benefit from being kept in a warm and humid climate in their early days of affinage. Those that host a *Geotrichum candidum* rind are best kept at temperatures of 20°C (68°F) and a humidity of 90 percent for 3 days after salting and draining. These warm and humid conditions are essential for the ecological developments of some of milk's most appropriate fungal microorganisms, those, like *Geotrichum*, that develop on the warm, humid animal.

To this end, wrinkly-rinded lactic cheeses, Stiltons, and other blue cheeses are often kept in what are known as hastening rooms in the U.K., *hastening* perhaps referencing the quick rind developments that result. These may be dedicated spaces within the already warm and humid dairy, or humidity- and temperature-controlled rooms that preserve the perfect environment for *Geo* to thrive. Tended to (flipped, protected from flies and drying air) in these right conditions for just 3 days, the cheeses are quickly covered by *Geotrichum candidum*, so long as they're made and salted right!

At home, you can easily create a hastening box from a wooden box or even a plastic container lined with wood, straw, or dry cloths and left at an ambient temperature of around 20°C (68°F). A wooden box lined with straw at room temperature will maintain the integral humidity of the cheese, and create an environment ideally suited to the development of *Geotrichum* on the rinds; the wood will breathe out excess moisture, helping to preserve the ideal 90 percent humidity. In a plastic hastener, natural materials like wood, straw, or cloth will hold excess humidity from the cheese and slowly release it into the cheese's environs, preventing the cheese from being overcome by humidity. Using a cloth to wipe out excess moisture from the sides and top of the plastic box every day helps keep humidity levels constant, as does placing a cloth under the lid.

Of course, these spaces must be resolutely fly-free—if even one fruit fly finds its way in, there's a risk of finding their larvae in the cheese! Known as vinegar flies in the U.K., *Drosophila* species should be called fermentation flies, as they are exceedingly attracted to the carbon dioxide and other gases given off by many forms of food fermentation: to bruised fruits, to the gases coming off of a fermenting wine vat, and, perhaps most significantly, to the *Geotrichum* rinds developing on young cheeses in a hastening room. Fruit flies love young lactic cheeses and Camemberts, as do their offspring; both

Some cheeses, like traditional cheddars, prefer to age in drier aboveground caves that limit the growth and development of ripening fungi.
PHOTO COURTESY OF QUEIJARIA BELAFAZENDA

eat the microbes growing on the cheeses, as well as the cheeses themselves, and can be especially hard to see on the white cheese rinds. Placing fine cheesecloth underneath the lids of cave containers and hastening spaces makes a tighter fit, which keeps out even the smallest flies and also can pull out excess moisture.

The preferred temperature for hastening is about 20°C (68°F). At higher temperatures, a successful hastening is still possible, but the cheeses must be flipped twice daily, as the fungus grows faster, and the appearances and flavors of the cheeses may not be as pleasing. If you're hastening in warmer weather, the whole process must also be hastened, preferably leaving the cheeses there for only 1 or 2 days until the *Geo* grows; leaving the cheeses longer will result in unwanted early curd breakdown and possibly a slipped rind. Conversely, in a colder climate, *Geotrichum* may not grow quickly enough; it is best to keep the temperatures as close to 20°C as you can to prevent cold-loving microbes like *Penicillium roqueforti* or *P. candidum* from establishing themselves on the rind first. Once the *Geo* covers the cheeses, they can be moved into the cave and will be protected by the established fungus against unwanted microorganisms.

Dry Caves

Many cheesemakers use dry aging rooms to age or finish certain styles of cheese. Kept at humidities around 70 percent, these spaces slowly pull moisture out from cheeses, and are only used in the hardest and the softest of cheese styles for a specific effect. Only a few cheeses benefit from being aged in such a dry environment, because humidity is essential to proper rind development, and proper rind development is key to the proper evolution of most styles. Drier caves have their origins in aboveground storage spaces like barns or granaries where environmental conditions were drier and warmer than those typically found in underground aging spaces.

Hard, Mediterranean-style grana cheeses and caciocavallos are often aged in drier conditions (this is a climactic consequence of being made in the Mediterranean) that restricts rind development. This dry aging results in cheeses that are fungus-free and consequently do not undergo a secondary, aerobic fermentation. In essence, these cheeses continue acidifying within as their rinds slowly dry, giving them a crumblier, drier texture. Grana cheeses, like Parmigiano-Reggiano, pecorino, and many Spanish goat and sheep cheeses, benefit from a drier aging that restricts de-acidification and curd breakdown. But to most benefit from this effect, the cheeses must be made as massive as possible, to restrict drying to the outermost layer. Too small a grana aged in such a dry environment will dry out to its interior and lose its best qualities. Size matters in cheese, especially when making cheeses like Parmigiano-Reggiano!

Clothbound cheddars can also benefit from a drier aging. The cloth binding helps to limit moisture loss and can restrict fungal development that can alter the flavor of the cheese near the rind. But again, like grana, they should be made massive, over 20 kg a truckle, to limit the drying effects to the rinds.

Some other styles of cheese are dried slightly after aging, to help firm their flesh, strengthen their rinds, and improve their transportability and their preservability. Soft cheeses that readily ooze, such as aged lactic cheeses, are often dried to prevent excess curd breakdown, keeping the cheeses solid but leaving a sort of dried broken-down layer beneath the rind. This drying has a flavor- and texture-intensifying effect on lactic cheeses and simplifies their sale and distribution. Lactic cheeses require much care and cooling to transport. A crottin that is aged in a drier environment will maintain its firmness and can be aged for a considerably longer time than the usual 1-month limit on aging lactic cheeses.

CHAPTER 9

Affinage Treatments

There are over a dozen different traditional ways that cheesemakers handle and treat their cheeses as they age in their caves. These various rind treatments encourage different developments on the cheeses and are the primary ways to encourage the cheeses' ripening or affinage and allow for the greatest expression of diversity within the world of cheese. In no particular order these include: flipping, rind washing, brushing, patting down, cloth binding, brine aging, oil rubbing, oil ripening, hastening, rubbing up, skewering, charcoaling, leaf wrapping, aging in pomace, and waxing.

Though some of these methods are applied for purely aesthetic reasons (like charcoaling), most are performed for the purposes of preservation, of slowing cheeses' decay or keeping unwanted ecologies from developing. The steps involved in such preservations have often evolved to become the identity of a cheese—for a well-preserved cheese is often an object of beauty, showing off its milk's most magnificent character when aged in that particular way.

Some of these methods, like aging in pomace, are most appropriate for hard cheeses; others, like oil ripening, are better for soft ones; still others, like rubbing up, are specific to certain cheeses. It's best to know, before you apply a treatment, what's best for the particular style of cheese you're making.

Some cheese types need a remarkable amount of care and handling over their long ripening period, and cheesemakers should be prepared to face the work required to ripen their cheeses after they put them in their cave. It should be understood that cheeses are like livestock, and need as much care and attention as animals when they're aged. Farmer-cheesemakers are first to neglect their aging cheeses if they only have enough time in the day for caring for their animals and transforming their milk into cheese. But regular care and handling of their cheeses in their caves helps assure that all the work of weeks or months past is able to develop its greatest worth.

Flipping

Flipping is a regular chore that cheesemakers of nearly all styles must do to assure their cheeses evolve in their most expressive way. Flipping assures a cheese's best ripening by encouraging airflow to the rind, regulating humidity, controlling cheese mites (discussed in chapter 10); keeping a cheese's shape, and limiting the tearing of rinds. Cheeses kept under brine or pomace, however, and those aged in oil do not need flipping. Plastic-bound cheeses and waxed cheeses can also be left unflipped, this being one of the time- and labor-saving advantages to these modern industrial rind treatments.

Flipping assures good airflow to both sides of a cheese, helping to create even ripening conditions all around. Naturally rinded cheeses need air on their rinds to develop their best fungal or bacterial ecologies. Regular flipping helps these microbial dynasties get their best footing all around the cheese.

Regularly flipping cheeses assures that their rinds don't adhere to their ripening boards or mats. Leaving a

Rind washing is a commonly practiced treatment that helps to control unwanted mites, flies, and fungi and encourages a washed-rind ecology to develop.

cheese unflipped impedes rind development altogether on the underside and causes the cheese to fuse with its supporting surface. Flipping a cheese whose rind has stuck to its board can cause the rind to rip. A soft, high-moisture, oozy cheese that isn't flipped regularly (every 2 days is standard care) could find its contents flowing out into the cave.

Flipping also assures that one side of the cheese doesn't get too wet. Leaving a cheese too long can give it a soggy bottom and can drive the ecology of that side of the cheese in the direction of a washed rind. A cheese that's only lifted regularly and not flipped could find differences in the microbiological regimes on its bottom and top. This happens intentionally with several styles that cannot be flipped, including the pyramid-shaped Valençay (the cheese must still be lifted and placed back in its place to keep its fungal rind from fusing with its boards).

Flipping can also keep cheese mites under control. Cheese mites take comfort and shelter on the underside of a cheese ripening on a board and feast most fiendishly on its dark side. Flipping the cheeses exposes mites, keeping them under control, and can help alert you to their presence so you can take effective action as needed . . . though to be quite honest, once the effects of mites are visible on a cheese, the little critters are too entrenched to stop.

Typically, cheeses are flipped every 2 or 3 days, with softer cheeses being flipped more often, and harder cheeses needing less flipping. Higher-moisture cheeses typically ferment faster and develop their rinds faster than harder, drier cheeses, and so need extra care and handling for their best ripening. But too much handling of soft cheeses can also interfere with rind development. For example, overhandling soft lactic cheeses can leave them susceptible to breaking apart and can impede rind development where they're touched. Warmer temperatures will necessitate more frequent flipping (during hastening of lactic cheeses, for example at 20°C [68°F], the cheeses must be flipped every day), while colder temperatures (like those of a refrigerator, around 4°C [39°F]) will slow the frequency of the flip.

Cheeses should be flipped regularly throughout their entire ripening. Consider flipping to be a regular check-in with your cheeses, to determine if they need care or attention, like the fungal control that comes from washing, or the mite control that comes from brushing.

Of course, the brushing or washing can be melded with the flipping. It's a task but also a meditation for the cheesemaker, handling and treating all the cheeses in their cave, all the good food they've put up to preserve over the past months. Ultimately, flipping regularly helps you to gauge the readiness of your cheeses and decide when it's their time to eat or sell.

There are some rare cheeses that are kept on their sides rather than their tops and bottoms. Roquefort, for example, is left on its side to assure the best passage of air and therefore improve the blueing through the piercings on the top and bottom of the wheel. And though the cheese isn't flipped, it is rolled roughly a quarter turn regularly to ensure that it keeps its round shape.

Saint-Marcellin, ripened in a pot, is a rare example of a cheese that isn't flipped or even lifted. As the cheese is left in the pot to ripen for almost 2 months, it becomes one with the pot: Its bottom rind essentially disappears within the cheese, and the cheese fills the pot as it ripens and liquifies. Any attempt to flip the Saint-Marcellin at this point would be disastrous, but that's beside the point, as the cheese is meant to ripen into a perfect puddle without any care or handling. Don't try this, though, with any other style, unless you want to scrape the cheese off the boards when it's ready.

Rind Washing

Rind washing is the forceful smearing of the exterior of a cheese, using a brine-soaked cloth or brush. Regular rind washing helps to prevent the growth of fungi and the proliferation of mites on the rinds of the cheese. It also establishes a yeasty and bacterially dominated washed-rind ecology that turns cheeses remarkable shades of orange and pink.

The cheeses should be washed with a light amount of force to reduce the rind to a creamy mess, often referred to as smearing; simply wetting the cheese is not sufficient for setting back unwanted fungal growth. Cotton cloths or soft-bristled horsehair brushes are best for this.

Regular washing beginning the day after salting is essential for encouraging the appropriate ecology; many washed-rind cheeses are monastic in origin, and cheeses to be brined are long-fermented stirred-curd cheeses, and lactic cheeses corresponding to the two brined cheese makes in this book—feta and sirene (Bulgarian feta). A lactic feta results in a cheese with a perfectly textured creaminess, whereas the stirred-curd feta is more crumbly.

But whether they're crumbly or creamy, brined cheeses should be allowed to ferment more fully. The acidity develops the cheeses' fullest flavor, gives them the most sought-after texture, and assures a better preservation. I recommend a full primary fermentation of the stirred-curd fetas (to below 5.0—the curd will disintegrate in the teabag test) before they are salted. Lactic fetas naturally achieve this acidity level.

The whey brine itself should also be fully fermented before it is prepared, to assure the best preservation of the cheese. Fermented whey is more acidic, which restricts microbiological development within the brined cheese, controls for the development of good microbes, and slows the aging of the cheese.

Cheeses are best kept in their brines at cool temperatures, like those of a standard cheese cave or refrigerator. Colder temperatures slow the aging, restrict the growth of surface fungi, and keep off flavors and gas from developing. They should also be kept in closed or covered containers, which limit evaporation of the brine and flavor exchange (as well as foreign objects) from the environment. Wooden barrels, the traditional vessel of choice for keeping brined cheeses, contribute their own flavor profile from their tannins and micro-aeration, much like wine or spirits.

Salt concentrations of the prepared brine are best between 5 and 7 percent by weight; with the added cheese salted to 2 percent, the salt content of the brine will fall when the cheese is added. A saltier brine will preserve a cheese longer, but the cheese will have to be eaten in smaller quantities or added to dishes in part for its saltiness (but that's what we do with feta anyway); or it can be left in water for some time before eating to pull out some of the salt.

Aging in a brine will change a cheese over time, but the changes will be slow, due to the limited air exposure and high salt content. Flavor complexity accrues over time as the cheese ages due to continued microbiological activity and enzyme action, but its crumbly or creamy texture and distinct acidity should be well preserved beneath the brine.

Depending on the salt concentration of the brine, and the temperature at which it is kept, fungus may grow upon the surface of the brine, typically a thin *Geotrichum*-y growth of wrinkly white fungus. It is best to leave the fungus undisturbed and not remove it—after all, *Geotrichum* belongs to a family of beneficial milk-fermenting fungi. And so long as the cheeses are submerged under the brine with the help of weights or stones, they will not be affected by any aerobic microbiological developments at the brine line.

Dry Brushing

Dry brushing as a rind treatment is meant to keep fungi from growing too thick, and to help limit the spread and damage of cheese mites. It is a technique best applied to the rinds of very firm alpine cheeses, to limit fungal growth after the first period of rind washing and restrict the establishment of mites. Essentially, brushing keeps hard cheeses clean.

Dry brushing should be performed with a large, natural-bristled, soft-textured brush. Horsehair works great for this purpose, as it is strong enough for repeated use, but soft enough not to cause damage to the cheese. You can also brush with a dry cloth.

Brushing is generally performed after the first period of an alpine cheese's ripening, when it is normally washed with a light brine to establish an orange/pink *Brevibacterium linens* rind. Beginning the brushing too soon (cheesemakers often wait a year or two on their Comtés) can restrict the growth and development of surface flora on a cheese and limit the flavor development and paste breakdown that gives cheeses their desired qualities. Brushing should then be continually practiced through the remaining life span of an alpine cheese, often over a year, to keep subsequent mold and mite developments in check. Brush the cheeses over the floor of the cave to limit the spread of mites and mess to other cheeses in their vicinity.

I generally don't recommend brushing the rinds of natural-rind cheeses like tommes. Brushing sets back the microbiological developments that lead to the ripening of these fine cheeses. Better to leave the fungal rinds of such semi-firm alpine cheeses unchecked, perhaps only patted down.

Patting Down

Unlike dry brushing, which restricts the growth and development of surface fungi, patting down a cheese enables it to grow its most beautiful fungal coat. This is often an inadvertent intervention that comes about from regularly handling and flipping naturally rinded wheels.

Patting down the cheese is a treatment that's used in the making of tomme de Savoie, a cheese that is encouraged to grow a thick rind of *Penicillium roqueforti*. Patting presses the fluffy mycelial growth of the fungus into the rind, resulting in a thick, waxy, and often colorful coat that envelops the cheese. The treatment is also applied to Camembert, Brie, and tomme de montagne, which grow thick white coats of *Penicillium candidum*. Patting down can also be done with naturally rinded cheddars, developing them into a style known in France as cantal. As you might notice by the examples given, this is a rind treatment most common in France, a nation filled with cheesemakers and cheese eaters who appreciate the improvement in flavor that comes from letting fungus grow—especially given that almost no extra effort is needed to achieve it.

As the name implies, the cheese is fully patted down all over. A good patting compresses the fungus growing on a cheese without rubbing it off, allowing the development of a thick rind. Regularly handling tommes by flipping them several times a week achieves this naturally.

This is a rind treatment most suited to semi-firm alpine cheeses that aren't aged too long. Allowing full fungal growth may improve the flavor character and texture of an aged tomme, but the treatment does not have the same beneficial effect on a harder grana style, which is best aged with a dry brushing, or on a full-alpine cheese, whose rinds are best washed with a light brine to encourage *Brevibacterium linens* and fend off fungi and mites. The lack of rind treatment can make these style susceptible to mites, as the long-aged cheeses can be slowly consumed by them over time.

The thick rind may be the most flavorful part of patted-down cheeses. Though many avoid this crust (*croute* is the French term for these thick rinds) for fear of fungus, the rinds present unique flavors and toothsome, crystalline textures. To appreciate these effects, eat a slice of a naturally rinded tomme slowly from the center to the rind, savoring the differences.

Oil or Fat Rubbing

Regularly rubbing cheese rinds with oil is a method used in the Mediterranean to restrict the development of fungi and other rind-ripening regimes on hard alpine-style cheeses aged in drier caves, to keep the rind of a cheese "clean."

This process is somewhat limiting, due to the instability of plant-based oils over time. Rancidity problems can arise in a cheese, especially if the oil used is not fresh. Unsaturated fatty acids from plant sources are more prone to destructive oxidations that can cause unwanted flavors in cheese (animal fats are saturated fats, which are less prone to these same effects). Some plant oils, however, are better for this purpose (and others) than others: Olive oil is monounsaturated and therefore less stable than animal fats, whereas sunflower oil, canola (rapeseed) oil, and other vegetable oils like soybean oil are polyunsaturated, and therefore significantly less stable and are extra prone to rancidification. Rancid oils are not just a threat to the quality of cheese; they are also a threat to our health. Animal fats, especially those properly preserved through cheesemaking or fermentation, do not develop rancidity, and should be considered more healthful. Olive oil is also a good choice for oil rubbing a cheese, but even better would be to apply ghee.

Ghee, whose fat is in its most stable form, are very resistant to oxidation and rancidity. As well, the delicious flavors of cultured butter or ghee (which can be recovered from the hard cheeses' whey cream) meld very well with an aged cheese, and only improve over time, unlike olive oil.

To have success with oil rubbing, the oil or fat should be applied to a cheese in several successions. Firmer ghee can be applied more thickly. The ghee can be applied in two or three applications over the first week of a cheese's aging. With oil, which does not form a thick layer on the rind of a cheese, the treatment must be repeated often and over the full course of a cheese's aging (though the frequency can be reduced over time) in order to limit the continued risk of fungal development on a rind.

But why use valuable ghee to rub the rind of a cheese when less valuable animal fats can also be used to rub the rind or make a cloth binding? Ghee-rubbed cheese has an advantage over a clothbound cheddar in that the cloth binding and its lard are ultimately discarded, whereas a ghee-rubbed cheese will have an entirely edible and delicious rind.

Oil Brining

Rather than rubbing cheeses in oil, small, soft cheeses can be preserved by submerging them in oil, much as feta is preserved in its whey brine. This technique works best with cheeses that are made more acidic, which controls the development of unwanted, toxicity-producing microbes like *Clostridium botulinum* in the anaerobic environment.

There are many traditions out of the Middle East and Mediterranean for aging cheeses in oil. Yogurt cheese or quark, salted, drained, and dredged in herbs, is preserved in oil to make shankleesh. And lactic cheeses can be aged under oil to make a sort of feta.

Like oil rubbing, this method is best performed with olive oil, which is less stable than animal fats, but not as unstable as seed oils. Even with olive oil, however, the cheeses ripened within might be subject to the oil's increasing rancidity over time. To restrict this effect, it is best to keep the cheese in oil in a cool environment to reduce oxidation, or to limit the preservation of cheese in oil to only a few months or a year. Too cold an environment or too long an aging can also cause olive oil to solidify and crystallize, which can make extracting the cheese challenging. Leaving the cheese at room temperature for several hours will cause the crystallized oil to melt.

Hastening

Hastening is a process of leaving cheeses in a warm and humid space for several days after salting to encourage certain microbiological developments, most notably the wrinkly white rinds of *Geotrichum candidum* and the gas bubbles of propionibacteria.

Though really just an environmental control, keeping a cheese warm and humid in its early stages of development to encourage the growth of milk's native *Geotrichum candidum* fungus can be considered a cheese treatment. This technique of developing the mammalian microbiology of milk by keeping the cheeses relatively warm and humid is commonly used in the making of wrinkly-rinded lactic cheeses, most notably Valençay, crottin, and Saint-Marcellin, as well as tomme Vaudoise and Stilton.

To encourage *Geotrichum candidum*'s best growth, cheeses should be kept at a temperature of around 20°C (68°F) and a humidity of about 90 percent with little air movement for 2 to 4 days post-salting. After salting then draining for 24 hours, the cheeses should be placed into a warm and humid hastening environment with the capacity to absorb excess moisture, as described in detail in chapter 8.

During this hastening period, you must handle the cheeses daily to ensure that the fungal development doesn't overwhelm them. Flipping infrequently can cause the cheeses to develop their rinds into their supporting surfaces, causing them to tear when eventually flipped. If the temperature is much warmer than 20°C (68°F), the cheeses should be handled twice daily.

Once the *Geotrichum candidum* visibly covers the cheese's rind and begins to initiate its wrinkling process (that's the *Geotrichum* running out of room to grow, modifying the soft rind of the cheese to create more surface area), the cheese's hastening can be halted. The cheese can now be placed into a cheese cave to age. The established *Geotrichum* rind will keep other unwanted fungi from establishing themselves on the cheese and continue the slower process of ripening it in the cave.

To encourage the development of propionibacteria and their eyes in alpine cheeses like Emmentaler, large, lightly salted wheels of alpine cheeses should be hastened in a warm and humid space, washing their rinds and flipping the cheeses daily, for about 1 week. The cheeses should swell visibly from the production of gas during this time.

In the making of Stilton, hastening the rind development by keeping the cheese warm and humid aids with a unique rind treatment known as rubbing up.

Geotrichum candidum is developed on the rind of lactic cheeses through keeping them warm and humid for several days after salting.

Rubbing Up

This technique is uniquely applied in the making of Stilton, a milled-curd cheese with blue veins. It involves encouraging the natural paste breakdown of the cheese due to yeasts early in the cheese's make to rub up the rind and seal it shut, almost as if you're icing a cake with the cake itself. This rubbing up restricts blueing to the interior of the cheeses, and only when you pierce them later in the affinage.

Stilton is given a more thorough fermentation than most milled-curd cheeses, partly to make it more buttery in texture and more complex in flavor (as is the case with most blue cheeses), but also to aid with this process of rubbing up. The half-lactic curd (the pH at salting for Stilton curd is around 5.1, when it melts loosely in hot water) becomes more significantly affected by a yeast-driven (*Geotrichum*-dominated) breakdown at its surface than cheeses salted at a higher pH, but not as much as a lactic cheese salted at 4.5.

To aid this process, a young Stilton is left in its tall cylindrical form to lightly press together and drain for 3 or 4 days, being turned in the form daily. When made with a diverse and fungus-rich mesophilic starter, kept in a warm and humid hastening environment, and made and treated right, the rind begins to soften after this time as a result of the yeast development. Once it's creamy and greasy to the touch, rub the rind of the cheese all around with a butter knife. The pressure from the knife turns the cheese to a creamy consistency and allows you to effectively seal shut the open crevasse-filled rind of the young Stilton. What once looked like a loosely adhered multitudinous collection of curds now appears monolithic, at least from the exterior. The interior, however, remains an open network of caves, sealed, for the moment, from outside air.

In preventing the passage of air to the cheese's interior, rubbing up permits the milled Stilton to age monolithically and therefore anaerobically without blueing for the first weeks of its affinage. Only after piercing does air finally penetrate to the open curd structure below the rubbed-up rind, and only then do the cheese's blue veins begin to develop. Were the cheese not rubbed up, air would naturally penetrate through to its interior without piercing, and the cheese would blue sooner, restricting its flavor potential.

Stilton's unique appearance illustrates the effect of rubbing up very clearly. Though it is a cheese made with an unpressed milled curd, the blue veining only shows on the interior of the cheese, not toward the exterior, because the rubbing up eliminates the exterior cracks between the milled curd in the cheese. The rind of the cheese is often wave-like in texture and creamy white due to the growth of *Geotrichum* on the rind. Some orange *Brevibacterium linens* also often grows on the rind, especially toward the middle of the cheese, as the Stilton often becomes wet and inadvertently becomes a washed-rind cheese in its midsection due to the extended period of time it spends wet in its form.

A milled-curd blue cheese can be made without rubbing up. The blueing will happen naturally as the air passes through the open structure of the curd to the interior of the cheese all on its own. But if a Stilton is rubbed up, it will have to be pierced to allow air to flow and *Penicillium roqueforti* to grow.

Piercing

The piercing of blue cheeses turns the interior into the exterior, and in so doing becomes an important part of the affinage of fungally ripened cheeses like Stilton.

Piercing introduces air to the flesh of the cheese and allows fungus to grow where it normally would not. Some blue cheeses can develop some *Penicillium roqueforti* on their interior without air being intentionally introduced, because air flows naturally through a cheese from the rind through its cracks. But to assure consistent blueing, a cheese should be well skewered. Piercing must be done right to assure the appropriate development of veins or pockets within a blue cheese. Of course, the cheese must also be made full of holes or full of crevasses to assure space for the *P. roqueforti* to grow within the cheese.

Piercing has some significant consequences to the flavor, texture, and evolution of cheese. Introducing air causes the fungal cultures that make the cheese blue to break it down much more quickly and uniformly than they normally would. This odd practice actually hastens the degradations of a cheese, and according to the traditions of cheese being a preserved food, this practice certainly goes against the grain. It's sometimes argued that blue cheeses were the first aesthetically driven cheese, made not from a cheesemaker's need to transform their milk into a more preservable form, but from desire for a more flavorful food.

For allowing air into a cheese causes the fungi to degrade the cheese in a most extraordinary way. Fungi typically ferment the components from the rinds of the cheeses they grow upon, so many of the aromatics they create evaporate into the cave air. However, in a blue-veined or -pocketed cheese, the fungi are much more numerous than when they grow simply on the rind, and they release their aromatics, including esters, ketones, and, most significantly, alcohol, into the flesh of the cheese itself, making these styles perhaps the most flavorful of all cheeses. In a strange way, the piercing of blue cheeses makes them most similar to wine, which is stirred regularly in order to mix in air and encourage yeasts to develop. If wine—or any other alcoholic fermentation—isn't stirred during its fermentation, it grows yeasts mostly at its surface, and doesn't develop alcohol the same.

Blue cheeses (aside from Stilton) are usually skewered in two series of piercings: the first the day after the cheeses are placed into the cave, and the second 1 month later. The development of the *Penicillium roqueforti* causes curd breakdown within the cheese, which may obscure the passage of air. Skewering twice assures a continued passage of air to the interior for the 2 to 3 months that it usually takes for the blueing to develop the best character in a cheese.

Cheeses should be skewered on the sides that are most exposed to air, not the sides that rest upon their boards. Stiltons, flipped between their flat tops and bottoms in the cave, should be skewered on their curved side to assure the most air passes to the interior. Roqueforts, which are rolled on their round side in the cave, should be pierced on their tops and bottoms to assure the best air penetration, and the most effective ripening.

A thorough piercing helps to assure an even blueing. Cheesemakers will typically pierce a blue cheese every 2 cm: A standard-sized Stilton could thus be pierced over one hundred times! Use a barbecue skewer or small knitting needle to assure sufficient airflow. An implement that's too fat compromises the integrity of the cheese.

Charcoal

Charcoal, often mistakenly called ash, is applied to the rinds of some cheeses as a purely aesthetic treatment. Charcoal is best applied to goat cheeses that grow white rinds, resulting in a striking contrast between the white flesh of the cheese, the black line of charcoal, and the white fungus growing through the black. Charcoal has no other effect upon a cheese; an ashed cheese is a cheese simply painted black.

Charcoal can be made by burning wood in an oxygen-deprived environment; this results in a carbonization of the wood, with only gases and other organic matter burning off, leaving behind the unburned carbon in the form of charcoal. Cheese charcoal is best prepared from softwoods like pine, which are easiest to grind after carbonization. Harder woods can be used but may be difficult to grind to a fine enough powder to apply to the cheese.

Some say that adding charcoal to cheese encourages the rind to develop the right microbiology sooner. My experience is that the charcoal only helps us see the fungal establishment on a cheese rind sooner: the blackened rind shows the cream-colored growth of *Geotrichum candidum* more than the white rind of an unashed cheese.

Charcoal should be applied the day after salting a cheese, just before the cheese is to develop its white rind. For ashed lactic cheeses, like Valençay, the charcoal is applied the day after salting, just as the cheeses are put away into their hastening space. For an ashed Camembert, as in the technique for Pont l'Évêque, charcoal should be applied after the first week of rind washing, just before the rind is left unwashed to grow its white rind.

Charcoal is best applied to the cheese as a sort of paint. Take the powdered charcoal, add clean water or fermented whey to make a thick ink (this is a traditional Japanese ink), then apply that ink, by hand or brush, to the rinds of the cheeses. The ink can be applied most evenly, quickly, and neatly in this manner. You can also dust charcoal on a cheese by sprinkling it through a very fine-mesh sieve.

Historically actual ash was added to cheeses. Ash, the mineral deposits remaining behind after wood is fully burned in air, is alkaline; applied to a cheese, ash

Painting cheese with charcoal creates a striking color contrasts and visual effects, especially with intensely wrinkled lactic cheeses. PHOTO COURTESY OF AGRICOLA FORADORI

can affect its microbiology and cause its paste to liquefy prematurely. Likely its addition completely hampers any mold development on a cheese and raises the pH chemically (like adding baking soda), causing the cheese to quickly deliquesce. But through inhibiting the microbes that grow upon the rinds, the ash may also keep flies from visiting a cheese. Morbier, made today with a line of charcoal through its center, was formerly a cheese made in two consecutive cheese makes, with the cheeses covered in ash after the morning make to keep out flies while waiting for the evening milking to finish the cheese.

Leaf Wrapping

Leaves and plant matter of all sorts are applied to the rinds of various cheeses for aesthetic and practical reasons, but they generally provide structural support to softer surface-ripened cheeses that liquify as they age. These leaves and plant materials were the first cheese packages, helping soft, delicate cheeses make the trip to market; but the plant materials also conferred their diverse flavors to the cheeses they wrapped or protected, helping define them as particularly delicious styles.

Large and tannic-rich chestnut leaves are used to wrap aged lactic cheeses, especially in the chestnut-growing regions of France. Cheeses are well protected by leaf wrappings, which can be any large and durable leaf of an edible plant, such as corn husks, maple leaves, vine leaves, et cetera. Leaves are generally applied late in the cheese's ripening, after the rinds are well established. Applying them earlier can cause a cheese to bind its rind to the leaf, which can lead to tears in its rind if it is mishandled. Still, cheesemakers can use this effect to advantage, by treating the leaf wrappings as a sort of container that allows the cheese to ripen to perfection like a Saint-Marcellin in its clay pot. Certainly the complex and flavorful tannins from leaves infuse into the flesh of the cheese better if the leaves are put in place before the rind has grown.

Ferns and other fine leaves, including leaves of grass, can be pasted onto cheeses to add visual appeal, making the cheeses look straight from the woods. Large white-rinded cheeses like Camembert and Brie, especially, can be made very dramatic with the careful addition of foliage to the rind just as the white rind is being established, when the cheese is still sticky from its washing.

Spruce strips are applied to the rinds of various soft ripened cheeses from the alpine regions of Europe, including vacherin Mont d'Or. The spruce bands, cut from the cambium of trees in the early spring when the bark strips easily, are dried, then rehydrated, preferably in fermented whey left overnight, before being applied to the rinds of the cheese with string. The cheeses should have their bark applied early in the ripening, preferably just after salting. The aromatics of the spruce bark penetrate this cheese beautifully, making it a most sought-after style.

Similarly, the reed juncus, which grows in wetland pastures, is used to wrap cheeses like France's little-known Livarot, a soft washed-rind cheese that would lose its shape upon ripening without the reed binding. Straw is used to pierce Saint-Maure, providing structural support, giving the otherwise brittle lactic cheese a degree of strength and preventing breakage of the cheese and its rind.

Leaves can be applied directly to the rinds of cheeses straight from trees, or fermented or preserved by other manners before application. Fermenting the leaves first helps them adhere to the rinds, and also helps protect the leaves with the right microbial ecologies. Leaves like maple, fig, and vine can be softened and fermented before pasting them to the rinds by brining them beforehand. Place the leaves into a large container with a bit of dry salt sprinkled between each layer of leaves, to create their own brine, or pour a liquid salt brine (3 to 4 percent salt) over the top. They will keep this way, under the brine, for many months if kept cool.

Leaves can also be preserved in brandy or other flavorful alcohols to preserve and soften them, but also to lend the essences of the alcohols to the cheeses they wrap. This technique marries especially well with blue cheeses, already rich in alcohol, and should be applied after the cheeses' ripening is complete.

Aging in Pomace

Pomace is a term used to describe the leftover fruit flesh remaining after the primary fermentation of wines or ciders. Cheeses have historically been aged in this agricultural waste product in order to impart unique characteristics to them and simplify their aging process.

Alcoholic pomace combined with high moisture and restricted air low dramatically alters the affinage of a cheese. For example, neither fungus nor any other microbial biofilms develop upon the rind of a cheese aged in pomace. The rinds maintain a clean finish, though of course they are colored brightly by the pigments and flavors of the fruit. Cheeses in pomace, therefore, need not be further treated—in fact, the less they're disturbed, the better they'll age. In these collective ways, aging a cheese in pomace is a natural equivalent to aging it in vacuum-sealed plastic, though the impact of pomace ripening is considerably more natural and delicious.

It is best to use pomace that is fermented during the primary fermentation of a wine or cider for this purpose. Using the unfermented pomace of grapes pressed to make a white wine or rosé can result in unwanted fermentations within the flesh of the cheese upon aging. (Though apples are rarely fermented on their skins and flesh, this technique makes for a much more flavorful and tannic cider and produces an extraordinary apple pomace for aging cheeses. More information for fermenting wine and cider and a technique for aging cheese in their pomace can be found in chapter 14.)

This technique of affinage is best applied to semi-firm cheeses, preferably medium-sized semi-alpine cheeses like tommes. A softer or smaller cheese can be easily lost in the lees.

Waxing

Waxing seals a cheese from the degradations of air and simplifies the regimen of aging. Waxed cheeses can be aged in a dry fridge, and no flipping, washing, or brushing is required.

But waxing seals off a cheese from air and the microbiological interactions that generally make most cheeses interesting. All cheeses sealed in wax tend to ripen in a similar direction, becoming more and more sour over time, due to continued primary fermentation. They never develop the complexity of a naturally rinded cheese, for they never have the opportunity to undergo a secondary fermentation.

Waxing can be an acceptable treatment for cheddars and Goudas, cheeses both appreciated in an extra-acidic state. But few other styles evolve into their recognizable selves when deprived of the air they need to breathe. Surface ripening ecologies, essential for the development of most aged cheeses, need air to develop and ripen right. Even other styles like grana, which ripen without rind ecologies and their second fermentation, need to slowly dry in the air to develop their best qualities—an evolution that a good coating of wax purposefully prohibits.

Before waxing a cheese, dry the wheel for 2 days after salting at room temperature. When it's dry to the touch all around, the cheese is ready to be sealed. Dedicate a pot to the process, filling it three-quarters of the way up with solid wax to be melted. Slowly and carefully heat it to its melting point, stirring occasionally (be very careful not to overcook—it's highly flammable!). Turn the heat off, and then grip and dip your cheeses, one half at a time, switching sides with every dipping, until three or four layers of wax have been applied and the cheese is no longer visible inside. Once it's cooled, the cheese is ready to be placed in a cool environment to age (the environment need not be humid, as the wax preserves the moisture within).

Cheese "wax," petroleum-based paraffin, is mixed with added plasticizers to improve its flexibility, and unnecessary colors are added to improve its appearance. Beeswax can be used as an alternative to paraffin that improves the flavor of cheeses ripened within, but it is quite brittle and hard to keep from cracking, especially in a cold cave. And when the wax cracks, air exposure happens, resulting in drying and or mold growth, depending on the humidity of the environment. If you're using beeswax, do your best to rarely touch the cheeses as they are aging. Fortunately, they do not need as regular a flipping; perhaps once monthly they can be turned to prevent their sticking to their supports.

Similar to waxing, vacuum sealing provides all the same benefits of waxing, and all the same drawbacks. Vacuum plastic doesn't come from a good place, is likely to contaminate the cheeses within, and ultimately is not recyclable—it's one use only. If the plastic seal breaks, air will enter, and a cheese will become a moldy mess.

CHAPTER 10

Ripening Ecologies

Cheese ripening, or affinage, is the realization of a cheese's potential as it is kept and tended to in a cave over time. It is the slow process of change in an aging cheese, typically resulting from complex microbiologies that grow upon the rind and feed on the flesh of a cheese below, allowing a fromage to fully express itself.

This chapter explores the many different ecologies of ripening cheeses, their preferences, and how to encourage them to work their particular magic. The most prominent surface ecologies of a ripened cheese are cream-colored *Geotrichum candidum*, blue-green *Penicillium roqueforti*, and orange *Brevibacterium linens*. Some other ecologies that inhabit certain naturally made cheeses include white *Penicillium candidum*, gray *Mucor miehei*, and cheese mites.

Rind-ripening cultures that live on a cheese are living breathing entities that have needs of their own that must be met for them to thrive. And any and all of them can be encouraged to grow naturally on the rind of a cheese given the right handling of the cheese during and after its make.

There are myriad ways of aging cheeses: A single fresh cheese can be transformed into well over a dozen aged cheeses depending on the techniques of affinage used (explored in chapter 9), each of which shepherds the cheese along a different microbiological pathway toward decay. Defining that pathway and staying the course for a particular cheese is key to a proper affinage. And though the process of change in a cheese happens largely unseen at a microbiological level, a cheesemaker can be in complete control of this beautiful and transformative process—a process that makes every cheese most flavorful and nutritious, but a process that ultimately also destroys it.

This chapter will also explore the developments of cheese acidification, cheese eyes, and cheese crystals, each sought after by makers and ripeners of certain styles of hard, long-aged cheeses.

Milk's Own Affinage

Milk, left to ferment undisturbed in a covered jar, preferably with an added culture like clabber or kefir, will evolve in a predictable, reproducible way through many stages of its own decomposition and ultimate decay.

The evolution of such a long-fermented milk embodies the nature of cheese as it ages, and in many ways can be considered the original affinage from which all aged cheese evolve. Milk does this entirely on its own when it drips from the udder to the ground, or if a cow dies with an udder full of milk. To best understand how cheese ages and decays, it is important to first recognize why and how milk itself ages and decays.

Such an evolution of milk can be witnessed by any cheesemaker by leaving a well-fed clabber culture to ferment undisturbed, with a loosely fitted lid, for months at room temperature.

In the early stages of milk's transformation into clabber, milk's lactose and other sugars fuels a frenzied fermentation, with dozens of species of lactofermenting bacteria feasting and reproducing exponentially. And

A single rennet cheese can evolve in many different directions, depending on its rind treatment; here a single batch of cheese was ripened four different ways to encourage (from top to bottom) *Penicillium candidum*, *Geotrichum candidum*, *P. roqueforti*, and the washed-rind ecology. PHOTO BY ALIZA ELIAZAROV

as they feed on the rich source of sugars in the milk, they rapidly transform them into lactic acid and other by-products of their metabolism. At their peak of activity, the milk reaches a certain level of acidity that causes it to curdle—at a pH of around 4.5. As the feeding frenzy exhausts its food, and by-products of the fermentation (lactic acid) accumulate, this primary fermentation slows, and the now fermented milk shifts to a secondary fermentation, equivalent to the affinage of a cheese. The *Lactobacilli*, *Lactococci*, and other lactofermenters run out of steam, stagnate, and, as they are overcome by their waste products, slowly die. But new microorganisms take over from the primary fermenters and bring about even more exciting and delicious changes.

At the surface of the clabber, with much oxygen and humidity, the secondary fermentation begins in earnest. Yeasts begin to thrive there, where there is abundant air, necessary for their growth and development. *Geotrichum candidum*, in its early stages very much a single-cellular yeast and very much a multicellular fungus in its later stages, reproduces wildly at the surface by budding with the abundant lactic acid, its favorite food. It quickly establishes a white rind that covers the ferment within a day, but only after the milk has curdled and lactic acid levels rise.

Once the *Geotrichum* covers the surface, it begins to function as a multicellular fungus and develops a wrinkled texture atop the milk. It seems to proclaim the clabber as its own and protects it from above; left to ferment for weeks or months, no visible manifestation of any other microscopic microorganisms develop. This is not a simplification or misperception; the culture of *Geotrichum* and its allies (it likely doesn't act alone) creates chemical compounds that protect its ecology, preventing unwanted microorganisms from developing. And it does this through continuously consuming the lactic acid of the fermented milk and transforming it into by-products like alcohol that keep other microbes at bay. Inevitably, however, the lactic acid supply that fuels the fungal growth dries up.

While the *Geotrichum* works its way through the clabber, below the surface, not much happens. With an entirely oxygen-free anaerobic environment and a high degree of acidity, all microorganisms are slow to develop within the curd; the change comes from above.

As the lactic acid runs out in clabber, the pH of its curd begins to rise. Concurrently, its moisture is evaporated by the ever-hungry fungus that's consuming it and respiring it into the air. Meanwhile, the proteins in the curd are consumed by the fungus and broken down into more umami amino acids; and as their pH is raised, they begin to deliquesce (more on this term soon) much like a Camembert. Attaining a gooey texture, this advanced state of broken down milk becomes a surprisingly flavorful cheese, especially when salt is added.

Such is the secondary fermentation of clabber. In many ways it is a microcosm of cheeses' ecological developments—the natural foundation of a microbiological process upon which almost all styles of cheese depend. For instance, if a naturally fermented and salted lactic curd is left to ripen in its typical fashion, white *Geotrichum* rind will develop that will defend the cheese against unwanted fungi and dominate its evolution. The lactic acid of the cheese will slowly be consumed by the *Geo*, and as the pH rises, the proteins of the curd will break down and liquify becoming what we know as a crottin.

Ecologies of the Rind

The three main ecologies that develop on the surfaces of naturally ripened cheeses are creamy-white *Geotrichum candidum*, pink-orange *Brevibacterium linens*, and blue-green *Penicillium roqueforti*.

On a well-made cheese, produced with good milk, invoking a good fermentation, made with good cheesemaking practices and ripened in the right aging environment (the cave), these are the only ecologies that will develop. Some other less significant but nonetheless important rind ecologies may grow on cheeses that are handled in certain ways, including *Penicillium candidum* and *Mucor miehei*. And cheese mites can also come to play an important role in cheese rind development. Other, sometimes colorful, surface bacteria and fungi can come to grow if the milk used is long refrigerated, if the primary ecologies are not well established, or if a cheese is ripened in too dry an environment to support the favored ecologies.

Establishing the right surface ecology is a matter of seeding the right microorganisms in the cheese in its infancy and creating the right circumstances for those microbes to grow on the ripening cheese. But the cheese has to be made right for those right microbes to develop. For instance, to make a proper crottin, the cheese must be started with a culture rich in *Geotrichum candidum*, the fungus that will establish its wrinkly white rind. It must also be aged in the right environment and handled the right way for that fungus to develop on the rind. But the cheese also has to be made with good milk and effective rennet, with just the right firmness, just the right amount of lactic acid, and just the right amount of salt, that altogether create the healthy "soil" upon which the fungus grows. Affinage is a continuum of cheesemaking and shouldn't be understood as a separate idea!

If the white fungus does not develop on the crottin and it instead turns blue, it is not because of contamination—for contamination is a state of mind! The crottin must have been made or ripened wrong in some way that discourages the *Geo*'s white rind's development, and encourages the *Penicillium roqueforti*'s blue. A bit of sleuthing should help you understand how it happened.

A naturally made crottin will not turn blue out of the blue! If the cheese is made to be white and it turns blue, there will be a reason. It should also be understood that there is not necessarily anything wrong with that blue crottin. However, if a cheese is meant to turn white, it should and it will if it is made and ripened right; if it is meant to turn blue, it will do that too, if it's made and ripened right.

Once those microbes are established on a cheese, they defend it—the first ecology to develop on the rind of a cheese wins! The established microbes create chemical compounds that prevent the establishment of other species of bacteria, fungi, or yeast, and in doing so claim the cheese for themselves and determine its destiny. All cheese-ripening fungi produce similar mycotoxins (fungal toxins) that protect cheese rinds from being infected by unwanted microorganisms. (Fungi from the same family as *Penicillium roqueforti* and *P. candidum* are producers of the well-known antibiotic penicillin.)

It should be noted that all these different surface ecologies can coexist within a cheese; the different microbial ecologies do not contaminate one another, but rather each grows best in certain circumstances and certain climates, and if each culture is given the respect and treatment it needs, the cultures can thrive side by side in the same cave—even on the same cheese. Many cheesemakers using modern methods keep blue cheeses out of their facilities for fear of contamination, and install HEPA filters to keep out *roqueforti* spores. And though such measures are needed for making a white-rinded cheese using modern methods, with natural methods these fears are entirely unfounded. Remember, contamination is a state of mind, and in a natural cheesemaking, cheeses will turn blue only if they're allowed to.

Stilton, for example, is made in a way that encourages all four of the sought-after rind-ripening microorganisms—*Geotrichum candidum*, *Penicillium candidum*, *Brevibacterium linens*, and *P. roqueforti*—to grow on the same cheese, each existing within its own niche where the right conditions are created for their growth, and none of them contaminating or compromising the others: The blue remains in the veins, the *Geo* grows in its place, and the washed-rind ecologies of *B. linens* and *P. candidum* thrive in their own space. So long as you make the cheese right, these right ecologies will develop right where they're meant to.

Now, it should be understood that my naming of species is a simplification of cheese ecologies; just like the description of a Douglas fir forest, say, is a simplification of a much more complex underlying ecology without which the Douglas fir cannot thrive and simply planting Douglas fir—as is often done—does not make a Douglas fir forest! For example, dozens if not hundreds of microbial species play an important interrelated role in the development of a *Brevibacterium linens* rind. Even cheese mites themselves likely play host to dozens, hundreds, or more species of microbes that are a part of their own microbial ecology. Rind ecologies are therefore too complex to understand or model on a microbial level. From a fermentation perspective, though, certain patterns within their development correspond to certain milk handlings and cheese treatments, and these can be completely controlled by a cheesemaker despite their complexity. In my own cheesemaking I do not concern myself whatsoever with the scientific identification of the species of microorganisms involved, for I don't believe that it matters. I simply make cheese the right way and watch the visual manifestation of the right microbes develop . . . for nature is exceptional in its ability to develop certain patterns when it follows certain pathways.

It is challenging not to subscribe to the monoculture approach to cheesemaking when describing ripening microorganisms. I have chosen to describe the microbes as an ecology rather than as individual species, and I apologize for any oversimplifications that might result from my choice of species. Perhaps another term to use would be *guild*, as used in the botanical sense, meaning a community of plants that thrive together in the right circumstances. We could call blue rinds a "*roqueforti* guild" and white rinds "*Geotrichum* and its allies."

Cheese Ripening and Paste Breakdown

Cheeses ripen as a result of two parallel biological processes: The dominant one, in most cases, is the metabolism of the aerobic surface ecologies, and the minor one, again in most cases, is the anaerobic metabolism or acidification within the flesh of the cheese.

All of the fungi, yeast, and bacteria that thrive on the rind of a cheese (or within its veins) consume that cheese in similar patterns, but each ferments the cheese in its own way due to its distinct metabolisms and produces unique signatures of compounds that develop distinguishing flavors in the cheeses it ripens. Each different ecology also leaves a different color on the rind, due to the different aspects of its biology that are visible to the naked eye. *Penicillium roqueforti* makes cheeses look blue, for example, because that is the color portrayed by the microscopic fruiting bodies that rise up above the surfaces or within the holes of cheeses upon which the fungus grows.

The dominant surface flora feed on the lactic acid left behind by the primary fermentation, and not the lactose, which is unfermentable by fungi. (Lactose is a tough sugar to crack! Even we probably can't do it.)

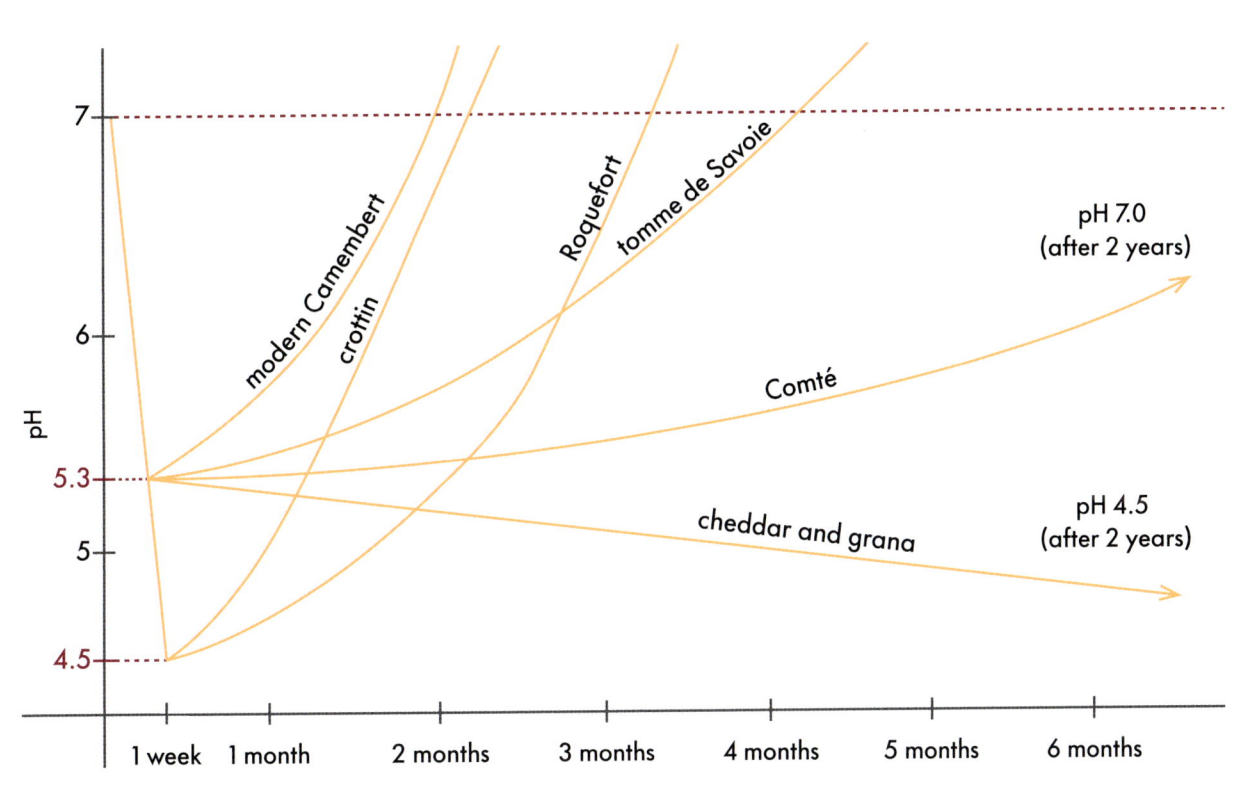

Most cheeses evolve through a combination of primary fermentation, which lowers the pH, and secondary fermentation post-salting, which raises the pH. Surface ripened cheeses typically are ready once the secondary fermentation has raised the pH back to around 7.0, when the curd deliquesces. But some rare cheeses, including cheddar and grana styles, continue to ripen without secondary fermentation, slowly becoming more acidic over the course of several years.

RIPENING ECOLOGIES • 181

And from what I understand and have experienced, it's only fermented by certain bacteria. They also feed on the fats and the proteins present within the flesh of a cheese and break them down into their component parts. Meanwhile, any remaining lactose continues to be very slowly metabolized into lactic acid by lactofermenting bacteria within the cheese.

As the surface flora feed on the nutritious elements of the cheese, the sugars, fats, and proteins are broken down by fermentation into simpler and typically more flavorful and aromatic components. The lactic acid is metabolized to alcohols, flavorful esters, and other less acidic substances that contribute unexpected aromas to the cheese. The proteins are broken down into umami-tasting amino acids and eventually ammonia in a process known as proteolysis. And the fats are lipolyzed into long- and short-chain fatty acids that contribute other elements of an aged cheese's most enticing aromas.

The metabolism of the proteins, fats, and lactic acid of the cheese by the surface bacteria and fungi all together results in the ripening of the cheese, witnessed by a dramatic effect on the cheese's texture. As the lactic acid is consumed and the pH of the paste rises, the texture of the proteins changes. As the proteins are broken down to ammonia, the pH rises more. And the broken-down fats contribute their changing character to the altered composition of the cheese. The combined effect is that of an oozing, liquified cheese, best observed inside a perfectly ripe and flowing Camembert.

This liquification of a cheese is often termed simply proteolysis, but I believe this in an oversimplification, as it results from diverse metabolic breakdowns that fit better under an umbrella of the fermentations that are occurring within the cheese. The liquification happens entirely because of a microbiological effect that raises the cheese's pH back to neutral, because cheese liquifies when it reaches a pH of 7. At this neutral state, the casein proteins attain the same liquefied state as they had when they departed the udder at about the same pH. From here on in I'll refer to this complex biochemical process as curd breakdown or liquification or, perhaps preferably, deliquescence, a term used by mycologists to describe the process of liquifying breakdown that happens in certain delicious mushrooms like shaggy manes.

The process of pH rising and its consequent effect on the texture of a cheese can be artificially reproduced by mixing sodium bicarbonate (aka baking soda) and a bit of water into a fresh chèvre or cream cheese. Within seconds, as the added alkalinity is absorbed by the cheese, the pH of the cheese rises dramatically to the neutral realm, and the paste of the cheese liquefies like a thick milk due to the return of its casein proteins to their original state. Similarly, if you add a teaspoon of baking soda to a cup of clabber or kefir, after a day the now-neutral mixture will cause the precipitated casein proteins to dissolve, and will return to liquid milk, but only in texture, not in taste.

Of course, adding alkalinity is not an acceptable shortcut to ripening a cheese. (But it is essentially the process of making American process cheese, melting cheddar cheese together into a paste with various alkaline salts that give a uniform texture and make the "cheese-food" shelf stable.) Because the chemical rise in pH that's responsible for the textural shift is not accompanied by a fermentation, the cheese tastes only of alkaline chèvre. And it is not a breakdown of the proteins that makes a Camembert flow like a Camembert, but a whole suite of diverse fermentations (including protein breakdown, which raises a cheese's pH through the production of ammonia).

Different cheeses deliquesce in their own distinct ways. Modern Camemberts ripen and flow from the rind to the core evenly as a result of the ideal conditions of firmness, moisture content, and acidity resulting from their make, which allows an even pH rise throughout the cheese and therefore an even ooze. A Camembert made with more moisture, like the traditional Camembert, will have more primary fermentation and a lower pH in its core, which counters the rising pH originating from the rind and thus restricts the ripening to its outer parts leaving behind a chalky center. Lactic cheeses, made even more acidic, show a more restricted ripening limited to the parts of the cheese just below the rind because it takes more of a pH rise to achieve deliquescence, and the intense acidity of the cheese buffers against this change. Alpine cheeses do develop a ripened flow from paste breakdown, but its effect is limited by the low moisture content and firmness of these cheeses. Tomme, an alpine cheese made with

Rind-ripening ecologies cause cheeses to deliquesce by consuming their lactic acid and converting it to alcohol, slowly raising the pH of the paste to 7.0, whence it turns to a most delicious ooze. PHOTO BY ALIZA ELIAZAROV

more moisture, shows much more breakdown than a firmer Comté, but neither compares to Camemberts.

All of the surface ecologies, major and minor, contribute to the breakdown of cheese. Each ecological regime, however, will contribute a unique character within a cheese due to the distinct way that each metabolizes the nutrients therein, contributing distinct lactic acid breakdowns, proteolyses, and lipolyses that result in unique flavor compounds, including diverse ketones, alcohols, esters, and aldehydes. The milks themselves provide the proteins, fats, and sugars, which differ from species to species, but the flavors of these remain locked up in the milk. In breaking down these elements into their more flavorful parts, it is the ripening ecologies that create the incredibly diverse flavors and aromas of ripened cheeses.

We salivate at the thought of an aged cheese due to the incomparable nutrition that these broken-down elements of milk provide (our bodies know it's good food!). Eating well-ripened cheeses may well be the finest way of feeding ourselves, for the pre-digestion of the cheese means this most important food is readily embraced by our bodies within our digestive tract.

Geotrichum candidum

Characterized by a velvety, creamy white growth, *Geotrichum candidum* tends to distort the rinds of cheeses it grows upon, often giving them an intensely wrinkled rind. Known best for growing on lactic cheeses, whose rinds are considered most beautiful when they show the fungus's intricate wrinkling, *Geo* is a culture that also establishes itself early on in the rinds of most naturally rinded cheeses: The fungus plays a pivotal role in the establishment of Camembert's and Brie's white rinds. It's a member of the washed-rind cheese ecology and is even present on the rinds of many blue cheeses.

Geotrichum is a yeast (a single-cellular fungus) that can also behave as a mold (a multicellular fungus) in certain circumstances. It reproduces by budding, dividing one cell into two much the way bacteria do. It is also understood to be a benevolent participant in our bodies complex ecologies, and a fungus that is significantly important to the preservation and development of many of our most important foods.

Geotrichum loves backslop and is most strongly encouraged by using a well-fed biodiverse mesophilic starter culture that includes it among its community of lactofermenting microbes. Because it grows by budding and not by spores, it reproduces quickly when the culture is fed milk. It is therefore the fastest growing of rind ecologies (at least in warmer weather), the one best propagated by backslop—it need not germinate from spores, nor grow them, both slow processes that together take several days to get most molds started—and is typically the first surface to develop on a naturally made cheese. But it will not grow until that medium is first fermented by its bacterial predecessors, and rich in lactic acid.

Geotrichum's favorite food is lactic acid. It therefore grows remarkably well in nearly all dairy ferments and lactic-acid-rich cheeses, as well as most other lactic-acid-rich foods. You'll find *Geo* and its allies wrinkling the surfaces of sauerkraut, sour pickles, sour beers (often called mixed fermented, or *Brettanomyces*), and some natural wines. It will even grow its interwoven wrinkles atop long-neglected sourdough starters left covered at room temperature (it may be that this same fungal phenomenon is responsible for raising sourdough breads!). And though common advice calls for the skimming of this so-called kahm yeast from atop the surface of sauerkraut and such, the *Geotrichum candidum* protects the ferments below from oxidation and microbial contamination, and slowly contributes fine flavors to the food below as it consumes the lactic acid over time. Ultimately, however, *Geo* does devour a ferment, just as it does a cheese!

Geotrichum candidum appears to be milk's intended rind ecology when left at temperatures just below body temperature, its preferred environment. Raw milk, left to its own devices, will typically grow a velvet layer of the fungus atop its surface after the primary fermentation is complete and the curd is full of lactic acid. Similarly, when cheeses are rich in lactic acid and left in a temperate and humid environment, the *G. candidum* coverage on the rind will be complete. Once the ecology is well established and the cheese is aged in a cave, no other culture will grow.

Using a regularly cultured (mesophilic) clabber will naturally seed the fungus in the cheese, as will a well-ripened whey culture, a healthy kefir starter, or a regularly used wooden cheese vat. I like to observe the slight signs of *Geotrichum candidum* growing atop a clabber or kefir culture before I add it to help assure its presence in the cheese. Look at the reflection of a window or an overhead light atop the ferment—if you see something atop the culture obscuring the light, that's the faint growth of *Geo* that typically develops atop the milk once it has curdled. A similar growth will be visible atop a mesophilic whey starter once its pH has dropped to 4.5 and it tastes mildly sour. The fungus will grow atop any of these ferments, every time they're fed, so long as the cultures are regularly fed (daily, preferably) fresh milk.

Geo's preferred conditions are warm and humid—the culture grows well on skin, where the temperatures are typically below body temperature. Keeping cheeses warm during their early stages is therefore essential for their best ecological developments. Place a *Geo* cheese in the cool cheese cave too soon, before its *Geo* rind is established, and it will likely develop a blue *Penicillium roqueforti* rind. Its preferred conditions for growth are 20 to 35°C (68–95°F) and 90 percent humidity. So to encourage cheeses to develop *Geo* rinds, I advise placing them in a warm and humid hastening space (see chapter 8) for 2 to 3 days after salting, before aging them in a cave. Once established on the rind of a cheese, *Geo* will stop *P. roqueforti* (among other fungi) from growing there, even when aged in cooler environs.

Because of *Geo*'s preference for lactic acid, lactic cheeses in particular are most inclined to grow *Geotrichum* rinds, as are cheeses left for a long ferment at room temperature before salting, like tomme crayeuse, and on the rinds of many blue cheeses like Stilton and fourme d'Ambert. If enough lactic acid isn't present in the cheese (and the pH is higher), the fungus can be slow to develop, leaving niches open for other cultures to gain prominence on the rind. This appears to be why Camemberts made with less lactic acid develop *Penicillium candidum* on their rinds, and Saint-Nectaire cheeses their *Mucor*.

Geotrichum evolves through several phases as the cheeses it grows upon age. *Geo* is, during its early stages of development, a yeasty culture, meaning it operates as a collection of single-cellular organisms. At this point, expect aromas typical of yeasts—grape, banana, apple,

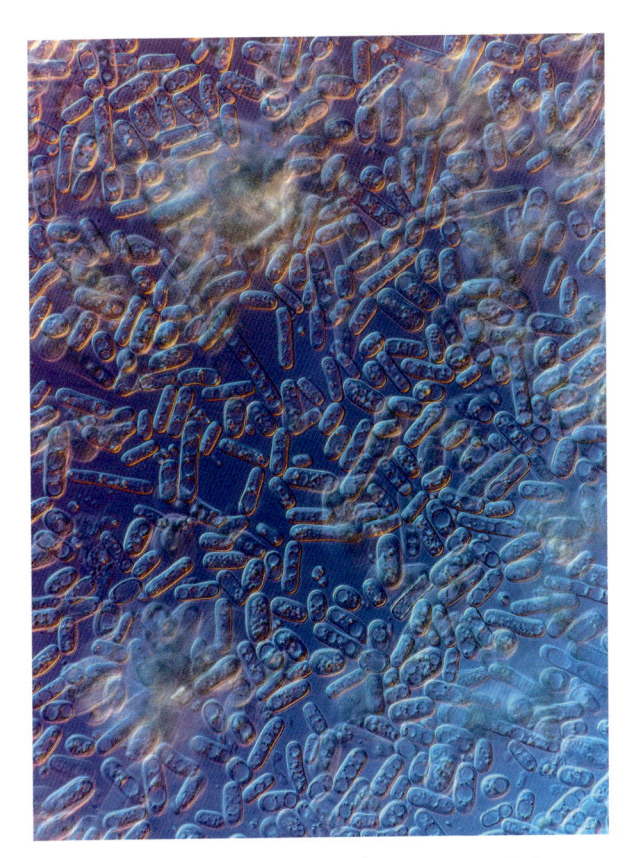

Geotrichum candidum is a yeast that sometimes behaves like a mold. It functions as a single-cellular yeast (as in this micrograph) when it first grows, but when mature it becomes a multicellular mold. MICROSCOPY BY CHLOÉ SAVARD

or sourdough smells are a good sign. The surfaces of young *Geo*-ripened cheeses will be slippery or greasy, and lightly covered with a slight cream-colored velvety rind that obscures the shine of the cheese and is easily disturbed when touched. A slight deliquescence might even be observed beneath the young rind.

As the cheeses continue to age, slight raised bumps rise all across the surface, indicating that the yeasty *Geo* has begun to run out of room to grow. As it continues to develop, the *Geo* begins to function as a multicellular mold; the raised bumps begin to connect as ridges, eventually forming an intricately interwoven wrinkling on the rind. The longer the cheeses are left to age, and the softer the flesh of the cheese is, the more intricate the wrinkles become, taking on a three-dimensional wrinkling if not overhandled. Indeed, the fungus seems to do this on purpose. Like a meandering

larger mechanical holes that become small pockets of blue by stirring a large-cut curd for a rather long time. The semi-firm curd that results knits together well, but not without large voids between individual curds. To enable the blue fungus to grow within these pockets or veins, the cheeses are pierced to permit the passage of air to areas it would not normally go. Piercing as a tool of affinage is explored in chapter 9.

Generally, wheels of blue cheese develop blue on their insides and not on their rinds: They give a surprising blue reveal when they're opened. Most blue cheeses do not develop *Penicillium roqueforti* rinds, but cheesemakers instead intervene to prevent the fungus from growing where it normally would, avoiding an unsightly blue rind. Gorgonzola has an orange rind, evidence of a regular washing, while Stilton is submitted to a process known as rubbing up that keeps blue from growing on the rind. Only a few techniques in the book explore the making of what are very rare cheeses with bright blue rinds: Camembert bleu and tomme de Savoie, which becomes gray-brown after it ages.

It is indeed a strange thing to turn a cheese blue, for encouraging this fungus by piercing a cheese hastens its decay. But there's something special about this degradation that draws us in. It's as if we've become transfixed (even trained) by the fungus, creating conditions that encourage it to consume our cheese. In exchange we're rewarded with one of the most flavorful foods in the animal kingdom.

Washed-Rind Ecologies

Washed-rind cheeses are often characterized by their pink-orange hues, sticky textures, and sometimes crystalline texture on the tooth. But what defines them most is a potent foot-like aroma referred to by the French as *au pieds de dieu*—of the feet of God!

Pungent and pudgy cheeses like Limburger, Époisses, taleggio, and Munster are the best known of this class of cheeses, but many harder alpine cheeses fit in here as well, including raclette, Gruyère, and Comté, which are washed in much the same way. To develop the colorful coats of these cheeses, cheesemakers wash their rinds with brine on a regular basis, through weeks or even months of ripening.

Keeping the cheese consistently wet and cool by regularly washing the rind in the cave shifts the dominant microorganisms that grow. Multicellular fungi do appreciate humid conditions but cannot get a chance to grow when constantly washed; thus, a consistently wet rind will develop very differently. Washing rinds encourages a unique microbiology that is dominated by bacteria and yeast (single-celled fungi). The particular species that grow give the cheese unique colors, ranging from orange to pink, and an aroma with a very particular stink.

The microbiology of washed-rind cheeses breaks down the cheese into aromatic compounds that are strangely body-odor-like. The ecology of these rinds is most complex, containing diverse species of bacteria, fungi, and yeasts including *Brevibacterium linens*, *Corynebacteria*, and *Geotrichum candidum*. Any part of the cheese left unwashed will develop the usual *Geotrichum* or *roqueforti* fungal ecologies. So the washing must be rather thorough—wash your cheeses like you're washing behind your ears and between your toes! Any spot left unwashed will grow other fungi like *Penicillium roqueforti*, *P. camemberti*, and *Geotrichum*.

The diverse members of the washed-rind ecology come from many origins. The particular yeasts and bacteria can originate from the fermented milk starter, or even from a source closer to home. You need not introduce these cultures intentionally to a cheese, for they're a part of your own microbial ecology.

The stinky-foot aromas they create are stinky-foot-like because the same ecology also grows between our wet and salty toes! And it is this washed-rind ecology that primarily gives our feet their stink. Now, I don't necessarily recommend washing between your toes with the same rag you use to wash your cheeses; simply using your unsterilized hands to wash the rinds introduces the appropriate microbes from your hands' microbiome that can establish the washed-rind ecology.

But what most helps these organisms get the best start is a backslop of the rind. In this case, that means washing the rinds of young cheeses with the community of microbes from an already established washed-rind cheese ecology of a mature washed-rind cheese. *Morge* is a term used in France to describe the messy liquid that transfers the culture from one cheese to another and perhaps the term *morgeing* should refer

The washed-rind ecology is grown by keeping cheeses consistently wet. After several weeks of washing, a sticky orange biofilm develops at the surface of a cheese that causes it to ripen. PHOTO COURTESY OF ALIMENTI FORADORI

to the act of washing the many generations of cheeses together. Morgeing hastens the development of the washed rind ecology, and helps cheeses ripen faster and develop their best flavors. It also helps keep the growth of unwanted microorganisms in check.

Though washing several batches of cheese together in a cheese cave is seen in modern cheesemaking as a food safety risk, the effect is, in fact, quite the opposite. Keeping cheeses in isolation and using freeze-dried *Brevibacterium linens* as the sole source of ripening organisms instead of a morge carries significantly greater risks—the packaged monoculture does not establish itself on the rind, resulting in a significant lag time and a cheese susceptible to the growth of wild species. Using a morge can get a washed rind to establish itself 1 week faster than it would otherwise, stopping unwanted microbes like *Listeria* from growing on a cheese sooner than more sterile techniques.

Softer, higher-moisture rennet cheeses like Limburger are considered the most characteristic of washed-rind cheeses, developing deep brothy flavors and just-fluid-enough textures. But many other styles of cheese are also washed in similar ways: Époisses is a half-lactic cheese like Camembert with a washed rind, while raclette and Comté are alpine cheeses washed with whey. But though they're all washed more or less the same way, each different class of cheese is affected in its own way by the washing. Higher-moisture, higher-acidity washed-rind cheese such as Époisses develop exceedingly bold flavors, and a two-textured interior like a traditional Camembert, even growing *Geo* wrinkles on the rind when ripe. Lower-moisture alpine cheeses aren't significantly broken down by the ecology and develop wonderful umami flavors, but remain untainted by the typical foot-like aromas of moister washed-rind cheeses.

The orange-pink color that defines cheeses with washed rinds is the result of carotenoid pigments produced by various coryneform bacteria (including *Brevibacterium linens*), a class of salt-tolerant bacteria that is usually present in the washed-rind ecology. Washed-rind cheeses tend to be pinker in color when they are being actively washed; the color fades to orange when the rind dries slightly, either between washings if the cave is too dry, or when the washing stops. If the washing is stopped long enough, often the orange rind will become colonized by filamentous fungi, especially *Penicillium candidum*, which can turn the rind white if washing is halted prematurely.

Crystals form at the surfaces of all cheeses due to the migration of minerals left over from the microbial processes that occur at the rind, which feed on proteins in the interior of the cheese, and leave behind waste minerals at the rind. But the wet conditions created at the rind of washed-rind cheeses make this style of cheese a crystal garden. Regular washing keeps the crystals from reaching a size that makes them gritty on the teeth; but prolonged periods without washing can make them grow to a visible size where they can sparkle in the light! More on crystals in cheese later in the chapter.

Penicillium candidum

Penicillium candidum, also known as *Penicillium camemberti*, is another white fungus that grows on the rinds of aged cheeses. It is considered the ideal fungus for the rind ripening of Camembert but can also be found residing on the rinds of alpine cheeses, Stiltons, and inconsistently washed washed-rind cheeses. Essentially an albino version of *Penicillium roqueforti* that develops white spores instead of blue, *P. candidum* grows in very specific conditions that encourage it to grow instead of its blue brother.

Penicillium candidum tends to obscure the surfaces of the cheeses it grows upon, covering them in a dense mat of pure white mycelium that rises above their rinds. Left untouched, the mycelium develops a cloud-like fluffiness, unparalleled for softness and whiteness in the world of food. If flipped or otherwise handled, the white rind is pressed firm, developing a denser and more toothsome texture.

If lactic acid bacteria are primary fermenters, and *Geotrichum candidum* a secondary fermenter, *Penicillium candidum* is a tertiary fermenter in cheesemaking and will normally only grow upon the rinds of a cheese if the conditions are altered by other ripening ecologies first. *Penicillium candidum* seems specifically to be friends with *Geotrichum candidum*, and getting the fluffy white fungus to grow on a cheese seems to require that *Geo* gets established first. This can be realized by a combination of leaving cheeses in a hastening space for 1 day after salting, and washing the rind of the cheeses during their first week in the cave.

Penicillium candidum prefers a less acidic rind to grow upon; encouraging *Geotrichum* to grow upon the rind, and consume lactic acid, raises the cheese's pH and sets the stage for the *P. candidum* to grow. Leaving the cheeses at room temperature in a hastening space for a day after salting helps get *Geotrichum* established on the rind of a Camembert, setting the stage for *P. candidum*'s growth.

Penicillium candidum also appears to grow well in cheeses that are washed while aging, once the washing stops, which raises the pH of a cheese, while preventing the *Geo* from getting fully established on a cheese (which would actually inhibit the growth of *P. candidum*). And a washed-rind cheese that's washed only during its first week in the cave will not have enough time to establish the typical pink/orange washed-rind ecology, and instead will find its rind overrun with white fungus after the washing stops, and the pH of the cheese has risen slightly.

For *P. candidum* to grow on Camemberts and other cheeses, it is best to practice a slightly altered rind-washing procedure. Prepare a washing brine of water or whey, salted to 2 percent at the time the cheeses are salted, and thoroughly wash the rinds daily during the cheeses' first week in the cave. Too little washing, or inconsistent washing, can result in the growth of *Penicillium roqueforti*, resulting in blue spots on the rind, while an overly extended washing can lead to excessive *Brevibacterium linens* and its pink or orange color. Washing the rind just right, preferably with morge (rind backslop; see above), seems to develop just the right white.

In addition, *Penicillium candidum* may also prefer a cheese that isn't too rich in lactic acid. Lactic cheeses, made more acidic through fermentation, do not grow *P. candidum* well, but rather support the growth of *Geotrichum candidum*, which thrives in the lactic-acid-rich cheeses, especially when hastened in a warm environment, which favors *Geo*. For a Camembert or other cheese to develop a *P. candidum* rind, it should be fermented less so that it has more lactose and less lactic acid. This can be achieved by salting the curd sooner at a higher pH (5.3, or when the curd stretches firmly when steeped in hot water), or by making the curd firmer through more stirring. The method for a modern Camembert takes this approach to develop its pure white *P. candidum* rind; the traditional Camembert is typically richer in lactic acid due to its half-lactic ladling, and often develops *Geo* instead.

Penicillium candidum is the Goldilocks of rind ecologies! It is perhaps the trickiest of rind-ripening regimes to develop, perhaps because its preference lies in the border of cheese styles, in an in-between ecology. Given the right ripening conditions, most washed-rind cheeses appear to develop some pure white fungi on their rinds once the washing stops; but the *P. candidum* will only dominate if the washing is stopped very prematurely, and only if the cheese has the right acidity.

Spores of *Penicillium candidum* may come from raw milk, though experiments that I've done cannot pinpoint its precise source. Likely it arrives, like most other important cheese microbes such as *Penicillium roqueforti*, from the pool of incredible beneficial microorganisms that exists all around us. But only under very certain circumstances does it evolve into the dominant ripening microorganism. And when it does, the results are dramatic and delicious!

Mucor

Mucor miehei is the name of a wild fungus that develops black-gray tufts upon the rinds of some cheeses. Easiest to identify, the fungal fruiting bodies, or conidiophores, of *Mucor* are visible to the naked eye: gossamer-thin grayish stalks topped with tiny black balls. Cheesemakers prefer, however, to call the fungus by its more relatable name, the ever-so-appropriate "cat's fur."

Cat's fur is a common wild species of fungus that often grows on decaying vegetables and flesh and sometimes even on dry-aging meat. It is not typically a desirable species in cheesemaking circles, as you might expect from its appearance and nickname. I encountered *Mucor* growing on my cheeses in my earliest cheesemaking days, when I didn't really understand what I was doing and was probably making my cheese wrong on many levels. But with a proper cheese make, using good milk, an effective fermentation, and an appropriate affinage, *Mucor* is one of many wild species of fungi and bacteria that's kept in check. *Geotrichum candidum*, for example, when well established on a cheese that's made right, effectively and completely prevents the development of cat's fur on the rind. In the past 15 years of cheesemaking I've yet to see it growing again on my cheese, unless I create the right circumstances for it to grow, which are essentially the wrong circumstances for making cheese.

Cat's fur develops when a cheesemaker does multiple things wrong in the make, like not adding an effective starter culture, salting a cheese too early, and drying the rind of a cheese before it is aged. But in making certain styles of cheese, this "wrong" ecological development is precisely what the cheesemakers aim to achieve. Such is the case with a cheese known as Saint-Nectaire, which is known for its gray *Mucor* rind. Saint-Nectaire may be the only cheese with a sought-after *Mucor* rind, and notably, this fungus develops the cheese into one of the most sought-after in France. So why not grow this fungus intentionally on your cheeses?

Mucor appears to grow best when a cheese is unfermented, and left unoccupied by the usual protective rind ecologies like *Geotrichum candidum*, *Brevibacterium linens*, and *Penicillium roqueforti*. Like *P. roqueforti*, it is a weed-like fungus that comes from the wild and finds a niche for itself in cheeses that are made in ways that do not develop the protective microbial ecologies of milk and cheese. Growing strong *Geo*, or getting good *P. roqueforti* growth, or washing a cheese's rind and developing *Brevibacterium linens* or *Penicillium candidum* all seem to keep *Mucor* in check! I never find it growing on cheeses made with the appropriate natural methods. It's as if the normal fermentations of a cheese must fail for *Mucor* to do well. And it seems as though it's mainly the lack of lactic acid in a cheese that assures mucor's best development.

RIPENING ECOLOGIES • 191

If you make cheese, they will come. Cheese mites reside on the rinds of well-aged, untreated cheeses. This cute little fella was found grazing on the rind of a 6-month-old tomme. MICROSCOPY BY CHLOÉ SAVARD

Cheese Mites

If cheeses with natural rinds are left untended for many months, they will inevitably develop macrobiological ecologies. Look closely at the rind of a long-neglected hard cheese left at the back of a cave, and invariably you'll see evidence of cheese mites, *Tyrophagus putrescentiae*, Latin for "putrid cheese eater."

If you make cheese, they will come. Mites reside in forest and farm environments, feeding on decaying leaves, hay and grain, and bits of dandruff from animals. Seeking out organic matter to eat, they'll hitch a ride on a cheesemaker's clothes, and will eventually find their way to a previously uncolonized cheese cave. But there's no reason to fear for their inevitable arrival. New cheesemakers should find their first mite sighting a cause for celebration, for the insects' only interest is a well-aged, naturally rinded cheese, and their presence equates to a certain level of success as a cheesemaker. As with *Penicillium roqueforti*, any efforts to keep out mites from a cave is overkill. Instead focus your energy on good cheesemaking practice and regular rind treatments, specifically rind washing and brushing, to limit mites' ability to reproduce upon a cheese.

Left to their own devices, mites will make short work of even a tall wheel of cheese. Like little goats that graze the rinds, they feast on fungus and nibble away at the surface of a cheese, eventually tunneling through. As the mites dive deep into a cheese, the passageways they create play host to *Penicillium roqueforti*, turning their tunnels into blue veins. A mite-y cheese that is not reined in will be heavily affected by these developments, and its paste will eventually become oxidized and discolored as it is exposed to introduced air, developing intense flavor as a result of the mites and the *P. roqueforti*.

A certain intensity of umami comes about from their activities in a cheese, sought after by some cheese lovers, disdained by others. French cheesemakers, especially, appreciate the gustatory effects of a mite infestation, and make a number of styles that just don't taste right without them—naturally rinded tommes that often show a brownish dust resulting from their activities, and mimolette, a harder alpine style that is riddled with mite-dug craters!

The US Food and Drug Administration once restricted (and perhaps still does) the import of cheeses into the US with more than a certain number of mites per square inch (six, to be precise). But mites do no harm, and they don't transmit illness. The FDA restrictions are perhaps understandable, given that the waste products of these often unwelcome arachnids accumulate within a cheese—but are these really any different from the waste products of fungi or bacteria that make most cheeses what they are?

Mites prefer cheeses that are firm, naturally rinded, and aged beyond 3 or 4 months. Picky eaters, they won't ever look at a Camembert or bother a blue cheese until it's way past its prime. But in the presence of a long-neglected clothbound cheddar or a naturally rinded alpine cheese like a tomme, they'll quickly make their mark, burrowing through the rind, leaving behind a brownish dust. To confirm the presence of mites, look for their brown dust on the underside of cheeses aging on their boards. To be sure the dust is a sign of mites, brush it into a line; come back the next day and the line will have moved!

To control mites, cheesemakers have their work cut out for them. Though French cheesemakers don't mind them on their rinds and appreciate the flavors they bring without them having to do any work, Swiss cheesemakers wash their alpine cheeses like clockwork to be sure they don't get established. Mites

The macro-effects of cheese mites inhabiting the rind of a tomme. Slowly but surely these creatures will turn an unwashed or unbrushed wheel of naturally rinded cheese to dust, making the cheese intensely flavorful in the process. PHOTO BY CHLOE GIRE

will completely devour a 12-month-aged Gruyère if not washed, so who can blame them. Their long-aged alpine cheeses are washed with brines several times weekly right up to maturity, which stops mites altogether. However, a slight lapse in treatments allows the mites to get below the surface, where they burrow in peace and cannot be treated. An established infestation is a significant challenge to remedy, but under no circumstances would such a situation have evolved if the cheeses were well cared for. The presence of cracks or excessive mechanical eyes in a cheese can make it easier for mites to enter, which is one of the reasons why full-alpine cheesemakers cook their curds to such high temperatures—to get a monolithic wheel without the faintest fissure.

Brushing can help reduce mites, but also tends to ruin some rinds. For cheeses that appreciate a natural rind, like tommes, cheesemakers generally let the mites take their share of the cheese (like the angel's share in whiskey). They'll let the mites feast until they decide they've had enough, and then sell their mite-riddled wheels. With clothbound cheddars, regular brushing and the cheese cave are the only natural options to control mites. Industrial cheesemakers keep their cheddars vacuum-sealed in plastic and their Goudas in wax in part to prevent mite infestations without the need for laborious rind care, but these cheeses miss out from the many beneficial fermentations that come along with air exposure.

Continued Acidification

Most styles of cheese continue acidifying to certain degrees as they age. After salting, the activity of the lactofermenting bacteria is significantly reduced, but not totally stopped. With the water mostly locked up by the salt, the bacteria are held back, and transformation of lactose to lactic acid continues at a snail's pace.

The effect of ripening microorganisms at the rind overrides this effect slowly reducing the acidity of the cheese, and the combination of the two contributes to the desired qualities of these cheeses. With firmer, drier alpine cheeses, the slow acidification over months or years of ripening works in concert with the rind-ripening microbes to help define the fullest flavor development these styles are known for. Lactic cheeses, however, see almost no continued acidification within their flesh, as the lactose is fully fermented to lactic acid in the early stages of their makes.

Hard and massive cheeses (with less surface area), like cheddars, grana, and even cheeses ripened under oil like shankleesh or in a whey brine like feta, have their ripening effects dominated by slower-acting anaerobic bacteria, which metabolize sugars as well as fats and proteins. It takes much more time for these cheeses to develop their most sought-after flavors, due to the slower nature of anaerobic microbes, but also due to their low moisture content. Ultimately, though, the physical effect of the anaerobic ripening on a cheese is different from the aerobic, for the diverse metabolic pathways that lead to pH rise (cheeses' sweetening) and curd breakdown (cheeses' deliquescence) are effectively paralyzed without oxygen.

Styles like grana and many milled-curd cheeses like cheddar benefit most significantly from this continued acidification. In fact, the more it occurs the more flavorful and sought after they become. Granas are salted at a pH of 5.3, but they are not considered ready to eat until the slow acidification within their core brings the pH of the flesh down to below 4.8, which is more or less the same as a lactic cheese, but it can take years for these low-moisture cheeses to achieve these high acidities. With cheddar, continued acidification over many years, especially in larger wheels with little to no air exposure, can yield strongly sharp and caramel characters resulting from a very low pH of 4.8. When fully ripened these two cheeses do not stretch when melted, but instead disintegrate like a lactic cheese. These styles have a great number of similarities. A crottin sec, especially when made with cow's milk, is comparable to a long-aged cheddar or a grana. Because the most sought-after character of these cheeses is this acidic and flavorful flesh, cheesemakers discourage rind-ripening microbes by ripening the cheeses in drier aboveground cellars.

Gas Development

Certain styles of cheese benefit from the growth and development of gas-producing microbes during their affinage. The most celebrated example of this is Emmentaler, known for its round eyes. Other cheeses that benefit from gas development include blue cheeses and some white-rinded cheeses like traditional Camembert.

It should be noted that gas development can also be a serious flaw in a cheese. Excessive yeast growth and coliforms cause gas that can be accompanied by odd flavors and can be indicative of the growth of opportunistic or pathogenic microorganisms, especially very early on in a cheese's development (in the first 24 hours). These are generally caused by the use of an overfermented starter and/or the use of older milk. It is important to distinguish between these good and bad gas producers, and appendix J helps a cheesemaker to do so.

Emmentaler's eyes are created with the help of lactic-acid-fermenting bacteria known as propionibacteria. These microorganisms feed on cheeses' residual lactic acid after the primary fermentation, transforming it into secondary metabolites including propionic acid, which contributes to the flavor of the cheese it grows in, and carbon dioxide, which develops the cheese's eyes.

But the propionibacteria will only develop eyes given the right circumstances within the flesh of a cheese. The firm and elastic paste of alpine cheeses permits the gases to accumulate and cheeses to swell just right. The cheeses must also be "improperly" salted to encourage the propionibacteria development, and they must be kept in a warm environment following salting to assure their development.

Emmentaler is among the taller alpine cheeses (often 20 to 30 cm high), which are generally made shorter, in part to avoid the development of eyes. The excessive height of the wheel slows the penetration of surface salt to its interior. As the salt takes its time to pull out moisture and slow the fermentation within the middle of the cheese, the propionibacteria get a chance to grow. The eyes of Emmentaler form toward the center of a cheese, where the salt is slowest to arrive.

Microbiologically diverse mesophilic starter cultures like kefir and clabber, considered heterofermentative due to the diversity of their microbes and their fermentation pathways, can help develop gas in a cheese (if they are made right; they won't grow if they're not given a chance) due to their natural populations of propionic cultures.

Cheese Crystals

Cheese crystals evolve as a result of a slow crystallization of various components of the cheese, usually amino acids, or various forms of calcium or phosphorus, which are by-products of various cheese fermentations. Though not exactly biological (though their compounds are created by diverse fermentations within the cheese), the crystals form in a slow, almost geological process and their crystallization cannot be hastened. Like "cheese diamonds," these crystals are a sign of quality and only develop in cheeses that are exceptionally well made and thoroughly aged.

Crystals result from minerals (in both cheese and stones) being chemically attracted to themselves, and slowly migrating toward one another, forming complex structures depending on their chemical makeup over time. To develop crystals, you must create the right conditions by making your cheeses well, and then patiently tending to the cheeses as they age in the right, underground-like environment.

Every cheese style that grows crystals, grows crystals with unique chemical signatures that result from their unique makes and distinct affinage. In high-temperature-cooked alpine cheeses, the crystals that develop are typically tyrosine (an amino acid whose name means "of cheese" in Greek)—a crystal made of the eponymous amino acid broken down from the protein by fermentation. The crystals that develop in cheddars are often calcium lactate, a salt of calcium and lactic acid. In plastic-bound cheddars, fine white calcium lactate crystals develop on the rinds where the moisture is trapped against the plastic for months or years. These white "rinds" are often trimmed by the cheesemonger or consumer in the mistaken belief that they're a fungus!

Washed-rind cheeses often develop crystals on their orange rinds due to the activity of their rind-ripening microorganisms; the minerals growing on the rinds of these orange cheeses include the unlikely crystals of ikaite and struvite, found in bladder stones. These minerals are excreted by cheese-ripening microorganisms that pull calcium and phosphorus from within the cheese for their nourishment, then excrete them as a waste product at the rind. Regular washing will keep the growth of these crystals in check, but if you leave your cheese unwashed long enough in a cool and humid setting, these minerals will grow out from the surface and can sometimes be clearly seen sparkling with the naked eye!

Other mold-ripened cheeses like Camembert and blue cheese form similar crystals where the fungi grow; in very long-aged blue cheeses a certain crunchiness can sometimes be noticed in the blue veins.

Keeping Cheeses Fresh

But what if you as a cheesemaker or cheesemonger wish to limit microbiological developments on a cheese, and keep it in its freshest possible state?

Unfortunately, to keep a cheese fresh is to work against nature. Cheese wants to evolve, to change, and we cannot, no matter how hard we try, ultimately keep a cheese at a certain stage of flavor forever. A fresh cheese will inevitably evolve into an aged cheese through the growth of fungus, either introduced or wild. For if a cheese is kept humid, fungus will grow, but if it's kept in a dry environment, the cheese will dry in turn.

If, however, you wish to extend the life of a fresh cheese, there are several things that you can do to slow its decay, mainly assuring that it doesn't begin its second fermentation too soon, or at all.

In making chèvre, for example, keeping a rigorous cheesemaking schedule and not letting fresh dairy products overferment can limit the growth of unwanted yeasts and fungi; draining lactic cheeses more quickly and thoroughly by hanging them can help as well. And salting to 2 percent well before any yeast growth happens helps assure that secondary fermenters don't get quickly established. Some fresh cheeses like chèvre and mozzarella can be frozen, but this is detrimental to their best tastes and textures.

You may consider propagating your cultures anaerobically, to stifle the development of surface flora: filling the culture jar right to the rim and sealing it tight will limit oxygen exposure and prevent aerobic surface cultures from developing in the culture and on a cheese made with it. Alternatively, starter cultures can be fed more frequently (twice daily) or placed under refrigeration as soon as they've curdled.

Refrigerating cheeses as quickly as possible after the make also helps impede ripening cultures in a fresh cheese or dairy ferment. Breaking a cheese down into smaller portions and chilling rapidly helps limit the growth of secondary fermenters.

As soon as a fresh or aged cheese is cut and repackaged (in either plastic, cheese wrap, or beeswax wraps), at the new interface between the cheese and air, fungus will inevitably grow. This is not dangerous, and it need not be trimmed. The cheese is simply growing a new rind that will eventually protect and consume the cheese below. Ideally, a cheese that is cut should be enjoyed or sold soon after cutting; otherwise it will either seal itself with a new rind if kept humid, or dry out if not. Though I don't generally recommend such actions, packaging fresh or aged cheese in vacuum seals keeps air out and prevents the degradations that depend on oxygen.

However, a cheese made with fungus-free methods will inevitably become exposed to air and wild fungi like *Penicillium roqueforti*, which will grow unchecked on the rind of the cheese if it is kept humid. If you make cheese naturally, fungi will invariably grow upon it. Many commercial cheeses are treated with preservatives and antifungal agents (such as natamycin) to prevent the inevitable fungal developments that come with long storage even under refrigeration. Mass-produced fresh cheeses like chèvre and cottage cheese often have potassium sorbate added to keep mold from growing once their air-sealed packages have been opened. Many industrially aged cheeses' rinds are treated with natamycin to prevent natural fungi from establishing.

And if you don't cultivate the appropriate milk-fermenting cultures in the right way, your cheese will grow cultures you *didn't* cultivate. Ultimately this leads to unwanted microbial transgressions and cheese's decay in unfortunate ways that don't preserve milk in its finest form. I prefer instead to follow nature's path and not restrict milk's inherent, inevitable fermentations.

Cheese's Inevitable Decay

The community of microbes and larger creatures that together form the rind of a cheese slowly consume the cheese, subsisting off the nutrients within and creating the complex aromas, flavors, and textures that we have come to know and appreciate in aged cheeses.

Cheese is of the earth, and if not eaten before its affinage consumes it, it will eventually return to the soil from which it came. For the microbiological (and macrobiological) breakdowns that are responsible for its ripening are the same microbiological phenomena that build our soil.
PHOTO BY CHLOE GIRE

Cheese-ripening organisms are also decomposers of organic matter that breathe air and create wastes as they eat a cheese. And though they are essentially causing the cheeses themselves to decay, they do so in a most extraordinary and flavorful way, in the process making them more nutritious, more aromatic, and generally more appealing.

The appreciation of such a cheese changes according to the culture, but in general the more broken down a cheese is the more flavorful it becomes. But only up to a certain point, where a strong ammonia smell—indicative of the final breakdown of the cheese's proteins into ammonia—sounds out the cheese's death knell.

Some extreme examples of eating decomposed cheeses, however, are worth noting: Some French cheese lovers appreciate the character of a Camembert or Brie so over-ripened that its flesh turns black and reeks of ammonia, referring to these past-their-twilight cheeses poetically as Camembert or Brie noir. And Sardinians turn their noses up at moldy cheeses but intentionally introduce maggots to their casu marzu cheeses to break them down and make them most intensely flavorful.

A cheese left to ripen too long will ultimately return to the soil from which it came. The rind-ripening regimes, be they *Geotrichum*, *Penicillium roqueforti*, *Brevibacterium linens*, or even mites or maggots, inevitably turn the cheese to dust. As cheesemakers, we must enjoy, share, or sell the cheeses before their inevitable destruction by their ripeners occurs. For either we or they will eat every last morsel of the cheese.

Cheese can thus be seen as a delicious metaphor for change. We cannot hold on to a certain stage of a cheese's existence forever; that cheese will ultimately transform if we do not enjoy it at that best moment (seize the cheese!). To truly appreciate cheese, we must understand this continuous evolution. And if we keep on making them in their most suitable seasons, we can always have a cheese to enjoy in its most exceptional state.

PART III
THE TECHNIQUES

PHOTO COURTESY OF MAX JONES

CHAPTER 11

The Families of Cheese

Though there's a remarkable diversity in cheese styles from around the world, all cheeses and fermented dairy products relate to one another in remarkable ways. This chapter will help you understand the complexities of each class of cheese, and how they differ and compare with the others; as well as how individual cheeses within each class are made slightly differently from their relatives.

Most styles of cheese fit into one of twelve or so major categories that vary from one another mainly in the way that the curd is handled as it is being made into cheese (though some cheeses blur these lines and inhabit more than one of these categories). These categories are:

Fermented milks   Milled-curd cheeses
Fermented cheeses   Washed-curd cheeses
Lactic cheeses   Half-lactic cheeses
Rennet cheeses   Spun-curd cheeses
Stirred-curd cheeses   Heat-acid-coagulated
Semi-alpine cheeses   cheeses
Full-alpine cheeses

Yet another variation in the cheesemaking process that yields an even greater diversity of styles is how the cheeses are aged. In essence, through affinage, each of the above-mentioned categories can evolve in numerous directions.

All styles of cheese can evolve from one single milk; what defines a cheese is the method that makes it. This chapter describes the many variations that a cheesemaker can bring about, and how each cheese differs from the next in its production.

Fermented Milks

Milk curdles through the action of its microbiological community alone, as it consumes the milks sugars and transforms them into lactic acid. This fermented milk is the first family of cheese.

Simple cultured dairy products are the foundation of all cheeses; for all cheeses are essentially evolutions of these first fermented milks. The makes of these cheeses are uncomplicated, and easy to reproduce with great precision.

Different milks (goat's, cow's, sheep's, buffalo's), treatments of milk (raw or boiled), as well as temperatures of fermentation (thermophilic or mesophilic) and conditions of fermentation (aerobic or anaerobic) can significantly change the texture and taste and even the color of the fermented milk in question.

> The category of fermented milks includes the following:
>
> - kefir
> - clabber
> - yogurt
> - ryazhenka
> - gros lait
> - kaymak
> - cultured cream
> - amasi
> - milkbeer
>
> All of these fermented milks are made by a variation on this basic process:
>
> Treat the milk (optional); ferment the milk.

FACTORS AFFECTING MILK'S FERMENTATION

type of milk	cow	strong curd, golden cream line (thicker for Jersey), simple flavor
	goat	no cream line, very soft white curd, complex & long-lasting flavor
	sheep	small cream line, thick, rich white curd, complex lamb-like flavor
	buffalo	white cream line, very thick curd, intriguing flavor
	yak	golden cream line, tastes of a yak
	horse	higher lactic acid content and alcohol
	human	smells of an infant
milk treatment	raw	strong curd (e.g., clabber)
	boiling	softer, but thicker curd (e.g., yogurt, gros lait), sweeter taste
	long-cooking	caramel color (e.g., ryazhenka)
	refrigeration	bad taste, rancidification, unwanted fermentations, strong animal notes
temperature of fermentation	thermophilic (37–45°C)	homofermentative: only a few strains of bacteria, mostly lactic acid production
		no *Geotrichum*
		more acidity, less complex flavor and lower pH (e.g., yogurt)
	mesophilic (20–35°C)	heterofermentative: great diversity of bacteria and yeasts, more diverse fermentation by-products
		Geotrichum growth
		more complex flavor, less acidity (e.g., gros lait)
amount of starter added (starter:milk)	no starter	wild fermentation (not recommended, except to begin a starter)
	1:1000	primarily bacterial fermentation, slower acidification
	1:100	balanced fermentation (e.g., kefir, clabber)
	1:10	faster fermentation, enhanced yeastiness (e.g., milkbeer)
aeration	aerobic	yeasts like *Geotrichum* develop at surface (e.g., clabber)
	anaerobic	no yeast development, only bacteria
	active mixing (or wooden vessels or gourds)	enhanced yeastiness and alcohol development (e.g., amasi)

Milk's fermentations can be quite diverse depending on many factors, including the type of milk, its treatment, the temperature of fermentation, the amount of starter added, and aeration.

Fermented milks are able to evolve in remarkably different directions depending on the milk, its treatment, the temperature conditions, and exposure to air. By re-fermenting boiled milk in aerated, mesophilic circumstances, milk can be made to alcoholize, effervesce, and develop a head of foam, much like beer! PHOTO BY CHLOE GIRE

Clabber and kefir are the most straightforward and formulaic of this class of cheese. To make them, simply add starter culture to milk, then leave the milk to ferment. Clabber is the simpler of the two, made by adding a small amount of active culture in the form of clabber or whey. Kefir is made by adding the polycultural kefir grain that causes the milk to ferment in a similar manner. Variations can be created by changing the milk; goat's, cow's, and sheep's milk all make wildly different kefirs or clabbers. Adding more kefir grains as a starter, or using a higher ratio of clabber, can also drastically change the texture and flavor of the finished dairy product.

Yogurt is another fermented milk—or more precisely, a fermented *cooked* milk. It is made by first cooking the milk to evaporate it and change its protein structure, then culturing the cooked milk at a higher, thermophilic temperature. The curd that results is smoother, thicker, and creamier than clabber or kefir, and more tart, too. This all results from the different way the milk is handled in its make. Kaymak, also known as clotted cream, is essentially prepared as the cream top of yogurt; a delicious and important dairy product in its own right.

Gros lait and ryazhenka are two variations of yogurt, one from Brittany, the other from Ukraine. Gros lait is essentially yogurt fermented at a lower temperature—a mesophilic yogurt—while ryazhenka is a caramel-colored yogurt made by fermenting milk that has been long cooked and caramelized. Both are excellent dairy products that deserve greater recognition outside their places of origin.

Cultured cream is yet another excellent cultured dairy product. It is cultured just like kefir or clabber, but with cream instead of milk. This yields the most flavorful and richest of milk's many ferments and is the precedent for making a traditional cultured butter.

Amasi and milkbeer are the most potent of the fermented milks. They're made like kefir but handled in a way to encourage milk's native yeasts to develop alcohol, in what is a rare intoxicating dairy drink. To create the right conditions, aerate the ferment, either by shaking or stirring as in milkbeer, or in a gourd, as for amasi. Larger amounts of starter are also intentionally added; both encourage the growth of milk's native yeasts.

Fermented Cheeses

Fermented cheeses, also known as acid-coagulated cheeses or sour milk cheeses, are simple styles of cheese prepared by curdling milk solely through fermentation, then draining, cooking, or churning the delicate and acidic curd. The typically very soft cheese that results can be manipulated in a number of ways, all of which make a fairly acidic cheese that is generally eaten fresh, though some are aged.

These are some of the most ancient of all styles of cheese. The many cultures that make them today are carrying on a cheesemaking tradition that dates back to the domestication of dairy animals. Most of the cheeses in this category evolved from the traditional practice of leaving warm milk to ferment without intervention to separate and curdle its cream for buttermaking—a most important fermented cheese; the skim milk curd below is then handled in different ways to make other fermented cheeses.

Draining the cheese removes its whey, concentrating the goodness of the fermented milk and focusing its texture, flavor, and nutritional qualities. It yields a cheese that preserves much better.

Depending on which fermented milk is hung, the amount of whey drained, whether or not the cheese is salted, and how the cheese is aged, a surprising complexity and diversity of cheeses results. Made entirely without rennet, this class of cheeses can offer a cheesemaker who chooses to avoid the ingredient a great selection of styles.

Fresh fermented cheeses include:

- skyr
- tvorog
- yogurt cheese
- butter

And some aged fermented cheeses are:

- shankleesh
- handkäse
- graukäse

The foundational process of making acid-coagulated cheeses is as follows:

Ferment the milk; drain the fermented milk.

Exemplifying the basic method, skyr is an Icelandic cheese made by hanging fermented skimmed milk in cloth to drain its whey. Tvorog is an Eastern European variation on the same technique that involves slightly cutting and cooking the fermented skim milk before draining the firmer curd that results into a drier cheese.

Yogurt cheese is an acid-coagulated cheese made by draining yogurt of its whey. The cooking of the milk prior to its fermentation gives the yogurt cheese an extra yield and improves its creaminess and flavor.

Shankleesh is an aged variety of yogurt cheese, submerged in oil to preserve it. It is one of dozens of different aged yogurt cheeses made across the Middle East and Asia, where yogurt, an important buttermaking by-product, is turned into a well-preserved food.

Handkäse is an aged acid-coagulated cheese made by salting skyr, then forming and ripening the cheese much like a Camembert. Perhaps one of the first mold-ripened cheeses, this ancient style of cheese is still appreciated among German cheese eaters. Graukäse is an Austrian variation, made much like a mountain cheese but with the sour milk curd left as a by-product of butter production.

Even traditional butter might be considered in this category—it's simply cultured cream, with its whey-like buttermilk removed through churning. Butter is more or less a fatty version of skyr, the skim milk cheese made with the butter's skim milk curd. The two cheeses evolve in parallel processes from a vat of fermented milk, as described in the technique for skyr and smjör.

Lactic Cheeses

Lactic cheeses are a group of softer cheeses made with rennet but fermented, typically overnight, to develop their lactic character. The long fermentation not only develops acidity and complex flavors, but also breaks down the semi-firm rennet curd into a soft and creamy textured cheese. Lactic cheeses can be either fresh or aged. Aged lactic cheeses are generally ripened with the help of *Geotrichum candidum*, as explored in chapter 10.

Fresh lactic cheeses include:

- cuajada
- faisselle
- chèvre / brebis / vache / bouflonne
- cottage cheese
- cream cheese

Some aged lactic cheeses are:

- crottin
- Saint-Maure
- Valençay
- le sein
- Saint-Marcellin
- sirene

The basic lactic cheese make can be summed up as follows:

Add starter culture; add rennet; ferment the curd; drain the curd; salt the cheese.

The lactic make is the most uncomplicated style of renneted cheese. And if you're taking milk straight from the animals, making this class of cheeses is perhaps easier than refrigerating the milk. Simply take the warm milk, add starter culture, add rennet, then forget about the milk until the next day, when the curd is ready to be ladled into forms or scooped into cloth.

Chèvre and its relatives vache, brebis, and bufflonne, are the poster cheeses of the fresh lactics. Their curd is made by adding active culture and rennet to milk at 20°C (68°F). The rubbery curd that first forms is left to ferment for between 12 and 24 hours, during which time it should shrink and sink beneath its whey. The sour, soft and brittle curd that results is drained in cloth for 6 to 12 hours, after which the curd is salted and drained for several more hours. Though typically made with goat's milk, similar cheeses can also be prepared with cow's milk, sheep's milk, and buffalo milk.

Cream cheese is a lactic-style fresh cheese, prepared like chèvre / vache / brebis / bufflonne, but with milk that is boiled first, which softens the texture, improves the yield, and gives additional flavor to the cheese.

Cuajada is just the fermented lactic curd itself, eaten in Spain by the spoonful. Faisselle is a French lactic cheese whose curd is ladled into forms to briefly drain its whey. The careful way that the faisselle curd is ladled yields a cheese with a soft and flaky texture. Lactic curd develops an exquisitely curdled texture that cleaves like chocolate when it is left to ferment undisturbed. Delicate handling of lactic curd, taking full scoops of curd without breaking it, preserves that texture and improves the character of the cheese.

Aged varieties of these cheeses simply call for ripening fresh lactic cheeses in the right environment to preserve them and encourage the right ecological developments on the rind. Most aged lactic cheeses are begun the way of chèvre, with slight variations that yield distinct cheeses. And a certain style of warm-weather affinage characterizes these cheeses, which is explored in chapter 9. To make crottin, goat's milk is cultured then renneted and left to ferment for about 24 hours. The curd is then carefully ladled into forms to preserve the delicate texture of the cheese, and then drained overnight; the following day, the cheeses are salted, then drained for yet another day. Once dry to the touch, the cheeses are placed in a curing chamber or hastening room with a high humidity and a temperature of 20°C (68°F). Flipped daily, the cheeses begin to bloom; within 3 to 5 days, they will be covered with *Geo*. Once the fungus is well established, the cheeses are placed in a cooler cave to age.

Saint-Maure is made in a similar way, but in a taller form and with a blade of dried straw positioned through the middle of the curd. Valençay, as well, is a lactic cheese much like crottin but in a pyramid-shaped form and coated with charcoal to give a contrasting black color to its white *Geotrichum*. Saint-Marcellin is a small lactic cheese left to age until completely consumed by its rind in a small terra-cotta pot.

The curd can also be drained before forming, as with le sein. Such pre-drained cheeses are made by first hanging the lactic curd in cheesecloth overnight. The soft curd is then salted and drained another day. Once dry to the touch, and modeling-clay-like in texture, the cheese can be formed into myriad shapes, each one of which will ripen into a distinct *Geotrichum*-ripened cheese. Pre-draining, however, curd gives the lactic cheese a consistent but grainy paste, in contrast with the beautifully cloven texture that results from hand-ladling.

Two other lactic cheeses can be made by encouraging different ecological developments on the rinds than *Geo*. With Bulgarian feta, the salted lactic cheese is left to age submerged in a brine made of its own whey. Without air, the *Geo* won't grow, and the cheese preserves its pristine rind. Lactic cheeses can also be made to turn blue by skipping the curing stage and placing them directly into the cool caves after salting to encourage *Penicillium roqueforti* on their rinds.

Rennet Cheeses

This curd is the foundation for making many styles of soft and hard cheeses, both fresh and aged. The method can be modified to make hundreds of different varieties of cheese. All the remaining categories evolve from this basic approach, and all of the other cheese classes begin the same way as these, by adding culture and rennet to warm milk, then making the cheese while the curd is still sweet. Unlike lactics, rennet cheeses are formed and drained before the curd has been affected by its fermentation. The result is a sweeter, semi-firm curd that's more dynamic and adaptable to many different styles.

To make a rennet cheese, warm milk to animal temperature, about 35°C (95°F). Add starter culture in

---

Fresh rennet cheeses include:

- primo sale
- crescenza

Some aged rennet cheeses are:

- modern Camembert
- Pont l'Évêque
- vacherin
- surface-ripened blue
- Gorgonzola

The rennet cheese make is:

> Warm the milk; add the culture; add the rennet; cut the curd to a walnut; stir the curd; drain the curd; salt the cheese.

the form of active kefir, clabber, or whey. Mix rennet thoroughly into the milk, then wait for the curd to set. Good milk will give a strong curd just beginning to give off whey within 45 minutes if you've added the right amount of rennet and the temperature is right. When you see a slight puddling of whey atop the curd, it's ready to cut.

Cut the curd to the appropriate size, typically the size of a walnut for rennet cheeses. Curd size defines the amount of moisture that the cheese will give off during the make, and determines to a great extent the moisture content of the final cheese.

Stir the curd very gently by hand until it firms. The length of this stir depends on the goal firmness of the cheese. The longer you stir, the firmer the curd becomes. Some milks, like sheep and buffalo, require much less stirring to achieve their goal firmness.

Leave rennet cheeses at a constant temperature while stirring. For when the temperature is increased, more whey is expelled from the curd and the cheeses become much firmer, entering the landscape of alpine cheeses.

Once you've achieved the right curd texture, leave the curd to settle in the whey for several minutes. The whey that rises to the top is then drained off with a scoop, or through a vat with a drain, in a process termed wheying off. Transfer the curd, now visible at the bottom of the pot, by hand into forms to drain.

Next up is hooping—the forming of the cheeses. Have the appropriate size forms or molds on hand in the appropriate number, and fill them with curd to their brims. The curds will drain slowly, eventually reaching a third or so of their original height. Do not overfill or underfill the forms, as the cheeses will not be the right shape, and will not evolve in the right way.

Flip the curds in their forms several times over the course of 12 or more hours. The first flip happens in the first hours, once the curds have knit together into a cheese and firmed somewhat. Two more flips help assure that the cheeses have an even shape all around. A neglected cheese will show an uneven shape, and this imperfection will stay with the cheese as it ages.

Leave the cheeses in their forms until they reach the appropriate acidity—a pH of about 5.3. Typically, this acidity is achieved after 8 to 12 hours, faster in warmer climates, slower in cool ones. At this point, when the curd stretches in hot water, the rennet cheese is ready to be salted.

Primo sale is a simple rennet cheese that is the embodiment of this basic method. "The first salting" in Italian refers to the cheese being eaten once it receives its first salt. Nearly all rennet cheeses begin as primo sale. The cheese at this point tastes of its milk, with only slight notes of fermentation.

Crescenza is another fresh Italian rennet cheese, though it's made in an odd manner. A proper crescenza is prepared without a starter culture, effectively halting the acidification that would otherwise influence its development. Instead, the cheese develops a runny texture much like a Camembert, though early in its development, and without the fungus that would normally raise the pH.

Modern Camembert and Pont l'Évêque are two aged rennet cheeses that begin their lives as a basic rennet curd, but whose final character is defined by the white fungus *Penicillium candidum*, which softens its paste to perfect flowing cream. All that really distinguishes these cheeses from one another is their shape: Camembert is made in a small round form, while Pont l'Évêque is made square. The shape, however, does significantly influence the development of each cheese, and though they're both made essentially in the same manner, each evolves in its own way. Both receive their white bloom through washing the rinds.

Camembert bleu is a rennet cheese made blue by encouraging the growth of *Penicillium roqueforti* instead of *P. candidum*. The make is simply a rennet cheese formed and salted like Camembert. But leave the cheese unwashed in its cave, to encourage the blue fungus usually kept in check with washing.

Vacherin is a washed-rind rennet cheese. It is made into a small round that evolves a flowing paste held together with strips of spruce cambium, infusing flavors of the forest to the cream. Vacherin is washed regularly with salted whey for at least a month to encourage its washed-rind ecology.

Gorgonzola is a rennet cheese, too, made with an extra step of drainage before forming that helps it to develop blue veins. To make a Gorgonzola, stir the rennet curd briefly to firm it, then drain it in cheesecloth bundles before you break it up and ladle it into a cheese form. The draining and breaking of the curd before forming helps further firm it, resulting in a curd with less capacity to knit together that develops fine blue veins when pierced.

Stirred-Curd Cheeses

The stirred-curd style evolved from the making of cheese in traditional wooden vats that couldn't be heated. Cheesemakers made as firm a cheese as possible by cutting the curd relatively small and stirring until the milk cooled, when the curd would be ladled into its forms to make the cheese.

In all of these cheeses, the make is a more intensive and involved than the standard rennet make and results in a cheese with less moisture and a longer life span. The curd is generally cut smaller, to evolve more whey, and stirred longer, to make the curd firmer still.

What distinguishes finished stirred-curd cheeses from rennet cheeses is the presence of mechanical holes resulting from a firm curd that does not quite knit together. Normally these cheeses are left unpressed, like rennet cheeses, but unlike alpine styles. They exist in a space somewhere between those two classes of cheese.

Roquefort and fourme d'Ambert are two stirred-curd cheeses that are turned blue. The mechanical holes created when fourme d'Ambert is stirred play host to *Penicillium roqueforti*. To maximize the size of the holes and make the most dramatic blue pockets, stir the cheese significantly—for up to 1 hour—before carefully draining the curds and forming. Roquefort is a sheep's milk blue made in much the same way, but the firmer sheep's milk curd requires only half the stirring of cow's milk fourme d'Ambert curds. Both these blue cheeses have a long primary fermentation before salting to develop lactic acid, contributing to a more thorough and flavorful second fermentation once they are pierced.

Feta is, in its make, much like Roquefort, but it differs in that it is submerged in a whey brine to age. The curd is cut small and stirred until quite firm. Pressed lightly into its forms, the curd still retains some small openings, but because the cheese is kept in salted whey, it does not develop blue. Like Roquefort and fourme d'Ambert, feta is long fermented before salting and brining.

Halloumi is a Cypriot stirred-curd cheese, that is made to withstand cooking without melting. The cheese is stirred until semi-firm before being formed. Once it's been drained, but before it has acidified, the curd is boiled in its whey to stop its fermentation, keeping the cheese sweet and prevent it from melting when grilled.

And torta is a Portuguese cheese, typically made with sheep's milk, as many of these styles are. But two practices set it apart from the flock: The cheese is renneted with cardoon; and it is made in a way that prevents fermentation, by omitting a starter and adding salt to the milk. The resulting cheese becomes broken down within weeks due to a foreshortened affinage resulting from its lack of lactic acid.

> Some stirred-curd cheeses covered in this book include:
> - feta
> - Roquefort
> - fourme d'Ambert
> - Halloumi
> - torta
>
> The stirred-curd make can be summed as follows:
>
> Warm the milk; add the culture; add the rennet; cut the curd to a medium size; stir the curd for a long period; drain the curd; salt the cheese.

Semi-Alpine Cheeses

Semi-alpine cheeses are those that are made in the foothills of the mountains, with a make halfway toward a full-alpine cheese. These styles were often made with the milk of smaller herds being moved to the alpine pastures, or in semi-mountainous regions where cheesemakers didn't perform a full transhumance but were still producing as large a cheese as possible with their milk. Often all the milk from the small herd would be put into a single pot and made into one medium-sized wheel of cheese—one pot, one cheese!

> Cheeses made in this semi-alpine style include:
> - tomme de Savoie
> - Tomme de montagne
> - raclette
> - tomme crayeuse
> - tomme de chèvre
> - Saint-Nectaire
>
> The semi-alpine make can be summed as follows:
>
> Warm the milk; add the starter; add the rennet; cut the curd small; stir the curd; cook the curd to medium temperature; form the cheese; press the cheese; salt the cheese.

These cheeses represent several possibilities for what is essentially the same lightly cooked alpine cheese. In all their makes, the curd is cut small and cooked until it just begins to press together (typically 45°C [113°F]), making a relatively high-moisture wheel.

The still-warm curds are then pressed together into a wheel under whey, then transferred into forms to make a medium-sized cheese with small mechanical holes. The higher-moisture semi-alpine cheeses that result ripen faster than full-alpine styles, which are cooked to an even higher temperature, but are still well preserved.

These styles are made to be aged and generally are not eaten fresh. Their low-moisture paste ripens slowly and becomes more complex and flavorful over time. These relatively small alpine wheels can ripen after 4 months to a year at most.

Tomme de Savoie is cut to a curd size slightly greater to a kernel of corn; it is less cooked than the others (to only 40°C [104°F]), making an even moister cheese that ripens relatively quickly. It is given a natural rind without treatment that turns brown-gray as it ages. Mites often play a role in its ripening.

Tomme de montagne is a firmer, larger wheel, cooked to 45°C (113°F). It is ripened with a white *candidum* rind resulting from a short period of washing. It can age almost up to 1 year. Raclette is made similarly to tomme de montagne, but with a fully washed rind that glows orange when ripe. It can be made by cooking the curd until firm, or by washing the curd with hot water, much like a Gouda.

Tomme crayeuse (chalky tomme) is a semi-alpine style that's left longer before its salting. The longer fermentation results in a cheese that's more lactic in its character. It has a chalky core and breaks down like a Camembert close to the rind. Because of the excessive fermentation, the cheese ripens like a giant crottin.

Tomme de chèvre is a semi-alpine style made with goat's milk. The special nature of goat's milk shines in this make, for the cheese can achieve a semi-firm almost alpine texture with only the cutting of its curd—no cooking required!

And Saint-Nectaire is a semi-alpine style made in wood, without cooking, that achieves its semi-firm curd through cutting and stirring alone. The stirred curd is left to coalesce in the vat, then transferred into forms and pressed into a relatively small wheel, essentially as big as an alpine cheese can get without cooking. The cheese is salted early and aged in a peculiar way to encourage *Mucor* to develop its rind.

Full-Alpine Cheeses

The alpine style is a most refined peak of milk's transformation into cheese. The philosophies behind the family are intertwined with the tradition of transhumance, and the constraints put in place on their production by the mountain. The method takes a large quantity of milk and transforms it into as manageable and movable a cheese as possible, at the limit of the human scale in cheese.

Made to be moved, these cheeses transform the abundant milk of the alpine pastures into massive wheels of cheese that are more easily carried down the mountain, and often shaped to fit on a cheesemaker's shoulders or the back of a donkey! The cheeses are also made to be preserved. The high-temperature cooking makes a monolithic wheel that prevents the entry of air, mold, and mites that would otherwise cause these long-aged wheels to decay.

Most often with this style, all the milk of one milking from a medium-sized herd is made into a single cheese, traditionally cooked in a copper kettle over a fire. This style fits with the philosophy of the semi-alpine cheeses—one pot, one cheese—but both the pot of milk and the cheese it makes are much more substantial!

Essentially an extreme version of the rennet cheese make, the alpine make includes several steps that differentiate it from the basic cheesemaking process, all of which help to make a firmer wheel.

The alpine cheese is started the same as a rennet cheese, by adding culture and rennet to warm milk. But once the curd has formed, the curd is cut to a much smaller size—around that of a lentil—to create much more surface area, which helps drain more whey and make a firmer curd when cooked.

The full-alpine cheeses covered in this book are:

- Comté
- Emmentaler
- grana

The alpine make can be summed as follows:

Warm the milk; add the starter; add the rennet; cut the curd small; stir the curd; cook the curd to high temperature; form the cheese; press the cheese; salt the cheese.

Full-alpine cheeses are made from curd cut small and cooked to a high temperature, with the entire vat's worth of curd transformed into a single, long-lasting wheel. PHOTO BY CHLOE GIRE

Next the curd is slowly cooked, to a higher temperature—toward 52°C (126°F)—which changes the texture of the cheese and gives it its alpine character. The pot is stirred nonstop as it is cooked, to distribute the heat throughout and prevent the curd from massing together prematurely.

Once it's fully cooked, the curd will form mountain peaks between your fingers and squeak loudly when chewed. Cooking the curd gives it this different character—it will stick to itself when pressed and have a certain degree of malleability, allowing the curds to be pressed together into a solid wheel of cheese. And it is this cooking, not necessarily making cheese on a mountaintop, that defines full-alpine cheeses.

Once you judge the curd to be at the right firmness, take the kettle off the heat. Leave the curd to settle for several minutes, so that all the small curds knit together. You then retrieve the curds from the vat full of hot whey with the help of cheesecloth and press them into the form, which often has a flexible shape to help achieve a cheese of a certain height—an important and defining characteristic of this style.

Comté is a cheese that's made to be moved down a mountain; cooked to 52°C (126°F) and formed into a massive wheel (about 50 kg). But the wheel is no more than 12 cm tall when made, and almost 1 m in width, to sit more comfortably on a cheesemaker's shoulders when carried down the valley below. The size and shape of the cheese influence its slow ripening. It's washed regularly to develop its orange glow.

Emmentaler is essentially a Comté made taller. It is a mountain cheese made instead in a mountain valley; unconstrained by its new geography, the alpine cheese could be made more massive, to be put on a cart to take to market. This taller cheese cannot be salted as effectively, which leads to the development of its characteristic eyes.

Granas, too, are alpine cheeses that descended from the mountain; they are among the most massive of cheeses and cooked to the highest temperatures of all the alpine styles—up to 55°C (131°F). This high-temperature cooking results in a tall cheese that doesn't swell with gas like Emmentaler but can be made just as massive and aged for many years. Granas are often aged in aboveground storerooms, which have drier conditions. They ripen solely from the interior, with a drying of the rind.

Other cheeses of this style include Gruyère—almost identical to Comté in its make (just about every mountain valley in the highest Alps has its own version of this style, with its own name). Abondance is made concave around its edge to be tied with a rope to the back of a donkey and taken down the mountain. Fontina, made on the Italian side of the Alps, is rennetted with a Southern European rennet, giving a stronger lipase character. Also in the family is pecorino, made like a grana, but with sheep's milk, and cooked only to 45°C (113°F). Parmigiano-Reggiano, too, is a grana, but one produced with 1,000 L milk to make two wheels.

Milled-Curd Cheeses

Milled-curd cheeses are a significant but often under-appreciated style (aside from the most famous member of its family, cheddar) made all around the world. This category includes some of the most ancient styles of cheese and may be the original way we made long-aged cheeses, for their production predates the invention of copper, which enabled the cooking of curds.

This family of fromages originates from the use of wooden vats that couldn't be heated and were the predominant cheeses in Europe before the development of copper vats. In wood, cheesemakers can only work

> The milled curd cheeses explored in this book are:
>
> - Lancashire
> - traditional cheddar
> - modern cheddar
> - Stilton
>
> The milled-curd make can be summarized as follows:
>
> > Warm the milk; add the starter; add the rennet; cut the curd; stir the curd; drain the curd; ferment the curd; mill the curd; salt the curds; drain the curds; form the cheese.

THE FAMILIES OF CHEESE • 213

The family of milled-curd cheeses includes (from left to right) cheddar, Lancashire, and Stilton, among others. PHOTO BY ALIZA ELIAZAROV

with the warmth of the animal, and they devised this fascinating technique to pull out moisture from within the cheese and improve its keeping capacity.

The purpose of the milled curd make is to create a firm cheese without added heat. Instead of cooking the curd to make a cheese firmer, as is the case with alpine styles, milled-curd cheeses are made firm by fermenting the curds, milling the curds, salting the curds, and then draining them before forming the final cheese. This method pulls out moisture from within the cheese, permitting a wheel to be made larger and kept longer than was previously possible with a make in wood.

The milled-curd make is typically performed in two stages, often over the course of 2 days. On the first day, the curd is cut, stirred, drained, and then left to ferment. On the second day, the now fermented curd is milled to smaller pieces, then salted, drained again, and assembled together back into a cheese.

If the milled curds are pressed firmly together, the cheese will take on a monolithic form, with little to no space between the curds, resulting in a long-keeping cheese like cheddar. If, however, the cheese is left unpressed, it will not knit together, leaving spaces within that will develop into blue veins, resulting from *Penicillium roqueforti* growing where air passes, especially if introduced through piercing.

Lancashire is a lesser-known but probably more traditional British territorial cheese that typifies the milled-curd make. Its method lies somewhere between that of cheddar and Stilton. It's a smaller, more approachable, more farmstead version of the milled-curd method.

Traditional cheddar is a very firm milled-curd cheese. This original cheddar technique makes a more "British" wheel, with more moisture, acidity, and a characteristic crumble. It is made with a variation on the milling method known as cheddaring, which uses the weight of the curd to press itself into a solid mass before it is milled, salted, and pressed into its final wheel.

The technique for modern cheddar is a more industrialized or "American" version of the cheddar make, which incorporates a slight cooking of the curd before draining, producing an even firmer and drier cheese curd that can be pressed into a more massive and longer-lasting wheel (or block), and that becomes arguably the most preservable of all cheeses. The make also yields cheddar cheese curds—the squeaky curds eaten for a snack or turned into poutine!

Stilton, a blue cheese from Britain, is a high-moisture milled-curd cheese that's left unpressed, giving it an open texture that results in *Penicillium roqueforti* growing throughout its crumbled curd. Curiously, the cheese is made somewhat like a Camembert. The curd is cut only vertically and ladled out of the vat, but instead of ladling into forms, it's ladled into cloth to drain. The following day that curd is milled, salted, and formed, making a large wheel that follows a fascinating journey to become a most esteemed blue cheese.

Other cheeses that fit into this category include: French Cantal, equivalent to cheddar but with a natural rind; Salers, made like Cantal but in a traditional wooden gerle; Italian castelmagno; French bleu de Termignon (more or less equivalent to castelmagno but made on the other side of the Alps and encouraged to turn blue); Mexican cotija; and the ancient Divle cave cheese from Turkey, one of a number of related cheeses made in the area from Bosnia to Turkey to Syria.

Washed-Curd Cheeses

Washed-curd cheeses are a class of cheeses wherein semi-firm rennet curds are washed with hot water to further firm them. Washed-curd cheeses are another traditional style that evolved from the historic use of wooden cheesemaking vats that couldn't be put over a flame. The contents of the wooden vat can only be heated by removing whey and replacing it with hot

---

The only washed-curd cheese in this book is:

- Gouda

The washed-curd make can be summarized as follows:

Warm the milk; add starter; add rennet; cut the curd; stir the curd; remove the whey; cook the curd with hot water; form the cheese; press the cheese; salt the cheese.

water. But, aside from this detail, the washed curd make is very closely related to the alpine cheesemaking.

To make a washed-curd cheese, milk is warmed, then cultured and renneted. Once the curd gives a clean break, it is cut into curds, and the curds are stirred until they firm to a certain degree. At that point, the curds are pitched and a portion of the cheese's whey is drawn off the top. That whey is replaced with an equal quantity of hot water at a certain temperature, and the curd is stirred until it reaches a greater firmness.

To make a Gouda, cut the curds to the size of a kernel of corn, and stir them until they firm as much as possible without cooking. The curds are then washed with water one or two times, bringing the temperature higher (to 45°C [113°F]) and making a firmer cheese. The firm curd is pressed into a mass at the bottom of the vat, and the whey removed. The curd is then placed into cheesecloth-lined forms to be pressed, then the cheese is submerged in brine to salt. Once thoroughly salted and drained, the cheese is typically aged in a dry cave, often sealed in wax.

Several other styles of cheese sometimes incorporate a washing process. Saint-Nectaire is occasionally washed with hot water to bring its temperature back to 35°C (95°F) if the temperature of the curds cools too low. This allows the curds to knit together better into a wheel when making the cheese traditionally in wood. Raclette is sometimes washed with warm water before cooking, or in place of cooking.

Half-Lactic Cheeses

Half-lactic cheeses are rennet cheeses that are more lightly handled, with less curd cutting and no curd stirring, resulting in a greater degree of fermentation and an almost lactic character.

These cheeses exist somewhere between the makes of rennet cheeses and lactic cheeses. Their name, half-lactic, precisely describes the methodology of the make. Prepare the curd much like a rennet cheese, but once you've achieved a clean break, cut the curd, then ladle it into cheese forms much like a more acidic lactic cheese. Half-lactic cheeses are often designated in France as *moulé à la louche* (ladle molded)—traditional Camemberts are very carefully ladled.

> Some half-lactic cheeses covered in this book are:
> - traditional Camembert
> - Époisses
> - Brie
>
> The half-lactic make can be summarized as follows:
>
> Warm the milk; add starter; add rennet; cut the curd; ladle the curd; salt the cheese.

Ladling the rennet curd without prior stirring changes the way the half-lactic cheeses evolve; the ladling is so important that they might also be called ladled-curd cheeses. With more moisture, the curd is slower to drain, and so it retains its higher moisture content. As a result, the cheese continues fermenting after salting, unlike rennet cheeses, where the salting nearly stops the acidification. The higher-moisture cheese achieves a lower pH—and this significantly changes the path of its ripening.

As a result of the greater acidity, the paste of these cheeses acidifies and softens in texture, and the ripening and breakdown changes. In this regard half-lactic cheeses again exist somewhere between rennet cheeses and lactic cheeses. Not softening from rind to core consistently like a modern Camembert, nor softening mostly at the rind like a lactic cheese, half-lactics take the middle road, often showing significant breakdown on the outer part of the cheese but retaining a small lactic core.

The softer curd also affects the development and appearance of the rind, and in this respect half-lactic cheeses again exist between rennet cheeses and lactics. Unlike lactic cheeses with their intensely wrinkled rinds, and lactic cheeses with their flat rinds, the rinds of half-lactics are often beautifully wavy, as epitomized by the washed-rind Époisses.

To make an Époisses, add starter culture and rennet to warm raw milk. Leave the curd to set until it just begins to show its whey, then ladle it directly into its forms. Allow it to drain for several hours until it is ready to salt; the cheese can then be flipped, drained overnight, then placed in a cave to age, washing with whey for 1 month to encourage proper ripening with a washed-rind ecology.

The make of a traditional Camembert is much like Époisses. It is post-make that the methods diverge and the two cheeses become distinct. After their make, the salted Camemberts are put into their caves then washed, but only for 1 week, to encourage the growth of fluffy white *Penicillium candidum* on the rinds.

Brie is a nearly identical make to Camembert, but the curd is ladled into a much larger form. The larger size changes the nature of the cheese and makes it one of the most challenging of all to execute because of the challenges of handling the sizable but delicate wheel.

Spun-Curd Cheeses

Spun-curd cheeses are often referred to as *pasta filata*, translated from the Italian as "spun dough." These celebrated rennet cheeses made around the world are fermented to develop their acidity, then plunged into hot water and stretched into shapes, much like bread dough.

The first stage of the pasta filata make is the standard rennet cheese make. When the rennet curd is fermented to the right acidity, it attains the magical ability to stretch endlessly when it's heated. The curd is brought to this state of stretchiness by submerging it in a quantity of very hot water. Lightly work the melted curd under the water to develop a smooth and plastic consistency. Then stretch the melted curd into myriad shapes and sizes, each of which becomes a distinct cheese.

At precise levels of acidity (a pH of 5.3) and at temperatures above 50°C (122°F), fresh curd loses its calcium. What was formerly a semi-firm cheese becomes a liquid and flowing curd that can be stretched into any shape, and it keeps that shape once it's cooled—a physical quality known as plasticity. The cooking not only changes the texture of the curd but also adds nuance and flavor and strips away acidity.

Mozzarella is the best known of the pasta filata; its make is also the simplest. To make mozzarella, milk is first transformed to curd through the addition of

> The pasta filata cheeses of this book are:
>
> - mozzarella
> - queso Oaxaca
> - stracciatella
> - burrata
> - caciocavallo
>
> The pasta filata make can be summarized as follows:
>
> > Warm the milk; add starter; add rennet; cut the curd; stir the curd; ferment the curd; melt the curd; shape the cheese; salt the cheese.

culture and rennet. The curd is cut to a walnut size and stirred gently until it can be easily held. The semi-firm curd is then left to sit in its whey and ferment until it develops its acidity, roughly 3 to 8 hours from the time of culturing, depending on the temperature. The curd is then drained of its whey, and hot water is slowly added to melt it. The curd is stirred gently with a wooden spoon or paddle for several minutes until it comes to temperature. Once it's supple and stretchy, the curd is ready to shape. To make the mozza, a handful of melted curd is stretched from the mass, and folded into its form, one after another, plunging each mozzarella into cold water to firm. Once cooled, the cheeses are transferred to an aging brine made of very lightly salted water.

Queso Oaxaca is a Mexican take on the mozzarella make. Much like mozzarella, Oaxaca is a rennet curd fermented to its ideal acidity and melted with very hot water. Once melted, the curd is shaped in a very different way: Instead of being formed into little buns like mozzarella, the curd is stretched into a long rope, after which it is heavily salted, then rolled into a ball. Though made in much the same manner as mozzarella, the different handling gives the cheese a different texture and flavor. The repeated kneading and pulling of the curd develops a stringy texture, unlike mozzarella, which is handled more delicately.

Stracciatella is a sort of Italian cottage cheese, made with pasta filata and cream. To make it, the melted

The spun-curd (or pasta filata) family of cheeses is made by fermenting curds to a certain degree, pouring hot water overtop to melt the cheese, and then stretching and shaping the curds into various shapes and sizes.

pasta filata curd is stretched like Oaxaca into lengths of rope. The cheese is then shredded and smothered in cream, fresh or fermented. Burrata is a delectable pasta filata cheese that is made essentially by stuffing a mozzarella with this stracciatella. An elaborate shaping technique, described in its technique, makes a perfect parcel of creamy curd.

Caciocavallo is a firmer pasta filata cheese. It is made through a more intensive, semi-alpine make that forces more moisture from the curd. The firmer curd is then fermented before it is milled, melted, then stretched into a cheese. With its firmness, the melted curd gives the shaping of caciocavallo more intensity and allows cheesemakers to create larger, more durable cheeses. These firmer pasta filata cheeses can also be aged for months or years, taking on a grana-like character because they're firmed during the many stages of the make. Caciocavallo's gourd shape is but one of many forms of a firmer pasta filata cheese: Variations include provolone (shaped like massive cylinders), Ragusano (massive bricks), and scamorza (small pears), which are often smoked.

Other cheeses of this style include the Armenian / Georgian majdouli / tenili, spun like a skein of wool (the technique is described in *The Art of Natural Cheesemaking*), and Venezuelan queso de mano, where the pasta filata cheese is essentially made flat in the hand, just like the region's arepas.

Heat-Acid-Coagulated Cheeses

These are the cheeses that don't fit into other categories. They are made with either milk or whey, and don't typically involve the use of rennet or culture for their make, but rather heat and acidity. Their curds are coagulated due to the sensitivity of milk's various proteins to a combination of heat and acid.

Sweet ricotta is the most famous of this group. It is prepared by taking whey fresh from a cheese, rich in albumin and cream, and bringing it to a boil. The thick curd made of the whey's solids rises to the top and can be skimmed off the surface and drained. It is best made with whey from an alpine cheese, still sweet, and with extra fat due to its small-cut curd.

> Some cheeses made with heat-acid coagulation include:
> - sweet ricotta
> - ricotta salata
> - paneer
> - mysost
> - kalvdans
>
> The heat-acid make can be summarized as follows:
>
> Boil milk or whey; add acid; strain off curds.

Fermented ricotta is prepared from fermented whey. It forms a softer curd that upon boiling can be separated from its whey only by straining all the whey through fine cheesecloth. The ricotta that results, however, is much finer and creamier in texture, and has a much more complex flavor and better keeping qualities than sweet ricotta.

Ricotta salata is an aged cheese made from ricotta. The ricotta is salted, dried, and placed in a cave, dry or humid, to age. The cheese can dry like a grana, or take on a mold-ripened character like a Camembert, depending on its treatment.

Paneer is a cheese made like ricotta, but with milk in place of whey. The yield, therefore, is substantially higher, and the texture much firmer, on account of the difference in amounts and types of proteins in milk and whey—primarily casein in milk, and only albumin in whey. To make paneer, milk is brought to a near boil, stirring all the while. A quantity of fermented milk is then added, and the acidity causes a destabilization of the milk's casein. The curd rises up, and the cheese can then be skimmed off the surface and drained until firm.

Mysost is a most rare and unusual cheese, unless you live in Scandinavia, where it is very common. It is made from whey, like ricotta, but cooked down until caramelized. To begin, fresh, sweet whey, preferably from an alpine cheese, is brought to a simmer; the whey is slowly cooked overnight, stirring occasionally. Come morning, the now-caramelized whey is cooked more intensely and stirred until it reaches the right consistency. The brown-colored, fudge-like cheese is cooled in brick-shaped forms and sliced extra thin to serve.

Kalvdans is also a dairy product from Scandinavia, though it's made in many other regions of the world with many other names. It's a fascinating "cheese" made only from colostrum, with added sugar and spice. The first milk is extremely high in albumins, much like eggs, and the colostrum curdles like an egg custard with only the slow addition of heat.

CHAPTER 12

Spring

Spring is the season of first cheeses. With animals finally freshening, milk begins to flow and cheesemaking can commence once again.

A maker can begin their cheesemaking year by conjuring a starter culture from these first drops of milk, leaving it to ferment until it turns to clabber. Kefir grains, dried to hibernate over the winter, can also be brought back to life. These cultures can be kept and fed through the year to furnish the fermentations of very nearly every cheese.

The first quantities of milk can be curdled with rennet to make junket, a springtime specialty; or fermented overnight into cuajada. The lactic cuajada can be then drained into a first fresh cheese, faisselle.

These lactic cheeses are well suited for springtime and ready to eat within days. They're nourishing and refreshing after a winter eating the long-preserved wheels of the previous summer. Fresh lactic cheeses also taste best when made in spring weather, and they work well in small batches made with the precious first milk, often shared with the calves, kids, or lambs.

The warming weather wrinkles the rinds of aged lactic cheeses like crottin, Valençay, and Saint-Marcellin, for *Geotrichum candidum* appreciates the growing daylight hours and the warmth they bring. But the still-cool temperatures are mild enough to assure these cheeses don't deliquesce prematurely and suffer from flies. Making these can be a challenge in the heat of summer!

Fresh eating cheeses like primo sale (and queso fresco and Halloumi) are the first rennet cheeses of the year, eaten just after salting. These rennet cheeses can also be aged, becoming bloomy white cheeses, like tomme Vaudoise, within weeks.

The blooming of the season's white rinds follows the first flowers in the pastures and hay meadows. *Penicillium candidum* grows well on cheeses made in the spring, gracing the rinds of Camembert, Brie, and Pont l'Évêque. These first rennet cheeses will be ripe in 2 months, ready to eat in the summer months ahead.

Clabber

Clabber is perhaps the simplest and most universally enjoyed dairy product—just milk, fermented. Everywhere on earth, milk transforms in this most delicious way; and though clabber is known by many different names around the world (filmjölk in Swedish, prostokvasha in Russian or Ukrainian, dickmilch—meaning thick milk—in German), all evolve from the same acidification and coagulation that result from milk's own fermentation.

Clabber, however, is actually a Gaelic word for this universal milk ferment, adopted into English from Irish immigrants to America who soured their milk instead of drinking it, as the Americans did. One of the few languages where the concept doesn't exist is English ("sour milk" doesn't quite fit) and it is indeed a strange, and probably misconceived, Anglo-American phenomenon to drink fresh milk, for the rest of the world ferments it.

It's a breeze to get into the habit of preparing this ferment daily, as your source of culture for the cheeses you make, or simply to eat as a delicious, sustainable, digestible, and nourishing food. You can prepare clabber by leaving milk to ferment spontaneously, but you'll get a much better result when you add an active starter culture, what I often refer to as clabber culture. This starter is a bit of clabber from a previous batch, or even some liquid kefir or active whey. The added culture initiates fermentation more immediately, helping suppress unwanted microbes and developing the most sought-after and balanced flavors in the ferment. Of course, you have to leave raw milk to spontaneously ferment to begin this clabber culture.

PHOTO BY ALIZA ELIAZAROV

Thinner than yogurt, clabber is not as easily enjoyed with a spoon. It can instead be shaken or stirred to break the delicate curd until its texture dissolves into a drink, much like strained kefir. When adding clabber culture as a starter to a cheese, it is best to break up its curd by shaking it before pouring it into the milk. Various flavorings or seasonal fruit, even a sprinkling of salt, can be added, too. Clabber can also be used in place of buttermilk (most commercial buttermilk is more akin in its production to clabber) for biscuits, scones, or pancakes.

Different animals' milks can ferment dramatically differently from one another. It's a lesson in milk's many possibilities to see the differences in aroma, flavor, and texture between cultured goat's milk and cow's milk, and even more so with cultured sheep's or buffalo milk, whose high solids yield an extra-thick curd with aromas and flavors evocative of the best aspects of those animals. And despite the differences between breeds and feeds and climates all around the world, cow clabber always tastes deliciously of cow, goat clabber of goat, sheep clabber of sheep, and buffalo clabber of buffalo. The inherent animal essence of a milk is best captured through its simplest fermentation.

You'll notice if you prepare clabber on a daily basis that a thin veil of velvety white fungus grows atop the culture, obscuring the reflection of light; if this is left for more than a day, it can grow quite thick and wrinkly. This *Geotrichum candidum* ecology growing atop is a sign that the culture is well tended and should not be a concern for the eater. Certainly, it's a sought-after quality for the natural cheesemaker, for it's an indicator of the microbiological diversity of the starter and a window into the endless cheesemaking potential of this incredible, universal culture.

Making Clabber

A fermented milk • 12- to 24-hour make

INGREDIENTS

milk	250 ml
starter	5 ml clabber (1:50) from previous batch

YIELD   250 ml CLABBER (1:1)

TECHNIQUE

Bring milk to culturing temperature, about 20°C (68°F).
Pour starter into a jar and mix to break it up.
Pour milk over the active starter and mix well.
Wait 12 to 24 hours at about 20°C, until the milk sets to clabber.
- Cover jar lightly.
- Leave undisturbed.

Reculture the clabber in fresh milk.
- Or keep refrigerated for up to 1 week.
- Or freeze for up to 1 month.

NOTES

The milk you use can be cow's, goat's, sheep's, or buffalo's; pasteurized or raw; low-temp-pasteurized, high-temp-pasteurized, and even ultra-pasteurized; homogenized or unhomogenized or even powdered. But the freshest raw milk, still warm from the udder, will give the best, most delicious results.

To begin a clabber culture, leave fresh raw milk to ferment for 2 to 4 days until thickened and sour, then re-ferment daily with raw or pasteurized milk until stable. Typically it will take about 3 days to 1 week to make a healthy and stable starter that can be used for cheesemaking. Clabber can also be begun with kefir or whey, or even a bit of sourdough starter.

Clabber can be fed cold milk, but the cold temperatures will slow the fermentation. Warm milk, straight from the udder, hastens the fermentation and gives better results.

If temperatures are warm (greater than 30°C [86°F]) and milk ferments quickly (within 8 hours), place clabber in the refrigerator within 12 hours of culturing to prevent overfermentation, or feed culture twice daily.

Kefir

Another form of fermented milk, kefir is a type of "clabber" prepared with the help of kefir grains, a mysterious and ancient culture that's been passed down from generation to generation for millennia. An extraordinary source of culture for fermentation, kefir grains can make the natural fermentation of dairy products accessible even to those without access to raw milk.

Strange but true, kefir grains are visible microbial manifestations comprising dozens (if not hundreds) of species of beneficial milk-fermenting bacteria and fungi that are cultivated and carried forward (or backslopped) by a daily feeding of milk. The grains themselves are a polysaccharide matrix constructed by certain members of the community that provides a home for the many other species in the culture. The ever-growing grains give the culture a distinguishing form, and this form helps cheesemakers better grasp the fermentation that makes all cheesemaking possible.

Kefir, Turkish for "good feeling," will also help you feel good all over. Perhaps the most diverse source of probiotics for gut health (though likely similar to clabber), kefir contains an incredible diversity of beneficial microorganisms. A daily drinking helps seed an abundance of microbes that survive the journey through the gastrointestinal tract intact and establish themselves better than laboratory-raised strains. There may be no better "starter" for your gut's own fermentation!

Known by many different names in many different places, kefir is called the milk plant in the Netherlands, *bulgaros* in Mexico, and *pacharitos*—little birds—in Chile, for the grains' tendency to float in the milk as they ferment it (if they're always at the bottom of the milk, they're not happy!). The culture seems to have a global distribution, and may have had significance in many cultures before the modern cheesemaking establishment nearly obliterated traditional cheesemaking practices.

There are many ways to make kefir, but I recommend a certain way to prepare the culture for optimal flavor, easy management, and best cheesemaking quality (and gut health!). The method of keeping kefir described here yields a ferment that cultures in about 12 to 24 hours at moderate temperatures; that grows a thin coat of *Geotrichum* atop; that does not separate or develop gas (two generally negative signs in fermentation, and problems that are not sought after in either cheesemaking or the gut); and that has a delicate but complex flavor. The flavor can be quite different if different species' milks are fermented, much like clabber—my, and most others', favorite is goat.

Kefir grains should be fed in a ratio of 1:50; 1 part grains to 50 parts freshest possible milk. The kefir grains should be left unwashed, to assure the best backslop and the fastest fermentation; and the kefir should be left undisturbed during its fermentation. Resist the urge to shake, which will mix in air and encourage yeasts and subsequent gas formation, unless that's what you want!

It's important to preserve this 1:50 ratio as the grains grow. If all the grains are preserved and still fed the same amount of milk daily, the culture will come to over-ferment daily; the result will be yeasty and gassy, and it does not make good cheese. As they double in size every 10 days or so, excess grains should be preserved by drying or freezing, or given away as gifts to those that might appreciate them. I happily give away excess kefir grains, but if you can't find a home for yours, don't hesitate to compost them—they're good for the soil! For most home kefir makers I recommend keeping only a tablespoonful—enough for between 250 ml and 1 L daily kefir—and about 20 ml of grains per liter of kefir to be prepared for commercial producers, preferably preparing it daily.

The kefir grains must also be active and regularly fed to ferment well. If the grains are new to you, or if they've been neglected, the first fermentation can be slow, but re-ferment them once the kefir has set and their second fermentation will be an improvement. Within three feeding cycles, waiting until the milk thickens and sours between feedings, the grains should be fermenting at their peak. When in their finest form they will sour milk in about 12 hours at 20°C (68°F).

A daily kefir feeding helps keep the culture in top form and assures a quick and effective fermentation every time. I consider my kefir grains to be like little goats, who appreciate their fresh milk feeding every morning. Leaving the kefir 2 days between feedings will result in overfermentation, a subsequent decline in lactose-fermenting bacteria, and a proliferation of lactic-acid-fermenting yeasts; using such a culture to ferment milk will result in an improper fermentation with slow acidification and gas production from coliforms and yeast. This condition can be remedied by re-fermenting on a daily basis or refrigerating (for up to 1 week) the kefir, once thickened, to prevent overfermentation. Freezing the strained kefir can also preserve the culture's best qualities without the need for a daily feeding.

Kefir can be intentionally put through a secondary fermentation to develop even more flavor and effervescence. After straining kefir of its grains, pour it back into its fermentation vessel, adding honey, maple syrup, or fresh fruit or preserves, mixing thoroughly and sealing the lid tight. Shake the kefir daily to encourage yeast development. After a week in the fridge, the kefir's yeasts will convert the added sugars to bubbly carbon dioxide and small quantities of alcohol, improving the ferment's feel-good factor. This can be done even more naturally, without the addition of sugar, through following the method for amasi or milkbeer in chapter 14.

Kefir can also be prepared by fermenting milk with only the liquid kefir as a starter, in a ratio of 1:50 to 1:100, instead of kefir grains. But the end result is closer to clabber, equivalent in every aspect to kefir, except for one thing I cannot fully explain: the mysterious kefir grain.

Making Kefir

A fermented milk • 12- to 24-hour make

INGREDIENTS

milk	1 L
starter	20 ml kefir grains (1:50)

YIELD

1 L KEFIR (1:1)

TECHNIQUE

Bring milk to culturing temperature, about 20°C (68°F). Place the kefir grains into a jar and pour milk overtop. Wait 12 to 24 hours, until the milk sets to kefir.
- Cover jar lightly.
- Leave undisturbed.

Pour the kefir through a strainer to separate grains.
- Shake or stir the kefir well to break up its curd before straining.
- Handle lightly so as not to break grains.

Reculture kefir grains in fresh milk.

Keep kefir refrigerated for up to 1 week to preserve.

NOTES

The milk you use can be cow's, goat's, sheep's, or buffalo's; pasteurized or raw; low-temp-pasteurized, high-temp-pasteurized, or even ultra-pasteurized; homogenized or unhomogenized. Kefir grains are less picky about their milk than we are and will tolerate any milk with lactose—but bear in mind that most of these processed milks will only work for making kefir or clabber. Most cheeses in this book are best made with raw milk or low-temperature-pasteurized milk, and those milks should be as fresh as possible. Best results for kefir, however, come from using milk still warm from the udder (or, boiled as for yogurt, and then cooled).

Kefir can be fed cold milk, but the cold temperatures will slow the fermentation. Warm milk, straight from the udder, will hasten the fermentation. I typically add my culture to milk as fresh as possible for best effect, and often perform this act of culturing as part of the morning milking process to ensure regular practice.

To preserve kefir grains between infrequent feedings, it is best to leave them submerged in their liquid kefir under refrigeration, or to dry them.

Junket

This is a delicious dish made by coagulating fresh milk into curd. An introductory lesson on renneting milk, this method is the jumping-off point for nearly all renneted cheese styles in this book. Its name is telling: *junket* derives from the Latin *juncus* or French *jonces*, referring to a type of marsh reed common across Europe (and that grows globally), which was once (and still can be) woven into baskets and used to drain fresh rennet curd into cheese.

Junket is made simply by adding the curdling enzyme rennet to milk, then waiting for the curd to set before eating. This simple step makes milk much more savory and eminently more digestible; this dish was formerly recommended as a healing and nourishing food to infants being weaned from their mothers' milk, to the elderly, and to those overcoming illness (many people of a certain generation have an aversion to it as they were forced to eat it in their youth!).

Adding rennet to milk completes it. Milk is made for this miraculous transformation, its casein proteins perfectly partnered to the chymosin enzyme in rennet. When added to milk in a warm and slightly acidic environment, rennet turns it to a semi-solid gel, aiding its breakdown and digestion within the gut, through the stomach's further churning and acidification. Junket is representative of the first stage of this digestion, an adjustment and rearrangement of milk's proteins that renders milk more healthful and less burdensome on our digestive tracts.

The milk to be curdled into junket is often first sweetened with sugar or honey and flavored with cardamom, nutmeg, cinnamon, or other milk-appropriate spices, turning the dish into a delicious pudding. But for the purists out there, junket is perhaps best eaten as is, with milk's natural sweetness accentuated by the slight fermentation brought on by the starter. And as the curd sits and continues fermenting, it slowly evolves its whey, becoming ever thicker and more delicious. Left to ferment overnight, it becomes cuajada, as described in the next technique.

Often considered too dangerous to make with raw milk, this dish can be safely made if the milk is as fresh as possible, the animals healthy, and a small amount of active starter added. Starter culture is not generally added to junket, but I call for its addition for its protective effects, as well as for its contribution to junket's character—the added acidity aids the formation of the curd, and the short fermentation improves the flavor of the junket.

Likely what Miss Muffet ate on her tuffet, this pudding was once eaten across the British Empire and in many other parts of Europe. But this former first food is now relegated to history books, Shakespearean tragedies, and nursery rhymes, as the good milk needed to make it is now largely inaccessible.

Making Junket

A rennet cheese • 1-hour make

INGREDIENTS

milk	1 L
spices	ground cardamom, cinnamon, and / or nutmeg to taste (optional)
sweetener	sugar or honey to taste (optional)
starter	20 ml kefir, clabber, or whey (1:50)
rennet	regular dose calf rennet
small clay pots	200 ml in size
Yield	**1 L junket (1:1)**

TECHNIQUE

Bring milk to cheese temperature, about 35°C (95°F).
Add spices and sweetener to taste, if using.
Add starter, mix in thoroughly.
Add rennet, mix in thoroughly.
Pour the mixture into clay pots.
  ▸ Don't wait too long, as the milk will begin to set!
Wait about 1 hour, until the milk sets to curd.
  ▸ Cover the pots and leave undisturbed.
Keep refrigerated to preserve, or ferment into Cuajada.

NOTES

As with rennet cheeses, the milk used should be the freshest raw milk or fresh low-temperature-pasteurized (bulk pasteurized) milk. But boiling milk then cooling before culturing and renneting will improve the junket's flavor and texture. Other rennet cheeses that are drained should only be prepared with untreated or lightly treated milk, as the process of syneresis is significantly hampered by milk's cooking.

Cuajada

Cuajada is the most basic rendition of a lactic cheese. It is prepared much like kefir or clabber, by adding culture to milk, but in addition, rennet is added. This causes the milk to first set like junket, then soften significantly over a daylong fermentation from the added starter. This soft and acidic curd is the first stage of the make for all lactic cheeses, but instead of ladling the cuajada to drain in forms, the fresh curd is eaten right from a pot, like yogurt.

Its Spanish name (the word means "curdled") refers to the texture of this dairy product, which is somewhere between yogurt and cheese, with a silky mouthfeel and a delicious sour tang. It can be prepared with goat, sheep, cow, or buffalo milk, but each will yield a different cuajada. Goat's milk and sheep's milk shine brightest with this make, for fermented without rennet their curds are disappointingly runny. But in cuajada, their curd is bright white, semi-firm, and intensely flavorful. Some goat yogurt makers decide to make this instead of their usual yogurt; if you do so, ferment it thermophilically like yogurt.

It should be quite easy to tell once the cuajada curd has gone lactic. It transitions from a rennet curd to a lactic curd at a pH of 4.5 when (like clabber or kefir) it tastes mildly sour and suddenly softens. The curd is stronger and more jiggly when it's not yet fermented, but once lactic it becomes brittle. The curd should also sink under a slight layer of whey, and a growth of *Geotrichum candidum* should be visible atop that whey, when the light catches it just right. To avoid the excess growth of fungus on this fresh ferment, get the cuajada refrigerated as soon as it has turned lactic, or consider a thermophilic fermentation instead.

Boiling the milk before curdling is an optional step, but it improves the character of the cuajada. Cooking the milk softens the texture of the curd through a breakdown of its casein proteins, and thickens the curd due to denaturing of the milk's albumins. The cooking also keeps whey from separating, creating a more stable product that is more consistent in texture and sweeter in taste.

This delicious "yogurt" is eaten all around southwestern Europe and Latin America. In Basque country it is known as mamia, and in Brazil and Portugal as caolhada. There are even versions of this cheese made in North Africa known as raïb, although they're commonly curdled with cardoon rennet in place of an animal enzyme. (Cuajada may have made its way to Spain and Portugal, along with the goats and sheep that make it best, from North Africa during the Iberian Peninsula's Muslim period.)

All of these curdled milks are sold traditionally in small, unglazed clay pots, and to make them right, you should leave your milk to set and ferment in clay, too. The clay cools the ferment, lowering the temperature through evaporative cooling, resulting in a more diverse microbiological expression with more elaborate flavors. And perhaps most important, the porous pots hold on to the culture from one batch to the next, enabling the ferment to succeed whether you recognize the microbiology of milk, or, like our ancestors, not!

Making Cuajada

A fresh lactic cheese • 12- to 24-hour make

INGREDIENTS

milk	1 L
starter	20 ml kefir, clabber, or whey (1:50)
rennet	¼ dose rennet
small clay pots	200 ml in size

Yield 1 L cuajada (1:1)

NOTES

Cuajada can also be eaten as a curdled but unfermented dairy product, like junket. But is more delicious, better keeping, and more nourishing once it is fermented.

TECHNIQUE

Bring milk to a boil, stirring nearly nonstop.
Cool the milk to culturing temperature, about 20°C (68°F), stirring nearly nonstop.
Add starter, mix in thoroughly.
Add rennet, mix in thoroughly.
Pour the mixture into pots.
Wait 12 to 24 hours, until the milk sets to a lactic curd.
- Cover the pots.
- Leave undisturbed at about 20°C.

Keep refrigerated to preserve.

PHOTO BY ALIZA ELIAZAROV

Chèvre / Brebis / Vache / Bufflonne

Certainly the most delicious of the fresh cheeses, chèvre, fresh goat cheese, and brebis, fresh sheep's milk cheese, are exceptionally easy styles to both make and devour. The lesser-known vache and bufflonne are equivalent cheeses made with cow's milk and buffalo milk, respectively, all of these referring to the French names of the animal that provided the milk. The French are adroit for simplifying cheeses' names as such, and, fittingly, these cheeses are perhaps the most expressive of each milk's fullest fermented flavors.

To make these simple soft cheeses, the fresh milk is renneted and cultured, then left to ferment and curdle overnight. As the curd increases in acidity it becomes lactic, developing a soft and chalky texture, and sinking beneath a puddle of its whey. To finish the cheese, the curd is transferred to cheesecloth and drained of its whey by hanging for several hours or overnight, before being salted then kneaded like bread dough. The salted curd is left to hang again to drain more whey for several hours more to finish the make.

A slow and undisturbed fermentation and coagulation at low temperature is what defines the make of these fine cheeses. Left to ferment for 12 to 24 hours, the milks take on complex flavors and sublime textures that are reflective of their great degree of digestibility and exceptional nutrition. The soft lactic curd is then concentrated by its draining and salting, resulting in a fresh cheese that is a distillation of all the best essences of its milk. This cheese can be prepared with cow's milk, but the flavor will never compare to the complexity of the goat or sheep versions due to its differing chemistry. The buffalo version may be the best of the beasts.

Renneting at 20°C (68°F) makes a more delicate lactic curd, so if you're using the freshest milk from the udder, it's best to let the animal warmth of the milk dissipate before adding the coagulant. The culture, however, can be added to the milk as soon as it is taken from the animal to assure its best fermentations.

You can take a slower pace with the draining and salting of these cheeses, but if you take too much time, they will begin to take on a strong yeastiness and show the growth of *Geotrichum*, especially in warmer weather. When making an aged lactic cheese, this is desirable, but to preserve the fresh taste of chèvre or brebis or vache or bufflonne, it's important not to let the fungus get too established, and a shortened make assures that. If the weather is warm, the make should be even shorter.

To prepare this or any other fresh or aged lactic cheese with cow's milk or buffalo milk, the make should be altered slightly because of the different nature of these milks. Goat's and sheep's milk do not easily separate their cream, so you can add a smaller amount of rennet, which causes the curd to set more slowly and delicately, without a loss of fat. With cow's or buffalo milk, a greater rennet dose must be added to encourage a quick set and reduce the cream loss. If you add a lesser rennet dose, you'll find a substantial layer of thickened crème fraîche atop a lactic skim milk curd, which you can skim from the top and mix back into the draining curd.

Making Chèvre / Brebis / Vache / Bouflonne

A fresh lactic cheese • made with goat's, sheep's, cow's, or buffalo milk • 24- to 48-hour make

INGREDIENTS

milk	4 L
starter	40 ml kefir, clabber, or whey (1:100)
rennet	¼ dose kid or lamb rennet for goat's or sheep's milk or regular dose calf rennet for cow's or buffalo milk
salt	

YIELD 600 G CHÈVRE / VACHE / BUFFLONE (7:1)
1 KG BREBIS / BUFFLONE (4:1)

NOTES

The curd can also be left unsalted, to keep its most delicate flavor, though its keeping capacity is reduced. This unsalted fresh lactic cheese is often known simply as curd.

TECHNIQUE

Bring milk to culturing temperature, about 20°C (68°F).
Add starter, mix in thoroughly.
Add rennet, mix in thoroughly.
Wait 12 to 24 hours, until the milk sets to a lactic curd.
- Cover the pot and leave undisturbed at 20°C.
- The curd will sink under its whey and become firm but brittle, and soft and acidic on the tongue.

Scoop the curd into cheesecloth.
Hang the curd in cloth to drain for 6 to 12 hours.
Knead salt into the curd to taste.
Leave to drain for another 1 to 2 hours (optional).
Keep the cheese refrigerated to preserve.

PHOTO BY ALIZA ELIAZAROV

Crottin

Emblematic of the French traditions of making aged lactic cheeses, crottin is a small lump of a raw goat's milk cheese, left to age until it's good and moldy. Though typically made with goat's milk because of the more intriguing flavors that develop in its relatively young cheeses, crottins can also be made with other milks; just increase the rennet dose if you're using cow or buffalo milk.

The make of this aged lactic cheese isn't all that different from a fresh lactic one: it more or less begins as faisselle. Fresh raw goat's milk, preferably still warm from the udder, is cultured and lightly renneted, then left to ferment for 24 hours. The lactic curd settles under the whey, becoming covered with a thin veil of *Geotrichum candidum* (just barely visible, obscuring the reflectivity of the surface of the whey), which will later ripen the cheese. The silky curd is then carefully ladled from the pot into forms like faisselle and allowed to drain for a day before it is salted on its surfaces.

It is the aging of the cheeses that sets the crottin apart. A careful two-stage regimen follows the make to assure the best growth of crottin's wrinkly *Geotrichum* rinds and the proper ripening of the cheese. First comes a hastening, in a warm and humid environment, which is followed by an affinage in a cool and humid cave. The first, warmer aging stage helps to establish the *Geo* on the rind and helps give the cheese its distinctive wrinkles. The second, cooler aging stage helps the fungus slowly break down the paste, giving the cheese its typical two-tone texture with a firm and chalky core and a deliquescent mantle. The acidic nature of the cheese resists the ripening and the changing pH and keeps the curd breakdown limited to the rind.

Most of the aged lactic cheeses are treated in this manner. Their simple make and foreshortened affinage—a result of their high moisture content—typically takes only a couple of weeks to 1 month, making these perfect cheeses for an early-season cheesemaking. This also makes them ideal for a beginning cheesemaker without much experience in ripening. I do recommend that cheesemakers perfect these simpler styles before attempting larger, longer, and more complicated makes and affinages.

To age crottins or other aged lactic cheeses longer, and develop even more exquisite textures and flavors, keep the cheeses in a drier cave after 1 week of a more humid affinage, and ripen them until they are dried crumbs of themselves and become crottin sec. (If they're kept longer than 1 month in a humid cave, these shaped cheeses will soften excessively and lose their shapes.) Taken to the extreme, some makers leave their crottin secs in caves for over a year, to be consumed nearly to the core by mites!

Making Crottin

A lactic cheese • made with goat's milk • aged with *Geotrichum* • 48-hour make • 2- to 4-week affinage

INGREDIENTS

milk	10 L
starter	100 ml kefir, clabber, or whey (1:100)
rennet	¼ dose kid rennet with lipase
salt	
forms	pots: 8 cm diameter × 12 cm deep

YIELD   EIGHT 200 G CROTTINS (7:1)

TECHNIQUE

Bring milk to culturing temperature, about 20°C (68°F).
Add starter, mix in thoroughly.
Add rennet, mix in thoroughly.
Wait 12 to 24 hours, until the milk sets to a lactic curd and *Geotrichum* veil grows.
- Cover the pot.
- Leave undisturbed at 20°C.

Ladle whey off from above the curd.
Carefully ladle the curd into forms, topping them up once or twice as they drain.
Leave the curd to drain in forms for 24 hours, until *Geotrichum* blooms.
- Flip cheeses after 12 hours.

Salt the cheeses on all sides.
Drain cheeses for 1 day on draining table, covered with cheesecloth.
- Flip cheeses after 12 hours.

Age cheeses for 2 to 3 days in hastening space until *Geo* wrinkles the rind.
- Flip cheeses once daily.

Age cheeses in cave for 2 to 4 weeks.
- Flip cheeses every other day.

Cheese may be dried at 70 percent humidity after 1 week of aging into crottin sec.

Valençay

A pyramid-shaped cheese from the Loire region in central France, Valençay is the most elegant of the many cheeses made in this celebrated center of goat's milk cheesemaking. It is a simple aged lactic cheese, but several special treatments make it stand apart.

What makes the Valençay is its unique shape and color. Valençay is ladled into pyramidal forms, giving it an elegant stature, and is then painted black with charcoal before its fungus grows to give it an even more formal appearance. As the *Geotrichum* blossoms on the faces of the cheese, it creates beautifully intricate black-and-white wrinkles.

The handling of Valençays as they are made and aged requires great attention to detail. Be careful to ladle the lactic curd without breaking it. Handle the cheese only by its corners; leaving the faces of the pyramid untouched allows the *Geotrichum* to fully express itself and develop the most intricate wrinkling. As the corners

PHOTO BY ALIZA ELIAZAROV

are repeatedly handled, the fungus is pressed, leaving straight black lines that define this breathtaking cheese. It is best to never flip the fromage, carefully lifting it off its bottom regularly instead, to keep it from fusing with its mats. The lack of flipping also helps the cheese develop very well-defined wrinkles on its upward-facing sides; a washed-rind ecology often develops on its soggy bottom.

Legend has it that Napoleon created Valençay on a trip through the Loire Valley, while returning to Paris after his failed military campaign in Egypt. Filled with rage upon seeing a pyramid-shaped cheese for sale at a market in a town called Valençay, he pulled out his lance and lopped off its top. More believably, though, the shape of Valençay evolves from the traditional use of pyramid-shaped pottery forms that couldn't be made with points, as they'd fall over! The pyramid shape is an elegant one that is carefully crafted using the slab method of construction. And unlike a perfect pyramid, the truncated shape of the clay Valençay form allows it to stand and receive its curd without falling—an extraordinary example of a cheese being defined by the natural materials that make it.

Making Valençay

A lactic cheese • made with goat's milk • aged with *Geotrichum* • 48-hour make • 2- to 4-week affinage

INGREDIENTS

milk	10 L
starter	40 ml kefir, clabber, or whey (1:100)
rennet	¼ dose kid rennet with lipase
salt	
fine charcoal	
forms	truncated pyramids: 12 cm deep
YIELD	SIX 250 G VALENÇAYS (7:1)

TECHNIQUE

Bring milk to culturing temperature, about 20°C (68°F).
Add starter, mix in thoroughly.
Add rennet, mix in thoroughly.
Wait 24 hours, until the milk sets to a lactic curd and *Geotrichum* grows.
- ▸ Cover the pot.
- ▸ Leave undisturbed at 20°C.

Ladle the curd into pyramid forms.
Leave the curd to drain in forms for 24 hours, until *Geotrichum* blooms.
- ▸ After 12 hours, remove cheese from form and put back in.
- ▸ And then press down lightly on the cheese to achieve a level surface.

Salt cheeses on all sides except pyramid top.
Drain cheeses for 1 day on draining table, covered with cheesecloth.
- ▸ Carefully lift cheeses after 12 hours.

Apply charcoal, moistened to an ink with water or whey, to all surfaces except pyramid bottom.
Age cheeses for 2 to 3 days in hastening space until *Geo* covers rind.
- ▸ Lift cheeses once or twice daily.

Age cheeses in cave for 2 to 4 weeks.
- ▸ Lift cheeses every other day.

Cheese may be dried after 1 week of aging.

Le Sein

This beautiful, breast-shaped cheese is instructive in a different way of handling aged lactic cheeses: pre-draining. The method is undesirable to some extent, because it changes the usual cloven interior texture that comes from ladling to a uniform crumbliness and exposes the cheese to the degradations of air, but this also frees cheesemakers from the confines of a form and allows them to shape a cheese to their will. Essentially a free-form cheesemaking, pre-draining lactic curd allows cheesemakers to incorporate an incredible artfulness to their craft.

Cheesemakers can form this pre-drained curd into nearly any shape, and age it as any other lactic cheese. In French cheese shops it is not uncommon to see pre-drained aged lactic cheeses in all sorts of shapes, from the patriotic to the agricultural and even anatomical, including fleurs-de-lis, Eiffel Towers, and little goats, sheep, and cows. Extra detail can be added with edible paints. I paint my little lactic cows with spots of charcoal to make them look like little Holsteins!

Free-forming lactic cheeses calls for a balance of speed and patience in their production so that they grow proper *Geotrichum* rinds and don't swell due to gas production in their curd. Goat's milk, preferably, as with all lactic, is slowly and mesophilically fermented, and curdled with a slight dose of rennet. The curd is left to ferment overnight until lactic, with the slightest growth of a *Geo* veil. Scoop the curd into cloth and drain overnight until the dripping stops, and a slight yeasty slipperiness can be felt on the cheesecloth. Then salt the drained curd to pull much of the remaining moisture out. The cheese that results is clay-like in texture, and can be shaped in all sorts of creative ways.

And what better shape to give the cheese than a breast, a symbol of the source of incredible nourishment and life force that is milk? As for those wrinkles: They're an important aspect of the microbiology of that milk, indicative of a creamy and flavorful curd within, not to mention an arresting addition to the finely formed cheese that should be appreciated by all as a sign of the good taste and quality of character that come with age.

Making Le Sein

A lactic cheese • made with goat's milk • aged with *Geotrichum* • 48-hour make • 2- to 4-week affinage

INGREDIENTS

milk	10 L
starter	100 ml kefir, clabber, or whey (1:100)
rennet	¼ dose kid rennet with lipase
salt	

Yield five 300 g Seins (6:1)

TECHNIQUE

Bring milk to culturing temperature, about 20°C (68°F).
Add starter, mix in thoroughly.
Add rennet, mix in thoroughly.
Wait 24 hours, until the milk sets to a lactic curd and *Geotrichum candidum* blooms atop whey.
- ▸ Cover the pot and leave undisturbed at 20°C.

Scoop the curd into cheesecloth.
Leave the curd to hang in cloth for 24 hours.
- ▸ Leave to drain until *Geotrichum* is just barely visible blooming on cloth.

Add salt into cheese to 2 percent or to taste.
- ▸ Knead with the help of cheesecloth until smooth.

Leave the curd to hang another 12 hours.
- ▸ Finished curd should have a modeling clay consistency.

Shape the curd into desirable form by hand.
Age cheeses for 2 to 3 days in hastening space until *Geo* covers rind.
- ▸ Lift cheeses once or twice daily.

Age cheeses in cave for 2 to 4 weeks.
- ▸ Lift cheeses every other day.

Cheese may be dried after aging 1 week to preserve.

PHOTO BY ALIZA ELIAZAROV

PHOTO BY ALIZA ELIAZAROV

Saint-Maure

This aged goat cheese pushes the limits of lactics, for it's a lactic cheese that's made longer than its soft and brittle texture usually permits. Just as alpine cheeses must be made massive to evolve right, ladled lactic cheeses must be made small so that the high-moisture curd drains well, and so the delicate young cheeses don't fall to pieces when handled. Such a lengthy lactic as Saint-Maure would normally break in the middle with the slightest of handling, but a cleverly pierced straw through the center of the long cheese acts as reinforcement, allowing it to be easily flipped, moved, and sold. As a bonus, the straw makes for a beautiful reveal when the cheese is cut, further distinguishing this beauty from all others.

Saint-Maure can be seen as a celebration of straw, a functional natural material undervalued in modern times or at least since the first telling of "The Three Little Pigs"! Straw, as a by-product of grain production, once served many roles in cheesemaking and beyond, including providing absorptive bedding to milking animals in the wintertime, aiding the forming and draining of cheeses on their draining tables, helping the developments of soft cheeses' rinds in the cave, and providing place-based packaging material for the shipment of delicate cheeses.

Saint-Maures are ladled into extra-tall cheese forms to create their long shape, and not removed from these forms until the draining is complete. The straw is carefully pierced through the center of the curd after its make, but before its salting. The straw itself should be hand-harvested and dried in the late summer, and cut between nodes to preserve its strength and aid the piercing.

Making Saint-Maure

A lactic cheese • made with goat's milk • aged with *Geotrichum* • 48-hour make • 2- to 4-week affinage

INGREDIENTS

milk	10 L
starter	100 ml kefir, clabber, or whey (1:100)
rennet	¼ dose kid rennet with lipase
salt	
forms	hoops: 6 cm diameter × 30 cm tall
pieces of dry straw	20 cm in length

Yield five 300 g Saint-Maures (6:1)

TECHNIQUE

Bring milk to culturing temperature, about 20°C (68°F).
Add starter, mix in thoroughly.
Add rennet, mix in thoroughly.
Wait 24 hours, until the milk sets to a lactic curd and *Geotrichum* grows.
  ▸ Cover the pot.
  ▸ Leave undisturbed at 20°C.
Ladle the curd into cheese forms.
  ▸ Top up forms as cheeses drain.
Leave to drain 24 hours without flipping.
Remove cheeses from their forms.
Pierce with straw through their centers.
Salt cheeses to 2 percent all around.
Drain 24 hours after salting.
Age cheeses for 2 to 3 days in hastening space until *Geo* covers rind.
  ▸ Lift cheeses by their straw daily.
Age cheeses in cave for 2 to 4 weeks.
  ▸ Lift cheeses by their straw every other day.
Cheese may be dried after aging 1 week to preserve.

Saint-Marcellin

Saint-Marcellin is the most spoonable style of cheese, for it is a soft lactic cheese that has the longest and most complete deliquescence. The lack of curd cutting keeps the full share of cream and moisture within the curd; combined with its high-moisture lactic style and the fullest extent of proteolysis possible, the result is a velvety and unctuous cheese with a delicate flavor, a mix of flowers and flesh.

Most lactic cheeses are eaten within 2 to 4 weeks, while they still stand tall and the curd hasn't broken down excessively from fungal fermentation at the rind. The delicate *Geo* doesn't quite hold the liquidy cheese together for long, and the paste turns to a puddle in the cave. Some lactic cheeses, like crottin sec, are allowed to dry to preserve them without excess deliquescence, but Saint-Marcellin is left to age in a ripening cave longer than other lactics, because the pot in which the cheese is ripened helps keep it from flowing away as it ages.

This is also the only aged lactic cheese that's characteristically made with cow's milk, which is suited to its make because of its longer ripening. Two- to 4-week-old lactic cheeses are generally more flavorful when made with goat's milk (or sheep's, or buffalo) as a result of the more flavorful fatty acids developed through the full lactic fermentation of those milks; cow's milk lactic cheeses eaten at the same young age are, frankly, lacking in character, and therefore not as valued. But age that same cow's milk lactic cheese about 2 months and something magical happens: The lactic cow's milk cheese achieves a fine flavor on par with its caprine counterparts.

The technique here uses a ladled curd—*moulé à la louche*. A small lactic cheese is made, hastened, and ripened for 2 weeks until it begins to soften. But before the cheese loses its shape, the small wheel is placed into a clay pot to help it stand tall and ripen fully. The cheese can also be pre-drained and salted, as per le sein, which makes the curd malleable, and pressed into any size of aging container, be it a mason jar (through which the ripening and breakdown can be observed) or an earthenware pot (as described in *The Art of Natural Cheesemaking*). But pre-draining the curd mixes in air, which degrades the quality of the cheese.

Like Valençay, this is a cheese defined by clay: The technique of aging the little lactic in a small pot is essential for realizing its great leap in flavor. A cow would jump over the moon if she could enjoy this incredible elaboration of her milk.

Making Saint-Marcellin

A lactic cheese • aged with *Geotrichum* • 48-hour make • 4- to 8-week affinage

INGREDIENTS

milk	10 L cow's or goat's milk
starter	40 ml kefir, clabber, or whey (1:100)
rennet	regular dose calf rennet for cow's milk; ¼ dose kid rennet for goat's milk
salt	
forms	pots: 8 cm diameter × 10 cm deep
ceramic pots	8 cm across

Yield: **eight 125 g Saint-Marcellins (10:1)**

TECHNIQUE

Bring milk to culturing temperature, about 20°C (68°F).
Add starter, mix in thoroughly.
Add rennet, mix in thoroughly.
Wait 24 hours, until the milk sets to a lactic curd and *Geotrichum* grows.
- Cover the pot and leave undisturbed at 20°C.
- Curd will sink under its whey and become firm but brittle, and soft and acidic on the tongue.

Ladle the curd into forms.
Leave the curd to drain 24 hours.
- Flip cheeses after 12 hours.

Surface-salt cheese to 2 percent.
Leave curd to drain another 24 hours.
- Flip cheeses after 12 hours.

Age cheeses in hastening space for 2 to 3 days, until *Geo* covers rind.
Age cheeses in cave for 2 weeks, until they begin to deliquesce.
Place cheeses into ceramic pots.
Age cheeses in cave for another 4 to 6 weeks.

Primo Sale

Primo sale is a rennet cheese made for fresh eating. Though fresh rennet cheeses are not commonly enjoyed in many modern Western nations because of their short keeping and tendency to change, they have a long history of consumption in traditional cheesemaking cultures around the world, like Italy, wherefrom this version hails. There, cheesemakers would eat or sell the cheese on the day of its make, just after it received its salt (its name means "at first salt").

Primo sale tastes mostly of milk, with slight hints of salt and fermentation. However, it has the capacity to evolve, if aged right, into extraordinarily diverse aged cheeses, including the *Penicillium candidum*–crowned Camembert, *Geotrichum candidum*–ripened tomme Vaudoise, orange-skinned Munster, and even a *roqueforti*-rinded Camembert. As a first taste of all fresh and aged rennet cheeses in this book, this is a style worth familiarizing yourself with!

The first step to making this and most other cheeses is coagulating milk. As was first seen with junket, fresh milk is warmed to just about body temperature to aid this transformation; an active culture (kefir, clabber, or whey saved from the previous day's make) and the right rennet dose are added one after the other to curdle the milk. Within 45 minutes, the curd should be firmly set to a clean break.

Upon achieving this full rennet set, the curd is cut to walnut-sized pieces, then slowly and gently stirred to help evolve its whey. Stirring helps the delicate curds firm and achieve their sought-after strength, but a certain slowness is needed to assure that they don't break. Depending on the species of milking animal, the particular breed, and the stage of lactation, this firmness can require more or less stirring. With a Holstein milk in midsummer, twenty minutes is typical; with late-lactation Jersey milk, 5 minutes is fine; with sheep's milk, no stirring is even needed. Drop a curd from 30 cm occasionally to judge its firmness: If it holds together on impact, it's ready.

Once the curds are ready, they're strained into forms that allow them to drain their whey while helping them fuse together. The curds are left to knit together for a certain amount of time to drain and ferment. The cheeses are flipped over and over to keep an even shape, with the first flipping as soon as the curds have just cohered, still warm from the make.

Upon achieving the right acidity, the cheeses are salted. The correct degree of acidity can be judged with the teabag test, submerging a small sample of cheese in hot water to test its stretch. With a pH of 5.3, the goal acidity for most cheeses (aside from most blue cheeses and lactic cheeses), the curd will spin an endless fiber. (Goat's and sheep's milk cheeses will only stretch slightly at this pH.) Salt preserves this acidity, preventing further fermentation, keeping the curd strong yet supple, with a mild flavor and sourness.

But as simple as the taste of this fresh cheese is, the subtle acidity hints at its capacity for extraordinary flavor if aged—it's this slight hit of lactic acid that fuels aged versions of this rennet cheese's second fermentation. Within this first wheel is the potential for Camembert, Munster, blue cheese, and more!

Making Primo Sale

A fresh rennet cheese • 8- to 12-hour make

INGREDIENTS

milk	4 L
starter	40 ml kefir, clabber, whey (1:100)
rennet	regular dose calf rennet
salt	
forms	baskets: 10 cm diameter × 8 cm deep

Yield two 300 g primo sales (7:1)

TECHNIQUE

Bring milk to cheese temperature, about 35°C (95°F).
Add starter, mix in thoroughly.
Add rennet, mix in thoroughly.
Wait for clean break, 45 minutes to 1 hour.
- Cover the pot to keep in warmth.

Cut curd to walnut size, 2.5 cm.
- Cut vertically; cut crosswise; cut horizontally or diagonally.

Stir the curds gently at about 35°C until they can be held: about 5 to 30 minutes.
Pitch curds for 5 minutes.
Fill forms to the top.
Drain 6 to 12 hours, until acidity develops.
- Goal pH 5.3—stretch test positive.
- Flip three times during the drain, with first flip at 1 hour.
- Salt 2 percent by weight on surface, about 2 teaspoons per cheese.

Or salt cheeses for 1 hour in saturated brine.
Drain cheeses for 1 day on draining table, covered with cheesecloth.
- Flip twice during draining.

Keep cheeses refrigerated to preserve.

PHOTO BY ALIZA ELIAZAROV

Queso Fresco

Queso fresco is a *queso tipico* (typical cheese) made across cultures from the US, Mexico, the Caribbean, through the length of Central America, and across the breadth of South America. The name may refer to different cheeses made differently in different regions, as it simply translates as "fresh cheese." But normally it refers to a simple, salted, unfermented cheese like this. It is one of the most significant cheeses of the world, with much milk in the region transformed on its farms into this nourishing and delicious fresh cheese by hand.

Fresco means "fresh." It's made typically in the morning with the still-warm morning's milk and in the evening with the evening's milk and eaten the same day, without being subject to fermentation or affinage. It's milk more or less transformed into meat; so much so that it is often cooked in a pan like steak. The fresh cheese holds itself together well and doesn't melt when heated, creating a beautiful, caramelized crust as it sizzles. Yum.

This cheese is one of the most quickly made. Most often it's produced on the farm, with milk still warm from the udder, as a continuation of animal agriculture. And it is usually prepared in simple containers, with no need for heating the curd, thanks to its low-tech method. The fresh milk is cultured and renneted; once it's set, the curd is cut, often with the hand, its fingers outstretched, to a small size. A few minutes of stirring suffices to firm and strengthen the curd; it is then wheyed off and strained into cloth to drain. Salt is mixed into the curd almost immediately, slowing its fermentation and preserving the cheese's sweetness. It's then pressed, again, by hand, while still warm, into its forms, to make the final cheese, which is ready to eat or sell less than 2 hours from the udder.

Queso fresco is normally made today without a starter, for it's a cheese that loses its best qualities for cooking when it develops its acidity (instead of holding together, a fermented queso fresco melts like mozzarella!). To that same end, the cheese is also salted early, and fermentation is suppressed. But making a cheese without a starter can cause problems, even when you're salting early, for the microbes in milk continue to change the cheeses, for better or worse, only more slowly. Traditional dairying materials like wood would have cultivated a strong starter naturally, but these are no longer in use. I recommend adding a starter for this and almost all other cheeses; even though the cheese is not actively fermented, the starter will have an inhibitive effect against the unwanted microbes that can still eventually grow in an unprotected cheese. This cheese will inevitably ferment, for better or for worse, as all fresh foods do, when stored more than a few days, so it's best to control that fermentation with a small amount of starter.

One of the pillars for a continent and a half's worth of cuisine, this is a style well worth making to re-create the best tastes of the region. It's a versatile cheese made for cooking over fire. And the fresh milky flavor and meaty texture are most satisfying.

Making Queso Fresco

A rennet cheese • 2-hour make

INGREDIENTS

milk	10 L
starter	100 ml kefir, clabber, or whey (1:100)
rennet	regular dose calf rennet
salt	
forms	baskets: 15 cm diameter × 10 cm deep

YIELD   **1 KG QUESO FRESCO (10:1)**

TECHNIQUE

Bring milk to cheese temperature, about 35°C (95°F).
Add starter, mix in thoroughly.
Add rennet, mix in thoroughly.
Wait for curd to fully set, 45 minutes to 1 hour.
   ▸ Cover the pot to keep in warmth.
Cut curd to hazelnut size, 1.5 cm.
   ▸ Cut slowly with a spino or by hand.
Stir curds gently at about 35°C until they firm, about 15 to 30 minutes.
   ▸ Curd should bounce from 30 cm.
Pitch curds for 5 minutes.
Whey off.
Drain curds into cloth for 5 minutes.
   ▸ Break the lump of curds occasionally to drain more whey.
Mix salt into curds to 2 percent or to taste.
Fill forms to the top with curds.
Press into a wheel by hand while still warm.
   ▸ Leave in the form for 1 hour until cool.
Eat the same day, or keep cheeses cool to preserve.

Halloumi

A famous product of the Mediterranean island of Cyprus, Halloumi is the perfect grilling cheese. More or less the opposite of a pasta filata, it is prepared specifically so that it doesn't melt when heated, by restricting its fermentation.

Cheese wants to ferment! This is one of the few that doesn't undergo any significant microbiological makeovers, and it takes much effort on the part of the cheesemaker to prevent it from doing so. But the work is worth it. Because it can hold its shape when heated, a consequence of a lack of fermentation and its acidification, Halloumi opens up all sorts of culinary possibilities for cheese.

Though not a pasta filata, the cheese is made like a firm mozzarella in two stages: You first make and form the cheese from milk, then boil it to halt fermentation and give it its desired shape and texture. It can be made from any milk, but the milk of goats and sheep makes a typically whiter and more flavorful Halloumi. They're also the more common milking animals in the cheese's Cypriot cradle.

Three specific treatments help to assure that Halloumi stands up to its grilling and doesn't melt as a result of acidity development. First, the cheese is boiled early in its make, halting its fermentation and preserving its sweetness. Second, the cheese is salted after boiling so that fermentation doesn't happen spontaneously. Finally, the cheese is preserved in a salted whey brine

PHOTO BY ALIZA ELIAZAROV

to age, to prevent any further fermentation or surface fungus growth.

Some fermentation is helpful in developing the flavor, working character, and digestibility of the cheese, but too much will cause the cheese to melt when it's boiled or grilled. The cheese retains much of its sweetness, as there is much unfermented lactose in it. This lactose darkens the cheese significantly when you cook it over a fire; the crust it forms on the grill is rich in caramelized milk sugars. The slight amount of fermentation also helps in the making of ricotta from its whey.

When you're boiling the cheese, it's best to cook it in the whey, for by pre-boiling this whey instead of water you also get a harvest of a ricotta known locally as anari. This is a traditional technique from around the Mediterranean (it's also used for caciocavallo) that makes the most of the efforts of the cheesemaker, the region's scant firewood, and the precious milk.

Making Halloumi

A stirred-curd cheese • 4-hour make

INGREDIENTS

milk	10 L
starter	100 ml kefir, clabber, or whey (1:100)
rennet	regular dose kid or lamb rennet with lipase
salt	
forms	baskets: 10 cm diameter × 10 cm deep

Yield: about 5 200 g cheeses (10:1)

TECHNIQUE

Bring milk to cheese temperature, about 35°C (95°F).
Add starter, mix in thoroughly.
Add rennet, mix in thoroughly.
Wait for clean break, 45 minutes to 1 hour.
- Cover the pot to keep in warmth.

Cut curd to hazelnut size, 1.5 cm.
- Cut vertically; cut crosswise; cut horizontally or diagonally.

Stir curds gently at about 35°C until they bounce from 60 cm, about 30 minutes.
Pitch curds for 5 minutes.
Whey off, preserving whey for making ricotta and boiling Halloumi.
Strain curds from whey into cheese forms.
Drain curds 1 hour, flipping every 30 minutes.
- Do not overferment the cheeses.
- Leave whey to ferment during this time.

Bring whey to a boil and remove ricotta when ready.
- reserve 1 L whey after boiling for brine and add 2 percent salt once cooled.

Cook cheeses in their whey.
- Simmer for 5 to 10 minutes in whey.
- Stir occasionally to keep from sticking.
- Cheese is fully cooked once it floats.

Drain cheeses on a draining mat until cool.
Salt cheeses to 2 percent.
- Leave to drain for 12 hours.

Brine cheeses in 2 percent salted whey brine and refrigerate to preserve.
- Cheese can also be kept dry.

Stracchino / Crescenza

Stracchino is the Lombardy word for "tired." This name reflects the cheese's tendency to relax and flow, growing in size—which presumably is the source of its other name: *crescenza*, Italian for "growing." But the name stracchino has also been attributed to the cheese made with milk from tired cows that had spent the summer up in the alpine pastures and had just made the long trip back to their valley farms in the fall. This raises an intriguing possibility that the cheese would have been made without a starter culture, resulting from the wooden cheesemaking vats being out of use for several moons. Perhaps what's tired about stracchino is its fermentation.

Against my own advice, to prepare this cheese right, do not add a starter. This results in a cheese that has a uniquely high pH when it is salted, and that is slow to develop its acidity. The higher-pH cheese tends to relax in its shape, losing its form quickly. And because little acidity develops within the cheese's paste, the microorganisms that usually grow on the rind aren't needed to raise the cheese's pH to the zone where the cheese flows. Only a few weeks of ripening are needed

PHOTO BY ALIZA ELIAZAROV

to get the cheese to ooze like a Camembert—crescenza is one of the only aged rennet cheeses that reaches its crescendo so quickly!

Because this cheese is free from added ferments, it is best to use the freshest possible milk—though this should, for many reasons, be a concern for every cheese in this book. The lack of fermentation causes various changes in the cheese's make and handling: The curd is slow to form and lose its whey due to its high pH. The salted cheese remains wobbly and never firms; it must be handled with extra care so it doesn't break. And it continues to expel whey as it ages, making for a messy affinage.

Although most washed-rind cheeses in this book are best if washed with salted whey, this one should be washed with salted water. Washing with a water brine causes the cheese to ripen faster, allowing its pH to rise faster without setting it back regularly with a lactic-acid-rich whey washing. Combined with the faster ripening, the water washing dilutes its flavor, making the cheese much milder than would be expected for a washed-rind cheese.

Stracchino makes a delicious breakfast cheese. It's a most mild but flavorful rindless formaggio with a milky taste and a smooth melting texture.

Making Stracchino / Crescenza

A rennet cheese • aged with a washed rind • 8- to 12-hour make • 2-week affinage

INGREDIENTS

milk	10 L
rennet	regular dose calf rennet
salt	
forms	baskets: 10 cm diameter × 8 cm deep

Yield five 200 g crescenzas (10:1)

TECHNIQUE

Bring milk to cheese temperature, about 35°C (95°F).
Add rennet, mix in thoroughly.
Wait for the curd to fully set, 45 minutes to 1 hour.
- ▸ Cover the pot to keep in warmth.

Cut curd to walnut size, 2.5 cm.
- ▸ Cut vertically; cut crosswise; cut horizontally or diagonally.

Stir the curds gently at about 35°C until they can be held, about 15 minutes.
- ▸ Drop test: Curds should bounce from 30 cm.

Pitch curds for 5 minutes.
Whey off.
Fill forms to the top.
Drain 6 to 8 hours.
- ▸ Flip three times during the drain, with first flip at 1 hour.

Salt cheeses 2 percent by weight on surface, about 1 teaspoon per cheese.
- ▸ Place cheeses back into forms after salting.

Age cheeses in cave for 2 to 4 weeks.
- ▸ Flip cheeses every other day.
- ▸ Wash rinds with 2 percent salted water brine every other day.

(Stracchino can also be packaged in containers and refrigerated instead of cave aging.)

SPRING • 251

Tomme Vaudoise

A small Swiss cheese, tomme Vaudoise is a pancake-thin, fungally ripened round of cow's milk cheese. It is a simple, fast-ripening, gooey cheese and a fitting first foray into white-rinded rennet cheeses. Vaudoise is perhaps the milkiest and creamiest of the aged rennet cheeses, and is one of the youngest aged cow's milk cheeses. It tastes of fresh cream, with slight hints of mushroom at the rind.

Vaudoise is a very simple aged cheese, made like a thin Camembert but ripened with a *Geotrichum candidum* ecology instead of *Penicillium candidum*. This first fungus-aged rennet cheese of the book exists between realms; its make is enzymatic like a Camembert, but its affinage falls in line with a lactic crottin. A short hastening period before aging helps establish *G. candidum* on the rind of this beautiful little wheel.

Gentle *Geotrichum* establishes a very tender and velvety rind that falls to pieces as it ripens. Unlike modern Camemberts, held together by stronger filaments of the *P. candidum* fungus, the rind of this cheese is soft as silk. All of the Camembert makes that follow can be ripened with *Geotrichum candidum* instead of *P. candidum* by following the technique for hastening and affinage outlined here, and not washing the rinds. And all may be improved by doing so because of the soft beauty and tender rind of *Geotrichum* that grows in its place. Still, the cheeses that result won't appeal to our ideal of what a Camembert should be.

Called a tomme, really it should be named tomette—a small tome or tile—because this is the thinnest of the rennet cheeses. Because it is so flat, it ripens quickly from its rind; its high surface-area-to-volume ratio means that its ripening is accelerated. The forms should be filled to only half their usual height. The final cheese should be only 1 centimeter thick.

Switzerland is not often considered a place where small soft cheeses are made, mostly because the production of the country's formerly extraordinary diverse cheeses was restricted, up until very recently, by the Swiss Cheese Union, which only permitted a few full-alpine styles to be made, hence the prevalence and popularity of raclette and Gruyère. Only with the dissolution of the marketing organization in the late 1990s did cheesemakers begin to re-create the historical cheeses of the region, like this extraordinary little tomme.

Making Tomme Vaudoise

A rennet cheese • aged with *Geotrichum* • 8- to 12-hour make • 3- to 4-week affinage

INGREDIENTS

milk	4 L
starter	100 ml kefir, clabber, or whey (1:100)
rennet	regular dose calf rennet
salt	
forms	hoops: 10 cm diameter × 5 cm deep

YIELD **FOUR 100 G TOMETTES (10:1)**

TECHNIQUE

Bring milk to cheese temperature, about 35°C (95°F).
Add starter, mix in thoroughly.
Add rennet, mix in thoroughly.
Wait for clean break, 45 minutes to 1 hour.
  ▸ Cover the pot to keep in warmth.
Cut curd to walnut size, 2.5 cm.
  ▸ Cut vertically; cut crosswise; cut horizontally or diagonally.
Stir curds gently at about 35°C until they firm, 5 to 15 minutes.
  ▸ They should stay together when dropped from 30 cm.
Pitch curds for 5 minutes.
Fill forms to 5 cm in height.
Drain 6 to 12 hours, until acidity develops.
  ▸ Goal pH 5.3—stretch test positive.
  ▸ Flip three times during the drain, with first flip at 1 hour.
  ▸ Salt 2 percent by weight on surface, about 1 teaspoon per cheese.
  ▸ Be careful not to oversalt the thin cheese.
Drain cheeses for 1 day on draining table, covered with cheesecloth.
  ▸ Flip twice during draining.
Age cheeses for 2 to 3 days in hastening space until *Geo* covers rind.
  ▸ Flip cheeses once daily.
Age cheeses in cave for 3 to 4 weeks.
  ▸ Flip cheeses every other day.

Making Traditional Camembert

A half-lactic cheese • aged with *Penicillium candidum* and *Geotrichum*
8- to 12-hour make • 6- to 8-week affinage

INGREDIENTS

milk	10 L
starter	100 ml kefir, clabber, or whey (1:100)
rennet	regular dose calf rennet
salt	
forms	hoops: 10 cm diameter × 10 cm deep
YIELD	**FIVE 300 G CAMEMBERTS (7:1)**

TECHNIQUE

Bring milk to cheese temperature, about 35°C (95°F).
Add starter, mix in thoroughly.
Add rennet, mix in thoroughly.
Wait for clean break, 45 minutes to 1 hour.
  ▸ Cover the pot to keep in warmth.
Cut curd to walnut-sized columns, 2.5 cm.
  ▸ Cut vertically, but do not cut horizontally.
Pitch curds for 1 hour.
  ▸ Leave curds untouched until much whey has separated.
Whey off from above.
Fill forms to the top by ladle (be sure cheese forms are on draining mats).
Drain 6 to 12 hours, until cheeses are firm and acidity develops.
  ▸ Goal pH 5.3—stretch test positive.
Salt 2 percent by weight on surface,
  about 2 teaspoons per cheese.
  ▸ Give cheese first flip during salting.
Drain cheeses for 2 days on draining table,
  covered with cheesecloth.
  ▸ Flip daily during draining.
Age cheeses in cave for 4 to 6 weeks.
  ▸ Wash rinds daily with water brine (2 percent salt) for first week.
  ▸ Flip cheeses every other day.

Modern Camembert

This is a technique for a different Camembert, with a bright white rind, a mild character, and an even proteolysis. Less flavorful and indulgent than the traditional, this more modern version of the cheese sits better on a shelf and is more suited to supermarket sales. Now, I realize I'm not selling this cheese well, and I suppose that's all right, for it really doesn't compare to a proper traditional Camembert. But it does provide an important lesson in milk's many ecologies, for this is also a method for growing *Penicillium candidum*—the fungus that grows the fluffy white rinds of the more modern version of this cheese.

Penicillium candidum has very precise and well-defined growing conditions, and it pops up naturally on the rinds of cheeses where these conditions are created. The fungus seems to live in partnership with *Geotrichum candidum*, but in the right circumstances this particularly fluffy fungus plays the dominant ripening ecology.

Penicillium candidum is a white variant of *Penicillium roqueforti*, the fungus that makes cheese blue. Unlike *P. roqueforti*, which has a weedy ecology and thrives in the cold environment of the cheese cave, *P. candidum* seems to be a part of milk's microbial ecology, for it lives in balance with *Geotrichum candidum*, which grows most readily in mesophilically fermented milk.

Penicillium candidum prefers a certain environment marked by the presence of lactic acid, its preferred food; relatively low acidity (high pH); and moderate amounts of moisture. It thrives in a niche just beside *Geotrichum*, which prefers more lactic acid, and seems to grow best when this other milk fungus gets established but fails to thrive.

So how do we get this fine fungus to grow? Simply by making the Camembert! The exact conditions for getting *Penicillium candidum* to grow are created by this very particular cheese make (it also can grow on Stilton or tomme de montagne where similar conditions are created). The Camembert must have just the right moisture content, just the right acidity at salting, and just the right amount of salt. It must be washed just right in the cave for 1 week, and then be aged in just the right environment for the white fungus to cover the rind.

And though it takes all this proper protocol to make this Camembert, really, it's the washing for a week that gets the *Penicillium candidum* to grow. The same Camembert could be made blue with *Penicillium roqueforti* by not washing at all (that's the technique for Camembert bleu) or orange by washing for 1 month to encourage the washed-rind ecology (that's the method for Munster). But by washing for just 1 week, the washed-rind ecology (including *Geotrichum candidum*) is allowed to just develop. This in turn raises the pH of the cheese slightly as its lactic acid is consumed. Once the washing stops, and the cheese is allowed to continue aging in the cave, the washed-rind ecology falters and *P. candidum* finally finds its footing. The fungus germinates and grows its mycelium over the rind, eventually defining this cheese as a fluffy white-rinded Camembert.

PHOTO BY ALIZA ELIAZAROV

Making Modern Camembert

A rennet cheese • made with cow's milk • aged with *Penicillium candidum*
8- to 12-hour make • 6- to 8-week affinage

INGREDIENTS

milk	10 L
starter	100 ml kefir, clabber, or whey (1:100)
rennet	regular dose calf rennet
salt	
forms	hoops: 10 cm diameter × 8 cm deep
Yield	**five 250 g Camemberts (8:1)**

TECHNIQUE

Bring milk to cheese temperature, about 35°C (95°F).
Add starter, mix in thoroughly.
Add rennet, mix in thoroughly.
Wait for clean break, 45 minutes to 1 hour.
- Cover the pot to keep in warmth.

Cut curd to walnut size, 2.5 cm.
- Cut vertically; cut crosswise; cut horizontally or diagonally.

Stir curds for about 15 minutes at 35°C, until they bounce from 30 cm.
Pitch curds for 5 minutes.
Whey off from above.
Fill forms to the top.
Drain 6 to 12 hours, until acidity develops.
- Goal pH 5.3—stretch test positive.
- Flip twice during the drain, with first flip at 1 hour.
- Salt 2 percent by weight on surface, about 1½ teaspoons per cheese.

Or salt cheeses for 1 hour in saturated salt brine.
Drain cheeses for 1 day on draining table, covered with cheesecloth.
- Flip twice during draining.

Age cheeses in cave for 4 to 6 weeks.
- Wash rinds daily with water brine (2 percent salt) for first week.
- Flip cheeses every other day.

Brie

Brie and Camembert evolve from very much the same method and have much in common. But most would argue that the two are very much distinct from each other. The two cheeses represent two very different techniques, from two distinct regions of France, with differing dairying traditions, different cows, different milks, and different microbes. But there are still an astonishing number of similarities between the two cheeses, and a number of common threads that bind them. And when all the politics and pageantry are removed, the only significant difference between the two is that Brie is big, and Camembert is small.

The origin story of Camembert confirms this. The cheese was supposedly created at the time of the French revolution, when religious orders were being persecuted for their connection to the Crown, and their land was being reclaimed by the peasantry. One monk, a Brie maker from the center of France, fled his monastery for Normandy and took refuge with a cheesemaker, Marie Harel, who lived in the village of Camembert. The priest, fearing for his life, nevertheless observed Marie making cheese from his hiding place, and criticized her cheesemaking practice, instructing her to make her cheese like the monks did in their making of Brie. Marie followed the monk's instruction but didn't have the large form that the monk suggested using; instead she ladled her cheeses into a smaller form used to make another Normandy specialty, Livarot. And so the cheese we know as Camembert was supposedly born.

Size matters in cheesemaking. The smaller cheese, with more surface area to its volume, ripened faster than the Brie and became something entirely different. Marie Harel named this new cheese for her village, but anyone, anywhere, even on the other side of the world from Camembert, can make both of these cheeses right, so long as they use raw milk fresh from the udder, enlist their milk's microbes to evolve the right character, and make the cheese according to centuries-old techniques.

Brie, however, is probably the most challenging of the classic white-rinded cheeses to make and may be one of the most difficult of all makes to execute. The curd must be ladled just right, to assure its amount of moisture, and drained right, with the help of draining mats underneath. The massive wheel is prone to breaking and must be handled delicately but firmly at all times, especially during its first flip! (Cheesemakers in France place a second draining mat on top of the cheeses in their hoops and flip the cheeses on their draining mats upside-down within their forms.) And the cheese must be washed just right to assure an even coat of *Penicillium candidum* atop its surface.

Making Brie

A half-lactic cheese • made with cow's milk • with a *Penicillium candidum* and *Geotrichum* rind
8- to 12-hour make • 6- to 8-week affinage

INGREDIENTS

milk	20 L
starter	200 ml kefir, clabber, or whey (1:100)
rennet	regular dose calf rennet
salt	
forms	hoops: 40 cm diameter × 10 cm deep (two nesting hoops)

YIELD   ONE 2 KG BRIE (10:1)

TECHNIQUE

Bring milk to cheese temperature, about 35°C (95°F).
Add starter, mix in thoroughly.
Add rennet, mix in thoroughly.
Wait for clean break, 45 minutes to 1 hour.
- ▸ Cover the pot to keep in warmth.

Cut curd to walnut-sized columns, 2.5 cm.
- ▸ Cut vertically, but do not cut horizontally.

Pitch curds for 1 hour.
- ▸ Leave curds untouched until much whey has separated.

Whey off from above.
Fill forms to the top by ladle.
Drain 6 to 12 hours, until acidity develops.
- ▸ Goal pH 5.3—stretch test strong.

Salt 2 percent by weight on surface, about ¼ cup per cheese.
- ▸ Give cheese first flip during salting.

Drain cheese for 2 days on draining table, covered with cheesecloth.
- ▸ Flip daily during draining.

Age cheese in cave for 1 week.
- ▸ Flip cheese daily.
- ▸ Wash rind daily with 2 percent salt brine for first week.

Continue aging cheese in cave for 4 to 6 weeks.
- ▸ Flip cheese every other day.

Pont l'Évêque

This technique is a tribute to the less appreciated, Camembert-like cheeses made elsewhere in Normandy and beyond. Many traditional cheeses from across the Continent are prepared in a similar manner to Camembert and achieve similar characteristics, but lack the celebrity of their more sought-after sibling. Likely these styles were once made more widely, for they were all conceived with original cheesemaking technologies of wooden vats, natural cheese forms, and rennet, all of which have been in use by cheesemakers since long before the supposed birth of Camembert in 1789.

As for the cheese of this method, Pont l'Évêque is a Camembert, only made square! The changed shape, of course, changes the cheese. The square shape increases its surface area, hastening its ripening, and also causes more deliquescence toward the corners of the cheese.

Though not normally covered in charcoal, I've taken artistic liberty with these cheeses and painted them black. The painting, which can be omitted, helps to define the cheese's most beautiful rind, giving the square cheeses beautiful black edges where the rind ecologies are set back through regular contact. Differing slightly in treatment from the charcoal-coated Valençay, made lactic, I recommend washing the rind of these cheeses for 1 week in the cave to develop their white rinds before applying their charcoal coat.

These diverse white-rinded cheeses exist on a continuum, with variations in shape, size, fermentation, and rind treatment. But the Camembert continuum should also extend to orange-rinded Munster and *roqueforti*-ripened Camembert bleu, cheeses that are remarkably similar in their makes and affinages, for mostly the difference between these is whether and how long they are washed. And though they're all related, each cheese is distinguished by a particular difference that makes it unique—*vive la différence!*

Some other small, soft cheeses that fit in this family include Livarot, another cheese from Normandy that is made in a shape like Camembert but washed regularly to give it an orange rind and tied together with juncus grass to keep it from slumping as it ripens. Farther away in France, reblochon, a medium-sized wheel with a washed rind that turns white when the washing stops, is made almost exactly like a modern Camembert, but in the Savoie. And Tomino, a small white-rinded cheese from the north of Italy, is made flat, just like the Swiss tomme Vaudoise.

PHOTO BY ALIZA ELIAZAROV

Making Pont L'Évêque

A rennet cheese • made with cow's milk • aged with *Penicillium candidum*
8- to 12-hour make • 6- to 8-week affinage

INGREDIENTS

milk	10 L
starter	100 ml kefir, clabber, whey (1:100)
rennet	regular dose calf rennet
salt	
charcoal	
forms	baskets: 10 cm square × 10 cm deep
YIELD	**FIVE 300 G PONT L'ÉVÊQUES (7:1)**

TECHNIQUE

Bring milk to cheese temperature, about 35°C (95°F).
Add starter, mix in thoroughly.
Add rennet, mix in thoroughly.
Wait for clean break, 45 minutes to 1 hour.
  ▸ Cover the pot to keep in warmth.
Cut curd to walnut size, 2.5 cm.
  ▸ Cut vertically; cut crosswise; cut horizontally or diagonally.
Stir curds for 15 minutes, until semi-firm.
Pitch curds for 5 minutes.
Whey off from above.
Fill forms to the top.
Drain 6 to 8 hours, until acidity develops.
  ▸ Goal pH 5.3—stretch test liquid.
  ▸ Flip once during the drain, after 3 to 4 hours.
  ▸ Salt 2 percent by weight on surface, about 2 teaspoons per cheese.
Or salt cheeses for 1 hour in saturated salt brine.
Drain cheeses for 1 day on draining table, covered with cheesecloth.
  ▸ Flip twice during draining.
Age cheeses in cave for 1 week.
  ▸ Flip cheeses once daily.
  ▸ Wash rind daily with 2 percent salt brine.
Apply charcoal as a paint on all sides of cheese.
Age cheeses in cave for another 3 to 5 weeks.
  ▸ Flip cheeses every other day.

CHAPTER 13

Summer

Summer is the season of sweat and salt; of rigorous warm-weather cheesemaking, with much milk to be transformed. It's the season of alpine cheeses, milled-curd cheeses, pasta filata styles, and brined cheeses, all the best preservations of this season of abundance.

In summertime, cheesemaking is finally in full swing. With the young animals weaned (or sacrificed), their mothers are providing milk to fill the vat. To best preserve the abundant milk during the warmer months, cheesemakers make their cheeses firm and massive. The biggest wheels, the milled-curd cheeses like cheddar and the alpine styles like Emmentaler, are best made in this season.

Summer is the season of cooking curds in kettles over fire, and is the time for tommes (tomme de Savoie, tomme de chèvre, tomme crayeuse, and tomme de montagne) and raclette. These long-preserving semi-alpine cheeses will provide good food through the winter; the full-alpine styles, like grana and Comté will last through to the next summer season.

With its high protein content, summer milk most elegantly and effectively transforms into these firmer alpine wheels. Likewise, firm milled-curd cheeses ferment their best in the warmth of summer, and preserve well even in the heat.

Summer is also the season of pickling, both vegetables and cheese. Salty brines keep feta and sirene free from mold and mites, which can consume a cheese rather quickly in the warmer weather.

Finally, summer is the season of pasta filata cheeses, when warmer weather encourages their best transformations and preservations. Indeed, mozzarella, Oaxaca, and caciocavallo may be the best cheeses for the heat. (Cheesemakers of the tropics, take note!)

Mozzarella

A first pasta filata, or spun-curd, cheese, mozzarella is made by melting curds that have been fermented to just the right degree so that they spin fibers when submerged in hot water. And it's just one of dozens of different pasta filata cheeses made in Italy and around the world that are all extraordinary elaborations of this most manipulable form of milk.

To prepare mozzarella, make a rennet cheese as usual from the freshest, least processed milk, but with a thermophilic starter. Cut the curd and stir lightly to firm, then leave it to ferment beneath warm whey until its curd spins fibers when submerged in very hot water. Test a small piece of curd for stretch every half hour or so while it is fermenting so that you don't miss the pasta filata stretching window.

Once the curd begins to show signs of stretch you will want to quickly prepare for the mozzarella making: Bring water to a boil, and prepare the cold-water bath and light salt brine. When the cheese achieves its full stretch, ladle the hot water over the curd in a melting vat a couple of ladles-ful at a time, stirring the curd gently between additions, until it melts into a stretchy mass. Do not add too much hot water too quickly, as the curd can overcook and be difficult to handle.

Once it's melted, gather together the curd with the help of a wooden spoon or paddle (steel sticks to melted curd) and lightly knead it to help work the melted cheese and develop its strength. Though the melting water is hot enough to burn, the curd should be just cool enough to touch and shape. Having a bowl of cold water handy to plunge your hands into helps with the high temperatures, as does using a wooden spoon to retrieve the curd from the hot water.

I'll do my best to describe the method of mozzarella's forming here. To start the process, lift the curd slightly out of the water with a wooden spoon and pull a lump of melted cheese from the mass of curd with your left hand (switch everything around if you're a leftie). Grab the curd with your right hand and pull it with your left one into a short length before folding it back onto itself accordion-like. Now stretch the curd in your left hand over the accordion with both hands to make a ball containing it all. Break this from the mass of curd between the thumb and first finger of your right hand with a very tight squeeze, while supporting it with the left. With your left hand, toss the mozzarella into the cold-water bath to cool and firm, before submerging it in the light salt brine to preserve.

It is best to work in a fluid fashion, keeping the curd in your hands so that you never have to retrieve the working end from the hot water in the melting vat. As you tear the mozzarella from the mass with your right hand, hold on to the torn end of the curd with that hand as the left hand takes the mozzarella and transfers it to the cold-water bath. This way, the working end of the curd is ready for the next tearing, and is transferred to your left hand before being pulled again into the next mozzarella by your right. As such the entire mass of curd can be spun into balls of mozzarella, one after the other after the other, until all the curd in the vat is used.

The name *mozzarella* derives from the Italian *mozzare*, to tear. Other shapes that are made with the spinning dough in Italy include: *nodini*, little knots, a pulled length of curd tied into a knot before being torn; *bocconcini*, little mouthfuls, small balls of mozzarella; *zizzona*, mozzarella made into a massive breast; *scamorza*, mozzarella shaped into little gourds to be hung for smoking; and *treccia*, a three-strand braid. It's even been known to be braided into elaborate shawls for festivals! Burrata and stracciatella are also made with a variation on this method, enriching the pasta filata cheese with cream (that's stracciatella) before encasing it with stretched curd into a dumpling (burrata). And a number of braided cheeses from Eastern Europe and the Middle East—often smoked and eaten in bars as beer cheese—are prepared by pulling the curd in hot water until it achieves a small fiber, then collecting the yarn-like curd and twisting it into a skein before smoking the salty cheese.

A most delicious and succulent fresh cheese, mozzarella can be prepared year-round with the milk of any animal. However, this method, along with the other pasta filata cheeses, works best with buffalo or cow's

PHOTO BY ALIZA ELIAZAROV

milk; if you're using goat's or sheep's milk, the stretch and workability of the curd won't be as developed, and it's harder to judge when the curd is ready to melt. For an improved stretch, consider blending these milks 50–50 with cow, as is often done for caciocavallo. Buffalo is considered best for the mozzarella technique, but a cow's milk mozzarella made using traditional methods, sweet and golden from its cream, will leave you questioning this reasoning (it may just be that water buffalo are more common in the swampy environs of Naples, the ancestral home of both mozzarella and pizza!). If you are using the milk of buffalo, though, be sure to handle the curds more lightly, barely stirring, if at all, and expect a roughly doubled yield—that's the big benefit of buffalo!

This technique calls for making mozzarella slowly, with a mesophilic fermentation (using a mesophilic starter, like clabber) and a fermentation at lower temperatures, 30°C (86°F), developing its stretch within 6 to 8 hours. But a faster, more thermophilic mozzarella (made with a thermophilic starter, like yogurt, fermented at 40°C [105°F], and ready in about 3 hours total) can be made instead. The thermophilic mozzarella retains more sweetness due to the straight lactic acid fermentation, but the mesophilic mozza develops more flavor in its flesh from a fuller heterofermentation (there's less lactose resulting when the more diversely fermented cheese reaches a pH of 5.3). You decide what works for you. And any pasta filata style can be prepared with either approach.

Making Mozzarella

A pasta filata cheese • made with cow's or buffalo milk • 6- to 8-hour make

INGREDIENTS

milk	10 L
starter	100 ml clabber, kefir, or whey (1:100)
rennet	regular dose calf rennet
salt	

Yield 1 kg mozzarella in brine (10:1)

TECHNIQUE

Bring milk to cheese temperature, about 35°C (95°F).
Add starter, mix in thoroughly.
Add rennet, mix in thoroughly.
Wait for clean break, 45 minutes to 1 hour.
- Cover the pot to keep in warmth.

Cut curd to walnut size, 2.5 cm.
- Cut vertically; cut crosswise; cut horizontally or diagonally.
- Stir curds gently over 5 to 10 minutes.

Ferment curds under whey at ambient temperature until acidity develops.
- Goal pH 5.3—stretch test spins.
- Perform teabag test every 30 minutes.
- Mozzarella usually takes 6 to 8 hours total to develop its stretch.
- If the temperature of the curds drops below 30°C (86°F), gently warm curds back up.

Prepare for mozzarella stretching as curd shows first signs of stretching.
- Bring 5 L water to a boil (half as much water as milk).
- Prepare a 5 L cold-water bath.
- Prepare a light salt brine with 2 L cold water and 1 percent salt (brine should taste good!).

Strain the curds from the whey.
Pour almost-boiling water over curds until curds melt evenly.
- Add hot water slowly, stirring curds gently between additions.
- Add more hot water as needed, but do not overcook.
- Gently bring curds together into a single mass under the water.

Shape curds into mozzarella.
- Stretch a handful of curd from the mass into a 15 cm rope.
- Fold rope into an accordion (or roll into a stout pinwheel).
- Stretch top part of the cheese over the accordion to make a ball.
- Tear mozzarella off the mass between thumb and forefinger.

Plunge mozzarella into cold-water bath.
- Leave in water until cool.

Transfer mozzarella to light salt brine and refrigerate to preserve.

Queso Oaxaca

Oaxaca is one of a number of traditional pasta filata cheeses made in the Americas. Latin America especially has an extraordinary diversity of spun-curd cheeses, likely brought over with the first Spanish colonizers, though these styles have by and large been lost from Spain. In Venezuela, pasta filata curd is stretched thin by hand and called queso de mano, hand cheese. And in Chiapas, an aged cheese called bola de Ocosingo is made by stuffing a round pasta filata cheese with a salted lactic curd. But nowhere else in the region is a pasta filata cheese so celebrated as in Oaxaca, in the south of Mexico. Oaxaca cheese is so important there that the name of the state has become shorthand for their incredible stretched-curd cheese.

The origin story of queso Oaxaca tells of a forgetful girl entrusted with the responsibility of turning the morning's milk into cheese. Neglecting her duties, the girl took the warm milk and began the cheesemaking process half-heartedly. Distracted by a beautiful butterfly that fluttered by, she forgot about the cheese altogether. It wasn't until later in the day that she remembered her responsibility. But on returning to the vat, she found that the curd had cooled and settled under the whey. Panicking that the now cooled curd would not form a cheese like her abuela had taught her, she warmed the pot on the stove but forgot about it yet again, and the curd overcooked! When she finally returned to the pot, the fermented curd had melted into a puddle. Not knowing what to do, she strained the cheese from the vat. Struggling to put the stretchy cheese into its form, she pulled it into a long string. She then wrapped the cheese up into a ball, hoping no one would notice. Her family, on finding the failed cheese, was upset at her irresponsibility, but they were enchanted by the flavor and texture of the new cheese, and named it quesillo, the local name for queso Oaxaca.

No other cheese (or other food, for that matter) has a more playful make. Working in pairs—the curd can only be stretched so long alone—cheesemakers pull and whip the spinning curd into great lengths, then salt the ropes and roll them together into great balls of cheese.

To begin, rennet curd is prepared and fermented until it spins, much like mozzarella (the two cheeses are in many ways the same). On achieving the right acidity, the curd is salted and melted stepwise with very hot water. The curd is then gathered together and kneaded under the hot water to develop its strength before being pulled. To shape the Oaxaca, a large lump of curd is removed from the water and worked into a thick rope; that rope is then stretched between two cheesemakers, with one maker whipping the curd to keep the rope even; both work quickly to preserve the cheese's warmth and stretchability. The stretched curd is then placed on a table to cool and flatten before being salted, drained, and rolled up into Oaxaca. The curd can also be pulled solo, like mozzarella, direct from the melting vat into a long rope right onto a draining table to cool before rolling.

The extended stretching of the Oaxaca gives this cheese a fascinating character: quesillo can be pulled apart into endless strings! Pulling the pasta filata curd in this unique way aligns the melted casein proteins; upon cooling, the parallel proteins firm and the cheese can be pulled apart along its microscopic seams.

Oaxaca is often sold in markets in massive yarn-like balls. A customer asks for a length, which is unraveled by the monger from the ball, and rolled into a smaller one to be taken home. To use Oaxaca the curd is shredded by hand (like stracciatella) into smaller and smaller strings then melted atop fresh tortillas, sprinkled on tacos, and stuffed into chiles rellenos. If you're making Mexican food, Oaxaca's a must; this delicious melting cheese envelops the region's cuisine!

Oaxaca can be made softer, like mozzarella, or firmer, like caciocavallo, depending on how the curd is cut and how long it is stirred. A lighter touch, like the one called for in this technique, makes a more delicate, creamy Oaxaca; a firmer Oaxaca can be kept longer, but is lighter in cream and in flavor.

And don't forget to save the stretching water to salvage its cream, known as crema in Mexico. After the make, leave the now cream-colored stretching water (the Oaxaca broth) overnight; the next day, its cream will rise to the top, wherefrom it can be easily skimmed. This crema, also especially important in

PHOTO BY ALIZA ELIAZAROV

Mexican cuisine, is half fermented; it's less sour than crème fraîche and maintains a texture between pouring cream and thickened sour cream. Its light cooking also brings out a nuanced flavor that's more floral than crème fraîche. Of course, this cream can be saved from every type of pasta filata cheese (I sometimes call it mozzarella-cream) and is essential for the making of a proper burrata (see next technique). A particularly flavorful Italian *burro* or Mexican *mantequilla* can be made by churning this pasta filata cream into butter!

Making Queso Oaxaca

A pasta filata cheese • made with cow's or buffalo milk • 6- to 8-hour make

INGREDIENTS

milk	10 L
starter	100 ml kefir, clabber, or whey (1:100)
rennet	regular dose calf rennet
salt	

Yield 1 kg Oaxaca (10:1) and 150 g crema (50:1)

TECHNIQUE

Bring milk to cheese temperature, about 35°C (95°F).
Add starter, mix in thoroughly.
Add rennet, mix in thoroughly.
Wait for clean break, 45 minutes to 1 hour.
Cut curd to walnut size, 2.5 cm.
- Cut vertically; cut crosswise; cut horizontally or diagonally.

Stir curds gently at about 35°C until they can be held, about 5 to 15 minutes.
Ferment curds under whey at room temp until goal acidity is reached,
- Goal pH 5.3—stretch test spins.
- Perform teabag test every 30 minutes.
- Curd is usually ready between 6 to 8 hours from adding culture.

Just before acidity develops, prepare for stretching and shaping Oaxaca.
- Bring 5 liters of water to a boil.

Strain curds from whey.
Pour hot water over curd slowly to melt.
- Stir curd gently until it melts into a dough.
- Add more water as needed.

Knead curd with a wooden spoon for a minute until strong.
Stretch curd into Oaxaca.
- Work the dough into a thick rope.
- Stretch the rope between two people until 2 cm thick.
- Whip as you stretch to keep the rope even.
- Submerge rope in hot water occasionally to keep it plastic.

Leave rope lying on table to cool and flatten.
Sprinkle lightly with salt, then wait 10 minutes.
Roll rope upon itself into a ball of Oaxaca.
Make crema.
- Leave stretching water to sit overnight, then skim off cream.

Burrata

Burrata ("buttery" in Italian) is a cheese make in two acts. First act: Make stracciatella. Second act: Stuff the burrata full with stracciatella! Essentially a cheese dumpling bursting with cream, burrata is the most decadent member of the pasta filata family.

So, for the first act: *stracciatella* means "shredded." It is a sort of Italian cottage cheese, with shredded bits of pasta filata smothered in thick and flavorful fermented cream. Prepare this by first stretching some spinning curd long (like Oaxaca), shredding the cheese fine, and chopping it into pieces before smothering them with a healthy quantity of crema (it's the same word—and the same concept—in Italian and Spanish).

Once the stracciatella is made, the burrata can be assembled. Break a handful of curd from the melted mass of pasta filata, then stretch it thin by hand the way a *pizzaiolo* stretches pizza dough. Slap a large lump of stracciatella into the middle, and bring the edges of the enveloping curd up and around the filling. Bring the ends of the curd together, twist them, pinch them closed, then tie around them with a small piece of string or a bit of curd stretched into a filament and cooled. Briefly submerge this delicate dumpling in cooling water and then a preserving brine, where the burrata can remain for several days or weeks.

What makes the burrata so buttery is the cream in which the stracciatella curds are drenched. My preference is to use the pasta filata cream, skimmed from the top of the stretching water of a previous batch of mozzarella or burrata left to stand; but to have this handy requires making a pasta filata cheese the day before! If you're into the rhythm of mozzarella making, this is simple to source, but if you want to pull off the burrata in 1 day, consider preparing the crema for the stracciatella with a 50–50 mixture of cream and cultured cream. But from a flavor perspective, and a food-waste perspective, burrata's best if made with mozzarella-cream skimmed from the top of the previous day's leftover stretching water. This crema is made more delectable through a combination of partial fermentation and cooking, and melds better flavor-wise and texture-wise with the shredded curd.

Before serving, it is best to warm this cheese ever so slightly. When cut, the burrata's buttery filling oozes out of the dumpling most provocatively.

Making Burrata

A pasta filata cheese • made with cow's or buffalo milk • 6- to 8-hour make

INGREDIENTS

milk	10 L
starter	100 ml kefir, clabber, or whey (1:100)
rennet	regular dose calf rennet
salt	
	skimmed crema or
	cultured cream and fresh cream

YIELD 1 KG BURRATA (10:1)

PART I: MAKE STRACCIATELLA

Bring milk to cheese temperature, about 35°C (95°F).
Add starter, mix in thoroughly.
Add rennet, mix in thoroughly.
Wait for clean break, 45 minutes to 1 hour.
- Cover the pot to keep in warmth.

Cut curds to walnut size, 2.5 cm.
- Cut vertically; cut crosswise; cut horizontally or diagonally.
- Stir curds gently over 5 to 10 minutes.

Ferment curds under whey until acidity develops, 6 to 8 hours from adding culture.
- Perform teabag test every 30 minutes when close.
- Goal pH 5.3—stretch test spins.
- Gently warm curds back to temperature if they fall below 30°C (86°F).

Before acidity develops, prepare for stretching and shaping curd.
- Bring 5 L water to a boil.
- Prepare a basin of cold water for cooling cheeses.
- Prepare a light salt brine with 1 percent salted water.

Prepare crema.
- Skim crema from broth of previous batch of pasta filata left to sit 12-plus hours.
- Or mix cultured cream 50–50 with fresh cream.

Melt the curd.
- Strain the curds from the whey.
- Slowly pour very hot water over curds to melt.
- Stir curds gently as water is added.
- Add more hot water as needed until it melts into a dough.

Prepare stracciatella.
- Stretch half of pasta filata curd into long ropes.
- Submerge in cold water to cool and firm.
- Shred the stretched curd into strings, and cut into 1 to 2 cm pieces.
- Smother shredded curd in crema.
- Add salt to taste.

Stuff into burrata or refrigerate to preserve.

PART II: MAKE BURRATA

Warm remaining pasta filata curd with additional hot water if needed.
Pull off a handful of curd from mass, and stretch into a flat circle.
Place a handful of stracciatella in the middle.
Close up dumpling and tie off with string or stretched curd.
Plunge burrata into cold water.
- Leave until cool.

Submerge burrata in 1 percent salt brine to preserve.

Sour Pickles

Before I became a cheesemaker, folks in my farming community used to call me Pickles, for I always had a jar of brined cucumbers or salted cabbage bubbling away on my counter, and invariably my contribution to a potluck was always a jar of pickles (before it invariably became cheese!). Certainly, my love of fermentation was piqued by my previous passion for pickles— itself inspired by the sour pickles and brined green tomatoes sold in the barrel by a shopkeeper named Simcha, at my old neighborhood grocer when I lived in Montreal. Simcha taught me to love the tang and tickle of a traditionally fermented pickle.

Every vegetable is made more digestible through fermentation. And our attraction to the flavors of fermented veg (fermented anything, really) completely crosses cultural lines. An important, biological reason for this is emerging: We evolved alongside, or perhaps even because of, our practice of fermenting foods to preserve them (our history of fermenting foods is believed to be more ancient than our history of cooking foods). Our large intestine, which supports fermentation and subsequent digestion, is significantly shorter in length than any other primate's, and the pre-digestion that occurs when our foods were fermented through their storage created a nutrient-rich and bioavailable diet that enabled us to live with a shortened large intestine, all the while supporting the energy demand of our growing brains. The practice of fermentation not only helped us best preserve and digest our foods, it literally expanded our minds.

Digestion is a metabolically expensive process. It taxes our bodies to produce the enzymes and acids needed to break down raw (and even cooked) foods. Ultimately what can't be broken down by our enzymes and acids needs to be fermented in our large intestines for us to make the most of the food. The more our foods are fermented before we eat them, the more nutrient-dense and assimilable they are, the more easily they're absorbed in our foreshortened intestines, and the more they nourish us. Our tastebuds tingle at the lactic tangs and umami notes of fermented foods, as our bodies recognize them to be the foods we need to thrive.

Our focus on fresh meat, fresh fruit, and fresh veg, like our focus on fresh milk, may be a modern fixation, one that's doing us and our world damage. Probably most of our foods—fruits, vegetables, grains, pulses, fish, meats, and dairy—should be fermented in some way before we eat them, and likely a combination of both cooking and natural fermentation, as in sourdough bread baking, yogurt making, and pasta filata cheesemaking, is best. As proof I present children, who are especially drawn to sourdough bread, yogurt, queso Oaxaca, salami, and, of course, sour pickles. I say, why force a child to eat their broccoli, when they'll gladly slurp up sauerkraut!

Very nearly any vegetable (or fruit) can be fermented this way; and the typical approach is very nearly always the same. A vegetable is preserved by submerging it in salty liquid, either a brine, like a cucumber pickle, or liquid extracted from the vegetable itself by salt as in sauerkraut. By keeping the vegetables submerged in the salty brine, a bacterial fermentation is initiated by the beneficial microbes on the vegetable. Lactofermenting bacteria convert the vegetables' sugars into lactic acid that protect the produce from various degradations, all while developing their flavor and nutrition.

From a cheesemaking perspective, this pickling lesson is indeed relevant, for there are a good number of brined cheeses that are preserved in a similar way. And for picklers, natural cheesemaking can be a wellspring of ideas for improving the quality of a vegetable ferment.

For instance, cheesemaking has taught me to source vegetables, be they cucumbers, cabbages, carrots, or kohlrabi, as fresh as possible, preferably harvesting them and fermenting them straight away, still warm from the sun. Vegetables right out of the garden are in their best state, both minerally and microbiologically, and serve as the best medium for pickling, much like milk still warm from the udder transforms into its best cheese.

Now, what about a starter? A cheese should probably never be prepared without the addition of a starter, but vegetable fermenters most often ferment without one. The use of starters is practiced mostly by larger-scale pickle processors, who take a more controlled approach and strive for a more consistent ferment. So, can a starter be kept naturally for making a perfect pickle? Of course! A small amount of the previous day's active pickle brine (or even kefir or clabber) can be added to the next to assure a more effective fermentation. And instead of it taking 2 days for fermentation to initiate, and all the microbiological uncertainty that brings, the ferment will be actively acidifying in 24 hours. This technique may necessitate a regular pickle preparation, but for best results in all realms of fermentation, especially for producers who sell to the public, this goal should be sought after.

Once the primary fermentation subsides, usually after 1 week at room temperature, the surface of the brine will often go moldy as a result of indigenous fungi like *Geotrichum candidum* (often called kahm), which create a beautiful white veil atop the brine. These yeasts are not dangerous—they play an important role in the ripening of many cheeses—but they do slowly change the character of vegetables preserved beneath the brine.

A big barrel of pickles will be less affected by this surface fungus than a small jar (size matters in pickling, as it does with cheese—and not just the size of the pickle!). And if you don't have enough at one time to make a larger batch, consider making a perpetual pickle, adding cucumbers to the barrel every day or two as they ripen throughout the summer and fall, covering each addition with its appropriate amount of salt and water. This traditional method also inadvertently creates the conditions for cultivating an effective pickle starter, assuring the best fermentation of each added vegetable.

A final way of integrating pickling and cheesemaking is to consider adding fermented whey to the brine, further improving the pickles' flavor and nutrition. I often even prepare my pickles with a full whey brine; if milk and water shouldn't mix, then why should vegetables and water? The savoriness, even meatiness, of such a pickle is unparalleled!

Making Sour Pickles

A vegetable ferment • 1 hour make • 1-week to 6-month affinage

INGREDIENTS

pickling cucumbers	2 kg (medium size)
	garlic, dill, caraway, pepper, mustard, coriander, and other pickling spices to taste
	vine leaves or other sources of tannins, like oak, maple, or blackcurrant leaves
pickle starter (optional)	40 ml (1:100) active pickle brine, kefir, clabber, or whey
salt	80 g or ⅓ cup (2 percent)
water, whey, or a mix	2 L
container	4 L jar or crock
YIELD	**2 KG PICKLES**

TECHNIQUE

Wipe cucumbers clean with a cloth, and remove spines.
Place spices and leaves into container.
Place cucumbers into jar, arranging for maximum packing.
▸ Fill jar to within 2 cm of top.
Pour in pickle starter, if using.
Dissolve salt in water to make brine.
Pour brine over cucumbers just to cover.
▸ Keep cucumbers submerged with the help of a plate or a weight.
▸ Cover to keep flies from congregating.
Leave to ferment 1 week at room temperature, until pickles change color.
Cover with a lid, and leave to age in cave or refrigerator 2 weeks before enjoying.
▸ Pickles can be enjoyed halfway through fermentation as half-sours.
Pickles can be preserved through the winter submerged under their brine.

SUMMER • 275

Feta

Feta is an aged brined cheese formed in large blocks. It's made to be firm but crumbly, with white flesh, small eyes, and a distinctive sour taste. Though feta is most often associated with Greek cheesemaking traditions, similar cheeses are made in many regions, especially around the Mediterranean and Middle East—climates where the warm weather made aging cheeses more challenging in certain seasons. In Iran it's known as Lighvan, in Palestine, Akawi, and in Turkey, beyaz peynir.

Feta is most commonly made from sheep's or goat's milk or a blend of the two, which develop its most sought-after flavors, textures, and colors. Sheep's and goat's milk evolve the best tastes when submitted to the cheese's longer fermentation. They also make a cheese that is whitest in color. They are also the most climate-appropriate animals in the dry and mountainous regions where this cheese is commonly made. The animals are often kept in mixed herds around the Mediterranean, and their milks often mingle in the vat, too!

As well, Southern European lamb or kid rennet is considered most suitable for this cheese. These more traditionally prepared stomachs, with their cheesy contents, give a sought-after lipasy character. Because the cheese is not ripened by flavor-inducing surface flora, this is especially important for creating a fuller-bodied feta.

The flesh of the feta is made firm and crumbly through a combination of a stirred-curd method and a subsequent long fermentation before salting. This is much like a blue cheese; indeed, the cheesemaker more or less uses the same technique as for Roquefort!

Cutting the curd to a medium size preserves the fat of the milk, helping to make the cheese creamy. Stirring for a long period helps firm the curd but also produces the characteristic mechanical holes. The cheese should be left unpressed to preserve the eyes within. After the make, fetas should be long fermented, preferably for 2 or more days before salting, to assure a well-acidified paste. A more acidic state assures a better crumble but also helps to keep the cheese safe and delicious for long periods under the brine.

Like pickled vegetables, brined cheeses are best preserved below salty and acidic brines, which limit the growth of unwanted yeasts and fungi, and create the best conditions for conservation. Feta brines should therefore also be made to be as acidic as possible before salting. Also, like pickled vegetables, you can make the cheeses through the dairying season, filling up barrels slowly with blocks of cheese and covering them with salty fermented whey.

An extra dose of salt helps keep feta fungus-free. If too little salt is added, the brine can become heavily covered with *Geotrichum*, which can alter the flavor of the cheeses below. The saltiness of the aged cheese can be quite high, so feta is often added to dishes both as a source of flavor and for salt!

Making Feta

A stirred-curd cheese • made with goat's or sheep's milk • aged in brine
72-hour make • 1- to 6-month affinage

INGREDIENTS

milk	20 L goat and/or sheep
starter	200 ml kefir, clabber, or whey (1:100)
rennet	regular dose kid or lamb rennet with lipase
salt	
forms	round: 20 cm diameter × 15 cm deep
YIELD	ONE 2 KG FETA IN BRINE (10:1)

TECHNIQUE

Bring milk to cheese temperature, about 35°C (95°F).
Add starter, mix in thoroughly.
Add rennet, mix in thoroughly.
Wait for clean break, 45 minutes to 1 hour.
- ▸ Cover the pot to keep in warmth.

Cut curd to hazelnut size, 1.5 cm.
- ▸ Cut vertically; cut crosswise; cut horizontally or diagonally.

Stir curds for 60 minutes at 35°C until they bounce from 90 cm.
Pitch curds for 5 minutes.
Preserve 4 L whey for aging brine.
- ▸ Leave to ferment overnight.
- ▸ Then add 7 percent salt by weight to make a medium brine.

Whey off.
Transfer curd to cheesecloth-lined form.
Drain 2 to 3 days until acidity develops.
- ▸ Goal pH 4.9—stretch test fails.
- ▸ Flip daily during the drain, with first flip at 1 hour.
- ▸ Keep cheese humid during draining.
- ▸ Rub down rind daily to limit fungal growth.

Salt cheese for 12 hours in saturated salt brine (6 hours per kg).
- ▸ Or surface-salt over 2 days.

Drain cheese for 1 day on draining table, covered with cheesecloth.
- ▸ Flip twice during draining.

Age cheese submerged in medium brine for 1 to 4 months at cool temperature.

Sirene

This is a variation on the brined feta technique that yields a softer, creamier cheese. Essentially a lactic cheese brined like feta, Bulgarian sirene (the name simply means "cheese") makes the most of its milk, conserving all the fat in its make and leaving the cheese even more flavorful. Perhaps the most perfect preservation of a lactic cheese, the brining allows the soft cheese to develop its most evolved flavors without exposure to air and its degradations.

To make this delicious and effortlessly preserved cheese, first prepare a lactic curd—preferably from goat's or sheep's milk, though cow's milk will do, too. The milk is cultured naturally, and preferably curdled with a more traditional Southern European rennet, giving this most flavorful cheese even more flavor. Leave the curd to ferment overnight, then ladle it carefully into forms to preserve the grain of the milk, essential to the fine texture of the finished cheese.

The curd is drained, then salted, and, once dried, left submerged beneath a brine of its salted fermented whey for weeks or months untouched. The brine, prepared with a generous helping of salt, keeps the little lactic cheeses free from fungus and allows them to develop their biggest flavors without interference from external rind ecologies. The result is a complex and creamy cheese, with a bright hit of acidity, that melts on the tongue.

A variation on this cheese involves submerging the salted and dried lactic cheese in olive oil. As with any cheese or food preserved in oil, the cheese must be fully fermented and very acidic to be properly kept without unwanted microbiological developments under the oil.

Making Sirene

A lactic cheese • made with goat's or sheep's milk • aged in brine • 48-hour make • 1- to 4-month affinage

INGREDIENTS

milk	10 L goat or sheep
starter	100 ml kefir, clabber, or whey (1:100)
rennet	¼ dose kid or lamb rennet with lipase or regular dose rennet for cow's milk
salt	
forms	baskets: 10 cm diameter × 10 cm deep
YIELD	TWO 0.75 KG SIRENE IN BRINE (7:1)

TECHNIQUE

Bring milk to culturing temperature, about 20°C (68°F).
Add starter, mix in thoroughly.
Add rennet, mix in thoroughly.
Wait 12 to 24 hours, until milk sets to a lactic curd.
- ▸ Cover the pot.
- ▸ Leave undisturbed.

Ladle curd into forms, topping it up as it drains.
- ▸ Whey off as the curd is ladled.

Preserve 1 L whey for aging brine.
- ▸ Add 7 percent salt to make medium brine.

Leave curd to drain in forms for 24 hours.
- ▸ Flip cheeses after 12 hours.

Salt the cheeses on all sides.
Drain cheeses for 1 day on draining table,
  covered with cheesecloth.
- ▸ Flip cheeses after 12 hours.

Surface-salt cheeses to 2 percent.
- ▸ Leave to drain 24 hours.

Age cheeses submerged in brine for 1 to 4 months
  at cool temperature.
- ▸ Leave undisturbed.

PHOTO BY ALIZA ELIAZAROV

Tomme de Savoie

Tomme de Savoie is a beautiful wheel of semi-firm cheese with a most natural philosophy in both its make and its aging. It is an uncooked, unpressed cheese made on the edge of the alpine style, originating in the foothills of the French Alps in a region known as Savoie. Similarly rustic styles known as toma are made on the Italian side of the mountain range in Piedmont.

Tomme and toma are examples of small alpine cheeses made by small farmers, and often in the early stages of the seasonal transhumance, while taking their animals up to their seasonal pastures before handing them over to the hands of hired summer cowherds. They are made with the smaller amount of milk of a smaller herd, sometimes mixed with a little goat's or sheep's milk. The smaller size of the resulting cheeses is not typical of alpine styles, but the making and aging still reflect the philosophy of mountain cheeses.

Tomme is a simple introduction to the alpine method. The cheese is made to a pared-down alpine technique, with less milk; a larger, corn-sized grain; no cooking; a light amount of pressing; and a hands-off affinage. Unlike other alpine styles, the cheese has a fairly moist, high fat paste that ripens relatively quickly and decadently, as well as a lightly broken-down curd. It is lightly pressed by hand, but because the cheese isn't cooked, the curd does not knit together perfectly, leaving small mechanical holes uniformly distributed throughout that sometimes develop into colorful eyes toward the rind. Because of its typical small size, it is also more affected by its rind ecology, which is left to grow untrammeled like a wildwood upon the cheese.

There are many different types of tomme (one of many terms used to describe a wheel of cheese in France), but the greatest distinction of the most famous "cheese of Savoy" is its gorgeous gray-brown rind. Wheels of tomme de Savoie are neither washed nor scrubbed and develop an extreme example of a rind. The farmers that make it have neither the time nor the desire to treat their long-aged cheeses. They thus leave their cheese to develop a natural bluish rind tinted with *Penicillium roqueforti*, which tends to brown and thicken over time due to age and develops intriguing fungal flavors. The cheese's only treatment is that it is flipped; this regular handling lightly pats down the fungus, turning it into a thick and almost waxy rind. This hands-off affinage helps develops substantial taste in the cheese that the French prefer over a more rigorously treated rind.

Due to the wild nature of the aging, cheese mites often inhabit the natural rind of well-aged tomme. This is the first cheese in the book susceptible to mites, which are attracted only to long-aged cheeses, especially the rinds of those that are left unwashed or brushed. And though many makers might turn in fright at the sight of a mite, French makers prefer the strong flavors they bring, a preference that contrasts with the Swiss style of controlling mites by rigorously washing their mountain cheeses.

Making *Tomme de Savoie*

A semi-alpine cheese • made with cow's milk • aged with a *roqueforti* rind
8- to 12-hour make • 4- to 6-month affinage

INGREDIENTS

milk	20 L
starter	200 ml kefir, clabber, or whey (1:100)
spores	*Penicillium roqueforti* (optional)
rennet	regular dose calf rennet
salt	
forms	round: 20 cm diameter × 10 cm deep

YIELD ONE 2 KG TOMME (10:1)

TECHNIQUE

Bring milk to cheese temperature, about 35°C (95°F).
Add starter and *roqueforti* spores, mix in thoroughly.
Add rennet, mix in thoroughly.
Wait for clean break, 45 minutes to 1 hour.
- Cover the pot to keep in warmth.

Cut curd to corn size, 0.75 cm, slowly with a spino.
Stir curds nonstop over 30 minutes at 35°C.
- Increase tempo of stirring as curds firm.

Pitch curds for 5 minutes.
Press curds together underneath the whey.
Transfer curds from pot into a cheesecloth-lined form.
Press cheese lightly by hand, or by stacking tommes.
- Keep cheese warm during pressing with whey or warm water.

Leave in form 8 to 12 hours, until acidity develops.
- Goal pH 5.3—stretch test spins.
- Flip four times during the drain, with first flip at 10 minutes.

Salt cheese two times over two days.
Or brine for 12 hours in saturated salt brine (6 hours per kg).
Drain cheese for 1 day on draining table, covered with cheesecloth.
- Flip twice during draining.

Age cheese in cave for 4 to 6 months.
- Flip cheese every other day.
- Never wash or brush.

Tomme de Chèvre

This is a study in alpine sheep and goat's milk cheese—extraordinary styles with unique characters that are deserving of their own methodologies. For sheep's milk and goat's milk behave differently than cow's milk, especially when transformed into alpine cheeses.

There's an extraordinary diversity of hard sheep and goat cheeses: cooked to various temperatures or not cooked at all, achieving all sorts of different paste textures, and pressed or unpressed to leave closed or open interiors. They can have rinds of all sorts: white *Penicillium candidum*, orange *Brevibacterium linens*, cream *Geotrichum*, or blue/brown *Penicillium roqueforti*. And sheep and goat's milk alpine cheeses can even be made to develop eyes, much like Emmentaler, by making them taller and keeping them warm after salting.

How do these two milks behave differently? Sheep's milk curds are very sensitive to heat and need only be lightly handled to respond as cow's milk curd does when handled more vigorously. To make an equivalent to tomme de montagne with sheep's milk (which yields a famous Basque sheep cheese known as Ossau-Iraty), the sheep curd should just be cut small and not cooked at all, instead being consistently stirred at its original temperature until it achieves its sought-out strong and

PHOTO BY ALIZA ELIAZAROV

elastic texture. To cook the sheep curd to the 45°C (113°F) of a cow's milk tomme de montagne would cause the cheese to achieve the equivalent texture to a Comté, which is normally made with cow's milk and cooked to 52°C (126°F). Goat's milk responds similarly to sheep, with very little effort from the cheesemaker, but is not as thick and rich and yields just over half the cheese that sheep's milk provides.

For this particular sheep's or goat's milk cheese, several steps make it stand out from the flock. First, the milk is best cultured with a sheep or goat's milk clabber, kefir, or whey, which provides sheep- or goat-specific microbes (though a cow's milk culture will work well enough); use a traditional lamb or kid rennet, which provides more diverse enzymes including lipase, resulting in a fuller flavor (though a modern cow rennet will work); cut the curd small; stir without cooking; and press the curd under whey into a cheese by hand. Dry-salt or brine the cheese, then leave it to develop a delicate white / orange rind through a regular washing in its early days. The cheese can be enjoyed young, unlike a cow's milk alpine cheese, which lacks flavor at that age but evolves even more extraordinary appeal when aged.

A more effortless and elegant alpine cheese just does not exist! It's almost as if goats and sheep that often graze the highest peaks (they can scale almost vertical cliffs!) provide a milk most suited to the alpine make. Indeed, alpine cheesemaking practices likely began in mountainous regions in the Middle East, regions very well suited to grazing goats and sheep.

Making Tomme de Chèvre

An alpine cheese • made with goat's or sheep's milk • aged with a *roqueforti* rind
8- to 12-hour make • 2- to 6-month affinage

INGREDIENTS

milk	20 L goat or sheep
starter	200 ml kefir, clabber, or whey (1:100)
rennet	regular dose kid or lamb rennet with lipase
salt	
forms	round: 20 cm diameter × 15 cm deep

YIELD ONE 2 KG TOMME DE CHÈVRE (10:1)
ONE 3.5 KG TOMME DE BREBIS (7:1)

TECHNIQUE

Bring milk to cheese temperature, about 35°C (95°F).
Add starter and roqueforti spores, mix in thoroughly.
Add rennet, mix in thoroughly.
Wait for clean break, 45 minutes to 1 hour.
Cut curd to corn size, 0.75 cm, with a spino.
Stir curds 15 minutes at 35°C, stirring nonstop.
Check for curd readiness.
- Curd will form peaks when drained in hand and squeezed.

Pitch curds for 10 minutes.
Form curds into a wheel and press under the whey.
Reserve 2 L whey to make light salt brine for washing.
- Ferment whey overnight, then add 2 percent salt.

Retrieve cheese from pot with cheesecloth and press into forms.
- Press by hand, or by stacking cheeses.
- Keep cheese warm by pouring whey overtop.
- Save whey for ricotta or for skimming cream for sheep or goat butter.

Drain 8 to 12 hours, until acidity develops.
- Goal pH 5.3—stretch test positive.
- Flip four times during the draining, with first flip at 10 minutes.

Surface salt cheese, covering the rind with salt daily for two days.
Or salt 8 hours per kg in saturated salt brine.
Drain cheese for 1 day on draining table, covered with cheesecloth.
- Flip twice during draining.

Age cheese in cave for 3 to 6 months.
- Flip cheese every other day.
- Wash rind with light brine every day for first week.

Tomme de Montagne

Literally "cheese of the mountain," this delicious semi-alpine cheese from France is lightly cooked and lightly pressed, with a light white rind—a sort of alpine cheese light. The curd is uniform and almost without mechanical holes, although propionic eyes may develop as in an Emmentaler depending on the application of salt.

A more elevated alpine cheese, this tomme is more reflective of the traditions of making cooked cheeses on mountain pastures. It is a relatively large alpine wheel, the first in the book, using heat to bring the curds together into a firmer form that is larger, transports more easily, and tastes more typically of mountain styles. If you want to make a long-preserving cheese with a good quantity of milk that is easier to age, this is your make!

Success with this style, and other cooked alpine cheeses, does demand a maker's full attention, however. Cooking curd requires constant care and tending: stirring without a break so that the curd doesn't knit and burn; watching the temperature so it doesn't cook too quickly; and judging when the curd reaches its best state for pressing so that it doesn't lose too much moisture. Imagining yourself alone, meditating on your milk on a mountaintop, helps to achieve this cheese enlightenment.

The cheese must be pressed right to become monolithic. Forming the wheels under their whey and retrieving with cheesecloth, pressing by hand, and flipping regularly, all the while warming with whey, all help to achieve a uniform, crack-free wheel. Especially with larger, longer aged cheeses, the finished wheel should be knit together perfectly, leaving no room for mites and fungus to burrow inside.

Because of the cheese's medium size and relatively high moisture content (due to being cooked only slightly), this semi-alpine cheese develops its flavor faster than fuller-alpine cheeses. Tommes are best eaten in their first winter, when they're less than a year old. More than a year of aging can cause the cheese to overripen, developing brown tints in its flesh, an overgrowth of cheese mites, and a strong reek of ammonia that doesn't cast the cheese in its best light. This is the limit of ripening a semi-alpine cheese. To age a cheese longer, cheesemakers must make the full-alpine style.

Washing the rind of the cheese during the first week of aging helps establish a white *Penicillium candidum* rind. The fluffy fungus is slowly patted down over time as it is handled and flipped regularly, to make a dense white rind. No other color will show on the rind if the cheese is made and handled right. But a diversity of rind treatments, or lack thereof, is acceptable for this style, because rind ecologies have less of an effect on the flavor of such a hard cheese, and because the degradation from mites is not as serious a concern compared with longer-ripened full-alpine styles.

Making Tomme de Montagne

A semi-alpine cheese • made with cow's milk • aged with *Penicillium candidum*
8- to 12-hour make • 6-month affinage

INGREDIENTS

milk	40 L
starter	200 ml mesophilic kefir, clabber, or whey (1:200)
	and 200 ml thermophilic yogurt, clabber or whey (1:200)
rennet	regular dose calf rennet
salt	
forms	round: 25 cm diameter × 15 cm deep
YIELD	**ONE 4 KG TOMME** (10:1)

TECHNIQUE

Bring milk to cheese temperature, about 35°C (95°F).
Add starters, mix in thoroughly.
Add rennet, mix in thoroughly.
Wait for clean break, 45 minutes to 1 hour.
Cut curd to lentil size, 3 mm, with a spino.
Cook curds to 45°C (113°F) over about 30 minutes, stirring nonstop.
  ▸ Increase tempo of stirring as temperature rises.
Check for curd readiness.
  ▸ Curd forms peaks when drained and squeezed in hand.
Take pot off heat and pitch curds for 5 minutes.
Form curds into a wheel by hand and press under the whey.
Retrieve cheese from pot with cheesecloth and place into form.
Press by hand into form, or stack cheeses two or three high.
  ▸ Keep cheese warm as they form by pouring whey from the vat overtop.
Drain 6 to 12 hours, until acidity develops.
  ▸ Goal pH 5.3—stretch test spins.
  ▸ Flip four times during the drain, with first flip at 30 minutes.
Reserve 1 L whey, ferment overnight, then make a light salt brine (2 percent salt).
Salt cheese, covering rind with salt daily for 2 days.
  ▸ Or brine for 24 hours in saturated salt brine (6 hours per kg), flipping twice.
Drain cheese for 1 day on draining table, covered with cheesecloth.
  ▸ Flip twice during draining.
Age cheese in cave for 6 months.
  ▸ Flip cheese every other day.
  ▸ Wash rind with light brine every day for first week (optional).
  ▸ Brush cheese weekly after 2 months to fend off mites (optional).

Raclette

Our first washed-rind alpine cheese, raclette is transformed from milk in much the same manner as a tomme de montagne, with a small curd size, a light cooking, and a light pressing. But the more careful aging regimen makes this a most attractive alpine cheese, for, as is typically practiced in Swiss alpine cheesemaking, raclette is a rigorously washed wheel of cheese whose rind glows orange-pink like an alpine sunset.

Famous for its meaty melt, raclette makes a rustic fireside "fondue" when the wheel is left with its cut end facing the intense heat of a fire to melt, so that the Dalí-esque ooze can be scraped up by a knife. (The word *raclette* in French refers to this scraping—it translates roughly as "squeegee.") The melted cheese is slathered over roasted potatoes, enveloping them in a fatty, velvety embrace. Its relatively high moisture content, for an alpine cheese, allows raclette to achieve its fullest flavor and best melt by winter, just in time for the fire to be lit.

Raclette has just the right acidity and moisture to slowly ooze its flavorful flesh from the wheel when heated. To achieve this just-right raclette melt, the cheese must be made just right, with the freshest raw milk, curdled with natural rennet, cut and cooked to just the right temperature where the curds form firm peaks when squeezed, and pressed just right into a wheel of just the right size so that it ages well. The wheel must be salted just right and at just the right acidity, where its firm curd spins. Once in the cave, the wheel is washed like clockwork for months until the rind glows, and the elastic paste just begins to break down; it's at this stage that raclette achieves the most magnificent melt in the panoply of cheese.

PHOTO BY ALIZA ELIAZAROV

Making Raclette

A semi-alpine cheese • made with cow's milk • aged with a washed rind
8- to 12-hour make • 3- to 6-month affinage

INGREDIENTS

milk	40 L
starter	200 ml mesophilic kefir, clabber, or whey (1:200)
	and 200 ml thermophilic yogurt, clabber, or whey (1:200)
rennet	regular dose calf rennet
salt	
forms	round: 30 cm diameter × 15 cm deep

YIELD ONE 4 KG RACLETTE (10:1)

TECHNIQUE

Bring milk to cheese temperature, about 35°C (95°F).
Add starters, mix in thoroughly.
Add rennet, mix in thoroughly.
Wait for clean break, 45 minutes to 1 hour.
- Cover the pot to keep in warmth.

Cut curd to lentil size, 3 mm, with a whisk or spino.
Cook curds to 45°C (113°F) over 30 minutes, stirring nonstop.
- Increase tempo of stirring as temperature rises.
- Curds are ready when they form firm peaks.

Take pot off heat.
Pitch curds for 5 minutes.
Reserve 2 L whey to make light salt brine for washing.
- Ferment whey overnight, then add 2 percent salt.

Retrieve cheese from pot with cheesecloth, and place into form.
Press by hand in form, or stack cheeses two or three high.
- Keep cheeses warm as they form by pouring whey from the vat overtop.

Drain 6 to 12 hours, until acidity develops.
- Goal pH 5.3—stretch test spins.
- Flip four times during the drain, with first flip at 30 minutes.

Salt cheese, covering rind with salt daily for 2 days.
Or brine for 24 hours in saturated salt brine (6 hours per kg), flipping twice.
Age cheese in cave for 3 to 6 months.
- Flip cheese every other day.
- Wash with light salt brine every other day for 3 months.
- Brush cheese once a week every week thereafter.

Tomme Crayeuse

The French name of this cheese translates to "chalky tomme," referring to the dense and chalky, lactic-like texture that this tomme contains at its core. But this smooth texture is in great contrast with the liquefied paste that develops on the tomme over time near the rind. It's a longer fermentation period prior to salting that gives this cheese its distinction in texture and taste.

This cheese is more or less a tomme de Savoie that's forgotten about on the draining table. When you wait 2 or 3 days before the cheese is salted, the paste become more fermented and brittle, and the rind develops a white hue due to *Geo* and *Penicillium candidum*, instead of the typical darker *P. roqueforti* rind.

But most importantly, unlike other mountain tommes, the outermost part of the cheese becomes broken down by fungus, resulting in a gooey rind contrasted with the chalky core, much like a traditional Camembert or crottin, but more massive and dense. This cheese in two parts comes about from the longer fermentation pre-salting, which increases the acidity of the cheese, softening the paste and restricting the effects of proteolysis to the exterior. The cheese even develops fat wrinkles on its rind, a result of the *Geotrichum*'s manipulation of the semi-soft cheese to maximize the surface area upon which it grows.

Now, this is not necessarily a sought-after condition for a cheese. Most alpine cheesemakers would consider such a softer-textured, proteolyzed, and shorter-aging wheel a to be a flawed one; after all, the goal in mountain cheesemaking is to make as easily transportable and long-keeping a cheese as possible! But the advanced fermentation causes the cheese to develop its best flavor sooner than a typical shorter-fermented and less acidic alpine cheese. This flavor is especially complex when the cheese is prepared with goat's or sheep's milk.

Because of the longer fermentation, this style is best prepared with small-ruminant milk, whose more flavorful fatty acids are rendered most expressive. With more exceptional flavors more quickly, and whiter, creamier paste, this style is a standout in the alpine family: more or less a lactic alpine cheese!

PHOTO BY ALIZA ELIAZAROV

Making Tomme Crayeuse

A semi-alpine cheese • aged with a *Geotrichum* rind • 48-hour make • 1- to 3-month affinage

INGREDIENTS

milk	20 L
starter	200 ml kefir, clabber, or whey (1:100)
rennet	regular dose calf rennet
salt	
forms	round: 20 cm diameter × 10 cm deep

YIELD ONE 2 KG TOMME (10:1)

TECHNIQUE

Bring milk to cheese temperature, about 35°C (95°F).
Add starter, mix in thoroughly.
Add rennet, mix in thoroughly.
Wait for clean break, 45 minutes to 1 hour.
- ▸ Cover the pot to keep in warmth.

Cut curd to corn size, 0.75 cm, with a spino.
Stir curds nonstop for 30 minutes.
- ▸ Increase tempo of stirring as curds firm.

Pitch curds for 5 minutes.
Press curds together under the whey for 30 minutes at 35°C.
Transfer curds from the vat into a cheesecloth-lined form.
- ▸ Press cheeses into forms by hand.
- ▸ Keep the curd warm and pressable with warm whey.

Leave in forms up to 2 to 3 days until acidity develops,
  and *Geo* grows on the rind.
- ▸ Flip four times during the first day, first flip at 30 minutes.
- ▸ Flip twice daily on subsequent days.
- ▸ Keep cheese covered in cheesecloth.
- ▸ Goal pH 4.9—stretch test fails.

Salt cheese.
- ▸ Surface-salt daily over 2 days.
- ▸ Or salt for 12 hours in saturated salt brine (6 hours per kg).

Drain cheese for 1 day on draining table,
  covered with cheesecloth.
- ▸ Flip twice during draining.

Age cheese in cave for 2 to 4 months.
- ▸ Flip cheese every other day.
- ▸ *Never* wash or brush.

Comté

Comté is a mountain cheese with a most challenging ascent. This famous fromage is considered a full-alpine cheese, with all the attendant tools and technologies. Many cheeses throughout the Alps are prepared in similar ways, according to similar mountain cheese-making philosophies, including Swiss Gruyère, German bergkäse, Italian fontina, and many lesser-known but equally extraordinary wheels that never leave their mountain valleys.

The most important consideration when making these traditional cheeses of the transhumance involves respecting the traditions of their mountainous makes. That means working with large quantities of pastured animals' milk still warm from the udder, cultivating the microbes of milk through backslop, curdling the milk with animal rennet, cutting the curd small, cooking the curd in a copper cauldron to 52°C (126°F), forming the curd into one wheel under hot whey, pressing and salting the cheese by hand, and aging slowly on wood with a regular washing of the rind.

To follow these traditions dictates that the cheese must be made massive, with the fresh milk of around twenty cows (about 200 L) per wheel. This style, made at the limit of the human scale, tests a cheesemaker's physical abilities. With full-alpine cheeses, the more massive they are (to a limit!), the easier they are to carry down the mountain, and the longer and better they age.

Comté is among the largest of the alpines. Today it is produced in collective dairies known as fruitières throughout the mountainous Franche-Comté region of France. The roots of the current incarnation of the cheese, like those of cheddar, are in the cooperatization of dairies in the nineteenth century, as many small herds pooled their milk to make more massive cheeses. A single small herd in the region cannot typically produce the milk needed for a single wheel of today's Comté, as written in its AOP—upward of 400 L!

To make these cheeses right, you should also consider making them in a copper kettle, which largely defines the way the massive cheese comes together and enables the proper cooking of the curd and its removal from the vat. Copper allows the cheesemakers to cook the curd over fire, permits the cheese to be pressed together into its wheel under whey, and enables the removal of the wheel from its whey within its cheesecloth to form a massive, seamless wheel of cheese.

And though these full-alpine cheeses are massive, when first made, their heights are usually less than 15 cm. This flattened shape is important for two reasons. First, these enormous cheeses were historically carried down the mountain on foot, and to be carried comfortably, the cheeses had to be supported above the hiker's center of gravity (a more rotund round of cheddar wouldn't easily make the trip). Second, if the cheese is taller, salt penetration is slower, and the cheese will develop propionic bacteria and their gas. The Emmentaler technique coming up calls for making a cheese taller, which naturally leads to the development of eyes.

So let's detail the lengthy alpine make. Most alpine cheeses adhere to a somewhat rigorous orthodoxy, which evolved from the unique circumstances faced by alpine cheesemakers in their mountain huts. Milk, still warm from the udder, is cultured with thermophilic whey and coagulated with traditional rennet, usually prepared by steeping a certain amount of a vell in the day's whey overnight to curdle and ferment the next day's milk. Upon reaching a full curdle, the curd is cut to a small size with a spino. It's then cooked in a copper kettle suspended over the coals of a fire and stirred nonstop as the temperature slowly rises to about 52°C (126°F).

Upon reaching the goal texture (the curd forms very stiff peaks and squeaks on the teeth), the pot is moved off the fire and the curds are left to settle to the bottom for some time; there they knit together under the weight and warmth of the whey into their cheese. Cheesemakers then bundle the cheese in cheesecloth with the help of a flexible bow that draws the strong cloth beneath it, so that the cheese can be maneuvered more easily from the vat.

With great strength (and often with the assistance of a second cheesemaker), the massive wheel is raised out of the whey within its cheesecloth bundle (there is no more birth-like cheese make—think of the stork and its baby bundle!) and placed in an adjustable cheese form known in German as a Järbe. The cheese is flipped

PHOTO BY ALIZA ELIAZAROV

several times and kept warm with whey to assure an effective pressing only by hand.

The cheeses are left overnight to ferment before salting (the low-moisture curd takes longer to develop its necessary acidity). Alpine cheesemakers typically salt the cheese slowly by hand, applying salt to each side every day for a week. This slow but thorough salting effectively pulls out moisture from within the massive cheeses, halting their primary fermentation, and eliminating gas production.

As the salting ends, the aging begins, with the wheel flipped and washed with a light salt brine every other day for months, even years. The washing develops the orange ecology and creates the most sought-out textures, flavors, and aromas in the alpine kingdom. And it takes a sustained washing for the life span of the cheese to maintain that beautiful rind; if the washing is stopped prematurely, white or blue fungus will grow above the rind while mites will burrow below, taking away from its visual appeal, altering its flavor, and slowly but surely consuming the wheel.

All of this work is worth it, for the most challenging ascent of a cheese makes for a most flavorful descent. With its intense concentration of good milk, its slow and patient affinage, and an even balance of surface and interior ripening, it is a most perfect preservation of milk.

Making Comté

An alpine cheese • made with cow's milk • aged with a washed rind
12- to 24-hour make • 6- to 24-month affinage

INGREDIENTS

milk	200 L
starters	1 L thermophilic yogurt, clabber, or whey (1:200)
	and 1 L mesophilic kefir, clabber, or whey (1:200)
rennet	regular dose calf rennet
salt	
forms	jarbe: 45 cm diameter × 15 cm deep
YIELD	ONE 20 KG COMTÉ (10:1)

TECHNIQUE

Bring milk to cheese temperature, about 35°C (95°F).
Add starters, mix in thoroughly.
Add rennet, mix in thoroughly.
Wait for clean break, 45 minutes to 1 hour.
- Cover the pot to keep in warmth.

Cut curd to lentil size, 3 mm, with a spino or whisk.
Slowly cook curds to 52°C (126°F) over 30 minutes, stirring nonstop.
- Increase tempo of stirring as temperature rises.
- Curd is ready when it forms peaks and squeaks.

Take pot off heat.
Pitch curds for 30 minutes.
Reserve 2 L whey to make light salt brine for washing.
- Ferment whey overnight, then add 2 percent salt.

Retrieve cheese from pot with bow and cheesecloth, and place into form.
Press lightly overnight 12 to 24 hours, until acidity develops.
- Goal pH 5.3—stretch test positive.
- Flip four times during pressing, with first flip at 10 minutes.
- Rewrap the cheese in its cheesecloth with each flipping.

Salt cheese every day in hastening space for 7 days.
- Flip cheese daily and cover only the top side with salt.

Age cheese in cave for 6 to 24 months.
- Flip cheese every other day.
- Wash the rind with brine every other day for 1 month.
- Rub rinds with cloth or brush every other day from 1 month onward.

Emmentaler

Emmentaler is a celebrated cheese of Switzerland that, because of its particular make, swells from the growth of gas-developing microorganisms. Famously reproduced as "big wheel Swiss" in North America, it is also sought after in South America (in Uruguay in particular), an alpine cheese introduced to the vast pampas by Swiss settlers in the nineteenth century.

You don't need to be in the mountains to make alpine cheese; and this particular alpine comes from the lowland Emmental Valley, not from the alpage. However, its make was adjusted to the circumstances of valley cheesemaking, which changed the cheese. Specifically, the wheels were made much more substantial, as they didn't have to be carried down the mountain. And because of the cheeses' significant size, they have the natural tendency to develop their eyes.

Emmentaler is made much taller than other alpine cheeses. This slows the passage of salt to the interior. This results in continued fermentation at the core and the growth of gas-producing microbes, known as propionibacteria, that thrive in the firm fermenting curd. Size matters for these cheeses: An AOP Emmentaler is typically made with 1,000 L milk and often measures almost a meter across and weighs in at over 90 kg, far too heavy for a person to easily carry! Most other alpine cheeses are, at most, half as big, and typically only 15 cm or so high.

PHOTO BY ALIZA ELIAZAROV

The big wheels swell on their own from the growth of the usually unwanted gas producers (even one propionic eye in a Comté or Parmigiano-Reggiano is an unacceptable flaw). But it's not just size that matters; the cheese must be cooked to 52°C (126°F) to achieve a fully elastic paste that swells with perfect round eyes, and the cheese must be kept warm after salting to encourage propionic growth (they don't like it cold, like most other milk microbes).

To get consistent gas development, it also helps to add a source of the often unwanted propionibacteria culture. Fortunately, raw milk, whey, clabber, and kefir all contain healthy populations of these friendly gas-producing microbes. My first Emmentaler (it's the one in the photo!) swelled without the addition of a specific culture. Altogether these acts create a cheese that is softer in texture than a typical alpine wheel, with a particular acidity and a very slight saltiness.

It's almost as if the Emmentaler has to be made wrong to be made right. Like a poorly pressed cheddar made more elegant and delicious with spontaneous blue veining, it is an alpine cheese made to have beautiful flaws.

Making Emmentaler

An alpine cheese • made with cow's milk • aged with a washed rind and propionic eyes
2- to 24-hour make • 3- to 12-month affinage

INGREDIENTS

milk	200 L
starters	1 L thermophilic yogurt, clabber, or whey (1:200)
	and 1 L mesophilic kefir, clabber, or or whey (1:200)
rennet	regular dose calf rennet
salt	
forms	hoops: 40 cm diameter × 30 cm deep
YIELD	ONE 20 KG EMMENTALER (15:1)

TECHNIQUE

Bring milk to cheese temperature, about 35°C (95°F).
Add starters, mix in thoroughly.
Add rennet, mix in thoroughly.
Wait for clean break, 45 minutes to 1 hour.
Cut curd to lentil size, 3 mm, with a spino.
Cook curds to 52°C (126°F) over 30 minutes, stirring nonstop, to achieve firm peaks.
▸ Increase tempo of stirring as temperature rises.
Take pot off heat.
Pitch curds for 10 minutes.
Reserve 2 L whey to make light salt brine for washing.
▸ Ferment whey overnight, then add 2 percent salt.
Retrieve cheese from pot with cheesecloth and place into form.
Leave overnight 12 to 24 hours, until acidity develops.
▸ Goal pH 5.3—stretch test spins.
▸ Flip four times, with first flip at 10 minutes.
Salt cheese on both sides daily for 7 days.
Or salt in saturated brine 7 days (8 hours per kg).
Age cheese for 1 week in hastening room to grow eyes.
▸ Flip cheese daily.
▸ Wash rind with light salt brine every day.
Age cheese in cave for 3 to 12 months.
▸ Flip cheese every other day.
▸ Wash rind with light salt brine (2 percent) every other day for 1 month.
▸ Wash rind or brush every other day from 1 month onward.

Grana

This is the most extreme of the alpine styles, so much so that it becomes a category of cheese all its own: an alpine cheese that has descended from the mountains but is ascendant, almost heavenly, in its taste.

The supposed inventors of this style were eleventh-century monks in the northern plains and hills of Italy. They used many of the alpine cheesemaking technologies from the nearby mountains (the make is more or less akin to Comté), but adapted their practices to their lowland monasteries, making more massive wheels that were more easily transported to market by cart.

The monks made their massive wheels not with milk of their own animals, but with quantities of milk coming from the tenant farmers on the vast landholdings of their monasteries. To enable the proper production of the larger wheel, the curd had to be cooked to a higher temperature than other alpine styles—55°C (131°F) instead of 52°C (126°F)—which limits moisture within the paste, restricting the development of propionic eyes that would otherwise grow in the taller, more substantial wheel (as is the case with the other alpine cheese that descended from the mountains, Emmentaler).

The word *grana* in Italian means "grain," referring to either the small curd size (described in Italy as a grain of rice) or the granular, often crystalline texture that this cheese achieves from its high-temperature cooking and long dry aging. The most famous cheese made in this style is Parmigiano-Reggiano, once known as grana Reggiano, a category which also includes many other Italian styles such as pecorino romano, whose name hints at the classical origins of this cheese. It is a style of extremes: they're made with the largest amounts of milk per wheel, with an alpine curd that's cooked to the highest of temperatures; the curd is formed into wheels weighing in at the top of their class, and they can be aged longer than any other style.

A unique aging gives grana cheeses their typical appearance and taste, with a polished skin and a dry, tough crust. The cheeses are aged in drier aboveground caves and regularly rubbed with a brush or cloth. The dry aging slowly pulls moisture from the rind and robs surface microbial ecologies of the chance to grow. Most other cheeses would not benefit from such a treatment, for rind development generally defines the flavor and character of smaller and higher-moisture cheeses (even full-alpine styles). But granas are massive wheels, and they continue to ripen from the interior despite the drying, and slowly become more and more complex and acidic within as the primary fermentation continues along at a snail's pace.

Diverse microbial and enzymatic processes break down the cheese as completely as possible without any aerobic decay over time, resulting in a cascade of flavor, making grana a cheese as delectable and delicious as milk permits. Almost too exalted to eat on their own, granas are best used as a boost of appeal and nutrition for a meal or enjoyed in the thinnest, most angelic slices.

Making Grana

An alpine cheese • made with cow's milk • aged with a rubbed rind
12- to 24-hour make • 12- to 24-month affinage

INGREDIENTS

milk	200 L
starter	1 L yogurt, thermophilic clabber, or whey (1:200)
	and 1 L kefir, clabber, or whey (1:200)
rennet	regular dose calf rennet with lipase
salt	
forms	hoops: 30 cm diameter × 30 cm deep
Yield	**one 20 kg grana (10:1)**

TECHNIQUE

Bring milk to cheese temperature, about 35°C (95°F).
Add starters, mix in thoroughly.
Add rennet, mix in thoroughly.
Wait for clean break, 45 minutes to 1 hour.
- ▸ Cover the pot to keep in warmth.

Cut curd to lentil size, 3 mm, with a spino.
Cook curds to 55°C (131°F) over 60 minutes, stirring nonstop.
- ▸ Increase tempo of stirring as temperature rises.

Take pot off heat.
Pitch curds for 10 minutes.
Retrieve cheese from pot with cheesecloth and place into form.
Press overnight 8 to 16 hours, until acidity develops.
- ▸ Goal pH 5.3—stretch test positive.
- ▸ Flip four times during pressing, with first flip at 1 hour.
- ▸ Remove cheesecloth during second flip.

Salt cheeses on both sides daily for 7 days.
Or salt in saturated brine 7 days (8 hours per kg).
Age cheese in dry cave for 12 to 24 months.
- ▸ Flip cheese every other day.
- ▸ Brush cheese every other day.

PHOTO BY ALIZA ELIAZAROV

Saint-Nectaire

Saint-Nectaire is a celebrated style from the renowned cheesemaking region of the Auvergne, a remarkable cheesemaking reserve where several techniques from this book originate. It is a cheese traditionally made by small-holding paysans, prepared with milk still warm from the udder, and curdled, historically, in wooden vats. To taste this style is to taste a cheese made across time, enjoyed by historical cheesemakers and their contemporaries, and still appreciated as a unique and especially delicious style. It is thus one that has become a favorite of mine to make and eat—and that's not to mention that its sought-after rind development is typically an unwanted microbial ecology in a cheese: *Mucor*, aka cat's fur!

No other cheese lets *Mucor* manifest. The steps taken to make white-rinded lactic cheeses prevent the usually unwanted fungus from taking hold; blue cheeses are quickly dominated by *Penicillium roqueforti*, which also keeps the *Mucor* from growing; washed-rind cheeses keep *Mucor* in check with a thorough and regular smearing. But Saint-Nectaire is left unprotected from this opportunistic fungus by its peculiar make, and the unwanted fungus grows rampant on its rind. But no worry—though considered a fault in other styles, it is the fungus of choice for the rare Saint-Nectaire and is the dominant force responsible for the cheese's particular flavor and texture.

Saint-Nectaire is a semi-alpine cheese made historically in a wooden vat without cooking the curd over fire. Instead, the curd is made as firm in the vat as possible. This is the longest-lasting cheese that can be made in wood without washing the curd or milling. The curd is made much like a traditional cheddar, but the slabs are pressed into a cheese without the cheddaring process. (In the Auvergne the cheese shares the first stages of its make with Cantal.)

It's not this, however, that makes the cat's fur grow. To encourage *Mucor*, the cheese is salted much too early (at a pH above 6.5) to prevent protective fermentation from occurring; then the wheel is left for several days in a cool and dry space before aging to fully dry its rind. These two acts combine to develop the conditions that lead to its characteristic fluffy coat. When that dry-rinded, unfermented cheese is placed into a humid cave to age, the cheese becomes a *Mucor* fur-ball!

This is considered another one of France's finest cheeses . . . but it is exceedingly hard to find anywhere else. I first learned to make it in America from Sister Noella Marcellino, the Cheese Nun, who makes the cheese in a wooden vat at the Abbey of Regina Laudis in Bethlehem, Connecticut. Noella herself learned the technique from a farmer from the Auvergne who visited her abbey after she prayed for someone to guide her in her cheesemaking endeavors. Was it coincidence or divine intervention? Like all things cheese, it's impossible to say!

Making Saint-Nectaire

A washed-curd cheese • made with cow's milk • aged with a *Mucor* rind
4-hour make • 8- to 12-week affinage

INGREDIENTS

milk	20 L
starter	200 ml kefir, clabber, or whey (1:100)
spores	*Mucor miehei* (optional—see appendix I)
rennet	regular dose calf rennet
salt	
forms	round: 20 cm diameter × 10 cm deep

Yield   two 1 kg Saint-Nectaires (10:1)

TECHNIQUE

Bring milk to cheese temperature, about 35°C (95°F).
Add starter, mix in thoroughly.
Add *Mucor* spores, if using.
Add rennet, mix in thoroughly, then still milk.
Wait for clean break, 45 minutes to 1 hour.
- Cover the pot to keep in warmth.

Cut curd to hazelnut size, 1.5 cm.
- Cut gently and slowly with a spino over 5 minutes.

Stir curds at 35°C until firm, about 60 minutes.
- Add hot water as needed to keep curd warm.

Pitch curds for 5 minutes, until they form a mass.
Press the curd firm under the whey by hand.
Whey off from above.
Cut curd into large slabs.
Fill cloth-lined forms with slabs, to a maximum of 10 cm.
Press heavily by hand or in a press, flipping cheeses
    after 15 minutes.
- Pour hot water (50°C [122°F]) or whey over cheeses to keep warm.

Salt cheeses after 1 hour, once cool.
- Apply salt all around.
- Leave cheeses to drain 12 hours in forms.
- Salt again after 12 hours.
- Drain an additional 12 hours.

Age cheeses in cool, dry cave (70 percent humidity)
    or refrigerator for 3 to 5 days.
- Flip daily.

Age cheeses in cool, humid cave (90 percent humidity)
    for 2 to 3 months.
- Turn cheeses every other day.
- *Never* brush or wash.

Gouda

This most famous Dutch kaas is made as firm as a mountain cheese, but in wood instead of copper. The styles are nearly identical in their makes, except that the Gouda's curds are washed with hot water to firm them instead of being cooked in a copper kettle. This hot-water washing allows the curds to be cooked to firm peaks in the wooden vat, enabling more whey expulsion and a firmer, longer-lasting wheel.

To form the wheel after washing, press the curds under their whey like a tomme, then transfer them to a cheese form to press by hand. To give the cheese its characteristic barreled shape, place it into a brine to salt when it's ready, or leave it out of its form during a surface salting. If you wish to develop round eyes, often a recognized feature of Gouda, add salt more slowly over a week to encourage propionibacteria to develop.

Gouda has many options for its affinage. If it's kept humid, it can be given a natural rind like a tomme or a washed rind like raclette. Kept in a dry cave, it can be dry-brushed like a grana if it's massive enough, or rubbed with fat or ghee. But the most characteristic appearances and flavors of Gouda come when the cheese is aged enclosed in wax.

Without the option for underground aging in the Netherlands (the country is almost entirely at or below sea level), these cheeses were historically aged in above-ground cellars or barns. To protect the cheeses from drying out in such spaces, the Dutch perfected the practice of waxing their Goudas.

Sealed in wax, a Gouda can be aged for years without decay, becoming tangy in taste, creamy in texture, and even crystalline on the tooth. Gouda's certain sweetness and caramel flavor come from the combination of cooking the curd and aging without the influence of air, and are unparalleled in other styles.

Though often considered a style inspired by the alpine make, it may be that washed-curd cheeses like Gouda inspired the alpine method. Being made in wood, a more ancient cheesemaking material than copper, it's likely that this make predates the alpine technique. And though it's commonly considered a Dutch cheesemaking style, semi-alpine cheeses like raclette, made in Switzerland and elsewhere, are often made with the washed-curd technique.

Making Gouda

A washed-curd cheese • made with cow's milk • aged with a waxed rind
8- to 12-hour make • 4- to 6-month affinage

INGREDIENTS

milk	40 L
starter	200 ml mesophilic kefir, clabber, or whey (1:200)
	and 200 ml thermophilic yogurt, clabber, or whey (1:200)
rennet	regular dose calf rennet
salt	
forms	round: 25 cm diameter × 20 cm deep
YIELD	ONE 4 KG GOUDA (10:1)

TECHNIQUE

Bring milk to cheese temperature, about 35°C (95°F).
Add starters, mix in thoroughly.
Add rennet, mix in thoroughly.
Wait for clean break, 45 minutes to 1 hour.
  ▸ Cover the pot to keep in warmth.
Cut curd to lentil size, 3 mm, with a spino.
Stir curds for 15 minutes.
Pitch curds for 5 minutes.
Whey off one-third of volume of vat.
Slowly pour in one-third volume of vat of hot water at 60°C (140°F).
  ▸ Stir slowly to distribute heat.
  ▸ Bring temperature to 40°C (104°F).
Stir curds for an additional 30 minutes, more quickly now.
Pitch curds for 5 minutes.
Wash curds again, bringing temperature to 45°C (113°F).
Press curds together into a wheel under the whey.
Retrieve cheese from pot with cheesecloth and place into form.
Press by hand in form.
  ▸ Keep cheese warm by pouring whey or warm water overtop.
Drain 6 to 12 hours, until acidity develops.
  ▸ Goal pH 5.3—stretch test positive.
  ▸ Flip four times during the drain, with first flip at 15 minutes.
Salt cheeses by hand over 2 days, or for 24 hours in saturated salt brine (6 hours per kg).
Leave cheese to dry for 2 days at room temperature.
  ▸ Flip daily.
Seal in wax.
Age in a cave for 6 months to 2 years.

PHOTO BY ALIZA ELIAZAROV

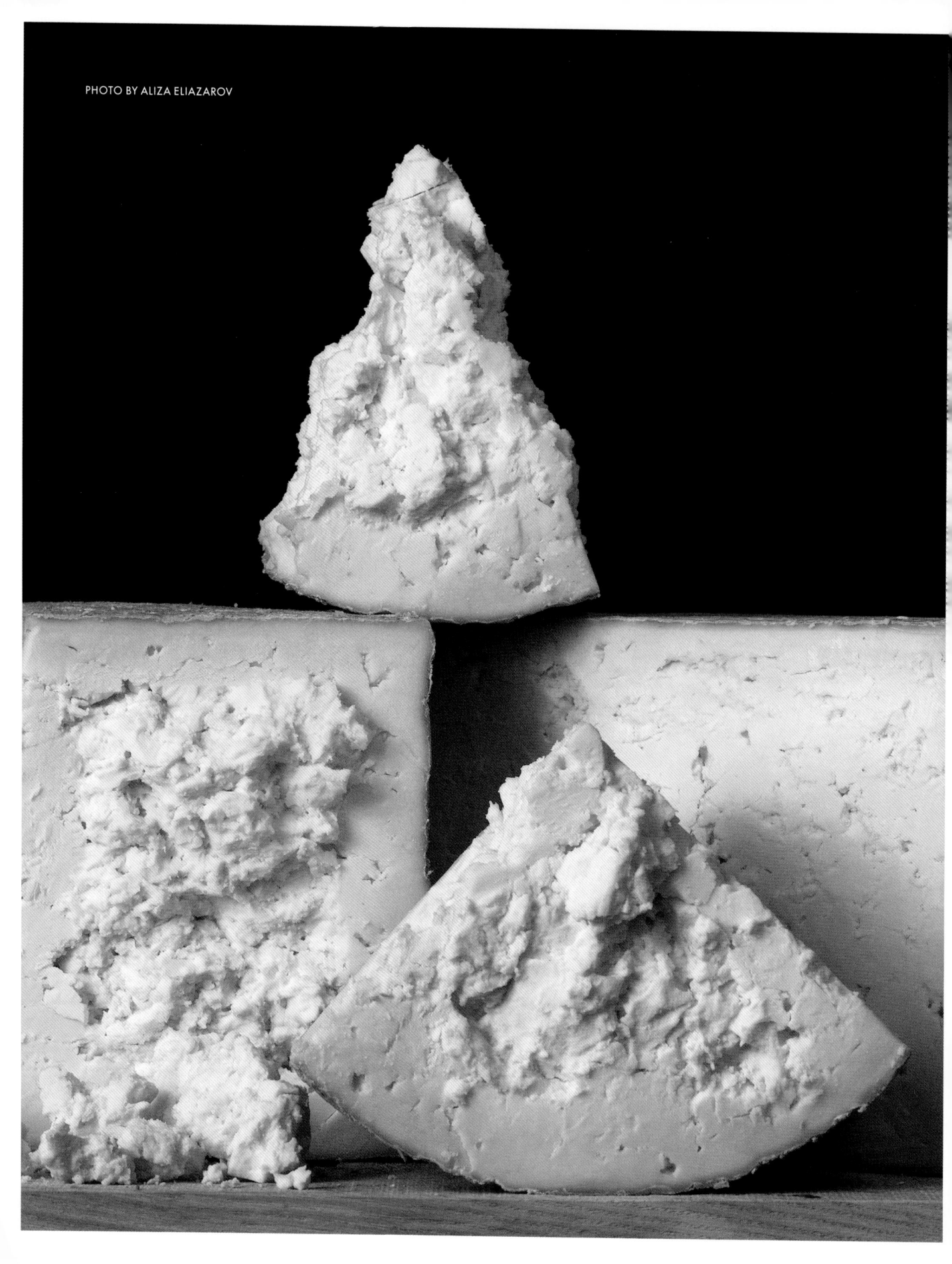

Lancashire

In my early cheesemaking days, I learned that my elderly neighbor Mae was once a milkmaid in the north of Britain during the war effort and had made Lancashire cheese. When I asked if she could teach me how to make it, Mae replied: "Of course you cannot make Lancashire—we're not *in* Lancashire!" She never even humored me with the recipe.

Though there may be some truth to her objection, I have to say I disagree with it. Lancashire is a style that may have evolved in Lancashire county, from particular agricultural and cultural practices, but it's a style that can be reproduced anywhere so long as the traditional aspects of its production are respected.

Lancashire is a medium-sized milled-curd cheese, with an acidic tang and crumbly texture. Unlike cheddar—historically made with larger quantities of milk on larger farms or collective dairies in the south of the country—Lancashire was made by smallholders in the north, who had fewer animals and therefore less milk to work with, and would often make their wheels of cheese from curd collected over the course of two or more days. The result is a sort of longer-fermented cheddar, often more acidic and with a fluffy chèvre-like texture, especially compared with its bigger and firmer milled-curd brother.

To make Lancashire, a batch of stirred rennet cheese is prepared. The curd is hung and left to drain overnight. The curds are then milled into smaller pieces, salted, and firmly pressed together into a wheel. Multiple makes of curd prepared over subsequent days can also be milled together to make a larger cheese. The cheese can be cloth-bound or washed, or even left to develop a natural rind. With a beautiful coat and a delightful mix of flavors and textures within the body of the cheese, this one's a keeper. It's a great option for making a large cheddar-style cheese if you only have smaller quantities of milk to work with or you don't have enough time to dedicate to cheddaring.

This is one of many related British territorial cheeses, usually made by milling the curd, that nearly disappeared from the country's dairying landscape during the modern era. Rationing and government control over cheese production during and after the Second World War eventually shut down farmstead and factory production of Lancashire, among many others, for it was deemed too soft and perishable. The technique was re-created only after its making had skipped a generation, as Britain entered into an era of cheese rediscoveries in the 1970s and '80s. Now several farmstead cheesemakers prepare Lancashire cheese according to traditional techniques, but it's one that's rarely known or made outside the country.

Making Lancashire

A milled-curd cheese • made with cow's milk • aged with a washed rind
24-hour make • 3- to 6-month affinage

INGREDIENTS

milk	40 L
starter	400 ml kefir, clabber, or whey (1:100)
rennet	regular dose calf rennet
salt	
forms	hoops: 20 cm diameter × 40 cm deep

YIELD   ONE 4 KG LANCASHIRE (10:1)

TECHNIQUE

Bring milk to cheese temperature, about 35°C (95°F).
Add starter, mix in thoroughly.
Add rennet, mix in thoroughly.
Wait for clean break, 45 minutes to 1 hour.
Cut curd to hazelnut size, 1.5 cm.
Stir curds for 30 minutes, until they bounce from a 60 cm drop.
Pitch curds for 10 minutes then whey off.
Transfer curds to cheesecloth-lined draining table and tie into a bundle.
Drain curds for 12 to 24 hours in cheesecloth bundle.
- ▸ Goal pH 5.1—stretch test very liquid.

Mill curds to 1 to 2 cm pieces (multiple makes of curd can be milled together).
Salt curds to 2 percent or to taste.
- ▸ Leave to drain 1 hour.
- ▸ Mix up curds occasionally to prevent matting and improve draining.

Fill cheesecloth-lined forms to the top with curds.
Press cheese in cheese press 24 hours.
- ▸ Flip twice during pressing, with first flip at 2 hours.
- ▸ Re-dress cheesecloth after every flip.

Age cheese in cave for 3 to 6 months.
- ▸ Flip every other day.
- ▸ Wash rind every other day for first month.
- ▸ Brush rind as needed from 1 month onward.

Traditional Cheddar

Cheddar is the firmest and finest of the milled-curd cheeses. Historically prepared in wood, this cheese, and close relatives of it, were made more widely across Europe before the technology of cooking curd in copper evolved. Despite its prevalence today, cheddar is an ancient style of cheese, and relatives that are made like it exist in France as Cantal, in Mexico as Cotija, and Italy as Castelmagno.

These firm milled cheeses, like firmer alpine cheeses, are best made with larger quantities of milk. These styles are, in some cases, even considered mountain cheeses—cotija and castelmagno owe their continued existence to their being made on isolated mountainous regions, unaffected by the developments of the dairy lands below. Ironically, cheddar itself likely only survived because it was adapted to become the darling of industrialized dairying across the English-speaking world, because its method of production was easy to scale up, and the firm cheeses were easy to ship from the colonies (especially the modern cheddar, whose technique follows).

Traditional cheddar makers employ different strategies than do cooked-curd alpine cheesemakers to get moisture out of their cheeses and help them last longer. The combination of stirring, cheddaring, milling, salting, and pressing helps to pull moisture from the curds, leading to a drier, firmer flesh that presses into a long-lasting wheel, all without the need for cooking. Only in an industrialized modernized cheddar make is the curd cooked, which sets it apart as a dramatically different cheese.

Now on to cheddaring. Cheddaring is a laborious method of using the curd to press the curd firm and expel as much moisture as possible from within it. This method is best executed with large slabs of curd. Though the process can be used with a small amount of milk, the cheese evolves best if made massive—an average-sized cheddar truckle is typically 25 kg—with up to 250 L milk in one single cheese, and easily four times that in the vat!

To make the slabs for this more traditional make, transfer stirred curd into large cheesecloths and drain in tied-up bundles. These slabs are left in their cheesecloth (it improves the draining), stacked atop one another, and regularly restacked, so each receives an even pressing. The bundles are rewrapped in cheesecloth as they press to accommodate their changing shape.

While cheddaring, the acidity of the curd should be carefully monitored, preferably through the teabag test. Take a small piece of cheese from the curd every hour and test it, as with mozzarella, by steeping in hot water. Once the curd achieves a full stretch, the cheddar is ready to be salted (further fermentation makes the cheese more crumbly, fluffy, and Lancashire-like).

Milling comes next, with the curd cut to a smaller size, often with the help of a mechanical aid—the Brits use a peg mill—though it can also be broken by hand, or cut with a knife. The curds are heavily salted, then drained, becoming like the squeaky cheese curds of the next technique (but a bit less squeaky because of the lack of cooking) before they're reassembled and pressed into their final form.

It takes much pressure to get the milled curds to press together. All milled cheeses are susceptible to blueing if they are poorly pressed, which results in air passing to the relatively open interior. To prevent this, cheesemakers have been known to take an iron to the rind, sealing it shut by melting the curd at the rind with its gentle heat! Some connoisseurs, though, consider the accidental blueing as a sign of the cheese's highest quality, indicative of a more traditional, handmade cheese.

The cheese can be left untreated to grow a natural rind (as in Cantal) or handled in various ways to keep the fungal rind and interior from developing. The cheeses can be washed with whey, smeared with butter, or clothbound, as this method instructs. The cloth binding, sealed with lard, makes a sort of sacrificial rind, allowing the cheese below to ferment without air, and preserves it much longer, even in a drier cave.

Cheeses like cheddar are made across Europe and even into to the Middle East. To make the closely related cotija or castelmagno, you need only make slight adjustments to the process: Curdle the milk with a more traditional rennet, preferably kid, with its extra lipase giving the cheese a bigger bite; and grind the curds up extra-fine, through a sort of meat grinder, before salting and pressing. While its cousin Cantal is made more or less just like the cheddar of this method (a variant, Salers, is still made today in the wooden vat), but given a natural rind or washed rind as it is aged.

Arguably the cheeses in a "sack" (sir iz mijeha, Tulum, Divle cave cheese) made in the Balkans, Turkey, and the mountains of Lebanon and Syria are the originators of this technique; all are ancient milled-curd cheeses, like cheddar, but made with goat's and sheep's milk (of course, the method works wonderfully with their milk). These fascinating cheeses are firmly pressed by hand into the skin of a goat to age, taking on the shape of the animal, and becoming, essentially, a goat-bound cheddar!

Making Traditional Cheddar

A milled-curd cheese • made with cow's milk • aged with a clothbound rind
8- to 10-hour make • 6- to 12-month affinage

INGREDIENTS

milk	200 L
starter	2 L mesophilic kefir, clabber, or whey (1:100)
rennet	regular dose calf rennet
salt	
forms	hoops: 30 cm diameter × 45 cm deep

YIELD   ONE 20 KG CHEDDAR (12:1)

TECHNIQUE

Bring milk to cheese temperature, about 35°C (95°F).
Add starter, mix in thoroughly.
Add rennet, mix in thoroughly.
Wait for clean break, 45 minutes to 1 hour.
Cut curd to hazelnut size, 1.5 cm.
- Cut vertically; cut crosswise; cut horizontally or diagonally.

Stir curds for 60 minutes at 35°C.
- Curds should bounce from 90 cm.

Whey off.
Transfer curds into cheesecloth and tie into large bundles.
- Leave curds to knit in cloth 1 hour.

Begin cheddaring process with curds in cloth.
- Stack slabs atop each other two or three high.
- Wait 30 minutes.
- Restack slabs, cycling each one through top and bottom of stack.
- Rewrap slabs in cheesecloth each time they are restacked.

Cheddar for 4 to 6 hours, until curd is strong.
- Goal pH 5.3—stretch test positive.

Mill slabs to 1 to 2 cm pieces.
Salt curds to 2.5 percent or to taste.
Leave to drain 1 hour, mixing curd every 15 minutes.
- Add salt again if need be, to taste.

Fill cheesecloth-lined forms to the top with curd.
Press cheese in a cheese press for 12 to 24 hours.
- Flip twice during pressing, with first flip at 2 hours.
- Re-dress cheesecloth after every flip.

Cloth-bind cheese (see chapter 9).
Age cheese in dry cave (70 percent humidity) for 6 to 12 months.
- Flip cheese every other day.

PHOTO BY ALIZA ELIAZAROV

Modern Cheddar (and Squeaky Curds)

Modern cheddar, made faster and firmer, is a product of the development of the first cooperative American cheese factories, which started churning out cheeses for export in the late 1800s. To achieve the dryness, size, and high degree of preservation required for overseas transport, cheddar making was changed: Instead of preparing an uncooked cheese in wood, cheesemakers scalded the curds, cooking them with steam in modern jacketed cheese vats in place of wooden vats.

This scald led to a different sort of cheddar—sweeter in taste, firmer in texture—that could be aged for ages in its cloth binding and even longer in a vacuum seal. It was also a process that ultimately also gave the world the squeaky cheese curd.

Cheese curds are the stepping-stone to making great cheddar. Essentially, they are the midway point in the production of a larger cheddar. After the curds have been milled and salted, but before they are pressed together into their wheels, these delicious curds can be devoured by the dozens.

To make them, and their modern cheddar, milk is cultured, rennetted, and when set cut into 1 cm curds. The curds are cooked and stirred to 40°C (104°F), then left to settle under the whey; the whey is then drained off while the mass is assembled into piles and allowed to knit. The knit curd is cut into thick slabs, and the slabs stacked and restacked, for hours, until the curd reaches its goal acidity. At this point the slabs of now cheddared curd can be milled—in America they are typically cut into finger-sized and -shaped curds, rather then milled, as in England. (They can be pressed by hand through a 1 cm × 1 cm square metal grid, or sliced in a french fry cutter.) They are then heavily salted to pull out moisture (enhancing their squeak) and left to drain before being eaten fresh or pressed into an even firmer wheel of "American" cheddar.

Cheese curds share a heritage across the North American landscape; they've been a part of the foodways of Wisconsin, New York State, Ontario, and Quebec since the establishment of cheese factories there in the 1800s. In Quebec, they spurred the creation of poutine, arguably one of the world's great cheese dishes (up there with raclette, pommes aligot, and tartiflette!). This (in)famous food likely evolved as a cheddar makers' lunch, prepared with loose curds left over from the day's make, often ready by noon. The fresh curds are sprinkled atop of french-fried potatoes and smothered in simmering gravy left over from the previous night's meal. The heat from the fries and the sauce melts the curds (they will have just the right pH of 5.3), enrobing the dish with a stretchy mantle, in the process becoming a sort of pasta filata cheese . . . much like the next technique for caciocavallo.

Making Modern Cheddar

A milled-curd cheese • made with cow's milk • 6- to 8-hour make • 1- to 2-year affinage

INGREDIENTS

milk	200 L
starter	1 L thermophilic yogurt, clabber, or whey (1:200)
	1 L mesophilic kefir, clabber, or whey (1:200)
rennet	regular dose calf rennet
salt	
forms	hoops: 30 cm diameter × 45 cm deep
YIELD	ONE 20 KG CHEDDAR (12:1)

TECHNIQUE

Bring milk to cheese temperature, about 35°C (95°F).
Add starters, mix in thoroughly.
Add rennet, mix in thoroughly.
Wait for clean break, 45 minutes to 1 hour.
Cut curd to hazelnut size, 1.5 cm.
- ▸ Cut vertically; cut crosswise; cut horizontally or diagonally.

Slowly cook curds to 40°C (104°F) over 30 minutes.
- ▸ Stir constantly, increasing tempo as curds firm.
- ▸ Curd should bounce from 90 cm when ready.

Pitch curds for 5 minutes.
Whey off.
Shovel curds into a pile, and leave to drain for 30 minutes into a large lump of curd.
- ▸ Curd can also be drained in cheesecloth bundles.

Slice curd into slabs.
Begin cheddaring process.
- ▸ Stack slabs three high, flipping every time.
- ▸ Wait 15 minutes.
- ▸ Restack slabs, cycling each one through top and bottom of stack.
- ▸ Keep cheese warm (35°C or higher) to speed fermentation.

Cheddar for 2 to 4 hours.
- ▸ Goal pH 5.3—stretch test positive.

Mill or cut slab to finger-sized curds.
Salt curds to 2.5 percent or to taste.
Leave to drain 1 hour.
- ▸ Mix every 15 minutes and add additional salt if needed.
- ▸ Cheese curds can be packaged and sold at this point in production.

Press cheese in a cheese press 12 to 24 hours.
- ▸ Flip twice during pressing, with first flip at 2 hours.
- ▸ Re-dress cheesecloth after every flip.

Cloth-bind cheese (see chapter 9), cover in wax, or vacuum seal.
Age cheese in dry cave (70 percent humidity) for 6 to 12 months.
Flip cheese once a week.

PHOTO BY ALIZA ELIAZAROV

Caciocavallo

To close off the summer chapter, we make caciocavallo. This most beautiful cheese is a coming together of the three main themes of the season: milled curds, spun curds, and alpine cheeses. Indeed, this cheese defies such categorizations, and proves to me that all cheeses are truly interrelated.

At the first stage of its make, the technique follows the milled-curd method and could easily become a cheddar. Milk is cultured, renneted, cut small, stirred, and formed into a first cheese; that cheese is fermented to the right acidity, and then milled, salted, and drained. But instead of pressing the curd together into a second wheel, like cheddar, the cheese is melted in hot whey, producing a firm pasta filata. The melted curd is then stretched and worked to give it a purposeful and firm shape that makes moving it down the mountain possible, evoking the philosophies of alpine cheese.

Working this warmest, most welcoming, endlessly stretchable dough is one of the most remarkable acts of cheesemaking. The melted curd is shaped in an intricate way to give it its form. Doing so turns the lump of dough into a shape like a child (or a gourd), with a body and head, so that it can be tied with string and more easily moved. Once the cheese has taken its shape, the cacio is chilled in water, and then salted in brine overnight to firm its rind before it's aged or eaten.

Versatile caciocavallo can be aged and eaten in many ways. The combination of the alpine make and the pasta filata method makes this among the finest of all cheeses, one that starts out intensely flavorful and only gets better as it ages. The cheese can be eaten on its first day as a firm and flavorful mozzarella, or, if made massive, aged to an almost grana-like texture over years. The cheese can be aged dry, to develop a clean rind and a drier paste; or aged in a humid cave to become mold- and mite-ripened like a tomme. If it's made and salted only from the rind, it can even grow round propionibacteria eyes like an Emmentaler! It's phenomenal as a slicing cheese, but my favorite way to eat it is the way Italian cowboys do: suspend the cheese by its string over a fire; let the flames warm it, raclette-like; and then scrape off the molten cheese from its bottom, to eat on bread or potatoes.

Caciocavallo is inextricably linked with the cultural and agricultural practices in the south of Italy, as explored in chapter 1. To make it best, you have to imagine yourself in these circumstances, following the extraordinary method as prescribed by the traditions of transhumance. The combination of milks and cheesemaking philosophies (it's more than the sum of its parts) makes this style among the most flavorful and aesthetically pleasing of cheeses, indeed of all foods, that our extraordinary world can make.

To me, caciocavallo is a most mystical cheese, a "cheese-child" made from milk that speaks to this integral fluid's capacity to make life possible, to morph and transform and build all parts of a body. Folks often ask about my favorite cheese; I'll let you know . . . it's caciocavallo.

Making Caciocavallo

A pasta filata cheese • made with cow's, goat's, *and* sheep's milk
6- to 8-hour make • 0- to 6-month affinage

INGREDIENTS

milk	12 L cow, 5 L goat, 3 L sheep (or 10 L cow, 10 L goat; or 20 L cow)
starter	200 ml kefir, clabber, or whey (1:100)
rennet	regular dose kid or lamb rennet with lipase
salt and brine	

YIELD: TWO 1 KG CACIOCAVALLOS (10:1)

TECHNIQUE

Bring milk to cheese temperature, about 35°C (95°F).
Add starter, mix in thoroughly.
Add rennet, mix in thoroughly.
Wait for clean break, 45 minutes to 1 hour.
Cut curd to corn size, 0.75 cm.
Stir curds for 30 minutes at 35°C, stirring nonstop.
Pitch curds for 5 minutes.
Press curds together beneath their whey.
Transfer curds into cheesecloth.
Drain curds in cloth until acidity develops, about 6 to 12 hours.
- Goal pH 5.3—stretch test positive.

Mill curds to 1 cm fingers, mix in salt to 2.5 percent, and leave to drain for 30 minutes.
Slowly pour boiling water or whey left over from ricotta over curds to just melt.
- Add hot water or whey stepwise, stirring between, until perfect melt is achieved.
- Stir curd gently until it melts into a dough.
- Add more boiling water as needed.

Shape curd into caciocavallo.
- Stretch skin of bottom of cheese over top to make a ball.
- Gather loose folds at the top together into a flower.
- Constrict cheese below flower to define nipple of the cheese.
- Cut flower off from atop nipple to make a ball.
- (Save excess for making more caciocavallo.)
- Pinch nipple to stretch and close it.
- Define neck of cheese by pressing into ball between thumb and forefinger.

Plunge caciocavallo into cold water to cool it.
- Tie a wide string around the neck to keep the cheese's shape.

Brine caciocavallos for 12 hours in saturated brine to firm and form a rind.
Hang cheeses for 24 hours to drain its whey.
Age cheeses hanging in cave for 1 to 6 months.

CHAPTER 14

Fall

Fall is the season of the richest milk, when butterfat tops the charts, and butter and its by-products are best and most bountifully made. With the end of summer and cooling temperatures comes this season of the harvest, of yeasty ferments and blue mold.

This is the time to harvest the fat of the land. It's the season to leave milk to separate its cream slowly and ferment in cooler weather, then churn that thickened cream into butter. Skyr and tvorog can then be made by draining the skimmed curd below. The leftover curd can also be aged into handkäse and graukäse, which preserve well for winter eating, alongside their butter.

Fall is the season of harvesting and milling wheat and rye, fermenting and baking sourdough bread, and making that bread moldy with *Penicillium roqueforti*.

Fall is the season of colorful trees . . . and colorful cheese! Blue-green *roqueforti* manifests best in the cool weather of the season, much like other wild fungi. The fat-rich milk of the late lactation makes the most flavorful Stiltons, fourme d'Amberts, and Roqueforts. When made in the fall, these beautiful blues are ready just in time for holiday feasts.

Fall is also the season of fallen fruit and the wild yeasts that consume them. It's the time of the vintage, of the gathering of grapes to ferment into natural wine. Milk, too, can be submitted to these alcoholic transformations, becoming amasi in a gourd or milkbeer in a bottle. Yeasts can also raise the rare cheese known as ambarees, and the alpine cheeses of the summer season can be aged beneath the pressed skins of the grapes.

Cultured Cream

Cultured cream is known by many names in many cultures: sour cream in America, sauerrahm in Germany, crema agria in Latin America, smetana in Russia, and crème fraîche in France. It is a culturally significant food around the world, being a stepping-stone in the making of traditional butter. Often some of the thickened fermented cream would be spared the churning and instead added to soups, sauces, pastries, fresh fruit, or anything else, really, for cultured cream gracefully adds flavor, texture, and richness to just about every dish.

The sweet cream we know today may be a modern invention, coming to us only in the era of the centrifuge. It is indeed peculiar that fermented cream is known in French as crème fraîche, meaning "fresh cream." But the name is a linguistic evidence of cream's cultured history; for in pre-industrial France (pre-centrifuge, pre-pasteurization, pre-stainless-steel), cream was left to rise and ferment in the wooden barrel and could only be taken off the top once it had been soured and thickened. In France today the term *crème fraîche* is used for two completely different dairy products: *crème fraîche liquide*, still-pourable unfermented cream; and *crème fraîche épaisse*—thick, fully fermented, and decadently creamy, the real deal. Dip a strawberry in and decide which you prefer. Invariably, it's the fermented cream that enrobes the berry best with a perfect coating of fluffy, flavorful fermented fat.

This is likely the simplest, most flavorful ferment that anyone can easily make at home, and a perfect one to start off the fall collection of fat-rich techniques. Simply add an active culture like clabber or kefir to cream, leave it to ferment at ambient temperatures, and voilà, your cream will thicken to a most luscious texture! It's even easier, and certainly more delicious (and probably more nutritious) than whipping that cream, for the culture does the work of the whipping for you. Chilling the fermented cream makes it really stand tall.

Crème fraîche is probably the first fermentation I ever conjured, as a simple kitchen experiment, long before I realized that I was either making cheese or fermenting. For crème fraîche is a common chefs' trick and is one of the few dairy fermentations regularly practiced in commercial kitchens. But why stop there, chefs? Any dairy product made in-house with fresh milk or cream is likely better than one purchased from larger industrial producers, who often don't respect traditions of fermentation. Consider this an opening act for a wider in-house cheesemaking program, and dive into dairying whole-hog!

Making Cultured Cream

A fermented milk • 12- to 24-hour make

INGREDIENTS

cream	1 L
starter	10 ml kefir, clabber, or whey (1:100)
YIELD	1 L CULTURED CREAM (1:1)

TECHNIQUE

Bring cream to culturing temperature, about 20°C (68°F).
Add starter, mix in thoroughly.
Leave cream to ferment until thick, 12 to 24 hours.
Chill to preserve.

NOTE

To make cultured cream from whole milk without a centrifuge, follow the technique for tvorog and smetana.

PHOTO BY ALIZA ELIAZAROV

Long-Cultured Cream

I'll admit that this is an odd make. Not exactly a traditional technique, this is something delicious that I stumbled across in my endless explorations of milk and its cream. It is essentially crème fraîche forgotten for a month (oops!), until the *Geotrichum candidum* that grows upon its skin consumes it. It is a full-cream cheese, if you will, left to age until it thickens—and deliquesces.

This technique permits the *Geotrichum candidum* to consume the cream, transforming it into a broken-down paste instead of being drained before aging like most cheeses, or churned into butter as cream usually is. The result is perhaps the richest and most luxuriant ripened cheese in the world: almost full butterfat, with a melting quality resulting from the small amount of casein proteins in the cream.

To make this fabulous ferment, the freshest cream is cultured, then poured deep into ceramic pots and left to its own devices. Once the cream thickens, *Geotrichum candidum* blooms at its surface; over subsequent days at ambient temperatures, that bloom transforms into a wrinkly *Geotrichum* rind. Once this is well established, place the cultured cream in a cave, where the *Geotrichum* continues consuming it, condensing the cream to less than half its height and developing delicate flavors and unctuous textures.

The technique could, theoretically, be performed with just milk, as described in chapter 9, but the concentrated solids content of the cream (five times higher than milk) allows a cheese to ripen that much faster. It would take the *Geotrichum* ages to transform a similar quantity of milk into cheese, and its texture and flavor would likely not compare. Really this is one of the important reasons that we turn milk to cheese by curdling it and separating its whey before its aging!

This cheese to me is representative of the serendipitous nature of natural cheesemaking. If you have fresh milk and good fermentation, anything you make with your milk (and especially its cream) will be become an extraordinary food. Most styles of cheese likely were discovered much the way this one was—it was forgotten about or made by error, and its makers tasted it and said, *Wow, let's make this again!* So if you make a mistake in your make, just roll with it. Likely what you've made will be something uniquely delicious.

Making Long-Cultured Cream

A fermented milk • aged with a *Geotrichum* rind • 24-hour make • 1-month affinage

INGREDIENTS

cream	1 L
starter	10 ml kefir, clabber, or whey (1:100)
salt	20 g (2 percent)
pots	5 cm deep
YIELD	500 ml LONG-CULTURED CREAM (2:1)

TECHNIQUE

Bring cream to culturing temperature, about 20°C (68°F).
Add starter, mix in thoroughly.
Cover tightly with a cloth.
  ▸ Cream should be no more than 5 cm deep.
Leave cream to ferment until thick and *Geo* grows, 12 to 24 hours.
Salt lightly at the surface of the cream.
Leave to age at room temperature for 1 week.
Transfer to cave and age for 1 month.

Butter

Good butter just melts into your body, and butter is better on every front when traditional methods like fermentation are followed. Natural methods improve every aspect of butter—fortifying its nutrition, simplifying its make, refining its flavor, increasing its fat content, improving its texture, lengthening its shelf life, and allowing it to be aged and preserved in traditional ways.

Natural butter is healthier. Fermentation improves the nutritional character, digestibility, and keeping quality of every food that is fermented. Well-made butter, especially from grass-fed animals, may be one of the most nourishing sources of fat out there, including an excellent supply of healthful fatty acids and a whole host of fat-soluble nutrients that are passed along from the milk. And culturing cream before its churning keeps butterfats from going rancid, so long as the cream used is at its freshest.

Natural butter is easy as pie—which, of course, is also made better with a cultured butter. Fermenting cream before churning breaks down the butterfat globule's protective envelope of lipoproteins, permitting the butterfat within to more easily agglomerate into butter. Churning cultured cream instead of sweet cream can reduce churning time by two-thirds!

Butter's flavor is fuller with fermentation. For the butterfat isn't fully flavored until broken down through fermentation into its fatty acids, and sugars including milk's citric acid are broken down into more flavorful compounds like diacetyl. The breakdown of the fats into fatty acids also makes the butter's fat more digestible.

Cultured butter is richer in fat than a sweet cream butter (I prefer to call it uncultured butter, for likely it's not a traditional or real food, but an artificial, industrial one) because the breakdown of the butterfat globule enables a better agglomeration of the butter, and a more complete expulsion of its buttermilk. Cultured butter is more buttery for this reason—its texture is richer, more in line with fat!

Culturing butter permits its preservation by eliminating fermentable sugars and fats and encouraging protective microorganisms. In addition, the slight acidity, along with other by-products of fermentation, significantly reduces the spoilage potential, so much so that it doesn't need refrigeration. And though cultured butter will change over time, it will not rancidify as uncultured butters do, due to the protective effects of fermentation on its fats. And with the addition of salt, cultured butters are so well preserved that they can even be aged.

Now that I'm done convincing you that natural butter is better, let's talk about the technicalities of its make. As soon as the cream thickens to crème fraîche, typically after 12 to 24 hours at 20°C (68°F), it should be chilled for 2 or more days. The cooler temperatures aid the cream's transformation into butter, and the long cooling period also seems to further break down the butterfat globules, making the churning more effective. Cooking the cream prior to its fermentation (as in the technique for Kaymak, page 366) also helps to break the butter.

Churning is best done slowly in a churn, allowing the butter globules to find themselves and agglomerate. For household churning, a large, sealable jar will do, filled no more than half full with cream to permit its movement when shaken. You can also simply whisk the fermented cream in a bowl. If you're scaling up, consider the gentle touch of stand mixers, dough mixers, or commercial barrel churns to agitate the cream. Plunge churns and paddle churns are also effective tools, especially with faster-churning cultured cream.

Cultured cream slowly thickens as it churns and becomes even thicker than a whipped sweet cream. Once enough churning has broken apart the butterfat's lipoprotein envelope, the butterfat within is released and begins to attract other small bits of butterfat around it. The butter suddenly breaks at the moment when the butter forms visible yellow globules and separates from its liquid buttermilk. A minute more in the churn grows the granules to their largest form, allowing the buttermilk to be most easily drained when strained. The butter should then be kneaded or worked slightly (with wooden paddles to keep it cool) after churning to increase its strength and smoothness, and to expel more buttermilk.

Skyr and Smjör

Here it is, the original butter make. For this is a butter (known in Icelandic as smjör) made alongside its curd (known in Icelandic as skyr); together they are a true traditional whole food made without the trappings of modern dairying. The previous butter technique takes an industrial shortcut, calling for sweet cream separated with the help of centrifuges or leaving milk to sit refrigerated (both of reduce the quality of a butter). If, however, you want to make the truest, healthiest, most elemental butter, a butter most in line with that of our buttermaking ancestors, this is the technique for you.

You could also call this technique barrel butter, for its original inception (Icelandic or otherwise) would have been inside a wooden barrel dedicated to buttermaking, wherein farmers would leave their fresh warm cow's milk (goat's and sheep's milk don't work well for this method, as their cream doesn't separate so easily) in tall vessels to sit and separate its cream. The cream, as it rose over the course of the day, would ferment thanks to the culture held in the wood; the thickened cream would be separated off the top by hand then churned into butter; and the skim milk below, now curdled, would be drained of its whey to make skyr.

This is an approach that I learned of in Iceland, a country with a particularly rich history of buttermaking that truly values this important fat, understood as the best food that they can produce in their northern climate. This buttermaking technology made its way to Iceland in the Middle Ages with the Vikings along with cows (you can still find the butter barrels used up until the early 1900s in museums there), and historical accounts from their journeys suggest that they first introduced butter from there to North America.

And though butter today is the most important product, cheeses made from the skim curd below were once, arguably, equally significant. All across Europe buttermakers employed similar techniques to produce their butters before the days of the centrifuge, and all across Europe buttermakers were also making sour milk cheeses with the fermented skimmed curd left below. Cheeses like German quark and French fromage blanc (among many others) are made in essentially the same way as skyr, and all these sour milk cheeses were produced because traditional buttermaking leaves behind a significant amount of skim milk curd that must be transformed into a cheese if the full value of a milk is to be realized and appreciated. To this day, for a farmer-buttermaker to have a financially viable operation, they must also value their skim milk. But Icelandic farmers in particular appreciated their skim milk cheeses, for they were forced by their Viking overlords to pay taxes for their protection in the more transportable butter, and so mostly ate skyr!

There is a very long history of buttermaking all across Europe, but typically it was most practiced in the northern regions and mountainous areas where the cool climates were most suited to its making and preserving. Similar techniques were likely also practiced among the Celts, who buried butter in bogs to preserve it across Scotland and Ireland (occasionally the long-preserved butters are dug up by turf cutters). But likely this most simplified buttermaking/cheesemaking technique came from the far eastern regions of Europe or Central Asia, where dairying began. This original dairying technology then made its way around the world with the milk cows that made it possible.

Making Skyr and Smjör

Fermented cheeses • made with cow's milk • 5-day make

INGREDIENTS

milk	20 L
starter	20 ml kefir, clabber, or whey (1:1000)

YIELD 1 KG SMJÖR (20:1)
AND 3 KG SKYR (20:3)

NOTES

Milk can be brought to a boil and cooled as per yogurt prior to fermenting, as is the tradition in much of the buttermaking world, improving the character and yield of both the butter and its skyr. The method for kaymak (page 366) explores this important dairying precedent.

TECHNIQUE

Bring milk to culturing temperature, about 20°C (68°F).
Add starter, mix in thoroughly.
Leave to ferment until milk thickens and cream curdles, 12 to 24 hours.
- Do not disturb.

Make butter:
- Skim thickened cream by hand into churn.
- Chill at 4°C (39°F) for 3 to 5 days.
- Churn until butter breaks.
- Strain butter through a colander or muslin to separate cultured buttermilk.
- Knead butter to remove excess liquid.

Make skyr:
- Carefully scoop skimmed curd into cloth.
- Hang to drain for 12 to 24 hours.
- Whisk curd intensely with a wire whisk or mixer until smooth and creamy.

PHOTO BY ALIZA ELIAZAROV

Smetana and Tvorog

Tvorog is a Ukrainian way of appreciating the skim milk curd left behind after skimming cream in the traditional way. In many homes across the region, milk is left to ferment, often spontaneously, until it curdles. The fermented cream, known as smetana, is then skimmed off the top by hand, for eating or buttermaking. But instead of simply draining the leftover curd in cloth, as in skyr, tvorog is made by cutting, cooking, and stirring the curd lightly before draining. The resulting cheese is more severe than skyr, because any remaining cream and moisture in the skim milk curd is cooked out in its make. But that's exactly the point of the cheese.

The barrel-butter method doesn't recover all of the milk's butterfat in the cream that rises; 5 to 10 percent of milk's cream remains in the curd below. Making tvorog with the skimmed curd and skimming the fat off its remaining whey after a day helps make up for the losses. And likely this cheese became popular in many cultures for this reason. As well, goat's and sheep's milk, whose cream does not easily rise for buttermaking, yield bountiful butters with the cutting and cooking of their milks' curd. My great-grandmother, who kept goats in a village in Poland, made her cheese and goat butter this way.

Cooking the acid-set curd also firms it and releases much more whey, resulting in a more solid and grainy cheese that also preserves longer than skyr. The cooking causes the curd to slowly rise—which eases its ladling—and to drain its whey much more quickly, in an hour, compared with more than 24 hours for skyr. The rising of the curd upon cooking is equivalent to the rising of a loaf of bread in the oven; the mesophilic milk fermentation makes a curd full of carbon dioxide, which expands when heated, resulting in a buoyant cheese.

Appreciated across Eastern and Northern Europe, tvorog is a foundational cheese for many culinary traditions; it's known as tvarog in Russia, twaróg in Poland, and skjørost in Norway. It's even pressed and aged into cheeses like graukäse in Austria. This is a significant cheese for many people, because it's made as a by-product of, and in much greater quantity than, butter, and though it's not that delicious, it is substantially nutritious! In English-speaking cultures it is rather rare, but it is sometimes known there as farmer cheese, bakers' cheese, or dry-curd cottage cheese.

In Ukraine, the dry texture of the tvorog is tempered with the addition of some smetana, almost always eaten together. It is essentially a version of clabbered cottage cheese, whose technique follows.

Making Smetana and Tvorog

Fermented cheeses • made with cow's milk • 2-day make

INGREDIENTS

milk	20 L
starter	20 ml kefir, clabber, or whey (1:1000)
salt	

YIELD 2 KG SMETANA (BUTTER) (10:1) AND 1.5 KG TVOROG (FARMER CHEESE) (20:3)

TECHNIQUE

Bring milk to culturing temperature, about 20°C (68°F).
Add starter, mix in thoroughly.
Leave milk to ferment 12 to 24 hours, until milk and cream thicken.
Separate smetana:
- ▸ Skim thickened cream from the top of the vat by hand.

Make tvorog:
- ▸ Cut curd to lentil size, 0.3 mm, with a wire whisk.
- ▸ Cook curd over 1 hour to about 50°C (122°F).
- ▸ Do not stir.
- ▸ Cook until all of the curd has risen.

Ladle curd out of the pot into cheesecloth.
Drain curd in cheesecloth for 1 hour.
Leave whey to stand overnight.
Skim cream off whey.

PHOTO BY ALIZA ELIAZAROV

PHOTO BY ALIZA ELIAZAROV

Goat and Sheep Butter

Goats and sheep don't give up their cream quite like cows. Because of their smaller fat globule, their creams aren't as buoyant, and tend to stay suspended in their milks. (With certain breeds and during later stages of the lactation, fat separation can be more significant.)

So what are goat or sheep butter makers to do? Because the cream doesn't rise so easily, they must resort to other technologies to make their butter. Centrifuges can help, but goat and sheep butter have been around much longer than centrifuges have, and their making doesn't require these unnecessary tools.

There are at least five ways to get the butter out of goat's and sheep's milk using traditional means: making whey butter from the whey of an alpine goat or sheep cheese; turning their fermented milk into to tvorog (farmer cheese) then separating the cream off its whey; churning goat's or sheep's milk yogurt, instead of churning their cream; making a pasta filata goat or sheep cheese and separating the cream off the pasta filata broth; and making chèvre / brebis with their milk, then kneading it and churning it in cold water.

Making an alpine sheep or goat cheese naturally cuts nearly half the fat out of a cheese, the result of a smaller curd size and a more vigorous stirring and cooking. The lost fat remains in the whey, giving it a milky appearance, but after a day of standing the whey cream rises to the top, fermenting along the way and thickening to an almost crème fraîche consistency. This risen cream can then be chilled and churned.

Goat's and sheep's milks can also be fermented and churned to separate their butter. This method, the most laborious of the five, was practiced in Central Asia, and still is to this day. Sheep's milk (often boiled first, making it into yogurt) is left to ferment inside stomachs hung from the rafters of yurts. And every time the stomach is passed by it is hit, bit by bit churning the fat out over the course of the day.

The pasta filata method can break out the cream from goat's milk and sheep's milk, too, although these milks do not stretch and make pasta filata cheeses quite like cow's or buffalo. Instead of stretching them to give them a shape, the curds are packed into forms and pressed. A famous caciocavallo-esque sheep cheese from the Carpathian Mountains of Poland known as oscypek is prepared this way, molded in exquisite hand-carved wooden forms. Oscypek is made this way for many reasons—but an unacknowledged one is that this method gives a significant by-product, sheep butter!

And finally, a Cypriot method of making goat or sheep butter involves making chèvre or brebis the standard way, then churning the cheese in cool water. The butter can be skimmed off the top of the water, and the water-diluted cheese can then be enjoyed as a refreshing and delicious yogurt-like drink.

Making Goat and Sheep Butter

A fermented cheese • made with goat's and sheep's milk • 3-day make

TECHNIQUE 1: WHEY BUTTER

Make an alpine goat or sheep cheese with 20 L milk minimum.
Allow whey to stand 24 hours to ferment and separate its cream.
Ladle cream off surface of whey.
Chill cream several days.
Churn cream into butter.
Separate butter from butter-whey.
Salt if desired.

TECHNIQUE 2: PASTA FILATA BUTTER

Prepare a caciocavallo-style pasta filata with sheep's or goat's milk, 20 L minimum.
Reserve whey for making whey butter as above.
Pour boiling water over salted fermented curd to stretch, and pack into forms to shape.
Leave pasta filata stretching water to stand 24 hours, then separate its cream.
Ladle cream off surface of whey.
Chill cream several days.
Churn cream into butter.
Separate butter from butter-water.
Salt if desired.

TECHNIQUE 3: TVOROG BUTTER

Ferment milk into clabber, with 10 L minimum.
Cut curd to a lentil size, 3 mm, with a whisk.
Cook clabber to 50°C (122°F), stirring all the while, until curds rise.
Strain the curds from the whey to make tvorog.
Leave whey to stand 24 hours to ferment, then separate its cream.
Ladle off the cultured cream.
Chill cream for several days.
Churn cream into butter.
Separate butter from butter-whey.
Salt if desired.

TECHNIQUE 4: YOGURT BUTTER

Boil milk and ferment into yogurt, with 10 L minimum.
Chill yogurt several days.
Churn yogurt until butter rises.
Separate butter from yogurt.
Leftover yogurt can be eaten as is, or drained in cloth to make yogurt cheese.

TECHNIQUE 5: CHÈVRE / BREBIS BUTTER

Make a batch of chèvre or brebis, with 10 L milk minimum.
Drain lactic curd in cloth 12 hours.
Salt curd to taste and knead well until smooth and creamy in texture.
Drain additional 12 hours.
Chill cheese for several days.
Dissolve chèvre in five times its volume of water, and mix around; churn if necessary.
Skim the white butter that gathers together at the surface.
Salt if desired.

Clabbered Cottage Cheese

This is a good ol' American cottage cheese with large and tender curds, smothered in cream. It's considered rather old-fashioned, because, well, it is. Cottage cheese predates the invention of the centrifuge, when milk was left to sit and ferment to separate its cream. As with skyr and tvorog, cottage cheese is a traditional buttermaking by-product. And as we come to recognize the damage done to our foods by industrial processing, and the need for appreciating all the elements of milk, it's about time we re-embrace one of our best skim milk cheeses.

To make cottage cheese, milk is first fermented until the clabber sets. It will form two layers: the thickened fermented cream on top, and the skim milk curd below. The layer of thickened cream can be skimmed off the top. Some of this will be reserved for smothering the curds; the remainder can be chilled and churned to butter.

The skimmed curd below is cut to a small size, then the curds are very lightly cooked and stirred to firm and round their edges. Once they've attained the right consistency, the curds are chilled in cold, salted water. They are then transferred to a vessel, and the crème fraîche is poured overtop to perfectly enrobe the curds.

This cheese is a delicious example of an acid-set curd handled like a rennet curd (a group that includes the less creamy tvorog, handkäse, and graukäse). But because the acid curd is so much softer than a rennet one, it must be very delicately handled.

A gentle, slow, but consistent handling is key to success with this method. To gently heat the curd, you might choose to use a warm-water bath, as excessive heat at the bottom of the pot can cause the delicate curd to break or overcook. When the curd is being very slowly cooked, it must be very softly but thoroughly stirred to prevent caking. Too much or too vigorous a stirring, however, can cause the curd to shatter into fines, which makes a poor-quality cheese. And only when the curd has reached the right firmness can the curd be strained into cold, salted water. The size and texture of the final curd is a measure of success: Though it starts as a kernel of corn, the ultimate size should be about that of a grain of barley.

Making Clabbered Cottage Cheese

A fermented cheese • made with cow's milk • 24-hour make

INGREDIENTS

milk	10 L
starter	10 ml kefir, clabber, or whey (1:1000)
salt	

Yield 1 kg cottage cheese (10:1)

TECHNIQUE

Bring milk to culturing temperature, about 20°C (68°F).
Add starter, mix in thoroughly.
Leave at room temperature for 12 to 24 hours, until milk sets.
Skim off fermented cream and reserve under refrigeration.
Slowly cut the curd to corn size, 0.75 cm, with a spino.
Slowly cook curds to 45°C (113°F) over 1 hour.
- Stir very gently at first.
- Pick up pace as temperature increases and curds firm.

Strain curds into a 5 L cold-water bath (salted to 2 percent) to chill.
- Stir occasionally.

Strain curds into colander.
Leave to drain for 1 hour.
Transfer curds to a pot and mix in cream (reserve some for butter if desired).
Keep cheese refrigerated to preserve.

Handkäse

Handkäse is another sauermilchkäse (sour milk cheese) from Northern Europe. It fits into a category of aged, fermented cheeses, little-known or celebrated cheeses today made from the skim milk curd left behind by traditional buttermaking. Because the cheeses are made of skim milk, the nearly pure protein curd leaves them lacking in flavor and texture. Nevertheless, if you're making traditional butter, you've got to do something with the leftover curd, and this is probably one of the more fascinating things to do with it!

The term *handkäse*, hand cheese, refers to the process of shaping the cheese by hand. The little wheels most closely resemble French aged lactic cheese in texture and flavor, though with something missing—that delicious fat. Oh, and the delicious goat.

To make this cheese, skimmed curd is lightly cooked, drained of its whey, and shaped by hand before being salted and aged with the help of our fungal friend *Geotrichum*. The curd goes through a proteolysis much like a lactic crottin, and when eaten fresh can be runny. Still, the cheese is often dried after aging (also like a crottin), and when it is, because it's free of fat, the broken-down curd takes on a strange translucent appearance.

Handkäse can be made with white *Geotrichum* or washed with whey to get a read-smear with a *Brevibacterium linens* ecology. It can even be made blue by omitting the hastening. Handkäse can also be aged in a drier, mite-infested cave and to become milbenkäse, the famous cheese-mite-ridden cheese of Saxony. My favorite way of enjoying it is to age it like a Saint-Marcellin, in a small ceramic pot, until it melts into a protein-rich puddle.

Though a traditional raw milk cheese of Germany, it is now made exclusively with pasteurized milk. Once made on a small scale by farmstead buttermakers, it is now produced almost exclusively by large industrial producers. It's become a sort of "utility" cheese, often sold dry, in six-packs. But handkäse can be something much greater if made in a way that's more in line with its handmade traditions.

There is an odd way of eating this odd cheese: Aficionados love to smother their handkäse in raw onions and vinegar, almost as if it were pickled herring. The iconic dish is known as *handkäse mit muzik*—hand cheese with music. The music apparently comes after the dish is devoured . . .

Making Handkäse

A fermented cheese • made with cow's milk • aged with a *Geotrichum* rind • 48 hour make • 2-week affinage

INGREDIENTS

milk	10 L
starter	10 ml kefir, clabber, or whey (1:1000)
salt	

YIELD **500 g handkäse (20:1)**
and 500 g butter (20:1)

TECHNIQUE

Make clabber:
- Bring milk to culturing temperature, about 20°C (68°F).
- Add starter, mix in thoroughly.
- Wait 12 to 24 hours, until set.
- Skim cultured cream, then chill, and churn to make butter.

Drain curd in cheesecloth for 24 hours.
- Break up curd occasionally to improve drainage.

Salt curd to 2 percent.
Drain for an additional 24 hours.
Shape into handkäse by hand.
Age for 2 to 3 days in warm and humid hastening space until *Geotrichum* grows.
- Flip daily.

Age in cave for 2 weeks.
- Flip every other day.

Dry cheeses to preserve.

PHOTO BY ALIZA ELIAZAROV

Graukäse

Another sauermilchkäse, this one from the Tyrolean region of Austria and Italy, is graukäse. This peculiar cheese is made with the leftover skim milk curd of traditional barrel-buttermaking, like the little handkäse, only made firmer, larger, and longer lasting, like an alpine sour milk cheese.

I include this cheese because it's an important one from historical and cultural perspectives. Graukäse is a sour milk cheese that submits butter by-products to alpine cheesemaking methods, cooking the fermented curd (made in wooden vats) in copper kettles until firm. But this cheese may predate alpine cheeses as we know them by many centuries, if not millennia.

This style is perhaps that of the original long-aged cheeses, made without the technology of rennet. When buttermaking is the main activity of alpine herders, as it

PHOTO BY ALIZA ELIAZAROV

still is today in Tirol, there is leftover curd to cope with, and these large cheeses are made with it to transport easily down the mountain.

Graukäse means "gray cheese," a reference to the grayish color of the rind that is most often a *Geotrichum*-dominated one. When aged, its rind is slowly broken down by the surface flora, but because of the high protein and low fat content, the flesh of the cheese takes on a unique translucent character. Toward the interior of the cheese, the curd remains tvorog-like, not surprisingly, as it is made a similar way.

Cheeses like graukäse inhabit lands where dairy products are still made along the path of transforming milk into one of its most enlightened and important products, butter. This style was likely much more common throughout Europe before rennet cheesemaking technology arrived. It has relatives from Scandinavia, like gamalost, through to Central Asia and evolved from the same original cultural practices of buttermaking, the earliest of dairying technologies. It could even be a cheese that the Beauty of Xiaohe, with her typical Tyrolean hat, made with her kefir.

Making Graukäse

A fermented cheese • made with cow's milk • aged with a *Geotrichum* rind
24-hour make • 2- to 6-month affinage

INGREDIENTS

milk	20 L
starter	20 ml kefir, clabber, or whey (1:1,000)
salt	
forms	round: 20 cm diameter × 20 cm deep
YIELD	1 KG GRAUKÄSE (20:1) AND 1 KG BUTTER (20:1)

TECHNIQUE

Make clabber:
- Bring milk to culturing temperature, about 20°C (68°F).
- Add starter, mix thoroughly.
- Wait 12 to 24 hours, until set.
- Skim cultured cream and refrigerate (after chilled for 3 days, churn into butter).

Cut curd into lentil-sized (3 mm) pieces.
Slowly cook curd to 50°C (122°F) over 60 minutes.
- Do not stir.
- The curd will rise as it cooks; it is ready when all the curd floats and firms.

Strain curd from top of whey and place into cheesecloth-lined forms.
Press curd lightly overnight.
- Flip after several hours.

Salt surface of cheese twice over 2 days.
Drain cheese on draining table for 24 hours, covered with cheesecloth.
- Flip once.

Age for 2 to 3 days in hastening space until *Geotrichum* grows.
- Flip daily.

Age in dry cave for 3 to 6 months.
- Flip every other day.

Sourdough Rye Bread

A recipe for sourdough bread is not out of place in a cheesemaking book, for the two realms of fermentation inspire each other, feed each other, in many fascinating ways. But this technique is not just here to complement your cheesemaking, it is here to help you grow your own blue fungus; for the *Penicillium roqueforti* that blossoms on blue cheese grows best on freshly baked sourdough rye bread.

Before I made cheese, I baked bread. I was regularly baking naturally fermented breads for sale at a farmers market when I made my first batches of naturally fermented cheese. Much of my own understanding of milk's fermentation comes from my prior experiences of keeping sourdough. And the more I examine sourdough cultures, the more they seem to parallel "sour-milk," and vice versa. They may well be the same.

Sourdough taught me that cheese cultures must be fed daily. Natural bakers religiously feed their starters, assuring that their cultures are in best form for their next day's bread bake. In no other realm of fermentation is the starter held in such high regard. Natural beer brewers or winemakers talk of their ferments preferring a spontaneous generation, and sauerkraut makers refer to a wild fermentation, but all of these would be improved with the addition of a starter kept much like sourdough. Certainly cheesemaking is a fine example of a fermentative process that can evolve its best through re-fermenting milk in the sourdough style.

The sourdough culture is an ecological equivalent to a clabber culture, but fed flour instead of milk. They may even have many microbes in common. A sourdough starter can quickly be started with clabber; in the absence of raw milk, a clabber culture can even be created from a sourdough starter as described in the appendices. Oddly, if a sourdough starter is made wetter (with a liquidy, milky texture), it ferments less like dough and more like milk, with the liquid souring like clabber and yeasty fermentation happening mainly at the surface of the liquid (for the water prevents air's passage through the ferment). Such a wet sourdough starter grows a wrinkly white fungal film at is surface that very much resembles milk's *Geo* (it probably is—both these film fungi are feeding on lactic acid!).

Even the rising of a properly baked sourdough bread can be related to cheese and its rind development. Like lactose's primary and secondary fermentations in cheese, sourdough makes bread rise through a two-step fermentation. *Lactobacilli* and other bacteria first break down the polysaccharide starches in the flour into lactic acid; the yeasts then feed on the lactic acid, producing alcohol and carbon dioxide, which gets trapped in the gluten network of the dough and causes it to rise. Cheesemaking and its affinage are equivalent—with bacteria breaking down lactose (also a polysaccharide) in the milk into lactic acid, which is then feasted upon by yeasts and fungi on the rind as the cheeses age, transforming the lactic acid into alcohol and carbon dioxide. It's not a stretch to suggest that sourdough and cheese are evolutions of the same fermentations. I personally believe that the raising of a bread is the same phenomenon as the growth of cheese rinds (at least the wrinkly *Geo* ones).

Finally, to bake sourdough well, particular attention must be paid to the qualities of flour and fermentation. To me these are equivalent to the priorities of good milk and good fermentation in natural cheesemaking; for if your flour is not freshly milled (whole-meal flour quickly rancidifies just like milk) or your fermentation is off, it doesn't matter how good your oven or your baking skills are, your bread will be a bust! And just like feeding a clabber culture regularly assures cheese's best white-rind development, regularly feeding a sourdough starter assures a bread's best rise.

Again, this technique is included partly to help us grow our *Penicillium roqueforti* fungus on sourdough rye bread. So then, why rye? *P. roqueforti* will grow on wheat bread, white or whole. But rye is the traditional grain used by bakers in the hills of Roquefort, and it's on rye bread that cheesemakers today still grow the fungus. Also, rye is like the goat's milk of bread baking: The sourdough process works especially well with this generally less appreciated grain, which lacks in working character but gives an extremely flavorful crumb when long fermented, making elaborate bread bakes impossible but also unnecessary. Simple bread bakes are best with rye, just like simple cheese makes best complement goat's milk.

Be sure the culture is well fed and active before the bake; that's what matters most (see appendix G for details on keeping sourdough). Don't feel you need to have a 100-year-old starter to bake good bread; there may be no difference microbiologically between a newer starter and an heirloom if they're both tended to the same. Research into the microbiome of sourdough shows that starters are remarkably similar the world over. What defines the microbiological makeup of the starter is primarily the type of flour it's fed, whether it's wheat, rye, or spelt. The provenance of a starter is much less significant than its feeding regimen. This is most likely the case in natural cheesemaking as well.

Also, be sure to add a more substantial amount of starter than in cheesemaking; most bread bakers add starter to their doughs in a ratio of about 1:5 to 1:10. If you add it 1:50 to 1:100 as in cheese, there would be an effective bacterial fermentation, but the dough would fail to rise (yeasts are slower to grow than bacteria, and need to be added in greater quantity in the starter to get the bread to rise).

Sourdough for growing *Penicillium roqueforti* should be fermented longer than normal (many bakers underferment their loaves) to assure sufficient lactic acid to fuel the growth of the *roqueforti*, and to suppress unwanted fungi. If fluffy gray *Mucor* grows on the moldy bread, that's a sure sign that the bread was not sufficiently soured.

This rye bread is especially delicious if it is long cooked, overnight, in a covered bread pan in a cooling brick oven. My regular bread bake is this one, a slow-baked, caramelized, dark rye bread in a Nordic style whose chewiness, color, and intense flavor are rarely reproduced by commercial bread bakers, even sourdough ones. For me, if there's one bread to bake, it is this one.

I see sourdough as a gateway fermentation—a daily practice that can bring great satisfaction and nutrition for the family, and can give you a better microbiological understanding of natural cheesemaking. I might say it helps to first get comfortable with this realm of fermentation beforehand, as I did.

Making Sourdough Rye Bread

A grain ferment • 8- to 12-hour make

INGREDIENTS

rye flour	3 cups	
salt	2 teaspoons	
starter	active sourdough prepared the previous day with ⅓ cup rye flour (1:9)	
tepid water		
cultured butter		
YIELD	**ONE 1 KG LOAF**	

TECHNIQUE

Mix flour, salt, and active starter together.
Add water until a soft, sticky dough is formed.
Let sit 1 hour or so, then mix dough again.
Leave to ferment at 20°C (68°F) about 6 hours, until dough has risen slightly and begins to sour.
Punch down dough, and leave to rise again for 1 to 2 hours.
Place in a loaf pan, greased with butter, or shape into a free-form loaf.
Leave to rise 1 more hour.
Bake at 450°F for 30 minutes, until browned.
▸ Or (my preference) bake at 100°C (210°F) for 12 hours in a covered pan.
Remove bread from pan and cool.

Camembert Bleu

This technique is for a Camembert made to be moldy with *Penicillium roqueforti*.

The first blue cheese of the book, this is also a great first blue cheese for a cheesemaker to attempt or a cheese eater to try, for the blue flavor is gentle and unintimidating. It is rather pleasant and mild, in fact, with the cheese tasting surprisingly akin to its white-rinded brothers. *P. roqueforti* and *P. candidum*, are, after all, close kin, and the two cheeses are made more or less the same.

This cheese is a lesson in respecting different ripening cultures, for the colorful *roqueforti* fungus that ripens this Camembert is generally not appreciated on the rinds of the esteemed white cheese—it's actually considered a contaminating fungus in the modern cheese make. If a spot of blue is found on the rind of a Camembert, it typically takes a beeline for the compost heap, despite its deliciousness. But the blue version of the cheese is in most respects the same as a Camembert ripened with *Penicillium candidum*, for the two closely

PHOTO BY ALIZA ELIAZAROV

related fungal cultures cause the cheese to break down in nearly identical ways.

The making of this blue Camembert and the classic white Camembert are essentially identical. But to encourage the blue rind to grow, two important steps are advised: *P. roqueforti* spores are added to the milk to help seed this sought-after fungus, and the rinds are left unwashed in their cave, which encourages *Penicillium roqueforti* over *P. candidum*.

As an exercise in the effects of affinage, specifically on the influence of rind washing, both a white-rinded Camembert and a blue-rinded Camembert can be made from the very same batch of cheese. To differentiate the two, the blue one is left unwashed while the white one is washed for 1 week to suppress the blue and encourage the white. Even an orange Munster can be made from the same curd, the cheese simply being washed with whey for 1 month to encourage a *Brevibacterium linens* ecology. And all three of these cheeses can even be aged in the same cave, for the *Penicillium roqueforti* spores will not "contaminate" the others. They will only get a chance to grow on a cheese if we allow them to, by leaving the rind unwashed! A single cheese can even be made with the three ecologies by treating each of the three parts of the rind differently—the bottom left unwashed to turn blue, the top washed for 1 week to turn white, and the sides washed for 1 month to become orange!

Making Camembert Bleu

A rennet cheese • aged with *roqueforti* • 8- to 12-hour make • 6- to 8-week affinage

INGREDIENTS

milk	10 L
starter	100 ml kefir, clabber, or whey (1:100)
spores	*Penicillium roqueforti*
rennet	regular dose calf rennet
salt	
forms	hoops: 10 cm diameter × 10 cm deep
YIELD	FIVE 300 G CAMEMBERTS (7:1)

TECHNIQUE

Bring milk to cheese temperature, about 35°C (95°F).
Add starter, mix in thoroughly.
Add spores, mix in thoroughly.
Add rennet, mix in thoroughly.
Wait for clean break, 45 minutes to 1 hour.
  ▸ Cover the pot to keep in warmth.
Cut curd to walnut size, 2.5 cm.
  ▸ Cut vertically; cut crosswise; cut horizontally or diagonally.
Stir curds gently for 5 to 15 minutes, until they can be held.
Pitch curds for 5 minutes.
Whey off.
Fill forms to the top with curds.
Drain 6 to 12 hours, until acidity develops.
  ▸ Goal pH 5.3—stretch test positive.
  ▸ Flip twice during the drain, with first flip at 1 hour.
  ▸ Salt 2 percent by weight on surface, about 2 teaspoons per cheese.
Or salt cheeses for 1 hour in saturated salt brine.
Drain cheeses for 1 day on draining table, covered with cheesecloth.
  ▸ Flip twice during draining.
Age cheeses in cave for 6 to 8 weeks.
  ▸ Flip cheeses every other day.

Fourme d'Ambert

La fourme d'Ambert is a classic blue cheese from the Auvergne still made today in the traditional way. It's a tall, cylindrical cheese with a rustic white rind, bright blue eyes, a bold blue flavor, and a buttery texture when aged. The name *fourme* derives from the Latin *forma*, or "form," and this blue is believed to be derived from Roman cheesemaking practices brought to France over two millennia ago.

Unlike Camembert bleu, which grows *Penicillium roqueforti* only on its rind, this cheese grows blue only on the inside. The big eyes are formed with the help of the stirred-curd method: The curd is cut to a medium size and stirred for a lengthy period of time to firm. The resulting curds don't knit together as well, leaving mechanical holes between them, which turn into blue eyes within the cheese only once air is introduced.

Three special treatment help accentuate the blueing in this beautiful cheese: Once the curds reach their desired firmness (one can be dropped from over a meter in height and not break!), the curd is drained as it is

PHOTO BY ALIZA ELIAZAROV

slowly ladled into the forms, to make it even firmer. The result is a cheese with a remarkably open texture, providing lots of habitat for *P. roqueforti* to grow.

The cheese is also long fermented before salting—3 days at room temperature—developing lots of lactic acid. This provides plenty of *Penicillium roqueforti*'s favorite food and gives the cheese its buttery texture. The extra primary fermentation provides for a longer affinage, which makes the blue cheese much more flavorful. It also helps *Geotrichum* grow in the early stages of the make—but only on its rind where there's air.

Finally, the young cheese is pierced to allow the *roqueforti* to blossom within. The piercing should be done with a needle all around the sides of the cheese, every 2 cm or so, penetrating deep toward the core. A second piercing 1 month later can help assure an even and consistent blueing. Only once air is introduced and the cheese is in a cooler cave does the *roqueforti* transform the open structure of the young cheese into a beautiful aged blue.

The cheese is eaten in 3 to 4 months, once the curd begins to soften from the fungal breakdown. The cheese ripens fast as a result of the piercing, which permits the surface ecologies to ripen the cheeses from the much greater surface area within; left unpierced, the cheese would age more slowly, like a tomme crayeuse, whose make it largely mimics. When ripe, and sliced in rounds, its cratered full-moon appearance will have you wondering if the moon is made of blue cheese, not green.

Making Fourme d'Ambert

A stirred-curd cheese • made with cow's milk • aged with *roqueforti* • 3-day make • 3- to 4-month affinage

INGREDIENTS

milk	20 L
starter	200 ml kefir, clabber, or whey (1:100)
spores	*Penicillium roqueforti*
rennet	regular dose calf rennet
salt	
forms	hoops: 15 cm diameter × 30 cm deep
YIELD	ONE 2 KG FOURME (10:1)

TECHNIQUE

Bring milk to cheese temperature, about 35°C (95°F).
Add starter and spores, mix in thoroughly.
Add rennet, mix in thoroughly.
Wait for clean break, 45 minutes to 1 hour.
- ▸ Cover the pot to keep in warmth.

Cut curd to hazelnut size, 1.5 cm.
- ▸ Cut vertically; cut crosswise; cut horizontally or diagonally.

Stir curds gently at around 35°C until they can drop from 90 cm, about 60 minutes.

Slowly scoop out curds onto a draining table, one small scoop at a time.
- ▸ Allow curd to drain whey as it is scooped.
- ▸ Continue to stir curds in the vat as they are scooped so they don't knit.

Carefully transfer curds to a cheesecloth-lined form, filling it beyond its brim.

Drain cheese for 2 to 3 days on draining table, covered with cheesecloth.
- ▸ Leave in form, flip daily.
- ▸ Goal pH 4.9—stretch test fails.
- ▸ Cheese will also grow *Geotrichum* on its rind.

Salt cheese.
- ▸ Cover cheese with salt.
- ▸ Leave cheese to drain overnight in form.
- ▸ Cover cheese with salt again.
- ▸ Leave cheese to drain overnight once more.
- ▸ Pierce cheese every 2 cm to allow air inside.

Age cheese in cave for 3 to 4 months.
- ▸ Flip cheese every other day.

Roquefort

This famous sheep's milk blue cheese from the south of France is made much like a fourme d'Ambert, but with a few refinements that make it stand out as the best of the blues. With the whitest and creamiest curd, and the biggest most beautiful blue eyes, this cheese is at the top of its class.

Roquefort's production is inextricable from its aging caves, the most famous aging caves on earth. And the celebrity of its caves is linked to the celebrity of its cheeses thanks to the way cold conditions encourages the growth of *Penicillium roqueforti*. Were the cheeses kept in warmer weather (20°C [68°F]) the wheels would have white eyes due to *Geotrichum candidum*, which takes the upper hand aboveground.

Very few cheeses in this book call for sheep's milk in their make, but it should be stated at the outset that it is essential for the desired Roquefort. Sheep's milk is exceptionally rich and creates curds that respond most effectively to this method. Goat's or cow's milk can be used in its place, but its curd must be cut to a smaller size—a hazelnut. If prepared with cow's milk, the cheese becomes Bleu d'Auvergne.

Sheep milk is also capable of remarkably aromatic transformations when submitted to the long fermentations—both primary and secondary—of blue cheese. Along the same vein, for this cheese it's best to keep everything sheep-specific. Fermenting with a sheep-milk-based starter, Roquefort will grow with its most appropriate ecologies that develop its best tastes. And curdling with lamb rennet, prepared without the lipase-rich cheese within, will give the truest flavor to the cheese, and contribute to the savoriness of the style.

When set, the curd should be cut larger than a typical curd for a blue cheese make, because the sheep curd begins nearly twice as dense as cow curd. With less cutting, it retains more of its cream, making this cheese the butteriest of the blues. Like the fourme before it, the curd should be slowly ladled into its forms to preserve the cheese's open structure, setting the stage for its best blueing. And like most blues, a fuller fermentation provides the lactic acid that fuels *roqueforti*'s growth.

As the cheese ages, it's stored on its side and regularly rolled. Few cheeses are oriented this way in their caves, but because of Roquefort's wide shape, it is pierced on its faces to maximize the flow of air to its interior and promote the growth of *roqueforti*. If it were flipped between its faces instead of rolled, air wouldn't flow through its holes when each face was against its boards. The brittle nature of the long-fermented curd helps keep the cheese from slumping in shape.

Modern Roquefort is scraped of the fungus that naturally grows on its rind, giving the cheese a clean exterior and a surprising blue reveal when the cheese is cut. But the wheel can also be left to grow a natural rind, like fourme d'Ambert, if desired.

Making Roquefort

A stirred-curd cheese • made with sheep's milk • aged with *roqueforti* • 3-day make • 2- to 3-month affinage

INGREDIENTS

milk	20 L sheep
starter	200 ml kefir, clabber, or whey (1:100)
spores	*Penicillium roqueforti*
rennet	regular dose lamb rennet
salt	
forms	round: 20 cm diameter × 10 cm deep

YIELD   ONE 3 KG ROQUEFORT (6:1)

TECHNIQUE

Bring milk to cheese temperature, about 35°C (95°F).
Add starter, mix in thoroughly.
Add spores, mix in thoroughly.
Add rennet, mix in thoroughly.
Wait for clean break, 45 minutes to 1 hour.
- Cover the pot to keep in warmth.

Cut curd to walnut size, 2.5 cm.
- Cut vertically; cut crosswise; cut horizontally or diagonally.

Stir curds gently at about 35°C until they bounce from 90 cm, about 60 minutes.
Slowly and carefully fill two forms, draining as the curds are ladled.
- Fill two forms, each to the top.
- Then flip one cheese overtop the other to unite.
- Keep curds moving in the vat all the while.

Flip cheeses after 1 hour and again after another 2 hours.
Leave cheeses in forms for 2 days in hastening space.
- Flip and rub down rind daily.

Salt cheeses once curd fails stretch test (pH 4.9).
- Cover cheese with salt and leave to drain overnight.
- Cover cheese again with salt and leave to drain overnight.
- Leave cheeses to dry for 2 days after salting.

Pierce cheeses every 2 cm on top and bottom.
Age cheeses in cave for 2 to 3 months.
- Rest cheeses on their sides to encourage air to flow to piercings.
- Turn cheeses every day.
- Scrape rind occasionally to remove mold if desired.

Gorgonzola

The creamiest and most indulgent of the blues, Gorgonzola is made with a curd most delicately handled and only lightly fermented, and is quite different in its make from other *Penicillium roqueforti*–ripened wheels. The result is a decadent blue with deliquescent texture and delectable *roqueforti* flavor that's deliciously *dolce*—"sweet," in Italian.

The cheese begins its life like most other blues, with only the freshest milk. The milk is cultured and renneted with a lipase-rich rennet, which contributes to the complex character of the style. But unlike the other blues, Gorgonzola curd is cut large to retain as much fat as possible, and stirred minimally to preserve its moisture. You then transfer the curd to cloth, pre-draining it before forming, and break up the drained curd and then pack it into forms to give the cheese its shape. This technique prevents the full knitting together of the curds, leaving hairline fissures between them, eventually providing the space for *Penicillium roqueforti* to bloom.

Then, unlike the other blues in this book, Gorgonzola is salted sooner—on the same day of its make—preserving the pH around 5.3. Together with the high moisture and fat of the curd, this creates a blue cheese that flows like a Camembert when ripened. Due to this foreshortened fermentation, the cheese doesn't have the strong blue flavor associated with the others. The earlier salting leads to an earlier ripening, and a preservation of the sweetness of the milk.

In the cave this blue is treated like all the others: It is pierced to permit the air to enter, washed or brushed to keep its blue on the inside, and flipped regularly to assure an even ripening. But within 2 months—sooner than the other blues in this book—Gorgonzola is considered ripe, sweet, and ready to eat.

Making Gorgonzola

A rennet cheese • aged with *roqueforti* • 8- to 12-hour make • 2- to 3-month affinage

INGREDIENTS

milk	20 L
starter	200 ml kefir, clabber, or whey (1:100)
spores	*Penicillium roqueforti*
rennet	regular dose calf rennet with lipase
salt	
forms	round: 20 cm diameter × 20 cm deep

Yield ONE 2 KG GORGONZOLA (10:1)

TECHNIQUE

Bring milk to cheese temperature, about 35°C (95°F).
Add starter and spores, mix in thoroughly.
Add rennet, mix in thoroughly.
Wait for clean break, 45 minutes to 1 hour.
Cut curd to walnut size, 2.5 cm.
- ▸ Cut vertically and horizontally.

Stir curds gently for 5 to 15 minutes, until they bounce from 30 cm.
Ladle curds into cloth to drain.
Drain on a draining table for 1 hour.
Break up the curd in the cloth into coarse pieces.
Fill form with drained curd.
Drain cheese until acidity develops, typically 8 to 12 hours total.
- ▸ Flip every couple of hours.
- ▸ Goal pH 5.3—stretch test positive.

Salt cheese over 2 days from the surface.
- ▸ Leave cheese in its form to keep it tall.

Pierce cheese on top and bottom faces every 2 cm.
Age cheese in cave for 2 to 3 months.
- ▸ Flip cheese every other day.
- ▸ Wash or rub rind occasionally to set back fungus.
- ▸ Pierce cheese again after 1 month.

PHOTO BY ALIZA ELIAZAROV

PHOTO BY ALIZA ELIAZAROV

Stilton

Stilton is a striking cheese. It stands tall, with a beautiful white-and-orange rind covering it all; only when cut does the cheese reveal its most distinct and vivid blue veining. It is the most fêted (and fetid) of the British cheeses, and a most extraordinary elaboration of the milled-curd cheesemaking method. And though complex in its production, Stilton is a cheese that excels in a small-scale operation and presents an achievable challenge and a great reward to a home-scale maker.

Stilton is made much like a cheddar, only wetter. The curd is cut larger, stirred less, drained less, pressed less (if at all); this results in a milled-curd cheese with a slightly higher moisture content and similarly higher cream content. This all helps the curd develop into a delicious fungally ripened cheese.

Because of the complexity of its make, the Stilton displays a complex ecology. More specifically, three distinct ecologies coexist within one wheel, one of the only cheeses to do so. The top and bottom of the wheel are typically white, the skin of its sides is often orange, and blue veins course through its flesh. *Penicillium roqueforti*, *P. candidum*, and *Brevibacterium linens* all display prominently on the cheese, each growing where the cheesemaker creates the specific conditions that allow the particular ecologies to flourish.

Penicillium roqueforti displays striking veins only within the body of a Stilton, perhaps the most beautiful blueing of all. The milled curd, made acidic through draining, and salted before forming, doesn't knit together well, leaving a tortuous open texture within the cheese that provides ample habitat for the blue fungus to grow. The curds are specifically left unpressed to leave the spaces between the curds for *roqueforti* to develop, unlike cheddars whose curds are pressed tightly together. However, Stiltons *must* be made tall to enable the acidic curds to knit together under the weight of the curds above. A Stilton made shorter would crumble to pieces.

Because the curd structure is open, *Penicillium roqueforti* should show itself at the rinds (the cheese, after its make, looks like a loose assembly of curds), but the unique treatment of rubbing up the rind prevents the blue growing on the cheese's exterior. Stiltons are hastened in a warm and humid room for several days after their make to encourage yeasts on the rind. Once the yeasts begin to soften the texture, the broken-down curd is smeared and rubbed into the rind, much like spreading icing on a cake. Rubbing up, described in detail in chapter 9, seals the cheese shut, preventing air from accessing the interior while encouraging the white and orange cultures to colonize the rinds.

The orange *Brevibacterium linens* and white *Penicillium candidum* ecologies (and occasional *Geotrichum*) dominate in different parts of the rubbed-up wheel. Typically, *P. candidum* grows toward the top and bottom of the wheel, while *B. linens* grows in the middle, giving the wheel a distinguished appearance. The treatment that allows the cheese to develop these two ecological zones is leaving the wheel in its tall cylindrical form during hastening. The sides of the wheel, smothered in their forms, remain wet, and are dominated by washed-rind ecologies; the tops and bottoms of the wheel breathe better, and tend to develop white *P. candidum*. *Geotrichum* wrinkles may form on the wheel, but only if the curds are left to ferment too long before salting, the more acidic curd favoring the growth of the lactic-acid-loving *Geo*.

Only after a couple of weeks of aging, and only after the rind ecologies are well established, are the cheeses first pierced. This delay allows the cheese to ripen more slowly, and develop more depth of flavor and lactic acid before the blueing and its resultant rapid ripening begins. Piercing allows air to finally flow to the open curd structure within the cheese and must be performed generously to permit air to flow through the rubbed-up rind. Piercing should be done horizontally or diagonally from the sides toward the center of the cheese to permit the most airflow to the interior. A second piercing, a month later, permits the continued passage of air through the cheese's softening and ripening rind. If left unpierced, the cheese becomes a white Stilton, a more crumbly, sour style.

Stilton can be made with varying degrees of creaminess, and different ripening times, depending on the handling of the curds. The draining curd can be made firmer or softer, or left to ferment shorter or longer, developing more acidity and an even more significant creaminess. This makes for an especially soft blue cheese, but one that doesn't last quite like a firmer, less primary-fermented wheel. The longer-fermented one will also feature a more wrinkled *Geotrichum*-dominated rind.

Jersey or Guernsey milk shines with this make for its golden color content, but even more flavor (along with a white color) can come with using goat or sheep's milk in place of cow's. For extra fun and flavor consider mixing the curds of different milks into one cheese for a colorful marbled effect. (This can also be done with any other milled-curd cheese, including cheddar.)

Stilton is a cheese made historically in the late summer or fall. It ripens better in aboveground "caves" in the cooler weather and is ready just in time for the holiday season. Its textures, flavors, and aromas are a perfect pairing for a celebratory winter gathering.

Making Stilton

A milled-curd cheese • aged with *roqueforti* • 48-hour make • 3-month affinage

INGREDIENTS

milk	40 L
starter	400 ml kefir, clabber, or whey (1:100)
spores	*Penicillium roqueforti*
rennet	regular dose calf rennet
salt	
forms	round: 20 cm diameter × 30 cm deep
YIELD	ONE 4 KG STILTON (10:1)

TECHNIQUE

Bring milk to cheese temperature, about 35°C (95°F).
Add starter and spores, mix in thoroughly.
Add rennet, mix in thoroughly.
Wait for clean break, 45 minutes to 1 hour.
Cut curd only vertically, to walnut size (2.5 cm) columns.
Leave curd to firm for 30 minutes.
Scoop curds into cheesecloth-lined colanders.
Tie up curds in cloth bundles.
Drain bundles on table for 12 to 24 hours.
- After 3 hours, slice drained curd into slices 4 cm thick to facilitate drainage.
- Drain until acidity develops.
- Goal pH 5.1—stretch test liquidy.

Mill curd to 1 to 2 cm pieces.
Salt curds to 2 percent, and leave to drain 1 hour.
Fill form with milled, salted curd.
Age cheese for 3 to 5 days in hastening space until rind develops *Geo* and softens.
- Leave cheese in form and flip daily.

Rub up the broken-down rind with a knife to cover up holes.
Age cheese in cave for 2 weeks before piercing.
- Flip cheese every other day.

Pierce cheese every 2 cm on sides.
Age cheese in cave for 2 additional months.
- Flip cheese every other day.
- Pierce cheese again every 2 cm on sides 1 month after first piercing.

Natural Wine

Fall isn't just the season for fatty cheesemaking, it's also the season of the vintage: the time of harvesting and fermenting grapes for wine, as well as apples for cider. So why not include a technique for wine in this book? Natural wine and natural cheese are even more perfect partners than sourdough bread and cultured butter!

Raw Wine, the largest community of natural winemakers and drinkers in the world, defines natural wine as:

- Wine made from organic grapes, grown without the use of artificial chemicals
- Wine prepared by hand using artisanal techniques
- Wine made using traditional winemaking processes that enable balance
- Living wine created with low intervention in the cellar
- Wine that promotes well-being in individuals and communities

Such is the new sought-after drink: Natural wine is poison-free and tastes of the sun, the earth, and the vintners' work instead of chemical intervention. All of these philosophies of natural winemaking are very much in line with those of natural cheesemaking, and an understanding of one can complement the practice or appreciation of the other. And surely the same can be said for natural cider, beer, and sake.

But what I'd like to focus on most is the common ground in fermentation between wine and cheese. Like traditional cheesemaking, natural winemaking is often criticized for its variability. Natural winemakers celebrate the spontaneity in their fermentation that comes with the exclusion of commercial strains of yeast—but should this be the most revered approach? Winemakers can gain more consistency in their fermentations and the final quality of their libations through the use of carefully cultivated natural starters. Though starters are rare and even controversial in the world of natural winemaking, vintners can use them to successfully initiate fermentation in their vats, just like a sourdough or a clabber culture.

A sourdough bread baker would never bake a bread without adding a starter; in cheese, a wild fermentation would invariably result in poor fermentation, wild inconsistency from batch to batch, and a possibly dangerous product. So why should a winemaker take such an approach? This is a challenging conversation to have with many natural winemakers, as many see their spontaneous fermentation as an antidote to the rigorously controlled industrial approach. Still, their wines will see fermentation hastened by the addition of a natural starter, which will transform their fruits into their truest, more consistent, and most expressive wine. Natural wines can be a high-fidelity affair, if practiced consistently, leaving nothing to be desired.

I see the addition of a natural starter, known in French as *pied de cuve*, as a middle ground between industrial and wild fermentation that helps yield deliciously consistent wine without commercial yeasts or the unpredictability of spontaneous fermentation. A vintner can begin one just before the vintage by leaving the freshly pressed juice of early-harvested grapes to ferment spontaneously (a similar effect can be achieved by adding kefir grains or clabber to pasteurized juice), stirring occasionally to encourage yeasts. Once bubbly, usually within 2 to 3 days at 20°C (68°F), the ferment can be fed fresh juice (fresh or pasteurized) at a ratio of 1:50 or higher. With the addition of a starter, the lag time is now reduced, and bubbling will often happen overnight. After just one or two feedings, usually within a few days, the culture will ferment within 12 to 24 hours at room temperature, reaching its peak fermentation period. You can now use this active culture as a wine starter, keeping it in top form through the harvest with a daily feeding of juice and a regular mixing of air to encourage its yeasts. A more thorough exploration of this starter can be found in appendix H.

The addition of the wine starter results in a microbiological protection of the fruit that keeps unwanted cultures from developing, by hastening the most desirable fermentations. The culture contains a microbiological diversity that initiates primary fermentation and provides diverse species that can aid fermentations even to a higher alcohol content. There are no stuck fermentations with this culture. And malolactic fermentation is assured.

The starter also contributes species that aid secondary fermentation and surface fermenters like *Brettanomyces*. If you don't want these microbes, often seen as "contaminants," stir, stir, stir! It is stirring that encourages the more sought-out wine/beer yeast *Saccharomyces cerevisiae* that can help a wine achieve many of its most sought out characters. (The French word for a brewery is *brasserie*, which translates roughly as "the stirring place.") In *Brettanomyces*'s case, this loved/loathed yeast is encouraged by leaving the ferment undisturbed without stirring, causing a surface yeast fermentation (it looks an awful lot like *Geotrichum*) and a subsurface, oxygen-free, lactic fermentation that gives a butteriness to wine.

This approach is relevant not just to winemakers, but also to cider brewers and mead makers, for the same philosophy of fermentation applies to all sources of sugar—even beer and sake (after the barley or other grains are malted and mashed or the rice is first converted to sugar with the help of koji). Though a generic starter can be preserved on apple juice, say, a more appropriate community of microorganisms can be cultivated by re-fermenting the particular fruit (or grain) that is to be fermented. For it's always best to keep the culture fed on the specific medium you're aiming to ferment.

Like cheese, wine, beer, and cider fermentation is generally regarded as a two-stage process, consisting of

PHOTO BY ALIZA ELIAZAROV

a primary fermentation and a secondary fermentation (I've adopted these terminologies for cheese from this realm). The first fermentation in wine is a yeasty one in which the fermented fruit is regularly stirred to expose it to air, initiating a yeast fermentation that converts sugars to alcohol and carbon dioxide (and other diverse metabolites). Once the primary fermentation has settled, after a week or so of vigorous stirring and bubbling, the brew is usually racked or transferred to a secondary fermentation chamber where air exposure is controlled, limiting or encouraging the development of anaerobic species that can affect the ferment, just like cheese.

If you're aging a wine in wood or clay, as opposed to constricted in stainless steel or glass, some slight air exposure continues to complement its evolution within the barrel, encouraging diverse surface yeasts to develop the secondary characteristics of the drink—much like aging a cheese. A most extraordinary complexity evolves if these ferments are exposed to some air in their secondary fermentation—one that for folks used to drinking industrial wine requires some time to appreciate. But for those who have become accustomed to the complexity of a more natural fermentation, there's little chance of turning back. It's like developing a taste for clothbound cheddar after a lifetime of eating only plastic-bound ones.

Finally, upon achieving a fine flavor, and a high clarity, after months or years of storage in the barrel, the wine is racked into bottles for finishing. There fermentation continues, albeit much more slowly now. The commonplace addition of sulfur is a chemical pasteurization that stops the bottled wine from re-fermenting with yeasts or vinegar bacteria once oxygen is mixed in, aiding its transportation without refrigeration. (Sulfur is also often added to halt wild fermentation when commercial yeasts are added to the fermentation vat.) But should wine be preserved in warm weather, or be aged for 50 years with the help of sulfur? Is there a need? Or has sulfuring wine enabled the stabilization and commodification of a food that probably shouldn't be stabilized and commodified? All things in life change—we can never hold on to the present state forever. This has always been the case with cheese; why not with wine? Let's drink to that!

Making Natural Wine

A ferment of the vine • 1-week make • 1-year affinage

INGREDIENTS

grapes	40 kg
starter	200 ml wine ferment (1:100)
vessel	40 L fermentation vat
	20 L carboy, barrel, or amphora

YIELD   FIFTEEN 750 ml BOTTLES

TECHNIQUE

Remove grape stems.
Crush grapes by hand, by foot, or machine.
- Leave juice, skins, and seeds in the vat.

Add active and bubbly wine starter (*pied de cuve*), mix in thoroughly.
Stir twice daily until fermentation begins.
Stir twice daily until fermentation slows, usually 1 week.
- Or omit stirring to encourage surface yeasts and a lactic fermentation.

Press the fermented grapes to separate wine and pour into the secondary fermenter.
- Save the dry pomace for pomace cheese.

Age wine in cave in glass, clay, wood, or steel.
- Fill containers right to the top.
- Leave undisturbed for up to 1 full year at 8 to 12°C (46–54°F).

Bottle the wine.

Amasi

Amasi (in Zulu and Xhosa) is an African technique for milk fermented in a gourd. It's one of many different fermented milks made similarly across the continent that are unfortunately unacknowledged and unappreciated in the rest of the world. It's time, of course, to change this.

Amasi may be the first dairy ferment. For when animals were first milked, their milk would have been milked into a vessel like a gourd. And as the gourd's natural materials interacted with the milk they held, some very special magic happened: The milk transformed to alcohol.

Lagenaria gourds (also called birdhouse-gourds, bottle-gourds, or calabashes) are known to be the original domesticated plant and are understood to be humanity's original container. From recent DNA analyses, we know that their seeds have been planted for over 12,000 years, beginning in Africa. Curiously, though, evidence of the gourds dated to over 10,000 years ago has been found in South America, hinting at possible proof of a pre-Columbian exchange (though they may have floated across the Atlantic, or traveled from Asia). The naturally hollow vessels are widely used for many significant cultural purposes across all continents: In India and Africa they are used as the resonant foundation of musical instruments like the sitar and kora; they're used as siphons and pipes in Mexico; Papuan New Guinea tribesman wear them as protective and decorative penis sheaths! But to me the most intriguing use of gourds is for fermentation. This is because any sweet liquid you put in the gourd, including milk, will be transformed by the vessel's very special magic into alcohol.

The gourds can be grown from seed (you can choose from varieties with a diverse array of shapes and sizes for many different purposes) or purchased from farms that grow them. They need a long growing season and plenty of heat to grow thick shells. They're often sold seasonally alongside pumpkins and decorative gourds and are even sold dried for various crafts by some companies online. The green gourds should be cured after harvest in a covered space outdoors over winter and will slowly transform to hollow, seed-filled gourds. With a saw or a rotary tool, cut a cork-sized hole in the top, remove the seeds, soak the gourd with several changes of water, clean its insides out with sand or gravel, pop in a cork, and you've made the perfect fermentation vessel.

The gourds hold a surprising amount of liquid, though it will slowly disappear through the porous shell via evaporation. This porosity creates different conditions for fermentation than a closed glass or steel container—it leads to a micro-aeration of the fluids as they ferment within. This shifts the fermentation in a highly aerobic direction, encouraging yeasts to flourish (alcohol starters for beer, wine, and cider can also be kept in gourds with a similar yeast-promoting effect). The wooden vessel also contributes vanilla essences to any liquids stored within; combined with the flavors of the yeasty fermentation, gourds make for a delicious brew. Alternatively, a low-temperature-fired, unglazed earthenware pottery will also work.

To make amasi, simply pour milk into the gourd, and leave the milk to ferment. Once it's thick and sour, pour out the contents and consume immediately (it will separate its whey and turn to a delicious cheese like ambarees—recipe below—if you wait too long) or refrigerate to preserve, then refill the gourd with fresh milk. Do so daily (twice daily in warm weather), and the amasi will hit its stride.

Do not attempt to clean out the vessel; doing so would wash out the all-important backslop that makes this special fermentation possible. I never wash my gourds, even when they're unused. If one is left to dry without washing, it will grow a rind of *Geotrichum candidum* in its interior; if it's washed with water first, though, you'll find *Mucor* growing within! To initiate a dry gourd, simply put fresh milk in it—with or without an added starter—and allow this to ferment. The first ferment will be slow and a bit odd, so discard it once it has set, and refresh with milk.

The amasi can be expected to take about 2 days of fermentation at 20°C (68°F) to develop its most important quality: its highest alcohol content! But by then the ferment may get too acidic, as it continues to lactoferment. As well, if you expose the alcoholic ferment to too much air over too long a time, it will sour into milk vinegar (as discussed on page 391), like all alcoholic ferments that aren't protected from air exposure. The method for the production of milkbeer, the following technique, solves this issue by bottling the ferment early and limiting air's influence, thus developing the milk alcohol to its fullest extent.

In many regions, like in Sardinia and Mexico, wine is still drunk from a gourd; while in China the gourds are recognized as symbols for both alcohol and good health. To me, these are traces on the landscape of a more ancient understanding; and a recognition of this plant's important contribution to our natural history of fermentation.

Making Amasi

A fermented milk • 1-day make

INGREDIENTS

milk	1 L
vessel	a *Lagenaria* gourd with capacity of approximately 1.25 L or equivalent-sized earthenware vessel

Yield 1 L amasi (1:1)

TECHNIQUE

Empty gourd of previous batch of amasi but do not wash!
Bring milk to culturing temperature, about 20°C (68°F).
Pour milk into gourd.
- ▸ Fill gourd about three-quarters full.

Leave milk to ferment.
- ▸ Seal with a cork.
- ▸ Shake occasionally as it ferments.
- ▸ Amasi is typically ready after 8 to 12 hours.

Empty vessel of amasi once lightly curdled and effervescent.
- ▸ Chill amasi to preserve.

Pour in more fresh milk to ferment again.

NOTES

To initiate an amasi culture, leave milk to ferment spontaneously in a gourd for 2 to 3 days until it's thick and sour, or add kefir or clabber to the first batch in a ratio of 1:10. Amasi must be made regularly to keep the gourd's culture active; if taking time between batches, keep the gourd in the refrigerator, filled with its amasi, for up to one week.

Milkbeer

Can the principles of natural winemaking be applied to milk? Yes! And the result is milkbeer, a modern adaptation of a number of historical milk alcohols, which include ayran, amasi, and koumiss. This ferment is proof that cheese and wine are evolutions of the same fermentations, and if all cheeses exist on a continuum, then that continuum should also include beer, wine, and cider!

This is my take on an ancient practice—making a smoother, more flavorful, more alcoholic and more effervescent ferment—that I developed during Covid lockdown.

To make alcohol from milk takes stirring, lots and lots of stirring. (The Scythians, a mare's-milk-fermenting civilization of antiquity, supposedly kept slaves to stir their alcoholic milk.) Doing so shifts the secondary fermentation in a new direction. Instead of developing a thick layer of *Geo* at the top as clabber and kefir do, the milk is fermented by air-dependent yeasts throughout the ferment's body. The lactic acid developed by the primary fermentation is thus converted into an abundance of alcohol and other flavorful compounds by yeasts found naturally in milk's mesophilic fermentation (including *Kluyveromyces*).

To make milkbeer, there are several modifications to the amasi technique that make a more quaffable ferment. First, the milk is boiled before fermentation, as with yogurt (technically I suppose this should be called a yogurt-beer), to make a softer curd that goes down smoother and develops a surprisingly foamy head. Second, the fermentation is controlled with the addition of a separate yeasty starter culture as opposed to a simple backslop within the gourd; the starter must be added in a larger proportion than I generally recommend—1:10 instead of the typical 1:50 to 1:100, to achieve a stronger yeasty ferment. (It's equivalent to the ratio for feeding a sourdough starter and encourages the growth of yeasts.) Third, the aeration is provided with a regular gentle stirring (of course this can also be done in a gourd). And fourth, once the ferment has reached a precise point, it is bottled and chilled in a cool cave for several days, with additional shaking to achieve a strong effervescence and maximize alcohol production, and minimal continuing lactic or vinegar fermentation (the cool temperatures stop the lactic acid fermentation; and the bottling stops the vinegar—but not the alcohol—from developing).

Historically, many traditional milk alcohols were prepared across Africa, Asia, and Europe, but few are still prepared today. Ayran is a bubbly milk from Turkey, likely originally produced in a similar fashion (these days it's artificially carbonated like soda pop). Koumiss was historically made by Mongol horsemen by leaving their mares' milk (warriors only rode mares for this reason) to ferment in skin bags hanging from their saddles. The regular shaking assured by their steed's gallop would have aerated their fermented milks, driving the ferment to become a rather powerful alcohol (mare's milk can have over 6 percent lactose compared with cow's milk's 3 percent) that reputedly fueled their rampages throughout Europe. This process can also be performed with whey; an alcoholic dairy product made this way was known as blaand in Viking circles (the term *blaand* may refer to the spoon used to stir the whey to make it alcoholic). And any of these diverse milk or whey beers can be distilled into a number of spirits like the Mongolian arkhi, that give new meaning to the term *milk-drunk*!

There's debate in anthropological circles about the original purpose for the domestication of grains. The "beer before bread" hypothesis posits that grains may have been first planted to provide an ample sugar source for brewing into alcohol for ritualistic (or enjoyment) purposes. Given the existence of amasi and other milk alcohols among many historic dairying cultures, it's quite possible that the domestication of milking animals followed a similar path: "beer before Brie"! I have a feeling that this is the case, for milkbeer is the dairy ferment that most satisfies me—the one I crave the most (I'm aching to drink one right now as I write). Besides, what else would incite a Neolithic human to milk a wild auroch?

Making Milkbeer

A fermented milk • 12-hour make • 1-week affinage

INGREDIENTS

milk	1 L
starter	100 ml milkbeer or amasi (1:10)

YIELD FIVE 200 ml BOTTLES OF MILKBEER (1:1)

TECHNIQUE

Cook milk to a boil, stirring constantly.
Chill milk in a cold water bath to 20°C (68°F), stirring constantly.
Add starter, mix in thoroughly.
Shake or stir milk as it ferments for 8 to 12 hours at 20°C.
- Lightly stir fermenting milk every 30 minutes.
- First fermentation is considered complete once milk has just thickened.

Strain ferment to remove cheese lumps and buttery bits.
Bottle in 200 ml bottles, and seal with caps.
Chill bottles immediately to 4°C (40°F), and condition for 1 week, laid on their sides.
- Shake bottles daily to mix, until whey separation stops.

Refrigerate to preserve for up to 1 year.
Shake well before opening—and expect contents to overflow like champagne!

NOTES

To make milkbeer best requires using an already developed milkbeer, or its whey, as a starter. This culture can be initiated with kefir or clabber, which have the appropriate yeasts but are undeveloped because the circumstances haven't been created for their growth. By aerating frequently and re-fermenting with a large ratio of starter, typically with two feedings over the course of 2 days, the fermented milk will develop its yeastiness, and create a culture that can be kept like clabber or kefir but with a full aeration.

Ambarees

Continuing with the theme of yeast in cheese, I present to you ambarees, a fascinating Lebanese cheese separated from its whey with the leavening power of yeast!

I sometimes refer to this style as cherry cheese, for there is no other cheese out there this fruity. The taste of a well-made ambarees is full of cherries and almonds—some of the most sought-after flavors you can find in the fold of fermentation—resulting from the complex secondary fermentation of milk's indigenous yeasts. The yeasts also give the cheese a distinctive effervescence when fresh and unsalted.

This technique calls for mixing air into a fermented milk to encourage an aerobic yeast fermentation—it's traditionally prepared in a naturally aerated clay pot—that causes a soft curd to firm and rise above its whey. The yeasty curd is drained in cloth, becoming a cheese rich in fruity esters from the yeasts' fermentation of the lactic acid. No other cheeses are separated from milk using such an ingenious trick. This simple method, rennet-free technique, and surprising flavor made me feel like it deserves a spot in these pages.

Normally, gas production from yeast is not sought after in a cheese, but this technique relies on the phenomenon and rolls with it elegantly. To make this delicious style, you first ferment milk, preferably in a clay pot with an already yeasty starter (think ripened milkbeer or amasi, though this can also simply be a clabber or kefir with its *Geo*). The set curd is then stirred, resulting in a full aeration that causes it to rise after an additional 12 hours of fermentation. Instead of yeasts just growing at the surface, they now thrive throughout the ferment, developing alcohol and carbon dioxide, causing the curd to rise like a loaf of bread, above its whey below. Strain the risen curd in cloth for a day and eat it fresh or salted. In essence it has become a milkbeer cheese.

The cherry cheese can also be aged. The texture of the salted cheese after draining is similar to a salted and drained lactic cheese. As such, the cheese can be shaped into various shapes, or packed into pots like a Saint-Marcellin, and hastened to develop *Geotrichum candidum*. The cheese can then be ripened in a cave to make a delicious full-fat, rennet-free aged cheese.

This is a simplified version of ambarees, which is actually a very long, complex, and fascinating make. The rare cheese is typically prepared in the mountainous regions of Lebanon with goat's milk, the most flavorful milk for the process. The clay vessel used to prepare it is tailor-made for the technique, with a spigot at the bottom to release whey. Day after day you add fresh milk and salt into the pot, mixing the milk into the yeasty ferment; every day the cheese rises up and you release the whey below from the tap. Repeat the process for several weeks until the clay jar is filled with risen cheese. Seal the jar and leave it in a cave to preserve through winter, when the labneh-like cheese is considered at its best!

Making Ambarees

A fermented cheese • 2-day make

INGREDIENTS

milk	4 L
starter	400 ml kefir, clabber, whey, or milkbeer (1:10)
salt	
clay vessel	4 L (optional)
YIELD	**600 g ambarees (6:1)**

TECHNIQUE

Cook milk to a boil, stirring constantly (optional, but recommended).
Chill milk in a cold water bath to 20°C (68°F), stirring constantly.
Add starter, mix in thoroughly.
Leave ferment to sit 8 to 12 hours at 20°C until thickened.
  ▸ Stir every hour if fermenting in a nonporous metal pot.
When thickened, give the curd a thorough stirring.
Leave cheese to ferment an additional 12 to 24 hours without stirring.
  ▸ Curd should rise and separate from whey due to yeast activity.
  ▸ Leave undisturbed and covered.
Strain curd in cheesecloth for 12 to 24 hours.
Salt to taste (optional), then drain again 1 to 2 hours (24 hours if it's to be aged).

Cheese on Lees

A fall phenomenon, cheese on lees calls for the season's cheeses to be left to steep in barrels full of freshly pressed pomace, allowing them to benefit from something usually only sought out for the fullest-bodied red and orange wines: skin contact.

This fascinating mélange of fermentation derives from the regions of Europe where winemaking and cheesemaking coexist—the foothills of the Alps and other hilly corners of the Continent. Farmers there who made both cheese and wine would often submerge their cheeses in barrels full of discarded grape skins and seeds, to shield them from taxation or hide them from the pillaging of enemy armies. The unintended consequence is a sort of drunken cheese (Italians call it formaggio ubriaco; the French, tomme au marc) enriched by its time in the pressings of the wine and altered by the unique environmental conditions created in the skins.

Keeping cheeses submerged in skins changes their nature considerably. The cheeses retain moisture, for they do not lose humidity as they age, but they do not develop a rind ecology due to the alcoholic nature of their aging medium, which restricts air and the subsequent development of the usual rind-ripening microorganisms. The cheeses therefore ripen in a much slower, anaerobic way. They do not break down as a similar cheese exposed to air might, and they take on beautiful flavors and colors on their rinds.

After aging for months or years, the cheeses can be eaten straight out of the lees, or left with a thick deposit of skins for several days in a drier space. This dehydrates them, giving them a colorful, crispy crust that contrasts deliciously with the creamy cheese within.

Cheese can be aged in the pressings of nearly any alcoholic ferment, not just the marc from grapes: the pomace from cider, sake lees, even the leftover barley from beer. Each gives the cheese a different cheer. What's important is that the medium be fermented, for an unfermented pomace is unstable and may ferment in an odd way. For best results, with barley the beer should be fermented on the grain; wine should be fermented on the skins; and cider should be fermented with the flesh of the apples, stems, pips, and all. And at the end of the fall, with the wine, beer, or cider mellowing in barrels and bottles, the final wheels of cheese can be left to ripen under their lees.

Making Cheese on Lees

A semi-alpine cheese • 8- to 12-hour make • 3- to 6-month affinage

INGREDIENTS

milk	20 L
starter	200 ml kefir, clabber, or whey (1:100)
rennet	regular dose calf rennet
salt	
	a bucketful of fermented pomace
forms	round: 20 cm diameter × 20 cm deep
YIELD	ONE 2 KG POMACE CHEESE (10:1)

TECHNIQUE

Bring milk to cheese temperature, about 35°C (95°F).
Add starter, mix in thoroughly.
Add rennet, mix in thoroughly.
Wait for clean break, 45 minutes to 1 hour.
- Cover the pot to keep in warmth.

Slowly cut curd to corn size, 0.75 cm, with a spino.
Stir curd nonstop at 35°C (95°F) over 30 minutes.
- Increase tempo of stirring as curds firm.

Pitch curds for 5 minutes.
Press curd beneath the whey.
Transfer curds from pot into cheesecloth-lined forms.
Press cheeses lightly by hand, or by stacking.
- Keep cheeses warm during pressing with whey or warm water.

Leave in forms 6 to 12 hours, until acidity develops.
- Goal pH 5.3—stretch test positive.
- Flip four times during the drain, with first flip at 5 minutes.

Surface salt cheese daily over 2 days.
Or salt cheeses for 12 hours in saturated salt brine (6 hours per kg).
Dry cheeses at room temperature for 24 hours.
Age cheeses covered in pomace for 3 to 6 months.
- Keep the barrel covered.
- Do not disturb cheeses as they age.

Yogurt

Considering the strange tendency to call attention to the nationalities of yogurt (Greek yogurt, Bulgarian yogurt, Icelandic yogurt)—given their common origins and parallels (they're all really the same)—I'll buck the trend and refer instead to this yogurt as a Hindu yogurt. For in Hindu India, with its prohibition on the eating of the cow (and consequently on the use of calf rennet), yogurt (*dahi* in Hindi) is most prominent, even holy (the sanctity of yogurt in India is prescient!). Its makers thus pay special attention to the preparation of this most exceptional food.

The yogurt in India, as such, is sublime. In many Indian homes and restaurants, yogurt making is a daily practice, and makers line up after the morning milking outside the nearest cow or buffalo shed (even in the city!) to pick up their milk. They carefully cook that freshest milk to a boil, evaporating as much moisture as possible with a unique method of ladling that pours milk from up high. Once sweet to the taste (the complex sugars in milk are made sweeter by the long cooking, and the caramelization and concentration that results), the cooked milk is cooled until just bearable to touch, then cultured with a bit of yogurt saved from the previous batch.

This technique calls for an active yogurt starter to ferment the cooked milk, which makes the perfect yogurt, if performed right. For a yogurt starter culture to remain active, it must be cultivated regularly (daily as many do in India) or refrigerated (for up to 1 week) or frozen (for up to 6 months) between uses. A neglected starter will ferment the yogurt unpredictably, resulting in odd flavors and textural inconsistencies, such as excess whey separation, ropiness (long viscous threads), and gas.

But if it takes yogurt to make yogurt, where does this traditional starter culture come from? You can borrow one from a neighbor as is tradition in India, or purchase one from culture houses that keep them, but such answers failed to satisfy my own personal desire to understand the origin of this particular community of microbes. As it turns out, milk is made to develop this most exceptional and important culture.

Raw milk, left to ferment in an animal-warm environment, will come to develop into a thermophilic clabber culture as part of its biological destiny. For the microbiological community that makes yogurt is meant to ferment milk in the young mammal's intestine. The microbes typically found in yogurt (species of *Streptococci* and *Lactobacilli*) are among the most common and important residents of both raw milk and milk-drinking infants' guts. Kept at around 40°C (104°F) for just 24 hours, fresh raw milk will reliably ferment itself, for the development of this community of beneficial microbes assures that infants make the most of their milk and have the best chance at life.

Nevertheless, that first fermentation is a slow one, which leaves opportunities for unwanted microbes to grow. What makes such a thermophilic starter stable (and what stabilizes the beneficial community in the infant gut) is regularly feeding the thermophilic ferment fresh milk (either in a jar or in the gut). This thermophilic clabber can thus be used to make a first batch of yogurt. And that yogurt then becomes the starter for the subsequent batch of yogurt, a process that can be repeated ad infinitum, for it is the biological imperative of the ferment to continue fermenting.

It's telling of our misunderstanding of milk that there is hardly a commercial yogurt maker anywhere that prepares yogurt with a backslopped starter. It's commonly understood that the traditional culture used to make yogurt is yogurt, yet hardly any commercial producers put their faith in anything but laboratory-raised cultures. This new biological perspective on the origins of our yogurt culture should help us gain faith in milk's exceptional capacity to develop starter cultures that we can depend on, and that improve the character and quality of our yogurt.

There are many, many reasons to make your own yogurt: to improve your personal health, to make good food for your family (perhaps the best food for both the young and the elderly) and community, and to enjoy its most amazing texture and taste (yogurt like this shouldn't need sugar!). But I believe the most important one is to reconnect with what sustains us.

When I make my yogurt, I consider it a meditation on milk. I cook it, stirring carefully until it boils, paying attention so it doesn't burn on the bottom, form a skin overtop, or boil over. I then cool the milk until it comes to a temperature that both our microbes and the milk's prefer. Next, I backslop my sacred starter, keeping the milk warm while it ferments in a protected place (an oven with a light on, a cooler filled with warm water) until thickened. Regular feedings and keeping the ferment warm, two elementally maternal acts, combine to create an extraordinarily important, indeed, a most divine food.

Making Yogurt

A fermented milk • 6- to 12-hour make

INGREDIENTS

milk	4 L
starter	40 ml active yogurt or yogurt whey (1:100)

Yield four 1 L jars yogurt (1:1)

TECHNIQUE

Bring milk to a boil.
- Stir nonstop.

Cool milk to 40°C (104°F) in a cold-water bath.
- Stir nonstop.

Pour starter into jars and stir with a spoon (10 ml in each jar).

Add cooked and cooled milk (1 L in each jar).
- Stir starter into milk thoroughly.
- Seal the jar with a lid.

Keep at 40°C for 6 to 12 hours to ferment.

Refrigerate once soured to your liking.

Yogurt will keep for up to 1 month.
- But it will work as an active starter for only 1 week.
- If frozen, the culture will remain active for months.

NOTES

To begin a yogurt culture, leave raw milk to ferment at 40°C for 24 hours until thick and sour, then re-culture (backslop) raw or boiled milk and ferment 1 or 2 times at 40°C until the culture sours and thickens in less than 6 hours. (A yogurt culture can also be made from mesophilic clabber or kefir through adaptive fermentation, refermenting the culture 1 or 2 times at higher temperatures.) At that point the culture is ready to make yogurt. And the culture can be preserved forever so long as the culture is regularly fed by regularly making yogurt. More information on keeping thermophilic starter cultures is provided in appendix E.

Yogurt can be incubated in an oven with the light on; in an insulated cooler filled with warm water; in an oven with an incandescent lightbulb turned on; or, in big enough quantities, wrapped in towels or baby blankets.

Gros Lait

"Fat milk" in French, gros lait (*grow-lay*) is a creamy and delicious dairy product from the celebrated butter-making region of Brittany, and is known in Breton as gwell. It's made much like yogurt, but with a mesophilic fermentation. This traditional ferment was likely once a by-product of historical buttermaking practices, but is now made expressly to be eaten.

Gros lait is appreciated as a natural dairy product, made with backslopping (it's one of the few dairy products or cheeses made in France whose identity is linked with natural fermentation), that evolves a delicious flavor distinct from yogurt. And what distinguishes the flavor of "fat milk" from yogurt is a lower-temperature fermentation that encourages a more diverse microbiological transformation.

The technique for this unique dairy product begins much like yogurt, by boiling the milk, which concentrates it and changes its sugars. The cooked milk is cooled and then fermented at a lower temperature, with a more diverse mesophilic starter. The result is a unique flavor profile that is less tart and acidic than a typical yogurt, but much more expressive and flavorful—as well as a bit moldy, in the best possible way.

It is best to use a small amount of a previous batch of gros lait as a starter, as is still traditionally done in Brittany. Using clabber culture or kefir works, but these ferments are adapted to the medium of uncooked milk. (Cooking milk changes its sugar profile, which, when fermented, selects for different microorganisms than raw milk.) Much like other starters, such a gros lait culture should be re-fermented as often as possible, preferably on a daily basis, in order to keep it in its top form. Gros lait can be refrigerated for up to a week between batches.

When fermented right, gros lait grows a healthy and delicious layer of *Geotrichum candidum* on top, which contributes a slight effervescence to the yogurt. The fungus is a natural part of the mesophilic starter used to make it, and though welcome on aged cheeses, it's not typically appreciated on yogurt. To make yogurt fungus-free, it's a matter mostly of fermenting at a higher temperature (above 40°C [104°F]), which encourages beneficial bacteria over fungi. Some biologists even theorize that mammals keep their body temperatures higher in part to keep potential fungal pathogens under control!

Trust me, though (along with the Bretons): *Geo* makes the gros lait that much better! So much so that traditional eaters of these room-temperature-cultured milks skim the fungus off their ferments and eat it first. Known as filmjölk in Scandinavia, matsoni in Georgia (it's made originally as a first step in making butter in many regions), mesophilic yogurt would be more universally appreciated if it weren't for the strange fear of fungus among-us!

Making Gros Lait

A fermented milk • *12- to 24-hour make*

INGREDIENTS

milk	4 L
starter	40 ml kefir, clabber, or gros lait (1:100)

Yield four 1 L jars gros lait (1:1)

TECHNIQUE

Bring milk to a boil.
- Stir nonstop.

Cool milk to 20°C (68°F) in a cold water bath.
- Stir nonstop.

Pour starter into jars (10 ml in each jar).
Add cooked and cooled milk (1 L in each jar).
- Stir in thoroughly.

Keep at 20°C (68°F) for 12 to 24 hours to ferment.
Chill to preserve.

NOTES

The gros lait culture is a preferred starter over clabber or kefir. You can begin a gros lait culture with clabber or kefir and adapt it to the cooked milk medium.

The white fungus on the top preserves this yogurt longer than others. Unlike thermophilic yogurts, which tend to develop shockingly bright-colored fungi—mostly *Penicillium roqueforti*—on the surface when they age because of a lack of protective surface ecologies, gros lait will only ever show the gentle white color of *Geotrichum candidum*. The *Geo* preserves the deliciousness of the ferment, until the gros lait is completely consumed by it.

PHOTO BY ALIZA ELIAZAROV

Ryazhenka

Ryazhenka is a famous Ukrainian specialty, known in some circles as baked-milk yogurt. It's a cold-weather classic, prepared historically in winter, by slow-cooking milk overnight in the massive masonry stoves that heat traditional Ukrainian homes (and are also used to bake breads and prepare other delicious dairy products).

A low and slow cooking, as the masonry stove cools overnight, distinguishes this unique yogurt from all others, for baked milk yogurt takes on an attractive caramel color and a rich and caramelized flavor that makes it the most beautiful and flavorful of the fermented milks.

The method for making ryazhenka is much like making yogurt or gros lait: Milk is first cooked, then cultured. But instead of being quickly cooked over fire, the milk is slowly baked, dramatically changing it. The long slow overnight cooking (you can replicate the effect with overnight cooking in a conventional oven at 100°C [212°F]) caramelizes the milk sugars, causing them first to sweeten. The milk also darkens in color through the caramelization and the Maillard reaction (which is a reaction between the milk's sugars and proteins).

The milk is usually left unstirred, developing a crispy brown milk skin that can be eaten on its own, still warm from the oven, or included in the final fermented dairy product. The cream of the milk, which rises and renders to a liquid ghee after 12 hours of cooking, is also taken off, preferably after fermentation. (Ghee, naturally, is significant to the region's cuisine.)

After baking, ryazhenka is best fermented under mesophilic conditions by the gentle warmth of the woodstove, though a thermophilic fermentation also works. The combination of the caramelization and the Maillard transformation with the diversity of the mesophilic fermentation makes this unique milk ferment one of the most flavorful of them all. The no-stirring-required method also makes it one of the easiest of yogurts to produce.

Like yogurt or gros lait made with a culture carried forward on cooked milk, it is best to ferment a baked-milk yogurt with a starter that is cultivated with and adapted to baked milk. You can still ferment an excellent ryazhenka with a milk-based kefir or clabber starter, but there will be a lag in the fermentation as the culture adapts to the caramelized milk medium. Of course, you can create a caramelized milk starter by adaptive fermentation from a milk-based mesophilic culture.

Making Ryazhenka

A fermented milk • 12- to 24-hour make

INGREDIENTS

milk	4 L
starter	40 ml kefir, clabber, or ryazhenka (1:100)

Yield   four 1 L jars ryazhenka (1:1)

TECHNIQUE

Bake milk at 100°C (212°F) for 12 to 24 hours.
- Bake milk in deep pot, covered or uncovered.
- Do not stir!

Remove milk skin and enjoy.
Chill milk to 20°C (68°F).
- Do not stir!

Pour starter into jars (10 ml in each jar).
Add baked milk (1 L in each jar).
- Stir in thoroughly.

Keep at 20°C for 12 to 24 hours to ferment.

NOTES

The best starter for fermenting ryazhenka is ryazhenka, already adapted to the medium of long-cooked milk. To develop this starter, you can begin with clabber or kefir.

Ryazhenka can also be fermented thermophilically with a thermophilic starter, and a 40°C (104°F) fermentation for 6 to 12 hours.

PHOTO BY ALIZA ELIAZAROV

Kaymak

Many dairying regions of the world, including those in Eastern Europe and beyond (Serbia, Turkey, Mongolia, and so on), focus their dairying mainly on their production of this one product, which is eaten on its own or serves as the basis for the production of butter and ghee. It's made by boiling milk for a long period of time, then cooling the milk and skimming off the cooked cream that rises to the surface and thickens.

Though not exactly a cheese (it typically lacks fermentation in its contemporary production), I've decided to include this technique for clotted cream because of its significance to the history of dairying. This practice may be the original inception of yogurt, because yogurt making may have originally evolved from the hunt for precious butterfat from the milk.

Kaymak's significance across a broad swath of Eastern Europe and Asia is not to be missed, but its production today has been affected by the specialization of the dairy industry. Once, this technique was but the first step in the production of an entire ecosystem of dairying; Mongolian pastoralists may produce fifty-plus dairy products that evolve from kaymak and its skim milk by-products. Nowadays kaymak is simply eaten sweet, and often the skimmed milk below is discarded.

To produce kaymak, also known as clotted cream, milk is very long cooked, which aids the separation and thickening of its cream. You'll want to stir nonstop to prevent burning at its bottom and the formation of a milk skin at its top. The cooked milk is then cooled, and once the temperature reaches about 5°C (40°F), the cooked cream solidifies on its surface and can be skimmed off as a sweet kaymak. This cooling can be easily enough performed outside in the winter, as is done traditionally, after warming the home with the cooking of the milk.

This pre-cooking of the milk primes its cream for its transformation into butter, by far the most important product of dairying in many parts of the world. Buttermakers get a greater yield of cream from milk through natural separation if that milk is boiled first, and a butter comes together much more readily from its cream if it's cooked. This is especially so if that cooked cream is then fermented.

In the more traditional places where kaymak is made, the milk is cultured once cooled. When thickened through fermentation (as opposed to just refrigeration), the kaymak that rises can then be ladled off and enjoyed as is (it's in essence the cream top of yogurt) or churned into butter. The milk can be fermented mesophilically or thermophilically; either way, the skim milk curd down below can be appreciated as a by-product of the fermented kaymak. That fermented skimmed milk would have been the original yogurt, eaten as a delicious and nutritious by-product of buttermaking!

Making Kaymak

A fermented milk • 24-hour make

INGREDIENTS

milk	20 L	
starter	20 ml active kefir, clabber (1:1000)	
YIELD	**1 KG KAYMAK (20:1)**	

MAKE A SWEET KAYMAK

Bring milk to a boil.
- Stir constantly.

Simmer milk for about an hour.
- Stir constantly.

Chill milk to 20°C (68°F) in a cold-water bath.
- Stir constantly.

Continue to chill milk to 4°C (39°F), and leave cold for 12 hours.
- Leave unstirred.

Skim off kaymak once thickened.

ALTERNATIVELY, MAKE A FERMENTED KAYMAK

Bring milk to a boil.
- Stir constantly.

Simmer milk for about an hour.
- Stir constantly.

Chill milk to 20°C (68°F) with cold water.
- Stir constantly.

Add starter, mix thoroughly.

Leave milk covered and unstirred at 20°C (68°F) for 12 to 24 hours.

Skim kaymak off yogurt once thickened.

PHOTO BY ALIZA ELIAZAROV

Yogurt Cheese

Yogurt cheese is among the simplest of cheeses to prepare, and one of the most delectable of the fresh cheeses in this book. It is made by simply draining yogurt in cloth overnight to thicken, then salting the creamy cheese that forms. The result is the smoothest-textured cheese in the entire realm of dairy, for the curd is softened and enriched by cooking the milk. A suitable substitute for cream cheese, it can be made with your own natural yogurt made by first boiling milk, then fermenting it thermophilically, or with yogurt purchased from the store.

What sets this cheese apart from quark—an equivalent cheese made by draining clabber instead of yogurt—is the cooking of the milk. Although many believe this was done historically for safety reasons, the real reasons for boiling milk in traditional dairying are more nutritional and gastronomical. For by boiling milk before fermenting it, the heat-denatured albumin proteins of the milk are coagulated into the yogurt; they do not end up wasted in the whey when that yogurt is drained into cheese. That means that more of milk's nutrition is preserved in the making of a cooked yogurt cheese than an uncooked quark. The resulting cheese is also more flavorful and better-textured than quark, two other benefits of boiling milk before fermenting it.

This technique can be performed with full-fat yogurt, or with yogurt skimmed of its cream. The thickened fermented cream can be taken off the yogurt before draining and eaten as a delicacy on its own (see the previous technique for kaymak), or churned into butter, as is done in many cultures in the Middle East. Or the cream can be left in the draining curd, making a fully creamy and colorful yogurt cheese. But it should be understood that this cheese is typically made, like yogurt, to separate the more important product from milk, its butter.

You can leave the yogurt hanging for just a few hours, or for up to 1 full day before it starts to molder. And then you can choose to salt the cheese or not, depending on how long you wish to preserve it. The draining time controls how much moisture is pulled out of the yogurt: Labneh, "white cheese" in Arabic, is made by leaving the skimmed yogurt (leben) to drain for only 12 hours, and not salting. Greek yogurt can be assumed to be the same dairy product, for such cheeses are made around the Mediterranean. Yogurt cheeses are also preserved in diverse ways around the region, as explored in a later technique, shankleesh.

Making Yogurt Cheese

A fermented cheese • 48-hour make

INGREDIENTS

milk	4 L
starter	40 ml active yogurt (1:100)
salt	

YIELD **600 G YOGURT CHEESE (6:1)**

TECHNIQUE

Make yogurt:
- Bring milk to a boil, stirring nonstop.
- Cool milk to 40°C (104°F) in a cold-water bath, stirring nonstop.
- Add starter, mix in thoroughly.
- Ferment at 40°C until thick and sour, 6 to 12 hours.

Pour yogurt into fine cheesecloth.
Hang 24 hours to drain.
Mix in salt to taste.
- Knead into curd thoroughly.

Hang 24 hours to drain again.

Shankleesh

This Levantine technique for a cheese known in Arabic as shankleesh calls for rolling yogurt cheese into balls and preserving them under oil, maintaining their freshness and acidity. In Palestinian cuisine, the cheeses are first rolled in a mixture of sesame, sumac, and thyme, known as za'atar, that complements the flavor and the tartness of the oil-preserved cheese.

Keeping cheeses under olive oil preserves them nearly eternally. The cheese is kept from air exposure under the oil, and never degrades as a result of aerobic fermentations—no fungus can grow upon the cheese if it is submerged. And the acidity of the cheese prevents unwanted anaerobic species (like *Clostridium botulinum*) from wreaking havoc. The best oil for this is olive oil, whose fats are more preservable than other oils; however, storing the cheese for longer than 1 year can result in rancidity.

The only other thing that can cause this cheese to spoil is if the balls rise above the oil. Take care to drain the yogurt cheese well and salt it effectively and regularly, to be sure yeast growth doesn't cause the cheeses to bubble and rise. It's always best to use a thermophilic yogurt, rather than mesophilic, whose yeasts can cause the yogurt to float.

This is an excellent first cheese to age as the simply executed cheeses can be just as easily preserved by keeping them submerged in oil. But there are other interesting methods for preserving them: Yogurt cheese can be preserved under a brine made from its whey, similarly to feta; or dried in the sun to make a cheese known in Persian as kashk, and in Turkmen as gurt. These dried balls of yogurt are often grated and rehydrated in water to reconstitute a yogurt, or added as a flavor and nutritional boost to soups, stews, and other dishes. Common across parts of Asia practicing an "original" cheesemaking—like Kyrgyzstan, where it's known as kurut—this preserved yogurt cheese may be one of the first long-preserved (and -preserving) dairy foods, and should be understood as a Central Asian superfood, having sustained nomadic shepherds over continents and through millennia.

You can even age a salted yogurt cheese into *Geotrichum* or *Roqueforti* ripened cheeses; however, the thermophilic ecology of the yogurt does not include many yeasts and fungi, making the rind development less predictable (moldy mesophilic yogurts like gros lait would work better for this reason). As such, add ripening cultures or spores to a thermophilic yogurt before fermenting if you're aging the yogurt cheese this way.

Making Shankleesh

A fermented cheese • aged in oil • 48-hour make • 2-month affinage

INGREDIENTS

yogurt	4 L
salt	
za'atar	
olive oil	600 ml
Yield	**600 g shankleesh (6:1)**

TECHNIQUE

Pour yogurt into fine cheesecloth.
Hang 24 hours to drain.
Mix in salt to taste.
▸ Knead into curd thoroughly.
Hang 24 hours to drain again.
Roll into balls 3 cm across.
Roll each shankleesh in za'atar (sesame, sumac, dried wild oregano, and dried thyme).
▸ Leave to air-dry for 1 hour.
Place into a jar one at a time, slowly adding olive oil to submerge.
Age in cave for 2 months.

PHOTO BY ALIZA ELIAZAROV

Cream Cheese

Forget Philadelphia; if you want to make a "schpectacular schmear" for your bagels and lox, or a creamy base for a cheesecake, this is the spreadable cheese of your dreams!

Cream cheese is a hybrid of yogurt making and lactic cheesemaking. This is the only technique in the book that calls for cooking milk before making a renneted cheese. No other cheese will respond well if made with such heat-processed milk, for cooking milk softens a curd's texture and slows its whey evolution. This, however, is exactly what you want for cream cheese!

Not necessarily produced with extra cream, cream cheese is all about making a cheese with a soft and creamy texture. Cooking the milk denatures its proteins, rendering them less effective at coagulating with rennet. Nevertheless, the boiled milk still curdles, but more softly, and without separating whey as it would normally. The boiling also denatures milk's albumin proteins, allowing them to incorporate into the curd of the cream cheese, improving the yield.

You can add more cream to the make if you wish, though making this cheese with straight milk leaves nothing to be desired. Be sure to use freshly centrifuged cream from as fresh a milk as possible. Add an amount equivalent to a tenth the volume of the milk, effectively doubling the cream content of the cheese, and making it a "double cream" cheese. Alternatively, fermented cream can be mixed into the drained cream cheese before salting.

Do not leave the curd to ferment too long, or *Geotrichum candidum* will develop and give a yeasty taste. Similarly, when you're draining the cheese, hang it to assure a quick evolution of its whey. A measured pace, not too fast, not too slow, ensures the freshest taste with this and other fresh lactic cheeses.

Making Cream Cheese

A fresh lactic cheese • 24-hour make

INGREDIENTS

milk	4 L
cream	400 ml (optional)
starter	40 ml kefir, clabber, or whey (1:100)
rennet	regular dose
salt	

YIELD 600 G CREAM CHEESE (6:1)

TECHNIQUE

Add cream to milk, if using.
Bring milk mixture to a boil, stirring nonstop.
Cool to 20°C (68°F) in a cold water bath, stirring nonstop.
Add starter and rennet and mix in thoroughly.
Ferment at 20°C until thickened and sour, 12 hours.
Scoop curd into fine cheesecloth.
Hang 12 hours to drain.
Mix in salt to taste—about 2 percent.
▸ Knead into curd thoroughly.
Hang 12 hours to drain again.
Refrigerate cream cheese to preserve.

Munster

Munster is a sumptuous but stinky washed-rind cheese. This fetid fromage from France—actually a German-speaking part called Alsace on the Rhine River—is an excellent small cheese to make and ripen, if you can get over the smell.

This is a first true washed-rind cheese of the book, with a sticky pinky-orange skin and a gooey, richly flavorful, broken-down paste. Unlike washed-rind alpine styles, which maintain their firmness as they ripen, the small size, relatively large surface area, and high moisture content of Munster results in a ripening that's dominated by the intensity of the washed-rind microflora.

This cheese, like all other washed rinds, can be washed with whey, water, or other flavorful liquids, each one bringing out a unique character in the cheese. My preference is to use fermented whey, as it imparts an extra level of fermentation within the cheese, which results in a longer ripening period; that in turn imparts richer flavors and aromas to the cheese. Fermented whey, rich in lactic acid, provides a continued source of nourishment to the cheese's washed-rind ecology, while washing with water dilutes the lactic acid within the curd, raising the pH, hastening the cheese's ripening, and enhancing the stink.

An original monastic cheese—a category that includes Époisses from Burgundy, Port Salut of the Loire, and even Canadian Oka, which evolved from these—Munster derives its name from the Latin *monastarium*, monastery. The cheese was originally made by shepherding serfs, who paid their tithes in cheese to the landowning monks, who in turn preserved the cheeses in the cellars of their monasteries. Following the Rule of Saint Benedict and its *ora et labora*—pray and work—they maintained their land and fed their brothers through their work in devotion to the Lord.

One of the monks' roles was to protect their cheeses from mold and decay by washing their rinds daily with a salted brine. What resulted from the washing was the development of a halo of protective microorganisms that preserved the cheese and gave it a particular monastic funk.

The cheeses' daily washing to remove fungus, maggots, and mites was done even on the Lord's day, for even one day's washing missed would cause the cheese to waver from its path. Monastic cheeses are most celebrated for their washed rinds, and it is the dedication of the monks that assures this style's best evolution. This is perhaps the key: To have success with this and other washed-rind cheeses, you must perform the washing with a monk-like fervor!

PHOTO BY ALIZA ELIAZAROV

Making Munster

An aged rennet cheese • with a washed rind • 8- to 12-hour make • 6- to 8-week affinage

INGREDIENTS

milk	10 L
starter	100 ml kefir, clabber, or whey (1:100)
rennet	regular dose calf rennet
salt	
forms	hoops: 8 cm diameter × 8 cm deep

Yield   five 200 g Munsters (10:1)

TECHNIQUE

Bring milk to cheese temperature, about 35°C (95°F).
Add starter, mix in thoroughly.
Add rennet, mix in thoroughly.
Wait for clean break, 45 minutes to 1 hour.
- Cover the pot to keep in warmth.

Cut curd to walnut size, 2.5 cm.
- Cut vertically; cut crosswise; cut horizontally or diagonally.

Stir curds for 5 to 15 minutes until semi-firm
(holds when dropped from 30 cm).
Pitch curds for 5 minutes.
Whey off from above.
- Reserve 1 L whey for washing brine, leave to ferment, then add 2 percent salt.

Fill forms to the top.
Drain 8 to 12 hours, until acidity develops.
- Goal pH 5.3—stretch test positive.
- Flip in form after 4 hours.
- Salt cheeses on the surface or salt cheeses for 1 hour in saturated brine.

Drain cheeses for 1 day on draining table,
covered with cheesecloth.
- Flip in form after 4 hours.

Age cheeses in cave for 6 to 8 weeks.
- Flip cheeses every other day.
- Wash rinds every day for 1 month or longer.

Vacherin

Vacherin, from the Latin for "little cow herd," is a cheese prepared in the alpine valleys of Switzerland and France during winter, when the cows are kept in small herds close to home. In contrast with the larger alpine wheels produced in the region during the summer alpage, vacherin is a small washed-rind cheese, bound by a spruce strip that holds the soft cheese together as it deliquesces, imparting aromatics of the upland forest to the paste.

In the summer, cheesemakers of the region send their animals to common mountain pastures, where the livestock of many small farms are milked together and the collective milk is used to prepare more massive alpine wheels. But in the winter, the individual herds are returned to their respective farms, where the farmers prepare this cheese at home.

Though they can be made in any season, many conditions make the wintertime a better time for vacherins: The demanding cheeses are made closer to the comforts of home, so cheesemakers can care for them more readily, as well as more easily taking them to market. Makers have less milk to work with, given the smaller herd size and lower production, which leaves the making of the

PHOTO BY ALIZA ELIAZAROV

larger summer-style wheels impossible. The milk is also extra fatty in the winter season, given the late (or very early) stage of lactation, making an extra-rich cheese. And the milk is whiter, given the lack of fresh green feed, resulting in a strikingly bright paste.

Unlike the summer alpine styles that beat the fat out of the curd with their cutting, vacherin preserves the calf's share of the winter milk's fat within the cheese. The curd is minimally cut and handled, and drained in small forms, before being salted and wrapped in spruce. The extra-high-moisture, high-fat, and low-acidity cheese that forms breaks down and flows readily, and must be kept contained. The spruce bands enable the cheeses to retain their shape as they liquefy fully with their ripening, without spilling out over the cave's shelves.

The spruce bands, or sangles, are prepared from the cambium of fallen spruce trees bucked up for firewood or lumber. They are most easily peeled lengthwise from the logs. The sangles are cut to size, usually 2 to 3 cm thick, and longer than the circumference of the young cheese, then dried before their rehydration in fermented whey as the cheeses are prepared. The cheeses are typically wrapped in the sangles early in their make (if not drained in cloth bags within the sangles, as in some vacherin traditions) to be sure that the cheese retains its height without slumping.

Vacherin comes in a few recognized styles, each a bit different from the others, but all made according to similar philosophies. Among them are vacherin Fribourgeois, from the region of Friboug, and vacherin Mont d'Or, which this technique re-creates. Vacherin can be made according to tradition, or prepared at any time of the year, but the high-fat winter milk makes this cheese shine brightest.

Making Vacherin

An aged rennet cheese • with a washed rind • 8- to 12-hour make • 6- to 8-week affinage

INGREDIENTS

milk	10 L
starter	100 ml kefir, clabber, or whey (1:100)
rennet	regular dose calf rennet
salt	
forms	hoops: 10 cm diameter × 10 cm deep sangles (spruce bands)

Yield three 300 g vacherins (10:1)

TECHNIQUE

Bring milk to cheese temperature, about 35°C (95°F).
Add starter, mix in thoroughly.
Add rennet, mix in thoroughly.
Wait for clean break, 45 minutes to 1 hour.
  ▸ Cover the pot to keep in warmth.
Cut curd to walnut size, 2.5 cm.
  ▸ Cut vertically; cut crosswise; cut horizontally or diagonally.
Stir curds gently for 5 minutes.
Pitch curds for 5 minutes.
Whey off from above.
  ▸ Reserve 1 L whey for washing brine, leave to ferment 24 hours, then add 2 percent salt.
  ▸ Leave dry sangles to steep in this fermenting whey.
Fill forms to the top with curd.
Drain 8 to 16 hours, until acidity develops.
  ▸ Goal pH 5.3—stretch test positive.
  ▸ Flip once during the draining, after 4 hours.
Salt cheeses for 1 hour in saturated salt brine.
  ▸ Or salt 2 percent by weight on surface, about 2 teaspoons per cheese.
Drain cheeses for 1 day on draining table, covered with cheesecloth.
  ▸ Flip twice during draining.
Wrap cheeses in soaked sangles and fasten with string or a rubber band.
Age cheeses in cave for 6 to 8 weeks.
  ▸ Flip cheeses every other day.
  ▸ Wash rinds with light brine every day for first month.

Époisses

Époisses is the most decadent of the washed-rinds. It, too, is a monastic style, but from Burgundy, France, a region celebrated for its wine and gastronomy—and Époisses is its most iconic cheese.

Époisses is instantly recognizable, either by sight—or by smell! Its distinctive rind, a dull orange / pink, is covered with vermiform wrinkles, giving it an aesthetically pleasing exterior. But from its interior arises an aroma most unpleasant—it's probably the most potent of the washed-rind cheeses. And yet despite its smell, the intense and meaty flavor of this cheese will draw you in. There's something about its complex fermentation that makes for an alluring juxtaposition: We're both attracted and repulsed at the same time—what Michael Pollan refers to as the erotics of disgust.

The cheese is more or less a traditional Camembert, made half lactic with a careful ladling and the right timing of the salt. However, instead of washing for a week as with Camembert, to establish a white rind, the monthlong washing called for in Époisses establishes a sticky orange washed rind. Époisses makers wash with marc de Bourgogne, the local brandy, to add a spirited flavor.

Once the washing stops, the wrinkling of the rind begins. This effect isn't found on the other washed rinds, for they all feature less fermentation. The extended primary fermentation of this cheese, resulting from the higher-moisture half-lactic curd, results in a softer-textured cheese whose rind is shaped by the rind ecology (likely the *Geotrichum*'s influence). The rind ecology forms wrinkles that maximize their growing area, resulting in more room for the washed-rind ecology to grow, and a maximization of the cheeses' stink!

As well, the extra lactic acid in the longer-fermented cheese leads to a longer and more flavorful ripening. The result is a positively putrid little wheel of cheese, with a remarkable resemblance to a living thing... indeed, it *is* a living thing, one going through its death throes.

Making Époisses

An aged half-lactic cheese • with a washed rind • 8- to 12-hour make • 6- to 8-week affinage

INGREDIENTS

milk	10 L
starter	100 ml kefir, clabber, or whey (1:100)
rennet	regular dose calf rennet
salt	
brandy	optional
forms	hoops: 10 cm diameter × 10 cm deep

YIELD THREE 300 G ÉPOISSES (10:1)

TECHNIQUE

Bring milk to cheese temperature, about 35°C (95°F).
Add starter, mix in thoroughly.
Add rennet, mix in thoroughly.
Wait for clean break, 45 minutes to 1 hour.
  ▸ Cover the pot to keep in warmth.
Cut curd to walnut-sized (2.5 cm) columns.
  ▸ Cut vertically, but do not cut horizontally.
Pitch curds for 1 hour.
  ▸ Leave curds untouched until much whey has separated.
Whey off from above.
Make washing brine.
  ▸ Add 2 percent salt to whey, and brandy to 10 percent, if using.
Fill forms to the top by ladle.
Drain 8 to 12 hours, until cheeses are firm and acidity develops.
  ▸ Goal pH 5.3—stretch test positive.
Salt 2 percent by weight on surface,
  about 2 teaspoons per cheese.
  ▸ Flip first time during salting.
Drain cheeses for 1 day on draining table,
  covered with cheesecloth.
  ▸ Flip once during draining.
Age cheeses in cave for 6 to 8 weeks.
  ▸ Flip cheeses every other day.
  ▸ Wash cheeses with brine every day for first month.

Torta

Torta—full name, torta del Casar—is representative of a fascinating family of cheeses from the Iberian Peninsula, curdled with the flowers of cardoon. These are typically made with sheep's milk, though they're also fine with goat or cow, with perfectly oozing curds and firm orange skins that holds their liquefied contents together. Calf or lamb rennet can be used in place of cardoon with a similar effect.

Oddly, torta, along with many of its Spanish and Portuguese relatives, is made without added fermentation (though in my technique I recommend its addition). Even more oddly, it is usually salted in the milk to prevent even a spontaneous acidification, resulting in a cheese with a uniquely high pH that ripens much more quickly than most. Torta and its cardoon kin are typically made semi-firm like a tomme, and aged for a short length of time with a washed rind. But despite their semi-firm texture and short ripening, these cheeses ooze like Camemberts quickly.

Several other considerations give tortas unique qualities. They are generally made small for semi-firm cheeses, as a similar style made larger would not be able to hold itself together as it softens. Also, the cheeses are wrapped around their circumference in cloth, to hold their shape as they relax with age, much like the similarly unfermented stracchino.

Members of the thistle family produce enzymes that cause milk to curdle. These are more generally proteolytic than chymosin, precipitating an enzymatic breakdown of the curd that's different from what animal rennet produces and leads to the development of unique flavors and textures in the aged cheeses. As well, these plants possess a slight but sought-after bitterness (like artichokes or the blanched stalks of cardoon) that contributes a pleasing character to cheeses; it especially complements young goat and sheep cheeses.

Thistles grow wild across pastoral lands around the world, and their seeds have followed cattle, goats, and sheep as they made their way across continents. They grow especially well in neglected pastures and overworked soil, and are a sign that minerals and fertility are lacking in the ground. These "weeds" repair neglected pastures and hay meadows by bringing up minerality from the soil below and building organic matter. In the process they protect themselves with prickles from the overgrazing that made the pastures deficient in the first place.

So if you see thistles, be they bull thistles (Scottish thistle or *Cirsium vulgare*), wild or cultivated cardoons (*Cynara cardunculus*), or other purple flowering species, consider leaving them to grow to help repair the soils. Let them flower, and pull off their purple stamens and save them for making this delicious cheese.

Making Torta

A stirred-curd cheese • aged with a washed rind • 12-hour make • 2- to 3-month affinage

INGREDIENTS

milk	10 L sheep
starter	100 ml kefir, clabber, or whey (1:100)
salt	100 g (1 percent of milk weight)
rennet	standard dose thistle rennet (2 g dried flowers steeped in 100 ml water overnight)
forms	hoops: 10 cm diameter × 10 cm deep

YIELD TWO 1 KG TORTAS (5:1)

TECHNIQUE

Bring milk to cheese temperature, about 35°C (95°F).
Add salt to milk, mix in thoroughly.
Add starter, mix in thoroughly.
Add rennet, mix in thoroughly.
Wait for clean break, 45 minutes to 1 hour.
- ▸ Cover the pot to keep in warmth.

Slowly cut curd to corn size, 0.75 cm, with a spino or whisk.
Stir curds slowly for 5 minutes, until curds form peaks when drained and squeezed.
Pitch curds for 5 minutes.
Whey off from above.
Fill forms to the top with curds.
Drain for 8 to 12 hours, until cheese is firm.
- ▸ Flip twice during the drain, with first flip at 4 hours.

Drain cheeses for 1 day on draining table, covered with cheesecloth.
- ▸ Flip twice during draining.

Age cheeses in cave for 2 months.
- ▸ Flip cheeses every other day.
- ▸ Wash cheeses with light water brine every other day for first month.
- ▸ Wrap cheeses in cloth after one month.

Ricotta Salata

This is a technique for an aged ricotta. Although the name translates from the Italian as "salted ricotta," really the ricotta is salted and ripened—*stagionata*—much like other aged cheeses. But because it begins as ricotta, this aged cheese has a uniquely textured and flavorful starting point that only gets better with age.

Ricotta salata is a true *paisano* (peasant) cheese, much loved by shepherds across the south of Italy. In the mountainous regions where it prevails, its makers produce both alpine sheep cheeses and fresh sheep ricotta. Their hard wheels of pecorino are brought down the mountain at the end of the summer grazing season. But the ricotta isn't as easily transported or sold, for it spoils quickly. During the spring and summer milking season, paisanos thus feast on their fresh ricotta; what they cannot eat or sell, they salt, put in their cellars, and age for their winter sustenance.

Ricotta salata can be made with either a sweet or a fermented ricotta, each of which will ripen in a different way due to the acidity in the curd, or the lack thereof. My preference is to use a well-soured ricotta, which is more flavorful, creamy-textured, and protected from unwanted microbes by its fermentation.

The fermented ricotta featured in this make can be replaced by a sweet ricotta, and can be similarly handled and aged. To make the fermented version, an alpine whey is first fermented for a day; the fermented whey is boiled, then drained through fine cloth to separate the ricotta. The ricotta is hung to drain and pressed into its shape in the cloth while still warm.

The cheese, once cool, will be fairly firm and ready for its salt. Salt the cheese to 2 percent from its exterior to help pull out its moisture and preserve it. After draining, the cheese will be ready to age in its cave.

Left in a dry cave, the cheese will slowly lose its moisture, becoming even more concentrated in flavor and crumbly in texture. If it's ripened in a humid cave, rind ecologies will slowly grow upon the cheese, fermenting it like any other mold-ripened cheese; but the cheese will maintain its solid texture due to its lack of casein protein, as only casein deliquesces as its pH rises.

Making Ricotta Salata

An aged heat-acid cheese • 24-hour make • 1- to 6-month affinage

INGREDIENTS

whey	20 L
salt	

YIELD 750 G RICOTTA SALATA (25:1)

TECHNIQUE

Reserve whey from an alpine cheese.
Leave to ferment 12 to 24 hours, until lactic.
Bring whey to a boil.
- Stir occasionally to prevent cream from rising and whey from scorching.
- Fermented whey will overflow when it boils!

Remove pot from heat and let stand 5 minutes to allow ricotta to coalesce.
Ladle off entire contents of pot through colander lined with very fine cheesecloth.
Leave ricotta to drain in cloth 24 hours.
- Scrape down and tighten cloth to form a cheese.

Salt cheese from exterior to 2 percent, about 2 tablespoons.
Leave to drain for 24 hours.
Age in cave for 1 to 6 months.
- Flip twice weekly.

Ghee

For all the alchemists out there, this is a recipe for gold. Perhaps the most exquisitely colored of all dairy products, ghee is a refinement of all the best in butter. A preserved and stable fat protected against rancidification, ghee is especially well suited to warmer climes, like its place of origin, India, where butter is quick to lose its best character.

Like rendering fat into lard, ghee making is the rendering of fat from butter. By slowly cooking butter over low heat for a long time, the elements of the butter are separated and refined. As the butter melts, it divides into three components: The liquid layer, which is densest, sinks to the bottom; the oil portion, the largest portion of the butter, sits in the middle; and the lighter foamy solids rise to the top.

To continue making ghee, the melted butter should be slow-cooked until the liquid portion evaporates and the solids crisp and sink. (These are delicious when fried, by the way, being essentially butter cracklings.) This prolonged cooking lightly caramelizes the fat, giving it a more golden hue and a nutty flavor. If the ghee is less cooked, the water portion has to be separated from the fat portion, which is more challenging to do, and the resulting product is a clarified butter, which, though delicious, pales in flavor, color, and keeping qualities.

Ghee should, preferentially, be prepared with cultured butter. This is the case traditionally in India, for in its original preparation, Indian butter, like all other cultures' butters, was a fermented dairy product. Ghee should also be made with an unsalted butter, the salt being unnecessary for the preservation of the pure fat ghee. When made right, ghee will last an eternity.

Other traditional ways to preserve butter include the Moroccan method of making smen, a cultured butter preserved for years in clay pots or animal skins; the French method of aging cultured butters in caves like cheese; the Italian method of preserving cultured butter within the skin of a caciocavallo, a cheese commonly called burrino; and the Irish / Viking method of burying it in bogs! Most of us are unfamiliar with anything but fresh butter thanks to refrigeration, but this important and revered food has a long history of preservation all around the world.

Making Ghee

A fermented cheese • 1-hour make

INGREDIENTS

cultured butter	1 kg
Yield	**800 g ghee (5:4)**

TECHNIQUE

Melt butter slowly in a small pot.
Continue cooking, stirring regularly, until ready:
- Liquid below solidifies and browns slightly.
- Solids at the surface descend.
- Ghee gives off a hazelnut-like aroma.
- Ghee is filled with tiny bubbles.

Strain out the butter solids (aka butter cracklings).
- Pour ghee through a fine-cheesecloth-lined strainer into heatproof jars.

Leave to cool.
- Ghee will solidify and turn golden when ready.
- Cover and leave at room temperature to preserve.

Lactic Cheese Sausage

To help us understand the "links" between fermented cheese and fermented meats, explored in the next technique, I'll first offer this method for making a lactic cheese aged in a sausage casing.

It's rare to find meat and milk mingling today in Western culture. But in Central Asia, many dairy products are packaged in animal casings. Picture butter stuffed in intestines in Mongolia; soft cheeses packaged in caul fat in Kazakhstan; and milled cheeses aged in goatskins in Turkey. And nearly all cheeses everywhere, even in Europe and North America up to the early twentieth century, were originally made with a piece of calf, kid, or lamb stomach. So why the concern?

This technique is a riff on the idea of mixing milk and meat, a lactic cheese stuffed into a sausage casing and left to go moldy with the fungus *Geotrichum candidum*,

PHOTO BY ALIZA ELIAZAROV

which grows on both fermented meat and fermented milk. The fungal culture grows on the casing in a cool and humid cave, and causes the contents to break down and deliquesce, but remain contained within the casing. When ripe, the entire contents of the casing can be gently squeezed out and enjoyed . . . no mess, no fuss, no rind!

Potentially any cheese can be stuffed into casings or other hollow animal parts, but the two most amenable are lactic cheeses and milled-curd cheeses, that can be more easily packed into a skin. Across the Middle East there are many variations of milled-curd cheeses being pressed into sheep or goat skins, then aged within the animal, fur and all.

To make these historical aged cheeses, leave the curds of one milking's rennet cheese make hanging in cloth overnight to ferment, then mill, salt, drain, and press them by hand into the insides of a specially prepared animal skin, with its legs and rear end tied closed. Over the course of several days, press several successive milkings' worth of curds into the same animal skin, until it's full. At that point you sew up the head end of the animal, making what amounts to an animal full of cheese. This can be aged within for many months or years, protected by the skin from mites and fungus, before being eaten or transported to market.

These are not the sort of cheeses that can be sold in North America, or even in Europe. To taste these original styles, you can either travel to the remote regions where they are made by fewer and fewer makers, or make them yourself, according to their long-standing traditions. You'll likely be pleasantly surprised, for milk and meat are made for each other.

Making Lactic Cheese Sausage

An aged lactic cheese • 48-hour make • 2-month affinage

INGREDIENTS

milk	4 L
starter	40 ml kefir, clabber, or whey (1:100)
rennet	regular dose calf rennet for cow's milk or ¼ dose kid or lamb rennet for goat's or sheep's milk
sausage casings	
salt	

Yield 400 g lactic cheese sausage

TECHNIQUE

Bring milk to culturing temperature, about 20°C (68°F).
Add starter, mix in thoroughly.
Add rennet, mix in thoroughly.
Wait 12 to 24 hours, until milk sets to a lactic curd and *Geotrichum* veil grows.
- Cover the pot.
- Leave undisturbed.

Ladle curd into cheesecloth.
- Reserve whey.
- Soak casings in whey until ready to stuff.

Hang curd to drain for 24 hours, until *Geotrichum* blooms.
Mix salt into the cheese to taste, and knead until smooth.
Hang cheeses for 1 additional day in cloth.
Stuff cheese into sausage casing.
Age cheeses for 2 to 3 days in hastening space until *Geo* covers casing.
- Flip cheeses once daily.

Hang cheeses to age in cave for 2 months until fully liquefied.
- Squeeze out contents to enjoy.

be kept inside of pasta filata cheeses; and, in the Icelandic tradition, meat can be preserved under sour whey without salt over winter. But to me the most exceptional integration of meat and milk is the use of fermented whey, and its *Geotrichum*, to cure charcuterie.

When I die, I'm planning for my body to be steeped in whey and consumed by *Geotrichum candidum* (and for the occasion, like the Beauty of Xiaohe, I'll be wearing a necklace with dried kefir grains!). In the meantime, I'm eating *Geotrichum*-cured sausages.

Making Salami

A fermented meat • 1-hour make • 2-month affinage

INGREDIENTS

meat	1 kg
starter	250 ml sour mesophilic whey (1:4)
salt	
cracked pepper (optional)	
sausage casings	
YIELD	**750 G SAUSAGE (5:4)**

TECHNIQUE

Grind meat.
Mix ground meat with whey.
Add salt to taste (2.5 percent).
Add cracked pepper to taste, if using.
Pack meat into sausage casings to desired length.
Hang sausages to ferment for 2 to 3 days in a hastening space (20°C [68°F], 90 percent humidity), until *Geotrichum candidum* blooms.
Hang to age for 1 month in dry cave (10°C [50°F], 70 percent humidity).

Milk Vinegar

If milk has the capacity to alter itself into alcohol, then it also has the vivacity to become vinegar!

This final act of milk's transformation is one that brings together nearly all the classes of fermentation into one full circle, having now already unified cheese with alcohol, bread, vegetable ferments, and koji. Thus the full family of dairy fermentation now includes the various types of vinegar (cider vinegar, wine vinegar, grain vinegar, and kombucha).

When milkbeer or amasi is left to ferment exposed to air, its alcohol continues in its decomposition to acetic acid, known also as vinegar. We cannot consider this stage of fermentation to be a transcendence; it is unquestionably a process of decay, another possible finale in milk's many pathways of decomposition.

Milk vinegar achieves a final low pH of almost 3.0, surprisingly acidic for milk, and much below any other dairy ferment. The acidity of this milk ferment is much more developed than kefir (pH 4.5) or even yogurt (pH 4.0); one taste on the tongue will assure you that the acidity is not of lactic acid but acetic acid. And no acetic acid bacteria need be added to achieve this vinegar fermentation: They are naturally present in all milk-fermenting cultures, whether clabber, kefir, amasi, or whey. But this common microbe generally isn't given the chance to develop in any other milk ferment.

Milk vinegar can be created via one of two pathways: either in a gourd, as with amasi, or by mechanical / manual aeration, as with milkbeer. It can be prepared with milk or its whey. Either way, the aerated fermentation must be allowed to progress for about 2 weeks, until all the lactic acid is transformed to alcohol, and then all the alcohol is transformed to acetic acid. The milk can also be boiled beforehand (my preference) to make the texture of the vinegar smoother and less cheesy. When the whey is boiled before transforming it into vinegar, its albumin protein is separated as ricotta, which makes for a residue-free cleanser.

There are many uses for this milk-based acid: It can be drizzled over greens as a delicious creamy salad dressing; it can be used to clean out copper kettles of their verdigris, or to scour stainless-steel equipment in a dairy without relying on harsh chemical cleaners (or questionably produced vinegars). It can even serve as an acid to curdle boiling milk in the production of paneer.

Making Milk Vinegar

A fermented milk • 7-day make

INGREDIENTS

milk	1 L
starter	100 ml milkbeer, kefir, clabber, or whey (1:10)

Yield 1 L milk vinegar (1:1)

TECHNIQUE

Cook milk to a boil, stirring constantly (optional, but recommended).
Cool to culturing temperature, about 20°C (68°F).
Add starter, mix thoroughly.
Leave milk to ferment 12 hours, until it curdles.
- ▸ Lightly mix and aerate milk regularly (or simply leave in a *Lagenaria* gourd).

Leave milk to ferment additional 7 days.
- ▸ Lightly mix and aerate once a day.

Bottle and refrigerate (optional) when extra sour.

WINTER • 391

Mysost

The name *mysost* translates simply as "whey cheese." It is a traditional Norwegian cheese, still made in isolated farms at the edge of fjords, where goats graze mountainsides seasonally and farmers make the most of every aspect of their milk.

This odd cheese is a sweet concentration of all the goodness left in whey. To make mysost, leftover whey, containing fats, sugars, and proteins, is slowly cooked to evaporate its moisture, resulting in a creamy, caramel-esque cheese. Indeed, the technique for making mysost is more like boiling syrup from the sap of trees than it is like any other cheese!

Mysost is, like ricotta, best prepared with the whey from an alpine or semi-alpine cheese, these wheys being much richer in fat due to the cheese's small curd size and higher-temperature cooking. And traditionally, it's made from the whey of an alpine or semi-alpine goat's milk cheese. Alpine-style goat cheeses are quite common in Scandinavia, being a part of the ancient seasonal transhumance tradition known in Norway as *sæterfjell*, that was once more widely practiced across the Nordic region. And as a part of the Nordic transhumance culture, the whey left over from these seasonally made alpine goat cheeses, cooked over a fire, would be slowly simmered over the fire's coals overnight until it thickened to a rich and delicious, golden-brown mysost.

To prepare mysost, fresh whey, right off the cheese and unfermented, is slowly brought to a boil. The cream-rich whey is slowly simmered and caramelized overnight, with an occasional stir. Come morning, the mixture is brought to a boil again. Now, with constant stirring, the caramelized whey is fully evaporated, and the thick, fudge-like paste that results is poured into forms to cool. Traditionally these are hand-carved, square-shaped wooden forms bearing distinct designs from the *sæter*—the seasonal farm—they were made at, identifying the cheese and its cheesemaker.

This cheese is best eaten, sliced paper-thin, on traditionally baked Nordic sourdough rye bread—highlighted in the technique for sourdough—baked low and slow until it caramelizes and darkens, much like mysost. With the bread warmed over a cheesemaking fire, stoked again after the morning milking at the sæter, the mysost melts most perfectly into its toast.

Making Mysost

A heat-acid cheese • 24-hour make

INGREDIENTS

alpine whey	20 L
forms	square 10 cm × 10 cm

YIELD 500 G MYSOST (40:1)

TECHNIQUE

Bring sweet whey to a boil.
- Do not wait more than 1 hour after make.

Continue to slowly simmer whey for 6 to 12 hours, until thick and caramelized.
- Stir occasionally.

Increase temperature and cook 1 hour or until a thicker, caramel paste is achieved.
- Stir nonstop.

Pour paste into square forms to shape.
- Lightly press paste into forms.
- Leave mysost in forms until cool.

NOTES

If you're using a non-alpine whey, it helps to add cream to give the cheese its best texture. Add cream to such whey in a ratio of 1:20, the usual proportion of cream in alpine whey (that's 1 L cream in 20 L milk). The whey should also not be from a cheese that is long-fermented, or left to ferment for more than 1 hour after the make, otherwise the cheese will lose sweetness, and the mysost will be overly acidic, though such a sour Mysost is also a regional specialty in Norway.

PHOTO BY ALIZA ELIAZAROV

PHOTO BY ALIZA ELIAZAROV

Paneer

A rennet-free cheese, paneer is another most appropriate food in predominantly Hindu India, with their reverence for the life-giving cow. It is also appreciated by Jains and their adherence to Ahimsa, nonviolence against all living things, as well as by their Muslim kin across India, Pakistan, and Bangladesh.

Paneer is prepared by bringing milk to a boil, then adding an acid to separate its curds without fermentation or adding an enzyme from a calf's stomach. The combined shocks of the heat and acidity cause milk's proteins to curdle in an instant, separating into a sweet, delicious, and nutritious cheese that's a staple food to over a billion people in southern Asia. There are more dairy farmers in India than the rest of the world combined—and likely more paneer makers than makers of any other style of cheese. For paneer is still prepared in a large number of farms, households, and restaurants in the region, the vast majority of which are small farmhouse producers (there are over eighty million dairy farms in the country, mostly very small in size, having between one and five cows or buffalo each).

Paneer is generally made by bringing milk to a boil and coagulating it with an added acid like lemon juice or vinegar without invoking a fermentation. But digging deeper toward a more traditional technique reclaims the fermentation of this South Asian staple, evolving a more holistic cheese, and one with more depth of flavor.

The more authentic way to curdle paneer is to enlist the help of a fermented milk or yogurt. Fermented milks, with their moderately high acidity, are readily on hand in India as household products or as by-products of cultured butter and ghee making; and at just the right temperature (just about a boil) and with the addition of just the right quantity, the hot milk separates its paneer instantly with only the addition of the soured milk.

With the addition of the acid, the curds quickly coagulate and rise up out of the hot milk to form the cheese. The milk's proteins, both casein and albumin, are sensitive to heat and acidity, and are denatured, or scrambled, out of the milk. Together they form a thorough curdle, incorporating the milk's cream as they separate from the now green-tinted, protein-less, fatless whey. The curds are then strained from the pot, drained in cloth, and pressed in a bundle to form a semi-firm cheese.

The incorporation of the albumin, along with the thorough denaturing of the proteins, and its low acidity, allow paneer to resist melting, staying firm as it's sauteed or grilled. Smothered in sauces rich in aromatic spices, with generous additions of both yogurt and ghee, paneer dishes are one of the most important foods of one of the most milk-loving cultures on earth.

Making Paneer

A heat-acid cheese • 1-hour make

INGREDIENTS

milk	4 L
fermented milk	1 L kefir, clabber, or yogurt (1:4)

Yield 500 g paneer (10:1)

TECHNIQUE

Bring milk to boil over high heat, stirring constantly, then turn off heat.

Pour in fermented milk.
- ▸ Break up curd before pouring.
- ▸ Stir into milk thoroughly.
- ▸ If curd does not begin to show, add more ferment and raise temperature.

Wait 5 minutes for paneer curd to firm.

Ladle the curd into fine cheesecloth.

Drain for 10 minutes.

Lightly press cheese for 30 minutes until cool and firm.

Kalvdans

Upon first tasting their mothers' milk, calves dance in delight—and you will too at your first taste of this incredible colostrum pudding from Scandinavia.

The first milk isn't milk, though—it's colostrum. When a calf first suckles at their mother's teat, their cheesemaking stomach (the abomasum) isn't yet developed. Their digestive tract isn't yet primed for cheesemaking, and so the young animal isn't ready to receive their mother' milk. Fortunately, the colostrum that comes first provides an incomparable boost for the infant and doesn't need chymosin to curdle it for its proper digestion.

Colostrum is low in casein, but very high in albumins and immunoglobulins (also known as antibodies): proteins that pass along immune function from cow to calf (and mother to child) and are absorbed directly through the infant's gut. The colostrum itself most resembles eggs in its composition and is likely a throwback to the common evolutionary origins of mammals and birds. The colostrum often has a yolky orange color, is sticky with lactalbumin protein, and contains other proteins and biological protection enzymes, like lysozyme, also commonly found in eggs.

Its peculiar proteins make colostrum unsuitable for cheesemaking but give it a whole other character in the kitchen. (These immunoglobulins also made colostrum a source of selective antimicrobial compounds for medicine before the development of antibiotics.)

Colostrum was and still is turned to pudding by slowly cooking it, thus denaturing the diverse proteins and giving the colostrum a thick texture—surprisingly thick, as the protein content in colostrum can be twice that of milk! The result is a custardy pudding that is often sweetened with sugar or honey, and spiced with cinnamon, nutmeg, and cardamom, much like junket, but thickened with heat instead of rennet.

Finding colostrum can be challenging off the farm, and its consumption is often frowned upon, for it is an essential food for the newborn calf. Newly freshened cows, however, often produce too much for their young; and to save the new mother from the agony of an overfilled udder, many modern dairy farmers milk out excess colostrum and unfortunately dump it. In industrial agriculture, colostrum is often considered unfit for sale because of its higher microbial loads as well as its white blood cell content (a large part of the somatic cell count), both of which, of course, benefit the newborn calves that drink it.

Colostrum is still celebrated in traditional dairy cultures, for it heralds in the beginning of the dairying season. The name kalvdans, from Sweden, is but one of many given to colostrum custards prepared the world over. Beestings pudding is the traditional name of it in English, though the dish is very rare in the English-speaking world today. It is still commonly eaten in India, where it's known as junnu.

Making Kalvdans

A heat-acid cheese • made with colostrum • 2-hour make

INGREDIENTS

colostrum	1 L
sugar	½ cup (to taste)
cinnamon	¼ teaspoon ground
nutmeg	¼ teaspoon ground
cardamom	¼ teaspoon ground

YIELD 1 L KALVDANS (1:1)

TECHNIQUE

Add sugar and spices to colostrum, and mix in thoroughly.
Pour into baking dish, filling 3 to 5 cm deep.
Bake in an oven at 100°C (212°F).
▸ Bake until curdled, 1 to 2 hours.

APPENDIX A

On Clabber Culture

All of the microorganisms needed to begin a clabber culture are present in all raw milk anywhere in the world. No one milk from any particular place or animal should be considered a better source of microbes, because all milks everywhere are meant to ferment, and may well develop the same climax microbial communities universally. (My experiences developing and using clabber cultures from diverse milks around the world show no differences in microbial character, flavor, or cheesemaking potential from one cow to the next; from one farm to the next; or even from one region to the next.) This process works equally well with cow's milk, goat's milk, sheep's milk, and buffalo milk (it develops a culture specific to every species of milk that best ferments that milk), and is even effective with human milk, for good reason. It can be elaborated with the milk of just one animal, or the milk from an entire herd.

To Begin a Clabber

To start a clabber culture, take milk—preferably fresh from the udder, still warm—place it into a clean (but not necessarily sterile) jar with a loose-fitting lid, and leave at ambient temperatures (around 20 to 30°C [68 to 86°F]). The teats should not be sanitized before milking to ensure there are no residual chemicals in the milk, and to have as robust a microbial population as possible. Refrigerated raw milk should be avoided if possible—exposure to cold temperatures can alter its microbial community—though it will work. Preferably the milk should come from animals who have not been recently treated with antibiotics—as should be the case with all milk for cheesemaking. My experience is that milk that comes out of a milking machine actually has a better capacity to ferment than does milk milked into a sterile jar, due to the inherent backslop that happens in milking systems' many impossible-to-perfectly-clean parts. But really, any milk will do.

The quality of milk's first ferment will vary depending on many factors. A clabber that ferments quickly will curdle evenly, with no gas development or whey separation. It should have a mild flavor, with perhaps even a slight development of *Geotrichum candidum* on top. If the milk is milked into a never-washed wooden bucket daily, this first fermentation will not be a spontaneous one, and will be perfect right off the bat.

The faster this first clabber fermentation happens, the better its quality; but it is typically a slow process due to the absolute cleanliness and absence of natural backslop in dairies today. If there are low numbers of microbes in the milk due to modern milking practices, the first fermentation will take longer, and it is more likely that unwanted microorganisms will take hold and cause off fermentations. A spontaneous clabber that ferments more slowly will usually taste off, with bitterness or acridness, and will likely show gas development and possibly even the rising of curd due to large numbers of coliforms or yeasts that get a chance to grow when fermentation is slow.

However, regardless of how this first fermentation turns out, once the milk has soured and curdled, it becomes clabber, and that first clabber should be re-cultured to help refine the culture before it is used as a starter. Even if the first fermentation shows signs of off fermentation, with regular feeding, the re-fermentation will select for the best microorganisms, and prevent the growth of the unwanted ones. That's how these microbes work.

To refresh or feed the culture the first time, place a small amount of the fermented milk into a fresh jar, cover with fresh raw milk (in a ratio of about 1:50), and leave to ferment again. Once the clabber is curdling milk within 12 hours at 20°C (68°F), it can be considered active, and it can be sustained through a regular feeding as described below.

A clabber culture can be kept with pasteurized or otherwise sterilized milk, but to establish a clabber culture, pasteurized milk will not suffice—only raw milk contains the community of beneficial microorganisms needed to kick-start the culture. If obtaining raw milk is too challenging, you can use kefir culture as a source of the same community of raw milk microbes. If raw cow's or goat's milk is illegal to obtain, you can instead leave raw breast milk to ferment to source this universal culture!

Alternatively, the culture can be initiated by re-culturing kefir without the grains (clabber and kefir are more or less the same); or by adapting a sourdough starter into a clabber culture by feeding it milk instead of flour.

Keeping a Clabber Culture

To keep a clabber culture, it must be regularly fed. A daily feeding is the simplest and most effective way of keeping the culture active and in its best state. That daily feeding will keep all the microorganisms in balance and assure a good growth of *Geotrichum candidum* atop the ferment. The clabber can be fed fresh raw milk or pasteurized milk; the microbiological quality of the pasteurized milk clabber shouldn't be significantly different.

A well-treated clabber culture fed the freshest milk on a daily basis will develop the same way each and every day. Fed warm milk, it will typically curdle within 12 hours at 20°C (68°F), once the pH has reached around 4.5. It will have even curd development from the top to the bottom of the jar with no gas development and no whey separation. Atop the curd will be a slight layer of cream, thicker and creamier with certain species and breeds, like Jersey cows, and thinner and whiter with breeds of goats like Saanen. And atop that cream, when it catches the light just right, a slight veil of *Geotrichum* can be seen that obscures what should be a shiny surface with a velvety fungal growth. The clabber will have a pleasant, mildly acidic taste, with a flavor of good milk fermentation, and not an excessive effervescence or yeastiness—you'll want to eat the whole jar! And you can, of course, so long as you save some to use as a starter for the next batch of clabber, and reserve the proper portion to use as a starter for the cheese.

If a clabber culture is not fed on a daily basis, it can overferment, developing too much acidity, which can reduce the numbers of lactose-fermenting bacteria and shift the culture into a secondary fermentation that is dominated by lactic-acid-fermenting yeasts. If a clabber culture is left to ferment for 2 days at 20°C (68°F) before it is refed, the culture will show thick—even wrinkly—growth of *Geotrichum candidum* atop its surface, a sign of that excessive yeastiness. If that older culture is then fed fresh milk, the milk will not develop acidity quickly because the lactose-fermenting microorganisms are no longer active; instead the surface yeasts will cause the culture to develop gas. That next batch of clabber culture fed with the older culture will almost certainly show points of gas and rising curd resulting from excessive yeast growth. That gas may also be indicative of the growth of coliforms, which can thrive when a lactic acid fermentation is slow. A cheese made with that older starter will ferment slowly and show the same unwanted gas development, resulting in a cheese with excessive eyes and off flavors.

Nevertheless, the clabber culture can be righted by bringing it back to its preferred daily feeding rhythm. The gassy cheese, however, is best suited for feeding chickens or the compost.

If the temperature is warm, above 30°C (86°F), these microbial communities all develop more quickly, and a daily feeding may not suffice, for the culture can quickly overferment and shift into a yeasty phase. In warmer weather, consider feeding your clabber culture twice daily to prevent it from developing unwanted gas in the ferment and in the cheese. Alternatively, once the culture has fermented and thickened, which can happen within 6 hours at 35°C (95°F), it can be placed into the refrigerator to prevent its overfermentation, then taken out and fed again the following morning.

To avoid the need to feed clabber daily, the culture can be slowed in the refrigerator. To best do this, feed the clabber and leave it to ferment until it has just set, typically after 12 hours, then promptly put it into the refrigerator to avoid overfermentation. In the fridge, the ferment will be slowed though not perfectly stalled, and the culture will slowly lose its activity over time. After a week the culture can be taken out, fed fresh milk anew, then left to ferment and set before refrigerating it again. To use a refrigerated culture, simply add it to warm milk and mix it in. The culture will show a slight lag time compared with a daily-fed clabber, but will still be quite effective.

To keep a clabber culture even longer without the need to feed it, consider freezing it. To freeze your clabber, leave it to ferment for 12 hours, until just curdled and mildly sour tasting, then freeze it in ice cube trays or freezer-proof containers in measured quantities for a batch of cheese. The culture will remain active for months that way. To use frozen clabber, simply leave it to thaw in the warm milk at the beginning of the cheesemaking session, stirring it in as it melts. As with a refrigerated starter, you should expect a slight lag time with a frozen starter.

Using Clabber Culture in Cheesemaking

For the absolute best results in cheesemaking, add clabber culture to milk as soon as the milk comes from the udder, and proceed with cheesemaking from there with the animal's warm milk. If you're using refrigerated milk, warm it to cheesemaking temperature and add the culture. Pasteurized milk should be cooled down quickly after pasteurization, and the culture added once the milk reaches cheesemaking temperature.

Clabber should be added to milk in a ratio of roughly 1:100—a little bit of starter to a lot of milk. And so, the day before a cheese make, the appropriate amount of clabber should be prepared to ferment an expected amount of milk.

Before adding the starter, mix the culture by either shaking the jar with a lid on or vigorously stirring with a spoon. This breaks up the thick curd structure and assures that the culture mixes well into the milk (this is most a concern with cow clabber and especially thick buffalo clabber). If the curd is not broken into small pieces, it can be strong enough that it won't dissolve into the milk, instead remaining in the cheese once it has set with rennet. One advantage of using a whey starter, as explored in appendix B, is that the curd does not need to be broken first—the whey is simply poured into the milk.

Always remember to feed the culture before you add it all to your cheese. I seldom wash out my clabber jar, to be sure that I never accidentally lose culture down the drain! And I add fresh milk back to that same unwashed jar to carry the culture forward—a simplification that a health inspector might not appreciate, but that actually assures a safer fermentation. Using the same unwashed jar too long, however, can lead to a buildup of *Geotrichum candidum* in the jar that can cause yeasty fermentation and gas development in the clabber or in a cheese, but with the added benefit of a more quickly established *Geo* rind.

APPENDIX B

On Whey Culture

Whey from one day's cheesemaking can easily be used as a starter for the next day's make without too many concerns or complications. After all, it is fairly simple to save a quantity of whey from one batch, leave it out at room temperature to ferment, then use it for the following day's cheese—you simply have to make cheese on a daily basis.

This is the truest form of saving culture from batch to batch, one that is most effective, and the simplest to incorporate into a regular commercial cheesemaking practice. It is also the one used most often today for natural cheesemaking around the world—it is the simplicity of its execution that makes it most effective and widespread.

To Keep a Whey Starter

To save a whey starter, simply hold back a quantity of whey at the end of the make, leave it to ferment, and use it the next day. Whey from lactic cheeses, already fully fermented, can be used the same day—if you wait another day, the culture will be overfermented.

It's best to get into the rhythm of reserving the whey at a certain point in the cheesemaking process—say, as the curds are formed—to be sure it is always done; for if you do not save the whey, you'll be at a loss to make your next cheese. In the making of washed-curd cheeses like Gouda, it is best to take off the whey before it (and its culture) is diluted with warm water. With high-temperature-cooked alpine and grana cheeses, take off the whey from the curd before it is cooked too high (say, once it hits 50°C [122°F]), as cooking above 50°C may hamper the whey's best fermentation.

To ferment, place the still-sweet whey into a jar or other container and, with a lid lightly fastened or a cloth tied atop, leave it to ferment at room temperature (20°C [68°F]) for between 12 and 24 hours (or 12 hours at 40°C [104°F] for a thermophilic starter). If you're making cheese twice daily, the whey saved from the cheese make at the morning milking should be ready in time for the cheese make of the evening milking. If you make cheese only once a day, the whey from one day's make should still be active for the next day's.

If 2 days have passed between cheese makes and the whey is left at room temperature for that period, that whey will have overfermented, and will not assure a good fermentation in cheese. Such a cheese will likely ferment slowly and develop gas from both coliforms and yeast. Warm temperatures can also cause whey to overferment in 24 hours. If the temperature is above 30°C (86°F), it may be best to preserve the whey in the refrigerator after 12 hours to keep it in its active phase and prevent gas from yeast and coliforms. Or you can feed it milk in a ratio of 1:50 and transform it into a clabber culture.

Unlike clabber, there are unfortunately not the same textural and visual reassurances that whey has properly fermented and is in its active phase: You have to trust your gut and know that the culture will ferment if it is handled right in the right amount of time. You can, though, also trust your tongue and taste the whey's developing acidity: It should taste sweet when taken off the curd on the first day, then taste mildly sour when used as a starter for the next make (a pH meter can also confirm this with a reading less than 4.6). Or you can look for telltale signs of *Geotrichum candidum*, a fungus that only grows well on fermented milk rich in lactic acid. You can usually observe the faint growth of *Geotrichum candidum* fungus atop a mesophilic whey starter at 24 hours of fermentation: It will look like slightly fuzzy islands growing atop what should be a shiny-surfaced whey. But if the whey is tasting very sour, or if the *Geotrichum candidum* has grown so thick that it forms wrinkles atop the whey, the culture is likely overfermented.

Whey from any style of cheese can be used as a starter for nearly any other style, so long as it's in the same temperature zone (meso or thermo). Some cheesemakers might be concerned with whey from a blue cheese contaminating a Camembert with the wrong fungus; others might say whey from a cow's milk cheddar couldn't possibly make a goat's milk chèvre, or that whey from an alpine cheese can't be used to make a mozzarella. To them I'd say that a whey starter doesn't just contain the microbes to make one style of cheese; rather, in the early stages of cheesemaking all styles of cheese are very closely interrelated, and if all of them can be made with a clabber culture saved on the side, then all of them can be made with whey saved from one cheese to the next. A whey starter contains within it an incredible microbiological diversity that allows it to adapt to nearly any cheese make. It is the work of the cheesemaker in handling the curds and making the cheese that selects for the right microorganism from the culture in the cheese. Sure, the blue fungus remaining in the whey in the blue cheese will end up in the curd of a Camembert, but if that Camembert is made right, the *Penicillium roqueforti* will never get a chance to grow, as the *Geotrichum* and *P. candidum* that are encouraged through the Camembert's handling keep the blue in check. And definitely, the chèvre will be better when fermented with a goat's milk specific whey, but the cow's milk culture still ferment the goat's milk, albeit a bit more slowly.

Whey can be saved endlessly, and it will ferment assuredly and consistently every single time, so long as you are consistent in your methods. Having faith in its fermentation, and being devoted to its care, is all you need to keep the culture for your entire cheesemaking career.

But if it takes whey to make whey, how can a cheesemaker begin a whey starter? The answer is simple, so long as you understand that all these dairy fermentations—whey, kefir, clabber—are essentially the same. To make the first batch of cheese without a whey starter, use an active clabber culture or kefir culture; then save the whey from that cheese and the whey will ferment and become the culture for the next day's cheese. That whey culture can then be preserved through a daily cheese make. Its microbial ecology will be good from the get-go.

To Use a Whey Starter for Cheesemaking

To keep a whey starter is to use a whey starter, for it is regularly making cheese that assures the whey culture's continued propagation. The culture need not be kept as a separate-lined starter from the cheese, like clabber or kefir. Rather, whey is kept in line with the cheese, like saving a bit of dough to use as the next day's sourdough starter, instead of keeping the sourdough starter in a separate jar. If you wish to have a backup, in case the whey does not ferment because of human error (it's easy to forget to save your whey at the end of a long cheesemaking day), you can always keep a clabber or kefir culture on the side, or freeze a batch of whey that fermented well to use in such circumstances.

Whey culture can be added to milk in a ratio of between 1:50 and 1:200, with 1:100 being ideal for both easy calculation and preparation in a commercial operation. Unlike clabber or kefir, whey need not be mixed before pouring it into milk, for it blends thoroughly into the milk on its own, saving a step and eliminating a curdy or grainy texture that can result from the addition of a fermented milk. For a commercial cheesemaker working with large batches of milk and large quantities of culture, this can save significant effort and make a better-quality cheese.

Whey can be preserved for up to 1 week in the refrigerator, though over time the culture will lose its activity. Whey can also be preserved for many months in the freezer. It is best to freeze or refrigerate it as soon as it has fermented to a pH of around 4.5, so that it does not overferment before it is preserved.

APPENDIX C

On Kefir Culture

Kefir culture is fairly simple to keep: Put grains in fresh milk, wait till the milk ferments and thickens into kefir, strain the grains from the kefir, repeat. But certain aspects of the culture need to be better understood to achieve the best results with this fermentation.

The surface of an active kefir culture, showing the velvety growth of Geo, and its kefir grains floating happily.

Finding Kefir Culture

Only kefir grains can beget kefir grains. From what I understand, you cannot create them from scratch. Fortunately, they're fairly easy to source in most locales if you ask around. You can often find a friendly local kefir keeper who is happy to share them with anyone who will give them a good home. You can also purchase them from kefir breeders online who will send them by mail. Once you've got the grains, if you treat them right you can keep them for life, and pass them on to your grandkids, who can pass them on to theirs.

When I first found out about kefir, I didn't know the culture could be so easily sourced. I was originally drawn to the drink, sold in supermarkets in Quebec, as it was packaged in a special container with a foil lid that rose due to its special "champagne" fermentation (and my mother always told me to avoid such bloated containers!). On the label of the package was a message that I still remember now, twenty years later: *Made with the original kefir grain, of the Caucasus mountains of Russia, under the exclusive authority of the Moscow Dairy Institute.* I thought that the culture that made kefir was controlled by some foreign agency. And that I could not easily gain access to the grains. I tried for years to make my own, by re-fermenting the kefir from the store, hoping to form grains spontaneously, but it never happened.

Eventually I found out that the grains are not controlled by the Russians, and that the culture can easily be obtained from breeders or other kefir fanatics. Many of my early ideas about kefir came from a website compiled by the late Australian master fermenter Dom Anfiteatro: https://myfermentedlife.com. I obtained my first grains in 2007 when I sent $20 cash to a breeder; she sent me back the grains in an envelope, dried. And I've kept kefir and given it away freely ever since. If you can't find some in your neighborhood, I might be able to send some dried grains to you in the mail, too.

To Keep Kefir Culture

Kefir culture is best fed daily (twice daily in warmer weather) in a ratio of 1 part culture to about 50 parts milk. Some recommend a ratio of 1:10 or even 1:5, but I find that such a kefir can be very sour and its fermentation can be challenging to manage. A 1:5 or 1:10 kefir quickly overferments and produces gas in a cheese; 1:50 keeps the culture balanced, and creates the ideal circumstances for using the kefir as a starter for cheesemaking.

At 20°C (68°F) and a 1:50 ratio it suffices to feed kefir once daily, with the kefir ready after 12 to 24 hours of fermentation. But the kefir typically tastes best after 12 hours of fermentation, as soon as it has curdled. With extra fermentation time, the kefir can become more sour, more yeasty, and less palatable to some. For the mildest-tasting kefir, put the ferment in the refrigerator once it has curdled.

In warmer weather (temperatures above 30°C [86°F]), it's best to feed the culture twice daily, or to put it in the fridge after 8 to 12 hours (once it has curdled) to prevent overfermentation and yeastiness.

I generally strain my grains in household strainers or cheese forms with relatively large openings. Removing the kefir grains from the top of the fermented kefir with your fingers or a spoon is an even gentler method of handling them The grains should be handled as gently as possible, to avoid breaking them into smaller pieces. I find that by shaking or stirring the ferment before straining it, the kefir grains are easier to strain and can be more delicately handled. If kefir is overfermented and separates into cheese, the grains can be challenging to remove from the curd, and tend to break into smaller-sized pieces.

Small-sized kefir grains, however, are really no different from larger-sized; they've just been handled more roughly, and have failed to grow as big. But there's something special about the beautiful lobed shape that kefir grains take when they are not heavily handled, and a big, healthy-seeming kefir grain seems to be more easily and enthusiastically cared for than a multitude of mini grains.

The grains should never be washed between batches, as this washes off the effective fermentation that the kefir grains carry from one batch of kefir to the next. Washing grains strips them of their backslop and significantly slows their fermentation. When washed, the kefir grains will not rise, and they seem to be unable to grow. The liquid culture they carry seems to be an integral part of the kefir grain itself.

Even if a kefir is fermenting off, because it has been long neglected, it is still best to not wash the grains. Doing so always sets back the fermentation, causing it to continuously lose its footing and allow unwanted microbes to grow. So no matter what the quality of your kefir, the grains should never be cleaned. They can, however, if long neglected, be dipped in an active clabber culture to quickly restore their activity, something I sometimes do when I've been regularly feeding a clabber, but have neglected to feed my kefir!

Similarly, a kefir grain should never be left exposed to air for too long. The complex microbial community of the grain will become imbalanced if it is not kept continually submerged in milk or its kefir. Exposure to air will shift the community toward a yeasty or even vinegary phase, and long exposure will cause a grain to dry.

Kefir grains grow, doubling in size every 10 days or so if fed daily, especially when fed fresh warm milk. A properly fed grain will grow and grow endlessly, so long as it's fed milk regularly. I've encountered kefir keepers with grains as big as softballs, and some with enough of the creatures to fill multiple 20 L buckets!

However, as kefir grains grow in size, we are less likely to feed them the larger and larger quantities of milk they need to ferment at their best. When the ratio of kefir grains to milk falls from 1:50 to 1:5, the culture will ferment very differently, exhausting its lactose, reaching peak activity in 3 to 4 hours, and becoming overfermented and yeasty shortly thereafter. As well, such kefir can become very sticky, due to the production of polysaccharides by some members of the kefir community. A 1:5 kefir is very thick and strongly flavored and is favored by some kefir keepers for its medicinal qualities, but from a cheesemaking perspective a kefir culture fed milk in a 1:50 to 1:100 ratio brings about a more balanced and controlled fermentation.

It is best to keep an amount of kefir culture that can be fed an appropriate amount of milk daily for your needs. I personally recommend keeping between 1 teaspoon (5 ml) and 1 tablespoon (15 ml) of kefir grains, which effectively ferments between 1 cup and 1 L kefir at around a 1:50 ratio. As the grains grow, you can gift excess to friends and relatives. Dry excess grains for a rainy day, or even eat them as some folks like to do (I do not!).

If you are a commercial producer looking to make large quantities of kefir, you do not necessarily need to have a large quantity of kefir grains. If you wish to prepare 100 L kefir, you can do so with just a tablespoonful of grains; that tablespoon can make a 1 L batch of kefir; then the 1 L kefir can be strained of its grains and used to prepare a 100 L batch of kefir. Though some may say that this is not kefir, as it is prepared without the grains, to me it is undoubtedly the same fermentation, as the grains themselves do not cause the fermentation to happen—it is the liquid kefir carried forward by the grains that ferments milk.

Kefir can be kept on raw milk or pasteurized milk; even UHT milk or powdered milk can sustain a kefir grain if no fresh milk is available. When traveling in parts of the world where good milk is hard to access, I prefer to feed my grains this overprocessed milk daily rather than neglect them, and I find the kefir to result from it to be surprisingly good, especially considering the origins of that milk (of course such milk cannot be used for cheesemaking, and that kefir will pale in comparison flavor-wise with a kefir made from fresh raw milk).

Kefir can be fed goat's milk, cow's milk, sheep's milk, and buffalo milk; each will ferment well, but each will ferment differently. My favorite is goat's milk kefir, which is the thinnest and most drinkable, and possesses the most exquisite flavor due to the unique and flavorful fatty acids of goat's milk. Cow kefir is still delicious, only much thicker. Buffalo kefir, though amazing, can be challenging to produce as it's so thick that straining the kefir grain is nearly impossible! It's best to keep the grains on a specific milk; transferring grains between milks is fine, but there is a slight lag time as the culture adapts to the sugars and nutrients of its new medium.

What matters almost more than anything with milk, though, is its freshness. Milk out of the udder and still warm yields the best kefir, and a long-refrigerated milk yields a strange brew. Try this yourself by using the same milk, pasteurized or raw, over the course of 2 weeks to feed your kefir, and note the declining quality as the days pass and the milk ages, due to unwanted microbial growth under refrigeration, and rancidification of milk's sensitive fats.

Using Kefir in Cheesemaking

To use kefir as a starter culture for cheesemaking, strain out the grains from an active kefir, and add the curdled liquid (the kefir) to warm milk in a ratio of around 1:100; add the rennet immediately thereafter. Continue on with the cheese make, being sure to feed the kefir grains fresh milk again to keep them happy, and to have starter ready for the next day's cheese.

If I'm going to be using kefir as a starter for 100 L milk, I'll prepare 1 L kefir the day before by feeding a tablespoon-sized grain 1 L milk and leaving it out to ferment 12 to 24 hours. The next day I'll strain the grains and add the starter to the milk, feeding the grains once again. Every day I'll feed the grains to that 1 L milk; but occasionally, about once a week, I'll have to pare back the amount of grains I feed back to 1 tablespoon, as they will continually grow. This will result in the most consistent fermentation both in the kefir and in a cheese.

Though kefir grains can be added as a starter directly to milk instead of the kefir, they generally do not create an even fermentation in a cheese. When I first started making cheeses with kefir, this is what I did, and conveniently, the happy kefir grains rose to the top of the curd as the curd set and were easily retrieved from there.

Neglecting Kefir Culture

Kefir culture does prefer a daily feeding, but it can also be preserved without feeding for many months or years with the right approach. The only things that will kill kefir grains are temperatures above 50°C (122°F) (they melt into goop at this point, and don't come back together when cooled); being accidentally washed down a drain (this is a good reason to label your kefir jar—and never to wash it out); and being eaten—unfortunately that's the end for them.

Kefir grains can be endlessly preserved in a refrigerator. Kept in their kefir, the grains will last as long as they remain submerged in the liquid. And though they won't remain active after weeks, months, or even years in a fridge, they can be brought back from their hibernation the same way as dried kefir grains, as described below.

Drying Kefir

Kefir grains can also be dried, and preserved that way, perhaps eternally. It's quite possible that the grains on the Beauty of Xiaohe's necklace (see chapter 3) could come back to life if put in milk! To dry kefir grains, simply place them on a dry cloth for 1 week, flipping them daily to be sure they desiccate evenly and don't stick to the fabric. Once dry, they can be kept indefinitely in a dry container, until the time comes to rehydrate them and bring them back to life.

Revitalizing dried kefir grains is as simple as refermenting them two or three times, waiting for the milk to sour and thicken until feeding, until they're fermenting reliably again. Place them into fresh milk (pasteurized or not) in a ratio of 1:50 and leave them at 20°C (68°F) for 2 to 3 days until the milk ferments and sours; strain out the grains from this first batch and discard the first kefir, then refresh the grains in fresh milk and ferment again. This second ferment should happen much faster—typically in 12 to 24 hours; refresh them one more time, and the kefir grains will be fermenting the same as they always were, and will have forgotten all about their long neglect. It will take about a week of regular feeding, however, before the dried grains grow again.

APPENDIX D

On Keeping a Wooden Vat

Wooden vats may be the most complex of all the forms that milk culture can take. But because the wooden vat is never washed out, simplifying both the make and the cleanup, the use of wood to culture cheese may also be the most simple form of cheese fermentation. To use a wooden vat as a starter culture, you simply never rinse it out. You just put in fresh milk every day and turn it into cheese, day after day after day, refreshing and feeding the culture that's kept in the whey-soaked wood.

Because of its porosity, wood holds on to whey saved from one batch of cheese to the next. It might be thought therefore that it functions like whey; however, its behavior appears to be more complicated. For yeasts will proliferate on the wood and develop a biofilm, as they will on the rind of a cheese. To this end, the wooden vat may function like a giant kefir grain, requiring the conditions of perpetual fermentation in milk in order to function properly. And for it to work right, a vat should never be left fully emptied.

Wooden vats, due to their porosity, are also permeable to air. So unlike glass containers, or plastic or stainless steel, some air passes through the walls of the vat, altering aspects of the fermentation. The result is a cheese with more microbial complexity, containing more secondary fermenting microorganisms, especially yeasts and fungi. The effect of wood can be realized in a small scale with the *Lagenaria* gourd used in the recipe for amasi. The gourd permits air to pass to the ferment within, causing a proliferation of yeasts, and a mildly alcoholic milk beer is the result.

Wooden barrels used for traditional cheesemaking in France, known as gerles, and the Tinas of Sicily are normally made of European chestnut. This wood is strong but lightweight, permitting a vat's transport between pastures in the summertime (cheese was and still is made out in the pastures with the animals to simplify the milk handling and assure its warmth for the cheese make). Chestnut is also porous, permitting the vat to hold on to the whey and better serve as a starter. Any strong and light wood can be used for a barrel, however. Tannins in the wood are not as significant a concern in cheesemaking as in winemaking, as the milk is not in contact with the wood for long; and the barrels are typically used over and over, thousands of times. Barrels can be used daily for many, many years with only minimal maintenance. They can be the most affordable vat option for a cheesemaking.

To Initiate a Wooden Vat

A wooden barrel without an active culture is just a wooden barrel. So how can you initiate the fermentation in a wooden vessel? To christen it and initiate the cycle of fermentation that's essential for cheesemaking, use active clabber, kefir, or whey as a starter for the first batch of cheese made in a wooden vat.

Before a barrel is initiated, however, it should be steeped with water for two days, for a thoroughly dried barrel will not be watertight. Add water to the barrel twice daily, until the staves have swelled, and no liquid is lost. Once the barrel is watertight, the water can be emptied, and cheesemaking can begin.

And as long as the barrel is in continuous, daily use, the barrel will continue to hold milk, and will continue to carry culture.

To Use the Wooden Vat as a Culture

To assure a successful continuous fermentation, the wooden vat must be used daily—twice daily in warm climates or weather—to prevent overfermentation of the culture within the wood. Ideally, all of the cheese's whey should be left to ferment in the vat after the make, right until the next cheese is made, to assure a full fermentation and limit air exposure and the resultant yeast development. And just as the wooden barrels best keep their form and hold liquid if they are kept filled with whey, wooden barrels are best for a culture if used daily and always left full with whey. Leaving the whey to sit after the make in the barrel also has the added advantage that the whey cream lost during the make rises to the top, and can be skimmed off to make whey butter, a tradition in the Auvergne of France where wooden gerles are still used for cheesemaking.

Much like kefir grains, the wooden vat should never be washed out; before the next make, it should only be scrubbed with the fermented whey left in the vat from the previous make. If the vessel is washed with water, the activity of the culture will be diluted, and the next cheese made in the vat will be adversely affected. This is what inadvertently happens with some traditional washed-curd cheeses made in wood like Gouda, where washing with hot water lessens the quality of fermentation in the wood. To use a wooden vat for such washed-curd cheeses, or to clean a vat with water, save a whey starter from the cheese before its curds are washed, then leave it to ferment apart from the vat and introduce it to the next vatful of milk. This whey starter should be fermented in the thermophilic zone to encourage an appropriate microbiology for the higher-temperature-cooked cheese.

If the barrels are used infrequently, the culture within its staves will lose its activity, causing a cheese to develop yeastiness or other unwanted consequences of a delayed fermentation. In addition, the barrel may have to have water added to preserve its milk-holding capacity. Fresh warm milk should be added to the vat as the animals are milked, and rennetted immediately after milking to work with the milk's warmth. (Some cheesemakers in France warm their milk with hot water coils or plug-in calf milk warmers if needed.) No starter need be added as the wood provides the appropriate microorganisms in top form if it is treated right, as well as the acidity that can help the milk curdle. When the curd sets, it is cut as usual, and stirred for the appropriate amount of time for the make. The curds can then be ladled off into their forms. In the case of semi-alpine cheeses traditionally made in wood without cooking, such as tomme de chèvre or Saint-Nectaire, the curds can be pressed together under the warm whey, then removed from the vat; leave the whey in the vat to ferment until you empty it just before the next milking.

The improper use of a wooden vat can result in a tendency to develop gas in cheeses made in it. Emptying the vat, for example, and leaving it exposed to air for long periods can result in excess yeast development in the culture of the wood. Leaving the whey in the vat too long can also cause it to overferment and develop excess yeast in a cheese. These two scenarios can cause curds made in a wooden vat to float. This can be seen when the rennet curd first sets in the vat; the curd around the edge is more buoyant, as it catches gas rising up from the yeasty culture fermenting within the wood.

For lactic cheeses, particular attention must be paid to be sure that the culture doesn't shift into a yeasty realm. The cheese's long fermentation in the vat can cause the delicate curd to be negatively affected by yeast development in the wood's culture if the vat isn't properly managed. When making lactic cheeses in wood, the curd should be left in the vat to ferment just until just before the next milking in 12 to 24 hours (the curd should be lactic within 12 hours, unless the cheesemaking environment is very cold, or the culture in the wood is inactive). The lactic curd can then be ladled into forms or cloths to drain just before the vat is filled again with fresh milk. After ladling the lactic curds, scrub the vat down with only the whey remaining in the wood to clean out remaining curd and set back fungal growth above the whey line, then empty it and fill it with warm milk from the next milking.

To Dry Off and Re-initiate a Wooden Vat

With a seasonal drying off, or a prolonged lapse in cheesemaking, the barrel should be properly dried off. It should therefore be expected that the culture in the vat will go dormant. The vat should be wiped down with whey, emptied, then left out with the lid on or covered with cloth (the wood will still dry just fine when covered) to keep out pests and dust. Washing with water can cause unwanted molds to grow, whereas leaving whey in wood will encourage the growth of *Geotrichum* upon the drying slats.

Before the cheesemaking season commences anew in the spring, the dried barrel can be re-initiated. First steep it with several changes of water over several days to swell the slats and seal it so that it holds milk. Meanwhile, you can leave the first small quantities of fresh milk from the animals to spontaneously ferment to develop a clabber culture. Add that clabber culture to the first batch of cheese you prepare in the wooden vat to assure an effective fermentation right from the get-go, to avoid the effects of a "tired" culture, as in the story of stracchino. That culture will then carry forward in the wood from batch to batch.

APPENDIX E

Thermophilic Starters

For some cheeses, like large alpine cheeses cooked to a high temperature, certain dairy ferments kept warm like yogurt, and a high temperature fermented mozzarella, it benefits you and your cheese to use a thermophilic (high-temperature) starter. *Thermophilic*, in cheesemaking, refers specifically to cultures that prefer a fermentation temperature at or just above animal temperature, 37°C (97°F).

Cheeses that ferment primarily in a hotter temperature zone appreciate a starter with microorganisms that are adapted to ferment in warmer weather. Refermenting raw milk or clabber at a higher temperature selects for the appropriate microorganisms that can aid a thermophilic cheese or yogurt make.

Thermophilic starters are simpler, microbiologically, than mesophilic starters fermented at lower temperatures, and consist primarily of lactofermenting bacteria. Considered homofermentative, they also develop simpler flavor profiles in cheese or yogurt than a mesophilic fermentation that is more heterofermentative. For this reason, cheeses are more flavorful when made in a mesophilic temperature range. However heat is needed for cooked-curd cheeses that rely on its effects on firming curd. Also, a hotter fermentation temperature is what gives yogurt its best qualities of tartness, rather than complexity.

Whey and clabber culture are suited to a thermophilic fermentation. But wooden vats cannot be kept as a thermophilic culture (unless the temperature of the make room is above body temperature, in which case the wood will be a thermophilic starter). Though kefir grains ferment fine at temperatures below 37°C (97°F), they cannot be kept as a thermophilic starter, as the grains tend not to function well at or above body temperature as their yeasts fail to develop. Yogurt, prepared weekly, and fermented in a thermophilic manner, can be used in place of a thermophilic clabber or whey in any technique requiring a thermophilic starter. Yogurt is essentially a thermophilic clabber, but fermented on boiled milk instead of raw.

Choosing to Use a Thermophilic Starter

Only if a portion of a cheese's fermentation happens above that temperature should a thermophilic starter be added. These conditions are usually only met with cooked-curd cheeses or high-temperature-fermented yogurt. Massive, full-alpine cheeses such as Comté cooked to 52°C (126°F) fit in this category, but smaller cheeses cooked to a medium temperature, such as tomme de montagne, raclette, and even Gouda can also benefit from a thermophilic starter, as the cheeses inhabit a high-temperature zone for an hour or so before they cool.

If cooked-curd cheeses made with a thermophilic starter are to be aged, it is best to also add a mesophilic starter, to assure that the cheese has the microbial community it needs to age well. Mesophilic starters contain more yeasts and fungi that can help establish protective rinds, whereas a thermophilic starter is completely lacking in that microbial department due to heat's hampering of fungi. Such cheeses will likely benefit in many other ways from the mesophilic starter; after all,

Cheesemakers should arm themselves with research papers documenting the effectiveness and safety of natural starter cultures in case they face resistance. There are a number of scientific papers out of Europe (referenced below) that explore the effectiveness of kefir culture as well as wooden cheesemaking vats for making safe raw milk cheese. However, the practices of using whey and clabber as starter cultures have not been greatly explored in academia. Their universal effectiveness has not been fully established, and may not be taken as proof that these cultures will make cheese safely in commercial production. Unfortunately there are no peer-reviewed studies as of yet that support the idea of natural starters as I propose in this book. Some studies in other realms of fermentation, though, like sourdough, can offer support for adapting ideas about natural fermentation into cheese.

Traditional methods of cultivating starters need more scientific inquiry before their effectiveness in the commercial production of cheese is proven. My stating their effectiveness in this book will not unequivocally convince inspectors that these methods are safe to use for making cheese. Nevertheless, despite being careful in your practice and referring to research proving their effectiveness, inspectors will want to know that the methods you're using are safe in your circumstances.

There are two main concerns to address with regard to the safety of starter cultures in commercial cheesemaking. They are:

- The effectiveness of the starter in initiating fermentation in a cheese
- The possibility of propagation of pathogenic microbes in the starter

The first is easier to prove than the second. The effectiveness of a starter's capacity for fermentation can be measured by the speed at which a cheese reaches a certain threshold pH, say 5.3 (4.5 in lactic cheeses), when most cheeses are salted. If the cheese reaches that goal pH relatively quickly (within 12 hours), it can be presumed to be effectively fermented, and therefore safely protected against the growth of many unwanted microorganisms that are controlled by fermentation and a pH drop. To document this pH change within a certain amount of time, the cheese's pH should be measured and recorded over the course of the make, either with a pH meter, or by recording that the cheese stretches endlessly when submerged in boiling-hot water (with cow and buffalo curd; less of a stretch with goat and sheep), which is a more reliable method of testing a cheese's pH for 5.3.

Proving that the starter is not propagating pathogens along with its beneficial microbes is more challenging. I know, and you should too by now, that in a properly propagated starter, made with fresh raw milk, and with a culture backslopped at its prime, an effective fermentation will prevent the growth and development of unwanted microbes. But an inspector is unlikely to take my or your word for it.

Regularly feeding a traditional starter should be understood as an effective way of keeping the culture clean and free of unwanted microbes. The starters themselves are endlessly propagable and self-cleansing, producing countless antimicrobial compounds, including organic acids (like lactic acid), alcohols, esters, ketones, and mycotoxins, that limit the growth of most if not all pathogens. Simply keeping the culture in its top form by feeding it the freshest milk daily should suffice to show its purity.

As well, there are visible (and gustatory) manifestations in starter cultures of both good and bad microorganisms. For example, an effective fermentation with beneficial microbes will result in a clabber with no gas development and a healthy growth of *Geotrichum candidum* at its surface. Seeing the clabber culture curdle within 12 hours is a sign that the pH dropped to 4.5 in an effective time frame, assuring that unwanted microbes are kept in check, as is the case in a well-made cheese. The growth of *E. coli*, salmonella, and listeria are all restricted by this low pH. Tasting the culture and finding it pleasing is another reliable way of determining that a good fermentation has taken place, one that can effectively control unwanted microbes, as less-desirable microbes will contribute off flavors. These developments can be documented as part of a food safety plan to show how care is taken to assure the proper progression of these fundamental fermentations.

As for unwanted microbes, sometimes the effects of these can also be seen. Coliforms produce gas that can be seen in a starter, or in a cheese that is similarly afflicted. It should be noted that coliforms are not necessarily dangerous—though certain strains such as *E. coli* O157:H7 can lead to severe illness and death—but their presence is a sign that other, possibly more dangerous pathogens are lurking. If fermentation is slow, or milk long-neglected, coliforms can develop, producing gas bubbles in the clabber's curd that are visible to the eye. A lack of visible gas in the fermented curd of clabber or kefir should be sufficient to prove the lack of coliforms to an inspector, though it is harder to observe this effect in a whey starter, as the gas bubbles rise to the surface. Performing a bench-type *E. coli* test on the starter occasionally will help build the inspector's confidence. It shouldn't have to be done daily so long as a protocol for feeding the starters is written and established—the results will always be the same so long as the starters are always prepared in the same way.

If you're using raw milk to make your cheese, a starter derived from the microbiology of raw milk is exceptionally well suited to fermenting that milk. This starter is representative of the best species present in your (and every) raw milk, and more effectively controls the growth of unwanted microbes from the milk than commercial starters due to the lack of lag time, and the exceptional nature of its diverse and balanced fermentations. Despite common food-safety concerns about its transfer of pathogenic microbes from batch to batch, a well-tended starter (fed the freshest animal warm milk daily or twice daily) propagates only the best players, and most effectively keeps unwanted species in check.

If you're using pasteurized milk to make cheese, using a starter derived originally from raw milk will likely raise alarm bells. However, an inspector can be reassured by the fact that after ten propagations of that starter, feeding that culture pasteurized or sterilized (boiled) milk in a ratio of 1:100 daily, only an infinitesimally small amount of raw milk is being added to the pasteurized milk cheese (about 1×10^{-20} parts raw milk to be precise). And effectively zero unwanted or pathogenic microbes from raw milk are being propagated by that culture because of the many successive successful rounds of fermentation. A well-practiced backslop does not propagate coliforms and other pathogens, and actually stops their growth better than pasteurization, in which these pathogens can still grow in the milk post-pasteurization. Testing the microbiological profile of this culture before its use can also help prove its lack of microorganisms of concern, but it should not be necessary to do so. It should be understood that an effective starter culture can be derived from raw milk in this manner for the making of pasteurized milk cheeses. If not, kefir culture can then be used safely in its place.

If an inspector is unfamiliar with or suspicious of kefir, then it's up to you to educate them. This culture, with its nearly 10,000-year history, its complex and protective microbiology, and its ability to propagate perfectly on pasteurized or raw milk, should assure any inspector of its capacity for beneficial fermentation. But the culture is often unknown or unrecognized by inspecting agencies, which may never have encountered or even heard of the thing.

Some cheesemakers propagate their natural cultures with sterilized milk, feeding their raw-milk-derived cultures on sterile milk. While this works, I do not believe this to be best practice, as it takes more effort, time, materials, and space to prepare the sterilized milk. As well, when milk is sterilized by cooking, its profile of sugars is changed, which changes the microbiological profile of the ferment. When such a sterilized milk culture is used on a raw milk cheese, there will be a lag time, with all its associated risks, compared to a starter propagated regularly with raw milk. It should also be understood that boiling milk in a small pot may not be recognized by an inspector as an effective pasteurization.

If animals are being milked daily, if cheese is being prepared daily, a starter culture can be most easily and most safely propagated daily on the freshest milk. The simplifications that come with feeding the culture with the freshest raw milk daily make the culturing technique more effective. If the procedure of propagating cultures is more challenging, or takes more time, then the method is more prone to be neglected or performed incorrectly, which reduces its effectiveness.

Using Natural Materials

Wood is considered an acceptable material in the cheese cave, for supporting the ripening of cheeses, but its use is typically restricted in the dairy itself. But why should wood or other natural materials be limited in the dairy, if, as we've come to understand, their use is responsible for the propagation of beneficial microorganisms that can make a cheese safer?

Proper use of wooden cheesemaking vats and cheese forms imbues them with the most protective microbiologies of milk, which create safer circumstances than the use of stainless steel and freeze-dried starters. A well-used wooden vat ushers in an effective fermentation that proceeds quickly and completely, creating just the right circumstances that good cheese needs to evolve and that restrict the growth of pathogens. If cheesemakers are permitted to use natural starters in their cheese makes (which they generally are), wood in the dairy should be seen as an effective natural starter (when it's never washed out, and used on a daily basis) and should be allowed by extension.

Using Natural Rennet

Traditional rennet stomachs can be used as a coagulating agent in dairies, both in Europe and in North America, so long as the product comes from an approved commercial source. Fortunately, traditional calf vells as well as kid and lamb stomachs (with or without their cheese) can be purchased from some commercial suppliers in Europe such as Cuajos Caporal (for lamb/kid stomachs—known as cuajares) in Spain, Laboratoires ABIA (for calf vells) in France, BHG in Austria, and Winkler Ag in Switzerland. The stomachs can be imported (to some countries legally, to some not) for use in cheesemaking.

The production of rennet in-house will, of course, raise eyebrows, for the mixing of such animal parts with raw milk hasn't been practiced in many locations in well over a century. But this should be understood to be a foundation of all cheesemaking practice, and fundamental to the sustainable production of good cheese with locally derived ingredients—and what is milk, again, but another animal part?

Natural rennet can be used safely, but current regulations around the origin of cheese additives would make the use of farm-produced rennet a near impossibility in most regions. Special allowances should be made for the inclusion of this ingredient in cheese; for when such naturally curdled cheese is also naturally fermented, the risk of propagation of unwanted microorganisms from the stomachs is effectively nullified.

Of course, the animals will have to be slaughtered in a certified slaughterhouse. But perhaps special circumstances should be permitted for on-farm slaughter of young animals from a dairy that are often considered unslaughterable (too small or too cute) by the slaughterhouses, and whose most ethical treatment and handling demands not being transported for hours or days by truck. If animals must be slaughtered in a slaughterhouse in your region, then do so, and ask for the slaughterhouse to return the animals' stomachs to you (they may be unaccustomed to doing so, but can nevertheless be trained) and to provide documentation of the stomachs' extraction. In certain regions, like the Canary Islands, stomachs of lambs, kids, and calves are certified and tagged, and returned to the farmers to transform into rennet in-house.

To create a rennet that is usable in commercial cheese production, the stomachs must also be prepared in an approved facility. They should be salted and hung, or cleaned and inflated to dry immediately, or frozen until the producer is ready to preserve them. This process can be performed in a specific space that is dedicated to the rennet preparation; nevertheless it should be permitted in a dairy, because, after all, this ingredient will be added to the milk. Proper protocol, including steeping the salted, dried stomachs in fermented whey to extract their enzymes effectively controls against possible pathogens. Their preparation and use is not all that different from the use of natural sausage casings in the making of sausages and salumi.

Remember, only a century ago cheesemaking depended wholly on the production of these rennets from animals slaughtered and butchered on the farm or in that region. And every dairy farm can source all the rennet for their cheesemaking from their own animals. If rennet production cannot be allowed in-house, we must collectively rebuild our capacity for producing these elemental enzymes in our respective regions, for food security, food safety, food sovereignty—and good cheese!

Using Warm Milk

Using animal-warm milk is a most important consideration when it comes to cheese flavor and microbiological quality. But milk must often be chilled to a certain cold degree (typically 4°C [39°F]) within a certain period of time (often 2 to 4 hours) to conform to the legalities of its production and sale. (In Canada, dairy farmers don't own their milk, and to make cheese with it, must have the milk diverted from a refrigerated bulk milk tank by an official, and only when the tanker truck arrives once every 2 days.) However, despite the near universality of such controls, it should be understood as dangerous to subject raw milk to any period of refrigeration.

If the milk is to be transformed into cheese within a couple of hours of milking, the cooling requirement should not be considered necessary. For this is good cheesemaking practice. Cheesemakers who are not dairy farmers should be able to transport milk to their facilities warm from the milking a certain distance within a number of hours. The energy it takes to cool and warm milk is entirely wasted. And the conditions creating by the cooling, and the delay of cheesemaking that results, encourage unwanted cold-loving microorganisms that can cause problems in a cheese.

In commercial production, natural cheesemakers should aim for a twice-daily production, avoiding the use of raw milk refrigerated for any amount of time. I suppose I should expand my tolerance to say that using milk refrigerated 12 hours is acceptable, if it is mixed with an equal quantity of fresher milk, as is commonly practiced, but my personal feeling is that this should be avoided if possible, and that milk should never be refrigerated in a raw milk cheesemaking. If milk is refrigerated for any time, cheesemakers should ideally pasteurize it (62°C [144°F] for 30 minutes in an approved pasteurizer), to significantly reduce the propagation of unwanted microbes caused by refrigeration.

Dairies should be designed to work with the milk produced daily or twice daily on the farm, not to store milk for up to 2 days, as they often are. Significant savings in systems costs and energy can be realized if milk goes directly from the milking parlor into the cheesemaking vat without storage and refrigeration. And though such designs necessitate a more regular cheesemaking, this is what it takes to make the safest, best-tasting, most nourishing, most sustainable cheese.

But cheesemakers should not feel bogged down by the idea of a twice-daily make. Some cheeses, of course, take much less effort for their production; cheesemakers can consider making more intensive rennet cheese with only the morning's milk, and making butter or less involved lactic cheeses with the evening milk; or making rennet cheese twice a day but realizing time and effort savings from natural methods. For example, not pasteurizing milk saves over 2 hours per batch; and adding culture and rennet simultaneously, as in natural techniques, can save an hour of cheesemaking time per make.

Once-a-day milking (not common in America, but fairly well known in France) can also relieve a farmer-cheesemaker of the need to produce cheese twice daily to work with only animal warm milk. Once-a-day milking followed by once-a-day cheesemaking works equally well with cows, goats, or sheep, which can be accustomed to almost any milking schedule. Yields of milk are reduced by around 20 to 30 percent, but yields of cheese do not suffer as substantially, because the milk concentrates its components slightly if it is taken less often. This time- and labor-saving choice for cheesemakers makes extraordinary cheese possible without the need for a twice-daily make. Ultimately, the udder (or the cheese) is the only safe milk storage vessel.

Scientific Papers and Protocols on Natural Methods

Bérard, Laurence, and Marie-Christine Montel. "La gerle, le vivant invisible entre traditions et normes d'hygiène." *Animal certifié conforme* (2012).

Bourrie, Benjamin C. T., Benjamin P. Willing, and Paul D. Cotter. "The Microbiota and Health Promoting Characteristics of the Fermented Beverage Kefir." *Frontiers in Microbiology* 7 (2016): 647. https://doi.org/10.3389/fmicb.2016.00647.

Didienne, Robert, et al. "Characteristics of Microbial Biofilm on Wooden Vats ('Gerles') in PDO Salers Cheese." *International Journal of Food Microbiology* 156, no. 2 (May 2012). https://doi.org/10.1016/j.ijfoodmicro.2012.03.007.

Farag, Mohamed A., et al. "The Many Faces of Kefir Fermented Dairy Products: Quality Characteristics, Flavour Chemistry, Nutritional Value, Health Benefits, and Safety." *Nutrients* 12, no. 2 (January 2020): 346. https://doi.org/10.3390/nu12020346.

Farmhouse and Artisan Cheese & Dairy Producers European Network. "European Guide for Good Hygiene Practices in the Production of Artisanal Cheese and Dairy Products." December 2016. https://food.ec.europa.eu/system/files/2017-12/biosafety_fh_guidance_artisanal-cheese-and-dairy-products_en.pdf.

Lortal, Sylvie, Giuseppe Licitra, and Florence Valence. "Wooden Tools: Reservoirs of Microbial Biodiversity in Traditional Cheesemaking." *Microbiology Spectrum* 2, no. 1 (January 2014). https://doi.org/10.1128/microbiolspec.cm-0008-2012.

Lortal, Sylvie, et al. "Tina Wooden Vat Biofilm: A Safe and Highly Efficient Lactic Acid Bacteria Delivering System in PDO Ragusano Cheese Making." *International Journal of Food Microbiology* 132, no. 1 (March 2009). https://doi.org/10.1016/j.ijfoodmicro.2009.02.026.

McKenney, Erin A., et al. "Sourdough Starters Exhibit Similar Succession Patterns but Develop Flour-Specific Climax Communities." *PeerJ* 11:e16163 (October 2023). https://doi.org/10.7717/peerj.16163.

APPENDIX G

Sourdough Starter

The philosophy of sourdough deserves its place in these pages for numerous reasons.

Sourdough is a medium that helps the development of an important cheese rind ecology. For *Penicillium roqueforti*, the fungus responsible for blue cheeses' evolution, grows upon sourdough bread. (*Mucor*, aka cat's fur, can be grown for rennet or spores atop non-sourdough bread as explored in appendixes H and I.)

Sourdough in its handling is very similar to dairy cultures such as kefir, and almost identical to clabber. For many readers, sourdough is likely the first fermentation they encountered, and one that can help them understand how dairy fermentation works. I often turn to sourdough fermentation for inspiration in my cheesemaking, having first understood how natural fermentation functions through its care.

Sourdough is also the medium in which natural fermentation is most evolved and universal. Its prevalence lends credence to the idea that natural methods in cheesemaking work. I find great hope that natural methods can be more readily practiced and adopted thanks to the growth that sourdough is experiencing and has been experiencing for decades. It's certainly a gateway fermentation, and one that has helped many gain a taste for creating a deeper connection to how they feed themselves, their families, and their communities.

Comparison of Sourdough and "Sourmilk"

As alluded to in chapter 3, there are significant parallels between the forms of fermentation in sourdough and in dairy that aren't often discussed. These may well be the same microbiological phenomenon.

Both sourdough and clabber/kefir feature two-phase fermentations. In both media this begins with a bacterial breakdown of complex sugars—either starches or lactose—into lactic acid (the origin of the sour in sourdough, as well as the sour in kefir and clabber). The lactic acid is then fermented by a yeasty or fungal fermentation into alcohol, carbon dioxide, and other metabolic by-products.

It may well be that the community of yeasts that cause sourdough bread to rise is the same community of microbes that establish white rinds on cheeses. Both cultures feature abundant lactic acid that supports the fermentation of yeasts; and both feature the growth of white surface yeasts that feast upon the ferment's abundant lactic acid. It's common for sourdough bread bakers to report the growth of white-film yeasts atop a long-neglected sourdough starter left at room temperature—this may be equivalent or even identical to the film yeasts that grow atop long-neglected kefir and clabber cultures.

When a kefir or clabber, with its abundant yeasts, is mixed into flour to form a dough, that dough will rise, like sourdough within 12 hours, and can be baked into a delicious bread. That dough can also be re-fermented like sourdough and become an effective sourdough starter culture more or less immediately, with no lag time. Equivalently, a sourdough starter can be used to begin a clabber culture, by adding a small amount of an active sourdough culture to milk. From what I've come

A well-managed sourdough starter showing good gas development. Unlike with kefir or clabber or most styles of cheese, such yeastiness is desirable. However, the yeast development in sourdough may be the same microbiological development of some cheese rinds.

to understand, they're more or less the same microbiological phenomenon.

Similarly, cheese can be made to go yeasty, rising like a loaf of sourdough, even featuring aromas reminiscent of bread and beer—as is the case in ambarees and aging lactic cheeses. Equivalent fermentations can even cause milk to develop a strong alcoholic character from the metabolisms of lactose to alcohol, as in my recipes for milkbeer and amasi.

This tendency of cheese to go yeasty is not a contamination as many believe. Many swear that sourdough bread and cheese should be kept in separate kitchens. Instead, such a yeastiness is an evolution of the yeast ecologies that are a part of milk's complex community. And the ways in which these yeasts can be encouraged are equivalent in cheese and bread and even wine!

Starting a Sourdough

Sourdough bakers always add a starter to their batches of bread. But how does that starter itself start? A clabber culture is begun by spontaneously fermenting milk, and you can similarly begin a sourdough through spontaneous fermentation—essentially by getting your wheat wet!

A sourdough starter can be started from scratch by getting flour wet and initiating its inherent tendency to ferment. It is best to source a natural flour, preferably whole wheat (or rye) and preferably stone ground, so as not to be submitted to the high temperatures of typical processing that essentially pasteurize the flour.

Add enough water (preferably unchlorinated) to 1 cup flour to make a paste—not too wet, not too dry. Leave this mixture in a jar with a loose-fitting lid or cloth covering (avoid metal, as it will rust over time—clay pots are often preferred by bakers for this reason) at room temperature (preferably 20°C [68°F]) for 2 to 3 days, until it bubbles. Interestingly, the speed and quality of this first fermentation can be improved by adding clabber or kefir in place of water as they share many microbes in common.

Once this first ferment is bubbling, refresh it with fresh flour. Take a small amount of this starter (1 tablespoon) and place it in a fresh jar (discard the remaining first ferment in the compost). Add fresh flour (½ cup) and enough water to make a dry paste; this constitutes the first feeding of the ferment. Now that a seed of starter was added, this second fermentation will proceed much faster, typically becoming bubbly in 12 to 24 hours.

Once this second ferment is bubbling, refresh again with fresh flour: Take a tablespoon of the ferment, add ½ cup of flour, and add enough water to make that paste. This daily rhythm can now be repeated, refreshing the starter in the morning with a tablespoon of the starter. Within 6 to 12 hours the ferment will be bubbly and ready to use for baking its first sourdough loaf.

Keeping a Sourdough

The most important consideration when keeping a sourdough starter is that it must be fed regularly to maintain its effectiveness. This regular, rhythmic feeding now conserves the best community of flour-fermenting microbes and will create as perfect a loaf as possible, with a strong rise in the right amount of time, and no off flavors. Keeping this culture reliably fed will keep it peaking within 6 to 12 hours daily, over and over and over. And this reliably fermenting starter can now be used to bake your daily bread.

The excess starter (made with about ½ cup flour) from each daily feeding will suffice as a starter to bake a medium-sized loaf of bread daily with around 3 cups flour (a 1:6 ratio, as below).

Or the steady starter can be preserved in its active state for up to 1 week in the fridge between feedings. If neglected longer, it will have to be refreshed once or twice before it is bubbly in 6 to 12 hours and ready to bake again.

In warmer weather (30°C [86°F]) a starter will have to be fed twice daily (at least) to keep in its finest form. The ferment will rise in less than 6 hours and will lose its peak fermentation after 12 hours.

Quantity Matters

When baking with sourdough, the second most important consideration is the ratio of starter culture to dough.

For years I failed in my sourdough baking because I added starter to my dough in the same ratio as my cheesemaking—1 part starter to 50 parts dough. Little did I know, but my sourdough style had devolved over the years into my cheesemaking style, and whenever I struck out to master my bread baking again, I'd always fall into this trap, adding too little starter and failing to successfully rise my loaves.

It was only in recent years, more than a decade after I'd started baking sourdough and had been fully distracted by cheesemaking, that I rediscovered why my sourdough had failed to raise my breads for so long: I wasn't adding enough starter. My doughs fermented too lactic and failed to develop almost any yeast fermentation at all.

The reason for this was fully related to the ratio of starter I added. I spoke with a professional baker, who told me that the ratio of starter to dough that he used in nearly all of his bakes was about 1:6, or ½ cup starter to 3 cups flour. It turns out that 1:50 compared with 1:6 makes a world of difference to the ecology of a bread and its balance of bacterial and yeast fermentation.

This comes down to the idea that bacteria grow and ferment much more quickly than yeasts, yeasts being more complex microbes with more substantial nutritional needs, and bacteria being simpler life-forms. Bacteria are able to double in number much more quickly than yeasts. Thus when low amounts of starter are added to a dough, yeasts are slow to grow, but the bacteria thrive and develop and acidify the dough, which then sours, but fails to rise significantly.

However, when a larger proportion of starter is added to a dough, larger populations of both bacteria and yeasts are added, both of which now can effectively ferment the dough in a more balanced manner. The yeasts have less growing to do to ferment that dough, so both the bacteria and yeasts ferment more equally. And such a bread features a pleasing and flavorful sourness (if fermented the right amount of time), the right amount of rise, and a good breakdown of the flour, resulting in greater digestibility.

However, when an even larger enough amount of ripe starter is added, a large amount of lactic acid is added directly to the dough, enabling the yeasts to develop and the bread to rise without any fermentation of the freshly added flour at all!

This occurs when you add starter to dough in a ratio of 1:2—a bread prepared with 3 cups flour, 1½ cups of which were in the prepared starter. A bread baked this way presents almost no sour notes or flavors at all, and the yeasts take off and cause the bread to rise almost immediately as the dough is already supplied with its favorite food, lactic acid.

In essence, a 1:2 dough isn't fermented, and the bread itself rises fully with the activity and ingredients in the starter instead of the starters' fermentation of the added flour in the dough. The paste of such a bread remains un-broken-down, showing the bran flakes distinct from the starch in a whole wheat loaf (similar to the appearance of a whole wheat bread prepared with a sweet starter). In a 1:6 dough, however, the bran breaks down due to a more significant fermentation and acidification of the dough. And in a 1:50 dough, there's a complete bacterial fermentation of the dough before almost any yeast development occurs at all.

Certain types of bread are made with different balances of yeast and bacterial fermentation. The same is true with cheese. Some styles of cheese need to be heavy on the bacterial fermentation and light on the yeast to avoid being overwhelmed by unwanted yeasty effects in cheese. Most styles, however—especially naturally rinded cheeses—need some yeast development in order to develop their best rinds (we add starter culture in a ratio of 1:50 to 1:100 for these). Some very special dairy ferments like milkbeer and Lebanese ambarees need to have even stronger yeasty fermentations in order to develop their most appropriate character (we add starter in a ratio of 1:10 for these).

APPENDIX H

Keeping a Wine / Beer / Cider Starter

I realize that this is probably going overboard for most natural cheese practitioners, but an alcohol starter is yet another culture you can keep alongside your other counter-cultures that can benefit your cheesemaking. There are several good reasons for a natural cheesemaker to keep a wine starter. Firstly, nothing pairs with natural cheese better than natural wine! Secondly, as will be explored in appendix I, such a starter can be used to bake a sweetdough bread that can cultivate Mucor fungus, which is helpful for ripening certain styles of cheese, as well as for growing your own fungal rennet.

Wines, and indeed all alcoholic ferments (ciders, beers, meads, pulque, palm wines, and so on), ferment in their most effective and complex way when a natural starter is added. Such a culture can be easily kept, like clabber or sourdough, by feeding a ferment daily. Referred to in French winemaking circles as pied de cuve, this ferment can be created from the crushed fruit or its juice, and can be refreshed daily with fresh juice during the harvest period to keep fermentation at its peak.

The starter can be begun with a spontaneous fermentation of the must or liquid to be fermented, be that grapes, apples, or the sweet wort of beer. The juice can be freshly prepared by pressing ripe fruit, and should preferably be raw (though a pasteurized juice can be used if inoculated with a small amount of clabber / kefir or even a sourdough culture). Leave the juice in a jar, with a cover or lid to keep out fermentation flies (it's not fruit they're after but fermented fruit and its yeast), and stir daily or twice daily to aerate, which promotes the growth and development of sugar-fermenting yeast. Once the ferment is actively fizzing (typically it takes 2 to 3 days at 20°C [68°F]), the spontaneous ferment can be fed fresh sweet juice again, in a ratio of about 1:50 or 1:100—a spoonful of starter to 1 cup of juice.

After the second feeding, this fermentation progresses much more quickly. Instead of 2 to 3 days, signs of fermentation should be visible within 12 hours. At 24 hours, the ferment will be bubbly and active.

With the ferment bubbling within 12 to 24 hours, the feeding can now be practiced daily (along with a once- or twice-daily stirring), and the culture will be active and bubbling again 12 to 24 hours later. Every day, the practice can be repeated, much like sourdough, much like clabber, keeping the yeast culture in its top form. It can now be used as a starter for a larger batch of booze, in a similar ratio: 1:50 to 1:100.

If you need to prepare a starter for a larger batch, scale up as needed. But I personally keep it small, feeding a small spoonful of my ferment a cup or so of juice daily during the fruit fermentation season. The culture can also be kept in the fridge once it's active, for up to 1 week. It can even be frozen for several months. If you can't feed it daily it's easy to preserve, just like a sourdough or clabber.

Stirring is essential to keep the alcohol ferment in an appropriate zone and prevent a shift to an anaerobic lactic acid fermentation. A regular aeration (twice daily, at least) helps fuel the growth of the most sought-after alcohol fermenters, including *Saccharomyces cerevisiae*. Some might consider constant stirring with the use of laboratory-bench-style stirring devices to achieve this. But an even better option is to ferment your alcohol starter in natural containers like *Lagenaria* gourds (as found in the recipe for amasi), which are porous and contribute a natural aeration that encourages yeasts without any stirring whatsoever. What natural winemaker would abstain from adding such a gorgeous gourd-based ferment into their wine?

If, however, winemakers, cider makers, and beer brewers are looking to make a more sour brew, they can consider adding instead a sour starter, fermented without air. This results in a mix of lactic-acid-fermenting bacteria that ferment below the aerobic surface, and surface-fermenting yeasts like *Brettanomyces* (which looks a heck of a lot like *Geo*) that consume the lactic acid in the brew. Such a sour starter can be prepared in an equivalent way to a wine starter, but without adding in air to encourage yeasts. Instead, bacteria will ferment the sugars below the surface where there's no air, and yeasts will grow only at the surface, where there's plenty, and will begin to form a visible veil, or flor, often within 24 hours (definitely within 2 days at 20°C [68°F] without aeration) when undisturbed. This sour wine / cider / beer starter can then be used to prepare a lactic / *Brettanomyces* wine, cider, or beer, which should be handled in a similar way, without stirring or mixing air into the vat.

Though the yeast culture thrives on sugar, it's best not to feed your wine ferment simple sugar-water. Sugar (or even honey) does not provide the nutrients (mainly the proteins) that help sustain the growth and development of the yeast in the fermentation. The ferment will inevitably sour into a bacterial one if there aren't enough nutrients to support the yeast in the fermenting liquid. For wine, I cultivate my starter on freshly pressed grape juice (or pasteurized juice); for cider, I ferment daily on freshly pressed (or pasteurized) apple juice. For my natural beer starter, I daily feed my culture milled barley malt covered in water (0.25 kg to 1 L); 24 hours later, I strain off the liquid and re-ferment 1:50, and / or pitch directly in the wort and get brewing (this technique creates a cold-mash which extracts enough sugars from the malt to support a fermentation).

This wine / beer / cider culture can even be used to raise a loaf of bread without the typical beneficial sourdough fermentation. Such a sweetdough bread, lacking lactic acid, is equivalent to a bread baked with commercial yeast and added sugar. It can be used to enable the cultivation of a certain species of beneficial cheese-ripening fungus, *Mucor miehei*, that can be a source of spores for the development of certain cheeses, like Saint-Nectaire, in which *Mucor* plays an important role. *Mucor*-fermented sweetdough bread can also be used as a source of homegrown microbial enzymes to use in place of animal rennet! The technique for both of these odd applications of fermentations is provided in appendix I.

Homemade commercial-style yeast for bread baking (and for beer brewing) can be derived from an alcoholic ferment in a related way. To do so, leave a wine or beer to ferment to its fullest extent, stirring daily to encourage yeast growth. Once the bubbling subsides, leave the ferment for another two days, undisturbed (this works best for a wine if it is first decanted off of its skins); the yeasts will settle to the bottom, leaving a brownish layer of sediment. Strain off the clarified ferment at the top into its aging vessels—the dregs at the bottom of the barrel are the live yeast, which now lay dormant, having exhausted their sugars. This live yeast can be used fresh for baking bread, or spread and dried in a dehydrator at a low setting for later use. When rehydrated and fed sugar, these yeasts will revive and raise a loaf of bread, though they leave the flour itself unfermented. Commercial bread yeast is raised today on sugar, with ammonia and other nutrients added to support its growth.

APPENDIX I

Cultivating *Mucor* for Rinds and Rennet

Mucor miehei is an unlikely fungus to grow at home, not being considered a desirable species today in cheese, nor really in any other realm of food fermentation. Generally recognized as a spoilage microorganism (even supposedly a causative agent of the human illness mucormycosis in immunocompromised individuals), it nonetheless has some fascinating applications in the realm of traditional and modern cheesemaking (as well as in the making of some traditional Eastern fermentations like tempeh).

This remarkable mold is the most like a mushroom. It's one of the few species of mold with visible sporangia (spore-producing bodies), which grow as tiny black spheres on the tips of long gray stalks. Its family is most easy to identify because of this peculiar appearance, one you've likely seen stalking leftovers in your fridge, or on old bread in a plastic bag.

Surprisingly, several traditional French cheeses are made with the help of *Mucor*. Though considered a fault in most styles for its fluffiness (its common moniker in cheese circles in France is *poil de chat*—cat fur!), *Mucor* helps cheeses like Saint-Nectaire establish their sought-after fluffy gray rinds and delicate sweet flavor.

Mucor's growth on a cheese is effectively kept in check through a good-quality fermentation and does not develop on the rind of a well-executed, naturally made cheese. A cheesemaker must intentionally make their *Mucor*-ripened cheese wrong in nearly every respect (as described in chapter 10), by restricting fermentation through using little or no starter culture; by salting the cheese before much fermentation has occurred; and by drying out its rind, as in the technique for making Saint-Nectaire. But just as importantly, it helps to add spores of the right fungus to get the *Mucor* established on a cheese before other unwanted fungi do. That's where growing your own *Mucor* spores comes in handy.

Mucor can even be grown as a rennet replacement. As strange as it may seem, perfectly acceptable cheese can be curdled from milk entirely with bread left to go moldy with *Mucor*. Certainly, I shocked myself the first time I managed to make mozzarella with moldy bread, but the technique has significant precedent in the cheese industry—nearly all certified organic cheese in North America is produced with rennet derived from *Mucor* species that naturally produce protein-digesting enzymes, because of restrictions on the use of genetically modified rennets, as well as avoidance of animal rennet due to the perception that organically minded consumers would be turned off by the ingredient.

So how does this fungus grow? Interestingly, the technique is somewhat related to the method for producing *Penicillium roqueforti* on sourdough bread. But instead of growing this culture on a freshly baked sourdough loaf, the *Mucor* is grown on a specially prepared slice of sweetdough bread that has effectively no lactic acid development. This sort of fermentation is referred to in microbiology as a solid-state fermentation (as opposed to a liquid-state fermentation, as in fermented milk). While *P. roqueforti* grows well in a lactic-acid-rich solid-state medium, *Mucor* prefers its solid state medium to be free from lactic acid entirely.

This can be done with a sort of spontaneous fermentation of a slice (or even a whole loaf) of sweetdough bread, a bread made very differently from sourdough, in which the loaf rises with a simple yeast fermentation of added sugar, not a more complex bacterial / fungal fermentation of the dough itself. The resulting loaf is the perfect breeding ground for this odd-looking but nonetheless friendly fungus.

Sweetdough Bread

A sweet, unfermented bread is the medium of fermentation for *Mucor*, which is a fungus that feasts primarily on proteins, unlike *Penicillium roqueforti*, which prefers lactic acid. To prepare the medium for the growth of this fungus, you must bake a bread without lactic acid production, which would favor the growth of blue fungus instead.

So, how to bake a natural bread without sourdough? With the help of a wine / beer / cider starter, a technique that's been used in bread baking for generations. It is perhaps not as old as sourdough but certainly predates the development of commercial yeast—which, of course, can also be used to bake the bread.

The idea behind this bread is to use added sugar as a medium for yeast fermentation. To do this, a wine / cider / beer starter can be used to start the bread's fermentation, and sugar must also be added to the loaf to give the sugar-fermenting yeast something appropriate to feed on for its fermentation. A bread prepared with dry or fresh yeast and sugar will ferment in an equivalent way and promote the growth of *Mucor*.

I add 250 ml (1 cup) of active wine starter for a medium-sized loaf of bread, as well as some sugar (15 ml or so) to give the yeast in the starter something to feed upon. When well fed, the yeast will produce the gas needed to raise the loaf, without a sourdough fermentation. Add to the starter 3 cups of flour and enough warm water to make a moist dough (no salt—it slows the *Mucor*). Leave the dough to rise 6 to 12 hours, until puffy, then bake in a hot oven. Once it's browned and hollow sounding, remove it from the oven and let it cool for several hours before preparing for the *Mucor* fermentation.

I have to say, I wouldn't normally bake bread this way. I find these loaves to be utterly without attraction, lacking various aspects—aroma, flavor, a mild sourness, digestibility—that make us want to eat them. But to get *Mucor* to grow, this is precisely what you'll need. A proper fermentation of the bread will fail the fermentation of the sought-after fungus.

Growing *Mucor*

Once it's baked and cooled, the sweetdough bread can then be used to grow *Mucor*. The fungal fermentation can be done spontaneously or—as is always better—with a bit of backslop from a previous batch if available. An eventual source of backslop can be obtained first from a spontaneous fermentation.

To begin, place a slice or whole loaf of sweetdough bread into a humid environment, like a hastening space for lactic cheeses (a wooden box or large plastic container, with a lid, lined with wood or cloth to absorb excess moisture from the fermentation). If backslop is available, add a slight smear of *Mucor* and its spores from a previous batch. Leave the bread in the container, flipping it daily to ensure good aeration and even fungal growth, and wipe down excess moisture from the sides of the container. Once the bread is fully covered with fungus, typically after 10 days at 20°C (68°F), dry it out to serve as a source of spores for ripening cheese (rub your finger on the *Mucor* bread, then introduce the spores to the milk at the beginning of the cheese make).

Wild *Mucor* spores, which seem to abound, are quick to germinate and will be the first to grow on the bread, claiming it for themselves. The first fermentation may show signs of some other wild species, but when re-fermented with a bit of backslop from a previous batch, the *Mucor* will not cede a centimeter to any other species of fungi. The *Mucor* wins the fermentation in a loaf that's lacking lactic acid.

This solid-state fermentation in a sweet, unfermented medium is ultimately what best supports the growth of *Mucor*. And re-fermenting with a cyclical feeding of fresh bread will eventually develop a pure culture of this fascinating fungus. *Mucor* can also be grown on rice, wheat, or barley, as in the production of koji, by leaving the cooked grains to spontaneously ferment in a humid environment for one week at 20°C and backslopping several times until the fluffy fungus dominates.

After one week, the slice of the *Mucor* bread can then be dried out, to preserve it; once dried, it can be kept for years in a closed container. When seeking spores for making a Saint-Nectaire, simply apply a fingerprint's worth of spores to a batch in the milk or on the rinds of the draining cheeses to inoculate the cheeses with the fungus.

Mucor Rennet

Mucor has another, even more intriguing, application in a nontraditional cheesemaking (though there's plenty of lore about traditional cheesemakers using bread to curdle their milk). This fungus can cause milk to curdle much like animal rennet, and can be used in its place for making any of the rennet cheeses in this book. And though these enzymes do not function precisely the same as the enzymes in ruminant rennet, they can provide a steady supply of a coagulating agent without the need to sacrifice a young animal or buy from multinational biotech companies.

Mucor fungus feeds on proteins. Its preferred medium seems to be fresh flesh (you'll often find it on a carcass after several days in the right circumstances), which it feeds on with the help of the proteolytic enzymes that they produce. These enzymes are not as specific in their activity on milk's main protein, casein, breaking it cleanly and efficiently like chymosin, but rather are more generally proteolytic. In breaking down the protein, these microbial enzymes cause the milk to curdle much like rennet, but the less specific enzymes cause further breakdown in the milk's protein structure that can possibly make bitter-tasting by-products including certain peptides (short chains of amino acids), though that has not been my experience. So it may not be a perfect substitute for animal rennet, but it will still do the job of curdling milk and making cheese.

Using *Mucor* Rennet

To use *Mucor* rennet, leave the *Mucor*-covered bread to steep in water overnight. In the morning, the steeped bread will have nearly dissolved into the liquid, which you can strain of its solids through fine cloth. Let the liquid sit for 1 more day, to clarify. Some solids (spores, crumbs) will rise to the top, while some will settle; the clear, enzyme-rich liquid in the middle (sans spores) is what you can use as a coagulant.

You can add the liquid to milk as if it were rennet, in a ratio that you calculate, much as with traditional rennet, by testing the strength of the preparation on smaller quantities of milk, determining the correct proportion to get the milk to curdle fully in about 30 to 45 minutes at 35°C (95°F), then applying the appropriate amount of the liquid to a larger vat of milk. Salted, you can preserve the liquid for up to several months in the refrigerator.

In my experience, one medium-sized slice of *Mucor* bread can curdle up to 20 L milk, so growing rennet on a small scale with this technique is more than doable—one loaf of *Mucor*-fermented bread would provide more or less enough enzymes for a commercial producer's daily make.

Now, is it safe to use this microbial preparation for coagulating your cheese? *Mucor* rennet should be as safe as or safer than animal rennet owing to the lack of animal microbes, though those are essentially controlled in traditional animal rennet preparation through salting and fermentation. *Mucor* fungus is not dangerous, and there is no link between the consumption of foods containing *Mucor* and the medical condition of mucormycosis. (This illness affects those with severe diabetes or others with weakened immune systems, and has no link to *Mucor* consumption in traditional French cheeses like Saint-Nectaire, which are understood to be perfectly safe.)

And though there is a live microbial character from the *Mucor* fungus in the rennet, this microbe doesn't get a chance to grow in the cheese so long as the cheese is made with an active starter culture that eliminates any opportunity for the unwanted microbe to grow. You don't have to worry about the *Mucor* fungus growing on cheeses coagulated with a *Mucor* rennet, for so long as you add an active starter like kefir or clabber to the milk and make the cheese right, the conditions created by the fermentation, including the development of lactic acid and the growth of *Geotrichum* and other rind-ripening ecologies, will effectively stop the *Mucor*'s development.

APPENDIX J

Flaws in Cheese

Nearly all flaws in cheese result from missteps on the cheesemaker's side. If a cheese doesn't turn out the way a cheesemaker expects, it's not because of some unexplained phenomenon or failure of natural methods. It might be *my* fault for writing a faulty recipe, but I hope that that isn't the case!

In most circumstances faults in a cheese can be traced to a problem with one or more of the four pillars of cheesemaking practice:

   I—fresh milk
   II—rennet
   III—good fermentation
   IV—a proper make (and ripening)

If a cheesemaker neglects but one of these, there will be opportunities for unwanted microorganisms to grow or a cheese to show a flaw. But if all of these pillars are considered, and each gets the respect it deserves, then a cheese will always develop as it should. Natural cheesemaking methods should be understood as perfect. There's always a reason why a cheese doesn't turn out the way it was meant to, but often it can be a combination of issues that come together to create the flaw, so diagnosing the defect can be challenging.

Cheeses simply do not fail on their own, for unwanted microorganisms don't just grow whenever or wherever they want to. It should be understood that there is no such thing as contamination in a cheese—again, contamination is a state of mind. If a cheese grows microorganisms that don't belong, it is because those unwanted microorganisms seized upon an opportunity provided by the cheesemaker in their practice. And the source of the problem will always return to one or more of these four pillars.

Cheesemakers can expect their cheeses to evolve essentially the same way every time, save for slight seasonal variations in the fat-to-protein ratios in their milk. Nothing ever goes wrong in my personal cheesemaking practice if I'm paying attention; no microbe ever grows out of place; no milk fails to curdle; no mozzarella fails to stretch. Just about all problems in cheese can be boiled down to human error.

Now, without a microbiologist it can be impossible to know for certain what unwanted microbe is growing on or in a cheese, but we shouldn't need a microbiologist to help us solve the problem. For good natural cheesemaking practice seems to keep all unwanted microbes in place: Unwanted fungi or gas-producing microorganisms will not grow on a naturally made Crottin, for example, so long as all aspects of the cheese make are properly conducted together. But if, say, the milk quality is neglected, then the cheese may bloat with gas, which can be a sign of coliforms proliferating in poor-quality milk. A bloated cheese can also be the consequence of a poor-quality, yeasty fermentation that develops carbon dioxide gas. Often it's both, for poor-quality fermentation permits the growth of unwanted microbes from milk. But a proper make will eliminate the development of both of those microorganisms in the cheese.

Below are some of the most common problems for cheesemakers and how to best address them. As you'll see

Kefir or clabber cultures can have gas in their curd, usually resulting from yeast overgrowth, if not correctly managed. However, gas development can also be a sign of *E. coli*, and often, it's both coliforms and yeasts. It is undesirable to see gas developing in your cultures or your cheeses, though there are some exceptions.

the issue is typically a question of milk quality, fermentation quality, rennet, or technique, not "contamination."

Clabber or kefir is full of gas

A small amount of gas development is acceptable in kefir or clabber to be used as a starter, but if so much gas is produced that the curd rises up and separates a significant amount of whey, then something is off in the culture's fermentation quality. This gas production will cause cheeses to bloat and develop unwanted flavors and textures.

Gas development can happen in cultures for numerous fermentation reasons that cause yeast to grow in excess and that can be righted by the cheesemaker: failing to adapt the culture to warm weather, adding too much starter, irregular feeding, or the mixing in of air. If the milk is not fresh, it can be a milk-quality issue resulting from the growth of coliforms, which are gas-developing microbes.

A gassy culture is typically overfermented, the gas being a sign of yeast development. The problem exacerbates if it is not remedied, as the unwanted fermentations snowball and yeast cultures overgrow.

Warm weather can cause this to happen. Keeping the usual daily feeding regimen of 20°C (68°F) when the weather gets above 30°C (86°F) can cause a culture to overferment in 24 hours, resulting in gas. Feeding a culture twice daily in warm weather, or putting the culture in the fridge once it has curdled, can remedy this issue.

Feeding with too much starter can also cause a culture to ferment faster, and shift into a yeasty phase sooner. I recommend keeping it at 1:50. With kefir, there is a tendency to add more kefir grains (especially so as they grow), but be sure to use only a small, consistent quantity, and preserve, give away, or compost excess grains. Too many grains leads to a faster fermentation, which results in faster yeast development and more gas, getting worse every day as the kefir grains grow.

Using an older starter with a lot of *Geotrichum* growing on its surface can cause the next round of fermentation to develop gas. (Note that the older starter may not show gas in its curd if it itself was prepared with an active culture.) Getting back into a daily (or twice-daily) feeding regimen will get a yeasty culture back on track. There is no need to discard a clabber or kefir if it's yeasty and start again—you'll only set yourself and the fermentation back. And don't ever wash kefir grains, for this same reason.

Alternatively, the kefir or clabber could have been shaken during fermentation. Always leave fermenting milk undisturbed, unless you're looking to develop gas, in which case, shake, shake, shake!

Gas development in a starter or a cheese can also result from the use of fairly old raw milk in which coliforms have grown. Milk stored under refrigeration is prone to the growth of unwanted microbes; even 2 days of cold temperatures can cause gas to develop. It's hard to know if the gas is from your culture or your milk, but if your milk is fresh (cow-warm), then it's a generally a fermentation issue and not a milk one. Typically, coliforms are accompanied by very odd aromas and flavors, and one whiff will tell you that something's off. Good fermentation and milk handling practices do the very best job of keeping them in check!

Milk fails to curdle with rennet

The likely cause is that too little rennet was used in the make to get the curd to set in a given time. This could be because of a simple miscalculation with the rennet dose, or because the rennet was older or kept in the wrong conditions. Next time perform a rennet test on a small amount of milk (250 ml or so) to see what it takes to curdle the small amount fully in around 30 minutes at 35°C (95°F), and then scale up from there. Unfortunately, if rennet has already been added, adding a second dose of rennet and stirring the pot disturbs the forming curd, and typically causes the curd to have a flawed formation.

It could also be that the temperature of the milk is too high or, conversely, too low, causing the curd to coagulate slowly. Once the milk has cooled, though, it's challenging to warm it; you'll have to stir it, and the curd will be damaged by the stirring once it begins to form.

It could also be that you simply forgot to add the rennet. I've never witnessed a good milk that has failed to curdle for an unknown reason! Just like milk's fermentation, milk's coagulation is a given.

Milk forms a very soft rennet curd

Again, this is likely due to a mistake in the rennet addition. If too low a rennet dose is added, the curd will develop slowly and take its time firming up. If there is cream that has risen to the top of the curd, that leaves a mark on your hand, this is a sure sign you have added too little rennet.

The soft curd could also result from using over-processed milk, like UHT milk, that fails to firmly curdle with rennet. Many brands of commercial milk are UHT, especially nationally distributed brands of certified organic milk in the US, and you may simply be unaware that is what you've purchased. Often store-bought milk is pasteurized to a higher temperature than cheesemakers typically pasteurize, to 72°C (162°F) for 10 seconds instead of 62°C (144°F) for 30 seconds, and this higher-temperature processing denatures milk's casein proteins and damages the curdling effect.

Cheese bloats in its first day or two

Any style of cheese can inflate in its early days for the same reasons that a kefir or clabber culture can be full of gas. It can be because of excessive yeast growth due to the use of an overfermented culture full of yeasts (either at the surface as a white *Geotrichum* veil or within the curd forming bubbles), and/or due to the growth of coliforms that result from a slow fermentation with an overfermented culture, or from using older milk.

If very bad smells are present as well, this is likely from coliform growth due to a combination of poor-quality (long-refrigerated) milk and poor-quality fermentation that together allows these unwanted microbes to thrive.

Alpine cheese bloats after several days

This is likely a sign of undersalting, causing fermentation to continue within the cheese, and propionic bacteria producing gas, as in Emmentaler cheese. Or perhaps the cheese was made too tall, which slows the salt penetration to the core. The cheese may also be cooked to too low a temperature, resulting in more moisture in the curd, and unwanted fermentations. This effect can happen in caciocavallo if the curds aren't salted sufficiently before stretching.

Clabber, yogurt, or kefir curdles without rennet but is not sour—it sets at a high pH

This problem can occur when a culture is long neglected, or when you're adapting a culture from one medium to another (such as when you're transitioning a mesophilic starter to a thermophilic starter), when pasteurized milk is fermented spontaneously, when milk is sourced from vessels washed with water but not sterilized, when milk is long refrigerated, or when you use a sourdough starter to begin a clabber culture. This scenario likely occurs because of the growth of milk-curdling microbes, like *Mucor*, which produce proteolytic enzymes that can cause milk to form a gel without acidification. (I sometimes call this "premature coagulation"!)

This is why it's important to taste your ferment to know that it's sour before you feed it or use it as a starter. If curdling of this type does happen, just be patient with the fermentation, wait for the curd to sour, then back-slop again, and you should find that the phenomenon resolves itself. Good fermentation practices eliminate the growth of unwanted microbes like *Mucor*.

Clabber or kefir fails to curdle in the usual time

Clabber or kefir should curdle within 12 hours when fed (at a ratio of 1:50) daily and kept at 20°C (68°F). If the culture fails to curdle within that standard time frame, this can be for several reasons.

Slight and unexpected variations in temperature can cause the fermentation to slow. If the culture is fed cold milk, this can slow the fermentation significantly.

The culture may be being fed irregularly. If you normally feed the culture in the morning, then instead feed it in the evening, it can get out of step and be slow to ferment. Combined with other factors, like feeding cold milk, or slightly cooler weather in the culturing space, and the culture may not be ready within the usual 24 hours.

A culture can also fail to curdle when you're transitioning from one species' milk to another. For some reason I cannot quite comprehend, a cow culture will ferment goat's milk (or vice versa) and cause it to sour, if a bit slowly, but the milk fails to curdle when it acidifies! This is likely due to the different sugar profiles in different species' milks, and their tendency to support different communities of microbes. Leave the culture a little longer at ambient temperatures, and it will eventually curdle (typically after 48 hours). That culture will then be adapted to the new milk, and will re-ferment normally from then on.

It's always a good idea to keep a backup culture in the freezer in case this scenario arrives and you do not have a starter to begin a batch of cheese. (I might recommend keeping a backup of frozen fermented whey for a rainy day!)

Fluorescent colors on cheese rinds

The presence of bright fluorescent colors—typically yellow or green—on a cheese rind is typically indicative of *Pseudomonas* development, and usually accompanied by off tastes. This unwanted microbe grows when milk, either pasteurized or raw, is left refrigerated too long before the make. Not much you can do about it once established. But you'll probably find that using the freshest possible milk will solve the problem the next time 'round.

Occasionally I find a similar yellow color on cheeses made with fresh milk, but aged in cold conditions, like a refrigerator at 4°C (39°F). It may be that *Pseudomonas* can establish itself on a cheese kept under too-cold conditions as well—conditions that discourage the growth of cheese's best microorganisms. Ideally, aging cheeses should be aged at a temperature of around 8 to 12°C (46–54°F).

Mucor "cat's fur" grows on a white-rinded cheese

Mucor grows when fermentation fails in a cheese, or when a cheese lacks a protective rind ecology, due to improper cheese handling after the make.

Good-quality fermentation will eliminate this typically unwanted microbe from cheese rinds, as will leaving the cheese to ferment for the right amount of time to develop its goal acidity at salting (typically a pH of 5.3 or less). An active meso starter culture full of *Geotrichum*, as well as an appropriate amount of fermentation, effectively control cat's fur.

Also, if cheeses are kept humid after the make, the conditions for proper rind development enable the appropriate microbes to get established. But if a cheese dries after the make and is then put into a humid cave, the cheese will lack a protective ecology, and *Mucor* will establish. I never find *Mucor* on my rinds with good natural cheesemaking practices, unless I want it to grow there and create the right conditions for its development.

Orange rinds fail to develop

Most likely, the washed-rind ecology failed to establish because washing didn't happen often enough (daily or every-other-day washing is essential for about 1 month to establish the culture); or the cave environment was too dry, and the culture failed to have the continuously humid environment it needed to develop.

Alternatively, the cheese may have been oversalted, slowing the development of the rind ecology. Or the brine you're using to wash the cheeses could also be too salty. Taste it to see; it shouldn't be so salty that your taste buds object.

Camemberts grow blue fungus

This can occur when the cheese fails to develop a *Penicillium candidum* rind in the first place, which ultimately will keep *Penicillium roqueforti* from developing.

Perhaps the cheese wasn't washed often enough, or long enough, or soon enough upon entering the cave; just a 1- or 2-day delay is enough to establish *P. roqueforti*. And once it's established, only washing can set it back; after the washing ends, a cheese rind will keep growing blue.

Perhaps the starter culture wasn't properly fed and lacked the appropriate microbes. If you're making a *Geotrichum*-ripened Camembert, perhaps the cheese wasn't left in the hastening space long enough to develop its protective *Geo* rinds.

Blue veins fail to grow

This can happen because the cheese was insufficiently pierced, or because it was treated in a way that sealed the holes, preventing the passage of air. (Keep cheeses sitting on their unpierced sides, and never wash their rinds or brush them.)

The cheese also may not have had much in the way of mechanical holes—there must be spaces within the curd for the *Penicillium roqueforti* to grow upon.

The cheese may not have been ripened long enough, or perhaps it was oversalted, slowing down the blue development.

As well, there must be sufficient lactic acid within the cheese to get the blue to grow, so be sure to leave the cheeses to ferment for several days before salting.

Rennet cheese develops *Geotrichum* and breaks down too quickly

This can be because of overfermentation—too late a salting, or insufficient salting, results in a more acidified, lactic-cheese-like curd, which deliquesces more from the exterior than a less acidic rennet curd. Or it can be due to an insufficient salting, which allows fermentation and curd acidification to continue on, resulting in the same problem. Warmer aging temperatures can also cause cheeses to overacidify and develop wrinkly *Geo*.

Aged lactic cheese fails to develop a *Geotrichum* rind

The cheese may have been left in an environment that is too dry, and the *Geo* fails to grow, resulting in the development of other rind microbes. Or the cheese may have been excessively salted.

Alternatively, the cheese might not have been left in its warm and humid hastening space long enough to develop *Geotrichum*, which can result in the growth of *Penicillium roqueforti* or even *P. candidum*. Or the cheese may have been salted to soon.

Patience pays off with this style. It's important to wait at every step in the process for the *Geotrichum* to grow—on the culture; on the whey; on the cheese in its form—before proceeding to the next step. For example, the salt may have been added too soon, before the *G. candidum* had a chance to establish itself on the cheese, which stops it from growing altogether (be sure to wait until you see *Geo* on the cheese before you salt it, which also might take longer in cooler weather). Or the starter culture was not regularly enough fed and lacked *Geo*, allowing other fungi to establish on the rind.

White- or blue-rinded cheese develops an orange rind

White-rinded cheeses can develop washed-rind ecologies inadvertently if they are kept in aging environments that are excessively humid. If a cheese is left in a suffocating plastic container or plastic bag without a natural

This tomme shows signs of excessive acidification in a cheese: the growth of velvety *Geotrichum candidum*, a wrinkled rind, and, inside, a chalky texture, with proteolysis toward the rind. The cheese, though flawed, was still delicious.

material to absorb moisture, or if excessive moisture is not regularly wiped out, relative humidities can reach 100 percent. When there is so much free moisture in the aging environment, the cheese can become consistently wet at the rind, developing conditions conducive to the growth of *Brevibacterium linens* and friends.

Mozzarella fails to stretch

When mozzarella fails to stretch it is typically because of fermentation or milk-quality issues. Only relatively unprocessed cow's milk will stretch properly in the traditional mozzarella make (in the fast mozzarella method I offered in my previous book, which I would love to redact, more processed milk will work). Only raw milk, or low-temperature-pasteurized milk (unhomogenized) will work for the mozzarella recipe. HTST pasteurized milk, even if it's unhomogenized, or UHT milk will fail for this traditional technique. If you are unsure of the pasteurization temperature used, call up the dairy you got the milk from and hopefully they'll divulge.

Only cow and buffalo milk will develop a perfect, infinite stretch at the right acidity and temperature. Do not expect goat's milk to behave like cow's milk—just like you should not expect goats to behave like cows! Sheep milk will respond slightly better than goat's milk to the stretching that the pasta filata method invokes, but still, not as well as cow or buffalo. With these other milks, you have to change your mozzarella expectations.

It could also be a timing issue. If all else has gone according to plan, you may have simply missed the short window in which mozzarella will stretch. Perform the teabag test regularly (every 30 minutes once the expected stretching time is near) to be sure that you don't miss the stretching window, which might be open for only 1 hour when the curd is warm and fermenting fast. A slow fermentation resulting from an old starter (or forgetting to add one) can cause it to take longer to reach the stretching window. Good cheesemaking practice with good fermentation practice will result in consistent mozzarella production every time. And if your mozzarella isn't stretching and it's past your bedtime, don't stay up late! Chill the mozzarella in the refrigerator, and revisit it in the morning, when it may be ready to stretch.

Mozzarella goes slimy in its brine

Mozzarella can sometimes go slimy at its rind and firm at its core if it continues to ferment in its brine. It's important to strip out as much lactic acid as possible from the cheese and halt further fermentation by using plenty of water in the make to prevent this from happening. Be sure to use plenty of very hot (but not fully boiling) water to melt the mozzarella. Leave the mozzarella in a water bath, changing the water once or twice, to remove any residual lactic acid.

Also, avoid overfermenting the mozzarella before stretching. Stretch it as soon as it achieves its infinite stretch; too much acidity in the cheese can yield a cheese that more easily and quickly becomes too fermented in its brine. If a mozzarella fails to melt properly on a pizza after it's made, this can be the same problem—too much fermentation after the make.

Fermented milk is viscous or ropy

This condition can arise when your milk ferment isn't fed regularly enough. This effect is caused by the growth of certain lactofermenting bacteria that ferment milk's lactose into long-chained sugars, giving the fermented milk a goopy character. Some traditional fermented milks like villi are intentionally prepared this way, but under proper fermentation conditions, milk will not develop such a quality, and should not be overly stringy. The condition can be damaging to cheese production; cheese made with a stringy starter will develop stringiness as well, which can slow the drainage of whey and result in cheeses with too high moisture content. Be sure to feed your cultures daily (or keep them active in the fridge) to avoid this circumstance.

Curds adhere excessively while stirring an alpine cheese

Small-cut and high-temperature-cooked alpine curds can knit together prematurely while stirring. This can happen due to overacidification (allowing too much fermentation to occur before the curd is cut and stirred) or when older milk is used that is slightly acidified. Using the freshest milk and following a fairly rapid cheese make can eliminate the problem. It can also be due to too slow a stirring—you need to pick up the pace of stirring as the curds firm.

Alpine cheese is full of cracks

Alpine cheeses should come together in a monolithic manner, with all the small curds adhering into a perfectly singular cheese without cracks or mechanical holes. Cracks or holes can allow fungi and mites to penetrate the cheese, causing the cheeses to be eaten away from the inside. This is a significant problem, especially if the cheeses are large full-alpine wheels that must be aged over a year to achieve their most sought-after qualities.

Insufficient cooking is generally the reason why an alpine cheese will show cracks. It could also be that the cheese has cooled during its production, resulting in the curd failing to fully knit. Ideally, form the cheeses into wheels beneath the whey to keep them as warm and malleable as possible, as is the traditional practice in alpine cheesemaking regions. Keep the cheeses warm during their forming and pressing by pouring warm whey overtop them as you put them into their forms.

Alternatively, it could be that the curd has acidified too much during the make, resulting from too slow a practice. These styles of cheese should be made quickly, before the milk (or its curd) has had a chance to ferment.

Curds float during the cheese make

Rennet curds can float during their make, when production of gas causes them to become buoyant. This has happened to me several times in my cheesemaking that I can recall, and I believe I have narrowed down the problem to a perfect storm of conditions.

This phenomenon can occur when too much time elapses between the adding of the culture and the adding of the rennet, and when an older, yeasty culture is added. In this case, excessive lactic acid develops within the curd and combines with the aeration resulting from stirring in the rennet. When this is done, the curd will form with active fermenting yeasts within its structure, causing gas development and rising curds. There is often a yeasty flavor in the resulting cheese. This condition can also result from adding too much starter culture to your milk—say, 1:10—or from using older milk that has already had a chance to ferment.

Fermented milk or cheese tastes strongly "cheesy"

This common problem can result from fermenting older milk. Unpleasant "cheesy" flavors can develop in a cheese due to oxidation of milk fats during milk's refrigeration. With cow's milk the consequence is a strong cheesy flavor (industrial cheeses have an unpleasant "cheesy" flavor for the same reason, that's why we use the term cheesy to describe this unpleasantness—well-made cheese shouldn't taste "cheesy"!). With goat's milk, strong goaty flavors will develop with older milk. And strangely, with sheep's milk, the flavor that presents with the same issue is a strong fishiness.

Cheese tastes awful

Typically this is the consequence of using very long-refrigerated milk; and the longer the milk has been refrigerated, the worse the problem. This effect results from the milk fat's rancidification under refrigeration, and because of the development of cold-tolerant spoilage microorganisms.

Certain styles of cheese, like aged goat's milk cheeses or stinky washed-rind cheeses, tend to develop the most unappealing flavors when prepared with older milk. Do your best to source and work with milk as fresh as possible to avoid this unfortunate flaw, so you don't have to throw away your cheese or force it down your throat!

Aged cheeses burn on the tongue

This can result from using relatively old (24 hours or more of refrigeration) raw milk when you're making hard long-aged cheeses, or other long-aged styles like blue cheese. Rancidification during the milk's cold storage, combined with aging in natural, aerated conditions (this effect doesn't occur as significantly in cheeses like cheddar aged in a vacuum), causes this problem, which isn't as easily recognized in young cheeses or fermented milks made with older milk.

The effect is often described as a histamine reaction—an unpleasant scratchy or tingling sensation on the tongue or throat that lingers after eating the cheese. In other words, it is an allergic reaction resulting from dangerous products created by milk abuse, a problem that can be completely avoided through using only the freshest milk. This problem likely presents some significant health issues for those eating such cheese, and should be understood as a great concern to public health.

This is a common flaw in long-aged commercial alpine cheeses prepared with older milk, both pasteurized or raw, that occurs even when older milk is mixed with fresh milk. Rare is the alpine cheese made in North America that doesn't have this effect, as very few cheeses there are made with exclusively animal-warm milk; they lack the healthfulness and good flavor they should inherently have. This is one of the more significant reasons to ban the bulk tank.

With proper techniques, animal-warm milk, well-tended starters, and good rennetting practices, cheesemakers can avoid all of these issues and more. Ultimately, it is you, the cheesemaker, that creates the circumstances that makes your cheese evolve into an extraordinary food. You can be in complete control of the process, and successfully transform your milk into countless delicious forms.

INDEX

Note: Page numbers in *italics* refer to photographs and figures. Page numbers followed by *t* refer to tables.

A

abomasum (fourth stomach), *84*, *94*
    age at slaughter and, 82–83
    for goat cheese, 82
    rennet preparation from, 77, 79, 85–88, *87*, *88*
abondance cheeses, 213
acetic acid bacteria, 391
acid-coagulated cheeses, 90. *See also* fermented cheeses
acidification, continued, 196
acidity monitoring, 115–18, *117*, 412
active vs. inactive cultures, 68
aeration. *See also* stirring curds
    of alcohol starters, 421, 422
    of ambarees, 354
    in cheese caves, 154–55, 156, 157, 158, 160
    of fermented milks, 206, 350, 352
    of milk vinegar, 391
    of natural wine, 349
affinage treatments, 163–176. *See also* ripening ecologies
    aging in pomace, 175
    brine aging, 167–68
    charcoal, 173–74, *174*
    cloth binding, 167
    dry brushing, 168–69
    flipping, 163, 165
    for Gouda cheeses, 300
    for grana cheeses, 296
    hastening, 170, *171*
    home scale, 159–160
    leaf wrapping, 174–75
    natural wine, 349
    oil or fat application, 169–170
    overview, 8
    patting down, 169
    piercing, 172–73, 188
    rind washing, *164*, 165–67
    rubbing up, 170, 172
    waxing, 175–76
aging. *See* affinage treatments; cheese caves; ripening ecologies
agricultural foundation of cheese, 1, 16–18, *17*

air exposure. *See* aeration
air quality, 154–55
alcoholic beverages
    aging in pomace from, 175
    amasi, *xiv*, 206, 350–51
    blaand, 120, 352
    leaves preserved in, 175
    milkbeer, *205*, 206, 352–53
    natural wine, 347–49
    starters for, 421–22, 424
    washing rinds with, 166
alpine cheeses. *See also* semi-alpine cheeses; *specific types*
    bloated, 429
    brining of, 143
    burning taste in, 433–34
    cheesecloth for, *123*, 129–130, *130*
    cheese forms for, 131
    continued acidification of, 196
    cooking curds for, 104–6, *105*
    copper kettles for, 125
    cracks in, 433
    crystals in, 197
    cultural foundation of, 21
    cutting curds for, 100
    dry brushing of, 168–69
    forming of, 103
    from goat's and sheep's milk, 325, 326
    hastening spaces for, 170
    making of, 291–92
    mesophilic and thermophilic cultures for, 70
    patting down of, 169
    premature knitting of curds, 432
    pressing of, 110
    ripening ecology of, 182, 184
    salting of, 140
    seasonality of milk and, 45–46
    from sheep's milk, 41
    slow movement of salt in, 136
    temperature needs, 112, 150
    thermophilic cultures for, 409
    traditional making of, 291–92
    washing rinds of, 194, 196
    whey from, 382, 384, 392, 400

alpine peaks tests, 113–14, *114*
amasi, *xiv*, 206, 350–51
ambarees, 354–55
ammonia development, in over-ripe cheeses, 200
anaerobic ripening, 196
anari, 249
Anfiteatro, Dom, 403
animal husbandry, 43–45, *44*
animal-warm milk. *See* fresh from the udder milk
ash rind treatments, 173–74
ayran, 352

B

backslopping
    of active cultures, 66
    fermentation strengthened by, 55
    in naturally fermented foods, 52–53, 54
    for secondary fermentation, 71
    as term, 24
    for washed-rind cheeses, 188–89
    of whey, 61
bacteriophages, 65, 74, 76
baked-milk yogurt. *See* ryazhenka
barrel butter (smjör), 320–21
barreling of cheeses, 104, 140
baskets, as cheese forms, 131, *132*
Beauty of Xiaohe (mummy), 62, *63*
beer
    milkbeer, *205*, 206, 352–53
    starters for, 421–22, 424
beeswax, for waxing rinds, 176
belief systems about cheese, 25–26
BHG, 414
Bifidobacterium, in milk, 33, 34
biological foundation of cheese, 15–16
blaand (whey beverage), 120, 352
Bleu d'Auvergne cheeses, 340
bleu de Termignon cheeses, 215
bloated cheeses, 429
blue cheeses
    crystals in, 197
    forming of, 103
    hastening spaces for, 160
    low temperatures needed by, 150

blue cheeses *(continued)*
    orange rind development on, 431–32
    Penicillium roqueforti role in color of, 186–87, *186*, 334
    piercing of, 172–73
    from primo sale, 244
    stirred-curd varieties, 40, 103, 126, 139
blue mold. *See Penicillium roqueforti*
blue veins
    failure to develop, 430
    piercing for the growth of, 154, 172
bocconcini cheeses, 266
boiling milk. *See* cooking milk
bread. *See* sourdough rye bread; sweetdough bread
breast milk, 42–43
brebis cheeses
    butter from, 325
    making of, 207
    springtime making of, 232–33, *233*
Brettanomyces species, 348, 422
Brevibacterium linens
    air exposure needs, 154–55
    on handkäse, 329
    on Munster cheeses, 337
    ripening ecology of, 177, 179, 180
    salt tolerance of, 137
    on Stilton cheeses, 172, 345
    washing rinds to encourage, 166, 337
Bries
    cheese forms for, 131
    leaf wrapping of, 175
    making of, 217, 260–61
    patting down of, 169
    scoops for, 129
brining
    description of, 142–43, *142*, 145, 167–68
    of feta cheeses, 276
    with oil, 170
    of pickles, 275
    of sirene cheeses, 278
    with whey, 120–21, 142, 143, 168
Brown Swiss cows, 38
brunost cheeses, 120
brushes, for dry brushing, 168
brushing rinds, 168–69, 196
budding, in yeast reproduction, 184
buffalo milk
    for bufflonne cheeses, 232
    for feeding kefir, 404
    for mozzarella, 266–67, 432
    uses for cheesemaking, 41–42, *41*
bufflonne cheeses
    making of, 207
    springtime making of, 232–33, *233*
bulk milk tanks, 47, 48
bulk pasteurization, 49. *See also* pasteurization
burning taste, in aged cheeses, 433–34
burrata cheeses, 266, 272–73
butter
    from cow's milk, 38
    ghee from, 385
    from goat's milk, 39
    from kaymak, 366
    making of, 207, 317, 319
    preservation methods, 385
    smjör, 320–21
buttermilk, 31, 319

C

cabrito (goat meat), 82
caciocavallo cheeses
    dry caves for, 162
    making of, 219, 310, 311–12
    story of making, 22–24, *23*
calves. *See also* abomasum (fourth stomach)
    digestion of milk, 16
    ethical considerations for rennet production, 79–85, *81*
camel's milk, 42
Camembert bleu cheeses, 209, 262, 336–37, *336*
Camembert cheeses
    acidity monitoring of, 118
    blue fungus on, 430
    Brie compared with, 260
    brining of, 143
    cave conditions for, 153, 157–58
    charcoal rind treatment for, 173
    cheese forms for, 131
    leaf wrapping of, 175
    low temperatures needed by, 150
    modern making of, 209, 257–59, *258*
    patting down of, 169
    Penicillium candidum for, 190, *191*
    from primo sale, 244
    ripening ecology of, 182
    salting of, 136, 138–39
    stirring curds for, 101, 103
    traditional making of, 217, 254–56, *255*
Canada
    cheese as value-added product, 21
    cheese curd popularity, 308
    material considerations, 133–34
    natural cheesemaking restrictions, xi
    Quebec cheesemaking, 308, 411
    tool use in, 123
Canadienne cows, 38
cantal cheeses, 169, 215, 305
cardoon rennet, 88–89, 210, 380
carrying forward. *See* backslopping
casein proteins, 29–30, *30*, 79
casings
    for lactic cheeses, 386
    for salami, *388*, 389
castelmagno cheeses, 215, 305
cat's fur fungus. *See Mucor miehei*
caves, 149–162
    design considerations, 156–57
    dry caves, *161*, 162
    hastening spaces, 160–62, 170, *171*
    home-scale affinage, 159–160
    natural materials used in, 153, 156, 157–59
    needs of cheese, 149–155, *151*, *152*, 159, 160
    original cheese caves, 155–56
    overview, 8, 149
    for Roquefort cheeses, 340
cellars, as cheese caves, 159, 160
charcoal rind treatments, 173–74, *174*, 236, *236*, 262, *263*
cheddar cheeses
    avoiding mites in, 196
    cloth binding of, 167
    continued acidification of, 196
    crystals in, 197
    dry caves for, 162
    modern making of, 215, *307*, 308–9
    patting down of, 169
    salting in the curd, 146
    temperature needs, 150
    thermophilic cultures for, 70
    traditional making of, *214*, 215, 305–6
    waxing rinds of, 176
cheddaring process, 305
cheese caves. *See* caves
cheesecloth, *123*, 129–130, *130*, 167
cheese crystals, 190, 197–98
cheese curds, squeaky, 308
cheese forms
    for Brie, 260
    for Comté cheeses, 291
    description of, 130–33, *132*
    for mysost, 392
cheese harps, 126, *127*
cheese mites
    control of, 165, 167
    ecology of, 194–96, *194*, *195*
    in tomme de Savoie, 280
cheese on lees, 356–57
cheese presses, 109–10
"cheesy" flavors, 433
chestnut, for wooden vats, 124, 406
chèvre cheeses
    butter from, 325
    making of, 207, 232–33, *233*
Christian Hansen company, 12, 91
Chrysosporium sulfureum, ripening ecology of, 193
churning butter, 317
chymosin, 30, 79, 83, 85, 88, 91–92. *See also* rennet
cider starters, 421–22, 424
clabber, 35, 222
    failure to curdle, 429–430
    kefir vs., 64
    lack of sourness when curdling, 429
    making of, 206
    sourdough starters from, 332
    springtime making of, 222–23
    transformation of milk into, 177, 179
clabber culture
    description of, 222–23
    feeding of, 398–99

gas development in, 398, 399, *427*, 428
making of, 397–98
mesophilic vs. thermophilic, 69, 410
overview, 3, 8
pH of, 115
sourdough compared with, 417, 419
as starter culture, 56, 57–58, *59*, 99, 399
for yogurt, 360, 361
clabbered cottage cheese, 327
clay pots
    for ambarees, 354
    cheese forms of, 131, *132*, 133
    for cooking milk, 119
    for cuajada, 228, *230*
    for Saint-Marcellin, 242
    for Valençay, 237
cleansers
    of milk vinegar, 391
    of whey, 121–22
cloth binding, 167
clothbound cheddar, 305, 306
clotted cream, 206. *See also* kaymak
coagulation tests, 89–90
coliforms
    bloated cheeses from, 429
    gas development from, 413, *427*, 428
color of cheeses. *See also* blue cheeses; white-rinded cheeses
    from buffalo's milk, 42
    concerns with, 193
    from cow's milk, 38
    from goat's milk, 39
    rind washing effects on, 166
    from sheep's milk, 41
    washed-rind cheeses, 188, 190
colostrum
    composition of, 46
    kalvdans from, 90, 396
commercial cheesemaking, *11*, 411–16. *See also* industrial cheesemaking practices
    milking schedules for, 415
    overview, 8
    rennets for, 90–92, 99, 414
    scientific papers on natural methods for, 415–16
    traditional starters used in, 411–13
Comté cheeses
    making of, 213, 291–93, *292*
    ripening ecology of, 188
cooked-curd cheeses
    brining of, 143
    from goat's milk, 39
    thermophilic cultures for, 409
cooking curds
    description of, 104–6, *105*
    for tvorog, 322
    for wooden vat makes, 97
cooking milk. *See also* pasteurization
    for cream cheese, 372
    for kaymak, 366

for ryazhenka, 364
in traditional cheesemaking practices, 30–31, 118–120, *119*
unwanted effects of, 49
for yogurt cheeses, 369
cooling milk, 120. *See also* refrigeration of milk
cooling systems, in caves, 156
copper kettles, *20*
    for Comté cheeses, 291
    for cooking curds, 105, *106*
    for cooking milk, 119–120
    description of, 125
    longevity of, 134
    whey starters used with, 60–61
Corynebacterium, in milk, 33
cotija cheeses, 215, 305
cottage cheese, clabbered, 327
cow's milk
    for caciocavallo, 22, 312
    clabber of, 58
    for feeding kefir, 404
    industrial starters limited to, 74
    for mozzarella, 266–67, 432
    for Saint-Marcellin, 242
    uses for cheesemaking, 37–38
    for vache cheeses, 232
cream. *See also* cultured cream
    composition of, 31–32
    fermented, 58
    from goat's milk, 39, 325
    from sheep's milk, 325
    skimming of, 122
    in whey, 120
cream cheese, 207, 372
crema, from queso Oaxaca stretching water, 269, 271, *272*
crema agria. *See* cultured cream
crème fraîche, 314. *See also* cultured cream
crescenza cheeses
    making of, 209, 250–51, *250*
    salting of, 138–39
crottin cheeses
    dry caves for, 162
    hay bedding for, 158
    making of, 208
    ripening ecology of, 180
    springtime making of, 234–35
crystalline salts, 148
crystals, cheese, 197–98
cuajada, 39, 207, 228–29, *229*, *230*
Cuajos Caporal, 86, 414
cultural foundation of cheese, *20*
    caciocavallo example, 22–24, *23*
    description of, 20–21
    seasonality of milk and, 45–46
culture, 51–76. *See also* starter cultures
    active vs. inactive cultures, 68
    benefits of natural starters, 75–76
    biological foundation of cheese and, 53–55
    clabber culture, 56, 57–58, *59*

cyclical aspects of fermentation, 52–53, *52*
dynamics of microbial communities, 65–68, *66*, *67*
faith in fermentation, 72–73
fermentation in cheesemaking, 56–57
industrial starters vs. natural fermentation, 73–75
initiating fermentation, 55–56
kefir, 61–64, *62*, *63*
mesophilic and thermophilic cultures, 69–70, *70*
overview, 7–8, 51–52
primary and secondary fermentation, 70–71
protective features of microorganisms, 64–65
quality of fermentation, 65
universality of natural starters, 71–72
whey starters, 60–61
wood as, 58, 60
cultured cream
    butter from, 317
    long-cultured, 316
    making of, 206, 314, *315*
curd. *See also* cutting curds; milling curds; stirring curds
    acidity monitoring of, 115–18, *117*
    from cow's milk, 38
    floating, 433
    from goat's milk, 39
    judging the curd for forming, 113–14, *114*
    overly soft, 428–29
    salting in, *144*, 145–46
    from sheep's milk, 41
curd knives, 126–27
curdling of milk
    for clabber culture, 57, 58
    description of, 15–16
    failure of, 428
    from rennet, 77
curd set, testing, 112–13
curd stirrers, 127–28
cutting curds
    by the abomasum, 79, *94*, 96
    for caciocavallo, 22, 24
    for Camembert, 254
    description of, 100–101, *100*
    for Gorgonzola, 342
    for Roquefort cheeses, 340
    testing the curd set prior to, 112–13
    tools for, 100–101, 126–27
    for tvorog, 322
    for wooden vat makes, 97

D

Danisco, 91
decay of cheeses, 198–200, *199*
deliquescence, 139, 154, 182, *183*
direct vat inoculants (DVIs), 73–75
Divle cave cheeses, 215
draining cheeses
    description of, 103–4

draining cheeses (*continued*)
    post-salting, 146–47
    for wooden vat makes, 97
draining curd
    cheesecloth for, 129
    lactic cheeses, 111, *111*
draining mats, 133
draining tables, 103, 133
dry brushing, 168–69
dry caves, *161*, 162
dry salting, 140, *141*
DVIs (direct vat inoculants), 73–75

E
Emmentaler
    gas development in, 197
    making of, 213, 294–95, *294*
entero-mammary pathway, 33
enzymes. *See also specific enzymes*
    in the abomasum, 83, 85–86, 89
    in colostrum, 46
    in milk, 33
    in rennet, 79
Époisses cheeses
    making of, 216–17, 378–79
    ripening ecology of, 188
ethical considerations
    commercial rennets, 90–92
    rennet production and use, 77, 79–85
    slaughter of animals, 414
Europe
    cultural imperative for cheesemaking, 21
    hegemonic cheese industry in, 6
    repression of traditional cheese culture, 4
    use of natural materials, 134
    use of traditional tools, 123

F
faisselle cheeses, 207–8, 231
faith in fermentation, 72–73
fall cheeses, 8, 313. *See also specific types*
families of cheeses, 8, 203. *See also specific families*
fat, in milk, 31–32, 48
fat and oil treatments, 167, 169–70
fermentation. *See also* primary fermentation; secondary fermentation
    in cheesemaking, 1, 56–57, 103–4
    of curd, 110–11, *110*, *111*
    cyclical aspects of, 52–53, *52*
    of fourme d'Ambert, 339
    four traditional types of, 56
    human digestion supported by, 274
    industrial views of, 51
    initiating, 55–56
    of kaymak, 366
    of kefir grains, 64
    of leaves, 175
    of meat, 389
    microbiological foundation of cheese and, 18, 20, 34

of milk into clabber, 177, 179
in natural cheesemaking, 51
of natural wine, 347, 348–49
paste breakdown, 181–84, *181*, *183*
for pickles, 274
primary and secondary phases, 70–71
quality of, 65
salt used for, 135, 136, 138
solid-state, 424
Fermentation Produced Chymosin (FPC), 92
fermented cheeses, 206–7
fermented milk culture. *See* clabber culture
fermented milks
    from goat's and sheep's milk, 325, 326
    making of, 203–6, 204t, *205*
    for paneer, 395
    sourdough compared with, 417, 419
    unwanted "cheesy" taste in, 433
    viscous or ropy characteristics, 432
fermented ricotta cheeses, 219
fermented whey, 120–21
feta cheeses
    brine aging of, 167–68
    making of, 208, 210, 276–77
flavor of the hand (sson mhat), 21
flavors of cheeses
    diversity of, 20
    flaws in, 433
    salt effects on, 137, 138–39
flaws in cheese, 8, 426–434, *427*, *431*
flies, protecting cheeses from, 147, 160, 162, 166
flipping cheeses
    description of, 103, 163, 165
    rind development from, 154
fluorescent colors, unwanted development of, 430
fontina cheeses, 213, 291
food safety concerns
    in commercial cheesemaking, 411–13
    from industrial food systems, 6
    Mucor-derived rennet, 425
forming cheeses, *102*
    judging the curd for, 113–14, *114*
    lactic cheeses, 111
    for wooden vat makes, 97
forms. *See* cheese forms
fourme d'Ambert cheeses, 210, 338–39, *338*
FPC (Fermentation Produced Chymosin), 92
France. *See also specific types of cheeses*
    author's teaching in, xi, xii
    calf vells available in, 414
    cultured cream made in, 314
    leaf wrapping used in, 174, 175
    mite-infested cheese styles, 194
    moulé à la louche label, *110*, 129, 216, 242, 254
    Mucor-based rinds used in, 191, 193, 298, 423
    patting down treatment used in, 169
    traditional cheesemaking practices, 86, 129
    washed-rind cheese treatments used in, 188
    wooden vats used in, 124, 406, 407

freeze-dried starters
    considerations in using, 10, 99
    description of, 73–75
    pH meters used with, 115
freezing
    of clabber culture, 399
    of milk, 48
    of whey starters, 401
fresh cheeses. *See also specific types*
    keeping freshness of, 198
    milling curds of, 107
    salting of, 136
    spinning curds of, 108
    springtime making of, 221
fresh from the udder milk. *See also* milk
    in commercial cheesemaking, 414–15
    cream integrated in, 31
    difficulties obtaining with machine milking systems, 49
    ideal characteristics for cheesemaking, 31, *47*, 48
    pH of, 115
    warmth of, 34, 37
fresh whey ricotta, 382, *383*
fruit juice, starters from, 421–22
full-alpine cheeses, 211–13, *212*. *See also* alpine cheeses

G
gardening uses of whey, 121
gas development
    in ambarees, 354
    in Camembert, 254
    in clabber culture, 398, 399, *427*, 428
    from coliforms, 413
    in declining cultures, 68
    description of, 196–97
    in Emmentaler, 294, 295
    floating curds from, 433
    in sourdough starters, *418*
    in wooden vats, 60, 407
genetically modified organisms (GMOs), 11, 92
Geotrichum candidum, 4, 70
    for aged lactic cheeses, 146
    air exposure needs, 154–55
    biofilms of, 133
    on brine-aged cheeses, 168
    on Camembert, 254
    cave conditions for, 153
    on clabber culture, 58, 179, 223, 398, 399
    on crottin cheeses, 234, 236
    on cuajada, 228
    failure to develop on aged lactic cheeses, 431
    on fresh lactic cheeses, 232
    on gourds, 350
    on graukäse, 331
    on gros lait, 362, 363
    on handkäse, 329
    hastening spaces for, 160, 162, 170
    on kefir, *402*

on lactic cheeses, 111
on lactic cheese sausages, 386–87
on le sein cheeses, 238
on liquid rennet, 88
on long-cultured cream, 316
original habitat of, 149–150
Penicillium candidum interactions with, 190
protective mycotoxins from, 64
rind color from, 167
ripening ecology of, 34, 177, *178*, 179–180, 184–86, *185*
on salami, 389, 390
salt effects on, 137, 138
secondary fermentation by, 68, 71
on Stilton cheeses, 172
temperature effects on, 69
on tomme cheeses, 252, 288
unwanted development of, 430–31, *431*
on wet sourdough starters, 332
on whey, 401
gerles (wooden vats), 97, 124, 406, 407
ghee
  making of, 385
  from making ryazhenka, 364
  rubbing rinds with, 169, 170
GMOs (genetically modified organisms), 11, 92
goatskins, for aging cheeses, 306
goat's milk
  for ambarees, 354
  brined cheeses of, 167
  butter from, 322, *324*, 325–26
  for caciocavallo, 22, 312
  for chèvre, 232
  clabber of, 58
  for crottin cheeses, 234
  for cuajada, 228
  for faisselle, 231
  for feeding kefir, 404
  for feta cheeses, 276
  for fresh whey ricotta, 382
  for Halloumi, 248
  for le sein cheeses, 238
  for mozzarella, 267
  for mysost, 392
  rennet doses for, 111
  for Saint-Maure, 241
  for sirene cheeses, 278
  for tomme de chèvre, 282–83
  uses for cheesemaking, 38–39
Gorgonzola cheeses, 209–10, 342, *343*
Gouda cheeses
  cheese forms for, 131
  making of, 216, 300–301, *302*
  washing curds for, 106
  waxing rinds of, 176
gourds
  alcohol starters in, 422
  fermenting milk in, 350, 351, 391, 406
grana cheeses
  brining of, 143
  continued acidification of, 196
  dry caves for, 162
  making of, 213, 296–97, *297*
  whey from, 400, 410
grass feeding. *See* pasture feeding
graukäse, 207, 330–31, *330*
gros lait, 206, 362–63, *363*
Gruyère cheeses, 213, 252, 291
Guernsey cows, 38
Gyr cows, 38

H

half-lactic cheeses, 216–17. *See also* Camembert cheeses
Halloumi, 210, 248–49, *248*
handkäse, 207, 328–29, *329*
handkäse mit muzik, *329*
Harel, Marie, 260
hastening
  for crottin cheeses, 234
  Geotrichum candidum development from, 170, *171*, 185, 190
  for Stilton cheeses, 345
  for tomme Vaudoise, 252
hastening spaces, 160, 162
hay bedding, in caves, 158–59
heat-acid-coagulated cheeses, 90, 219–220. *See also* paneer; ricotta cheeses
heterofermentation, 69
High Temperature Short Time (HTST) pasteurization, 49–50. *See also* pasteurization
Hindu cheesemaking traditions, 82
historical background
  buttermaking, 320
  Camembert cheeses, 260
  cheese caves, 155–56
  cheese forms, 130–31
  cheesemaking, 5, 16–18, 112
  cheese vats, 124–25
  grana cheeses, 296
  kefir, 62, 63, 76
  ladles and scoops, 129
  milkbeer, 352
  Munster cheeses, 373
  Penicillium roqueforti, 334
  queso Oaxaca, 269
  relationships among cheeses, 25
  rennet production and use, 80, 82, 83
HMOs (human milk oligosaccharides), 54
Holstein cows, 38
homogenization, unwanted effects of, 47, 50
horns of milking animals, *44*
horse's milk, 42, 352
human milk, 42–43
human milk oligosaccharides (HMOs), 54
humidity
  of cheese caves, 150, *152*, 153–54, 157, 159, 160
  in dry caves, 162
  for *Geotrichum candidum* growth, 185
  for *Penicillium roqueforti* growth, 187
  for washed-rind cheeses, 188
Hundsbichler, 12

I

inactive vs. active cultures, 68
India
  ghee making, 385
  junnu making, 396
  paneer making, 395
  rennet-less cheesemaking, 82
  yogurt making, 360
industrial cheesemaking practices
  in France, xiv
  Geotrichum candidum concerns, 186
  packaged starters, 73–75
  pH meters, 115
  rennets, 90–92, 99
  square vats, 125–26
  widespread nature of, xiv
industrial food systems
  cream separation practice, 122
  dairy farming practices, 5, 10
  detriment from, 18
  food safety concerns, 6
  kefir production, 63
  limited definition of fermentation in, 51
  scientific approach to, 72
  yogurt making, 360
isoelectric point of milk, 90
Italy. *See also specific types of cheeses*
  cardoon rennet used in, 88
  Tinas used in, 22, 24, 124
  traditional cheesemaking practices, 24, 25, 86, 266
  whey starters used in, 60

J

Järbe (German cheese form), 131, *132*, 291
Jersey cows, 38
juncus reeds, 175, 226
junket, 226–27
junnu (colostrum custard), 396

K

kahm yeasts, 184, 275
kalvdans, 46, 90, 220, 396
kappa-casein, 29–30
Käsesäbel (cheese saber), 128
kaymak, 206, 366–67
kefir
  antiviral properties of, 65
  failure to curdle, 429–430
  lack of sourness when curdling, 429
  making of, 206
  as mesophilic culture, 69, 361
  as part of clabber culture, 3
  pH of, 115
  springtime making of, 224–25
  as starter culture, 56, 61–64, *62*, *63*, 99, 404

kefir culture, *402*
    in cheesemaking, 404, 413
    gas development in, *427*, 428
    keeping of, 403–4
    overview, 8, 402
    preserving of, 404–5
    research papers on, 412
    sourcing grains for, 402–3
    sourdough compared with, 417, 419
kefir grains
    drying of, 405
    feeding of, 224–25
    microorganisms in, 61–62, 63–64, 224
    origins of, 62
    preserving of, 405
    sourcing of, 402–3
    straining of, 403, 404
kitchen uses of whey, 121
koji (*Aspergillus oryzae*), 89
kosher salt, 148
koumiss, 352
kuhwarm, as term, 37

L

labneh, 369
Laboratoires ABIA, 414
lactalbumins, 30
lactase, 32, 96
lactase persistence, 54
lactic acid
    fermentation of lactose into, 71, 181–82
    Geotrichum candidum propensity for, 184, *185*
    metabolization of, 182
    Penicillium roqueforti propensity for, 187
lactic cheeses. *See also* half-lactic cheeses
    acidity monitoring of, 116
    brine aging of, 168
    cave conditions for, 157–58
    cheese forms for, 131
    draining curd, *111*
    failure to develop *Geotrichum* rind, 431
    fermenting of curd, 110–11, *110*
    flipping of, 165
    from goat's milk, 39
    hastening spaces for, 160
    ladles for, 129
    low temperatures needed by, 150
    making of, 207–8
    mesophilic cultures for, 70
    rennet doses for, 12, 99
    ripening ecology of, 182
    salting in the curd, *144*, 145–46
    wooden vats for, 407
lactic cheese sausage, 386–87, *386*
Lactobacillus bacteria, 33, 34
Lactococcus bacteria, 33
lactofermenting bacteria
    dynamics of microbial communities, 65–68, *66*, *67*

milk sugars metabolized by, 18, 71, 177, 179
    in thermophilic starters, 409
    for vegetable fermentation, 274
lactoferrins, in milk, 33
lactoglobulins, 30
lactose
    digestion of, 32, 54, 96
    fermentation into lactic acid, 71
    washing curds and, 106
lactose intolerance, 54
ladles, 111, 128–29
ladling
    of Camembert, 254
    of faisselle, 231
    of lactic cheeses, 111, 129, 216
    moulé à la louche, *110*, 129, 216, 242, 254
Lagenaria gourds
    alcohol starters in, 422
    fermenting milk in, 350, 351, 391, 406
Lancashire cheeses, *214*, 215, 303–4
lard, for cloth binding, 167
Latin America
    diversity of spun-curd cheeses in, 269
    quality of milk in, 56
    rennet production in, 83
    repression of traditional cheese culture, 4
leaf wrapping, 174–75
legal considerations. *See* regulations
le sein cheeses, 208, 238, *239*
lipase, 31, 85–86, 87–88
lipolysis, 31
lipoproteins, in milk fat globule membranes, 31
liquification, during ripening, 182
Livarot cheeses, 175, 262
local food systems, importance of, 6
long-chain fatty acids, 32
long-cultured cream, 316
low-temperature pasteurization, 49. *See also* pasteurization
lysozymes, in milk, 33

M

machine milking systems, milk damaged by, 48–49
the make, 95–122
    boiling milk, 118–120, *119*
    cooking curds, 104–6, *105*
    culturing milk, 98–99
    cutting curd, 100–101, *100*
    draining and fermenting cheeses, 103–4
    fermenting curds, 110–11, *110*, *111*
    flexibility in, 8–9
    forming cheeses, *102*, 103
    list of steps for, 95
    milling curds, 106–8, *107*
    origins of, 96–98
    overview, 8, 95–96
    pressing cheeses, 108–10, *109*
    renneting milk, 99–100
    salting cheeses, 95, 104
    skimming cream, 122

spinning curds, 108
stirring curds, 101, *101*, 103
testing cheeses, 111–18, *114*, *117*
warming milk, 98
washing curds, 106
whey use and reuse, 120–22
Marcellino, Sister Noella, 298
mare's milk, 42, 352
mascarpone cheeses, 382
materials. *See also* wooden vats
    for cave equipment, 157–59
    in commercial cheesemaking, 413–14
    for tools, 133–34
meat, fermented, *388*, 389–390
mesophilic cultures
    description of, 69–70
    for gros lait, 362, 363
    for mozzarella, 267
    primary and secondary fermenting microorganisms in, 71
    for ryazhenka, 364
    thermophilic cultures combined with, 409–10
Mexico
    goat cheesemaking in, 82
    queso fresco from, 246
    queso Oaxaca from, 269
micelles, 29–30, 79
microbial communities
    in clabber culture, 398
    in cultures, 53–55
    diversity in mesophilic vs. thermophilic cultures, 69, 410
    dynamics of, 65–68, *66*, *67*
    in kefir grains, 403
    in whey starters, 401
    in wooden vats, 406
    in yogurt, 360
microbiological foundation of cheese, 18–20, *19*
microorganisms. *See also* microbial communities; unwanted microorganisms; *specific types*
    controlling effects of salt on, 136
    digestive role of, 32, 54
    diversity of, 20
    faith in the goodness of, 72–73
    growth of bacteria vs. yeasts, 420
    in hay and straw, 158
    in kefir grains, 61–62, 63–64
    in milk, 33, 64–65
    overview, 3
    primary vs. secondary fermentations, 71
    protective purpose of, 20, 53, 64–65
    in re-fermented milk culture, 56
milbenkäse, 328
milk, 27–50. *See also* buffalo milk; cow's milk; fresh from the udder milk; goat's milk; pasteurized milk; raw milk; sheep's milk
    aging and decay of, 177, 179
    clabber from, *35*
    colostrum, 46

complexity of, 3–4
cultural aspects of handling, 21
curdling of, 15–16
digestion of, 79
drinking of, 4–5
fat in, 27, 31–32
for feeding kefir, 404
fermentation and coagulation properties, 4–5
generation of, 28–29
ideal characteristics for cheesemaking, 1, 9–10, *47*
microbes in, 33–34
overview, 7, 27–28
pasteurization of, 49–50
problems of milk processing, 28, 39, 46–49
proteins in, 27, 29–31, *30*, 33
seasonality of, 45–46
species producing, 37–43, *40, 41*
sugars in, 27, 32–33
udders, *28*
warmth of, 9–10, 34, 37
well-being of milking animals, 29, 43–45, *44*
milkbeer, *205*, 206, 352–53
milking animals, care of, 43–45, *44*
milking robots, milk damaged by, 48, *49*
milking systems, milk damaged by, 48–49
Milklab, xv
milk mothers. *See* clabber culture
milk processing, problems of, 46–49
milk replacers, for young ruminants, 43
milkstone deposits, 49
milk vinegar, 391
milled-curd cheeses
    in casings, 387
    cloth binding of, 167
    continued acidification of, 196
    diversity of, 107
    making of, 213–15, *214*, 305–6
    pressing of, 103, 108–10, *109*
    salting of, 139, 145, 146, *147*
    scoops for, 129
    wooden vats for, 124
milling curds
    for caciocavallo, 311
    for cheddar cheeses, 305, 308
    description of, 106–8, *107*
    for Lancashire cheeses, 303
    for Stilton cheeses, 345
    for wooden vat makes, 97
mimolette cheeses, 194
mites, cheese. *See* cheese mites
morge, 121, 166, 188–89
mother's milk, 42–43
moulé à la louche (ladled curds), *110*, 129, 216, 242, 254
mountain cheeses. *See* alpine cheeses
mozzarella cheeses
    acidity monitoring of, 116
    failure to stretch, 432
    making of, 217, 266–68, *267*

slimy, 432
temperature guidelines, 112
Mucor miehei
    conditions for growth of, 58
    overview, 8
    for rinds and rennet, 89, 91, 423–25
    ripening ecology of, 177, 179, 191–93, *192*
    for Saint-Nectaire, 298, 299
    sweetdough bread starter from, 422
    unwanted growth of, 430
Munster cheeses
    Camembert compared with, 262, 337
    making of, 373–75, *374*
    from primo sale, 244
mysost cheeses, 219–220, 392, *393*

N

natural caves, 155–56. *See also* caves
natural cheese, 15–26
    accessibility of, 5
    agricultural foundation of, 16–18, *17*
    biological foundation of, 15–16
    cultural foundation of, 20–21, *20*
    microbiological foundation of, 18–20, *19*
    overview, 7
    PDO designations, 24–25
    as religion, 25–26
    story of making a caciocavallo, 22–24, *23*
    universality of, 25
natural cheesemaking
    author's first batch, xiv
    four foundations of, 1, 15
    growing practice of, xiii–xv
    importance of practicing, xiii
    mystical aspects of, 1
    philosophy of, 7
natural materials. *See* materials
natural wine, 347–49, *348*
nodini cheeses, 266
North America. *See* Canada; United States
Northern European rennet, *84*, 85–87
Nuevo León, Mexico, goat cheesemaking in, 82

O

Oaxaca, spun-curd cheeses in, 269
oil brining, 170
oil rubbing, of rinds, 169–170
oligosaccharides, in milk, 32
olive oil
    brining with, 170
    rubbing rinds with, 169
    shankleesh preserved in, 370
opportunistic microorganisms. *See* unwanted microorganisms
orange rinds
    from *Brevibacterium linens*, 172, 177, 179, 345
    dry brushing for, 168
    failure to develop, 430
    rind washing for, 165, 166–67, 188–190, *189*, 292

rubbing up for, 172
unwanted development of, 431–32
organic agriculture, 1, 74–75
osmosis, 136, 142
oxcypek cheeses, 325

P

packaged cultures, 73–75
packaging of cheeses, 198
paddles, for stirring curds, 127–28
paneer, *394*
    making of, 219, 395
    temperature guidelines, 112
paraffin waxes, 176
Parmigiano-Reggiano cheeses
    brining of, 143
    dry caves for, 162
    making of, 213
pasta filata cheeses
    from goat's and sheep's milk, 325, 326
    making of, 217–19, *218*
    mesophilic vs. thermophilic cultures for, 70
    salting in the curd, 145
    spinning curds for, 24, 108, 311
paste breakdown, 181–84, *181, 183*
pasteurization
    cooking milk vs., 119
    incomplete effectiveness of, 411
    of refrigerated raw milk, 415
    unwanted effects of, 49–50
pasteurized milk
    in commercial cheesemaking, 413
    considerations in using, 6, 10
    soft curd from, 429
pasture feeding
    benefits of, 32
    management of, 43, 45
pathogenic microorganisms. *See* unwanted microorganisms
patting down, 169
PDO (Protected Denomination of Origin) designations
    author's concerns with, 24–25
    copper kettle requirements, 125
    fresh milk requirements, 47
    unreliability of, 4
pecorino cheeses, 213, 296
pêle à Brie (perforated scoop), 129
Penicillium candidum
    air exposure needs, 154–55
    on Camembert, 254, 257
    cave conditions for, 153
    low temperatures needed by, 150
    on rennet cheeses, 209
    ripening ecology of, 177, *178*, 179, 190–91
    salt effects on, 138
    salt tolerance of, 137
    on Stilton cheeses, 345
    on tomme de montagne, 284
    white rinds from, 166, 167

Penicillium roqueforti, 340
    air exposure needs, 154–55
    in Camembert bleu cheeses, 209, 336, 337
    cave conditions for, 153
    cultivation of, 187, 334–35, *335*
    in fourme d'Ambert, 338, 339
    in Gorgonzola, 342
    in lactic cheeses, 208
    low temperatures needed by, 150
    in mite-derived passageways, 194
    piercing for the growth of, 172, 173
    rind color from, 167
    ripening ecology of, 177, *178*, 179, 180, 186–88, *186*
    on salami, 389
    salt tolerance of, 137
    secondary fermentation by, 71
    on sourdough bread, 332, 333, 417
    in Stilton cheeses, 345
    in stirred-curd blue cheeses, 210
    on tomme de Savoie, 280
pepsin, 83
pH. *See also* acidity monitoring
    for brining, 143
    changes during ripening, 181–84, *181*
    commercial cheesemaking documenting of, 412
    for salting, 139–140
    scale of, 115
philosophy of natural cheesemaking, 7
pickles, sour, 274–75
piercing
    description of, 172–73
    of fourme d'Ambert, 339
    of Stilton cheeses, 173, 345
pitching curds, 103
plastic
    for cheesemaking tools, 133–34
    concerns during rind development, 153
Pollan, Michael, 378
pomace
    aging in, 175
    cheese on lees from, 356, 357
Pont l'Évêque cheeses, 209, 262–64, *263*
poutine, 308
pre-draining curds, 208
presses, 109–10
pressing cheeses
    cheddar cheeses, 305
    description of, 108–10, *109*
    overview, 103
    tomme de montagne, 284
primary and secondary fermenting microorganisms
    in milk, 177, 179
    salt effects on, 136, 138
primary fermentation. *See also* fermentation
    description of, 70–71
    of natural wine, 348–49
    pH lowered by, 181

    pomace from, 175
    of stirred-curd feta cheeses, 168
    timing of salt application and, 136, 137, 138–39, 140, 146
    of waxed cheeses, 176
primo sale cheeses, 209, 244–45, *245*
probiotics, 65
propionibacteria
    bloating from, 429
    for Emmentaler, 197, 294, 295
    for Gouda, 300
    hastening to encourage, 170
    in milk, 33
Protected Denomination of Origin. *See* PDO (Protected Denomination of Origin) designations
proteins, in milk
    denaturing of, 50, 119, 120
    types of, 29–31, *30*, 33
proteolysis, 182
proteolytic enzymes, 89
Pseudomonas, fluorescent colors from, 193, 430

Q

quark, 369
Quebec cheesemaking, 308, 411
queso fresco, 145, 246–47
queso Oaxaca, 217, 269–271, *270*

R

racking, of natural wine, 349
raclette cheeses
    making of, 211, 286–87, *286*
    ripening ecology of, 188
    washing curds for, 106, 216
rancidity concerns
    refrigerated milk, 48, 122, 433
    vegetable oils and fats, 167, 169
raw milk
    accessibility of, 6
    in commercial cheesemaking, 411, 413
    procurement of, 48
    quality considerations, 56–57
    refrigeration concerns, 22, 37, 39, 47–48, 415
    restrictions on, xiii, xiv, 411
    as term, 6, 46–47
reblochon cheeses, 262
re-fermentation. *See also* backslopping
    mesophilic and thermophilic cultures, 69
    as term, 52–53
refined vs. unrefined salts, 148
refrigeration of cheeses, 198
refrigeration of milk
    considerations for, 9–10
    unwanted effects of, 22, 37, 39, 47–48, 415
refrigerators, as cheese caves, 159–160
regulations
    cheese caves, 12
    commercial cheesemaking, 12, 411–15
    raw milk, xiii, xiv, 411

rennet, 77–92
    from the abomasum, 12, 77, *78*, 79–82, 83–85
    adding simultaneously with natural starters, 76, 99
    adding to milk, 99–100
    biochemical effects of, 77, 79
    for caciocavallo, 22
    in commercial cheesemaking, 414
    commercial sources of, 90–92
    considerations in choosing, 10–12
    dose for lactic goat and sheep cheeses, 111
    failure of milk to curdle, 428
    for feta cheeses, 276
    as foundation of natural cheesemaking, 1
    genetically modified, 11
    from *Mucor*, 425
    multiple pathways of ripening of, *178*
    overview, 8, 77
    preparation of, 85–89, *87*, 414
    slaughtering of young animals, 82–85
    testing of, 89–90
rennet cheeses. *See also* specific types
    acidity monitoring of, 116
    brining of, 143
    cheese forms for, 131
    forming of, 103
    fresh, 244
    Geotrichum candidum development on, 430–31, *431*
    making of, 208–10
    mesophilic cultures for, 70
    milling curds for, 107–8, *107*
rennet curd. *See* curd
rennet-free cheeses, 90
research papers, 412, 415–16
ricotta cheeses
    from fermented whey, *2*, 120
    from fresh whey, 382, *383*
    Halloumi made simultaneously with, 249
    making of, 219
    whey leftover from, 120
ricotta salata cheeses, 219, 384
rind-ripening microorganisms. *See also* ripening ecologies
    air exposure needs, 154–55
    cave conditions for, 149, 150, 153
    overview, 71
    salt effects on, 136, 137
rind treatments. *See also* brining; flipping cheeses; rind washing
    brine aging, 167–68
    cloth binding, 167
    dry brushing, 168–69
    patting down, 169
rind washing, *164*
    of alpine cheeses, 292
    of Camembert cheeses, 254, 257, 337
    description of, 165–67
    of Époisses cheeses, 378

of Munster cheeses, 373
of raclette cheeses, 286, 287
of Stilton cheeses, 345
of stracchino cheeses, 251
with whey, 121, 166
ripening ecologies, 177–200. *See also* affinage treatments
  Brevibacterium linens, 177, 179, 180
  cheese caves, 190
  cheese crystals, 197–98
  cheese mites and, 194–96
  continued acidification, 196
  ecologies of the rind, 179–181
  flaws in, 430
  gas development, 196–97
  Geotrichum candidum, 177, *178*, 179–180, 184–86, *185*
  inevitable decay of cheese, 198–200, *199*
  keeping cheeses fresh, 198
  in mesophilic starters, 75
  milk's own affinage, 177
  miscellaneous fungi, *194*, 195
  Mucor, 177, 179, 191–93, *192*, 423–25
  multiple pathways of, 177, *178*
  overview, 8
  paste breakdown, 181–84, *181*, *183*
  Penicillium candidum, 177, *178*, 179, 190–91
  Penicillium roqueforti, 177, *178*, 179, 180, 186–88, *186*
  salt effects on, 139
  washed-rind ecologies, *178*, 188–190, *189*
robot milkers, 48, 49
rolling of cheeses, 340
Roquefort cheeses. *See also* blue cheeses
  air exposure needs, 154–55
  making of, 210, 340–41
  Penicillium roqueforti for, 334
  piercing of, 173, 188
  salting of, 139
  side aging of, 165
rubbing, of rinds, 169–170
rubbing up, 170, 172, 345
ruminants, care of, 43–45, *44*
ryazhenka, 206, 364, *365*
rye sourdough bread, 332–33

S

Saint-Marcellin cheeses
  Geotrichum candidum in, *4*
  making of, 208, 242–43
  pot ripening of, 165
  salting of, 138–39
Saint-Maure cheeses
  making of, 208, 241
  straw for, 175, *240*, 241
Saint-Nectaire cheeses
  making of, 211, 216, 298–99
  Mucor rind of, 191, *192*, 193, 423
salami, *388*, 389–390
Salers cheeses, 215

salt, 135–148
  acidity monitoring, 116, 118
  affinity for water, 136
  for alpine cheeses, 292
  for brining, 142–43, *142*, 145, 168
  for butter, 319
  for cheddar cheeses, 305
  effects of, 104
  for feta cheeses, 276
  for Gorgonzola, 342
  making from seawater, 148
  overview, 8, 135
  post-salting, 146–47
  for Saint-Nectaire cheeses, 298
  salting in the curd, *144*, 145–46, *147*
  surface salting, 140, *141*
  timing of application, 103–4, 136, 137, 138–140, 146
  two percent application of, 137, 140
  types of, 147–48
  for wooden vat makes, 97
salt brines. *See* brining
sanitation
  fermented whey for, 121
  square vats and, 126
saturated fat, in milk, 31–32
sauerrahm. *See* cultured cream
sausage casings
  for lactic cheeses, 386
  for salami, *388*, 389
scalding of curds, for modern cheddar, 308
scale of cheesemaking, 9
scamorza cheeses, 266
scientific papers, 412, 415–16
scoops, 128–29
Scopulariopsis, ripening ecology of, 193
seasonality of milk, 45–46
seawater, making salt from, 148
secondary fermentation. *See also* fermentation
  of clabber, 179
  description of, 70–71
  of kefir, 225
  of milkbeer, 352
  of natural wine, 348–49
  pH raised by, *181*
  salt effects on, 136, 137, 138–39
  of sourdough bread, 332
semi-alpine cheeses. *See also* alpine cheeses
  aging of, 284
  cooking curds for, 112
  cutting curds for, 100
  making of, 210–11, 265
  stirring curds for, 106
  whey from, 120
  wooden vats for, 407
settling curds, 103
shankleesh, 170, 207, 370, *371*
shape of cheeses
  caciocavallo cheeses, 311
  Comté cheeses, 291

dry salting effects on, 140
Roquefort cheeses, 340
sheep's milk
  for brebis cheeses, 232
  for brined cheeses, 167
  butter from, 322, *324*, 325–26
  for caciocavallo, 22, 312
  clabber of, 58
  for cuajada, 228
  for faisselle, 231
  for feta cheeses, 276
  for fresh whey ricotta, 382
  for Halloumi, 248
  for mozzarella, 267
  rennet doses for, 111
  for Roquefort cheeses, 340
  for sirene cheeses, 278
  for tomme de chèvre, 282–83
  uses for cheesemaking, 39–41, *40*
  wool cushioning for cheeses of, 159
sirene cheeses, 278, *279*
skerpikjøt, 389
skim milk curd
  cheeses from, 320
  cottage cheese from, 327
  graukäse from, 330, 331
  handkäse from, 328, 329
skimming cream, 122
skyr, 207, 320–21
slaughter of animals, 79–85, 414
smear-ripened cheeses. *See* washed-rind cheeses
smetana, 314, 322–23, *323*
smjör, 320–21
soft cheeses. *See also specific types*
  cave conditions for, 157–58
  cutting curds for, 100
soil amendments, of whey, 121, 122
solid-state fermentation, 424
sour cream. *See* cultured cream
sourdough rye bread, 332–33, 392
sourdough starters, 417–420
  fermentation of milk compared with, 417, 419
  making of, 332, 333, *418*, 419
  microbial communities in, 53
  overview, 8
  ratio of starter to dough, 419–420
sour milk cheeses, 206–7
sour pickles, 274–75
Southern European rennet, *84*, 85–86, 87–88, *88*
Spain
  cuajada from, 39, 207, 228
  lamb/kid stomachs available in, 86, 414
spinning curds, 24, 108, 311. *See also* pasta filata cheeses
spinos, for cutting curd, 22, 24, 101, 127
splat tests, 113
spring cheeses, 8, 221. *See also specific types*
spruce bands, for vacherin cheeses, 175, 376, *376*, 377

spun-curd cheeses, 217–19, *218*. *See also* pasta filata cheeses
square vats, 125–26
squeaky cheese curds, 308
sson mhat (flavor of the hand), 21
stainless steel, for cheesemaking tools, 124, 126, 133–34
starter cultures. *See also* culture
    backslopping into fresh milk, 66
    benefits of, 56–57
    for caciocavallo, 22
    in commercial cheesemaking, 411–13
    four traditional types of, 56
    frequency of propagation, 75
    freshness of cheese and, 198
    for *Geotrichum candidum* growth, 184–85
    ideal characteristics for cheesemaking, 10
    industrial starter limitations, 73–75
    industrial vs. natural, 73–75
    microbiological foundation of cheese and, 18, 20
    microorganisms for, 34
    for milkbeer, 352, 353
    for modern milks, 55–56
    for *Mucor* growth, 193
    natural starter benefits, 1, 75–76
    for natural wine, 347–48
    for *Penicillium roqueforti* cultivation, 334
    for pickles, 274–75
    protocol for, 98–99
    for queso fresco, 246
    for sourdough bread, 8, 53, 332, 333, 417–420, *418*
    universality of, 71–72
    for yogurt, 360, 361
sterilization of equipment, 48
sterilized milk, in commercial cheesemaking, 413
Stilton cheeses
    air exposure needs, 154–55
    blue veins in, 187
    hastening spaces for, 160, 170
    making of, *214*, 215, *344*, 345–46
    piercing of, 173, 345
    pressing of, 109
    ripening ecology of, 180
    rubbing up, 170, 172
    salting in the curd, 146
stirred-curd blue cheeses
    forming of, 103
    salting of, 139
    from sheep's milk, 40
    square vats for, 126
stirred-curd cheeses
    brine aging of, 168
    cutting curds for, 100
    making of, 210
stirring curds
    for cheddar cheeses, 305
    description of, 101, *101*, 103
    for fourme d'Ambert, 338

for primo sale cheeses, 244
    tools for, 127–28
    for tvorog, 322
    for wooden vat makes, 97
stracchino cheeses, 250–51, *250*
stracciatella cheeses
    burrata stuffed with, 272, 273
    making of, 217, 219, 266, 273
straw
    bedding of, 158–59
    for Saint-Maure, 175, *240*, 241
Streptococcus bacteria, 33, 34
stretched-curd cheeses. *See* pasta filata cheeses
sugars, in milk, 32–33
sulfuring, of wine, 349
summer cheeses, 8, 265. *See also specific types*
super-saturation of brines, 142–43
surface salting, 140, *141*
sweetdough bread
    alcohol starters for, 422, 424
    Mucor cultivation on, 423–25
sweet ricotta cheeses, making of, 219
Switzerland
    calf vells available in, 414
    diversity of cheesemaking in, 252
    Emmentaler cheese from, 294
    traditional cheesemaking practices, 86
    vacherin cheese from, 376
syneresis, 79

T

taleggio cheeses, 133
teabag tests, 116–18, *117*
temperature
    of brines, 145, 168
    of cheese caves, 150, 159
    for *Geotrichum candidum* growth, 185
    for hastening, 162
    lactofermenting bacteria affected by, 66, *66*, *67*, 68
    mesophilic and thermophilic cultures, 69–70
    monitoring of, 112
    for *Penicillium roqueforti* growth, 187
    for washed-rind cheeses, 188
    when renneting milk, 99
terroir, 24
testing
    cheeses, 111–18, *114*, *117*
    rennet, 89–90
thermization, 6
thermophilic cultures
    choosing to use, 409–10
    description of, 69–70
    feeding of, 410
    for mozzarella, 267
    overview, 8
    primary fermenting microorganisms in, 71
    for shankleesh, 370

    starting, 410
thistle rennet, 88–89, 380
Tinas, 22, 24, 124. *See also* wooden vats
Tomino cheeses, 262
tomme cheeses
    brining of, 143
    mites in, 194, *195*, 196
    patting down of, 169
    ripening ecology of, 182, 184
    salting of, 138
tomme crayeuse, 211, 288–290, *289*
tomme de chèvre, 211, 282–83, *282*
tomme de montagne, 211, 284–85
tomme de Savoie, 211, 280–81
tomme Vaudoise, 252–53
tools, 123–134. *See also* clay pots; copper kettles; wooden vats
    cheesecloth, *123*, 129–130, *130*, 167
    cheese forms, 130–33, *132*
    curd knives, 126–27
    curd stirrers, 127–28, *127*
    for cutting curds, 100–101, 126–27
    draining tables and mats, 103, 133
    human hands, 126, 128
    ladles and scoops, 111, 128–29
    materials considerations, 133–34, 157–59, 413–14
    overview, 8, 123–24
    vats, 124–26, *124*
torta, 89, 210, 380–81
traditional cheesemaking practices
    agricultural foundation of cheese and, 16–18
    as foundation of natural cheesemaking, 1, 3, 4
    regional variety in, 5–6
transhumance, 22, 25, 211, 392
treccia cheeses, 266
triglycerides, in milk fat, 31
troubleshooting guide, 426–434, *427*, *431*
turning cheeses. *See* flipping cheeses
tvorog, 207, 322–23, *323*, 325, 326
two percent salting guideline, 137, 140
Tyrophagus putrescentiae. *See* cheese mites

U

udders, *28*
    concerns with sterilizing, 34, 48
    description of, 28
    whey for cleansing, 121
UHT (ultra-high-temperature) pasteurization, 50, 429. *See also* pasteurization
underground natural caves, 155–56
United States
    cheese as value-added product, 21
    cheese curd popularity, 308
    material considerations, 133–34
    tool use in, 123
units, 12
unpasteurized milk. *See* raw milk

unrefined vs. refined salts, 148
unwanted microorganisms
    commercial cheesemaking concerns, 412–13
    controlling effects of salt on, 136
    fermentation as defense against, 64–65
    flaws in cheese from, 426, 430
    growth in refrigerated milk, 47
    growth when no starter is added, *67*
    natural starter benefits, 76
    susceptibility of industrial starters to, 74
US Food and Drug Administration, 194

V

vache cheeses, 207, 232–33, *233*
vacherin cheeses, 209, 376–77, *376*
vacuum sealing of cheeses, 176, 198
Valençay cheeses
    charcoal rind treatment for, 173, 236, *236*
    lifting of, 165
    making of, 208, 236–37, *236*
vats, 124–26, *124*, 143. *See also* copper kettles; wooden vats
veal
    decline in the market for, 91
    ethical considerations, 80, *81*
vells, 86–87. *See also* abomasum (fourth stomach)
vinegar, from milk, 391
viruses, probiotics for protection against, 65

W

Walcoren tablets, 12
warm and fresh milk. *See* fresh from the udder milk
warming milk, for the make, 98. *See also* cooking milk
washed-curd cheeses. *See also* Gouda cheeses
    making of, 215–16
    whey from, 400
    wooden vats for, 124, 407
washed-rind cheeses
    cave conditions for, 157–58
    crystals in, 197
    failure to develop, 430
    Penicillium candidum for, 190
    ripening ecology of, *178*, 188–190, *189*
    secondary fermentation for, 71
    in winter, 359
washing curds, 106, 300
washing rinds. *See* rind washing
water buffalo milk. *See* buffalo milk
waxing rinds, 175–76, 300

wax types, 176
whey
    from alpine cheeses, 382, 384
    brines from, 120–21, 142, 143, 168
    cleansing uses of, 128
    description of, 61
    fermentation of, 61
    Halloumi boiled in, 249
    meat preserved with, 389
    for mysost, 392
    for rennet preparation, 86, *87*, 88
    use and reuse of, 120–22
    washing rinds with, 121, 166, 373
    washing udders with, 48
whey butter
    from cream skimmed from wooden vats, 120, 407
    from goat's and sheep's milk, 325, 326
whey cheeses
    benefits of, 30–31
    fresh whey ricotta, 382, *383*
    ricotta salata, 219, 382
    types of, 120
    in winter, 359
whey cream, 120, 325, 407
wheying off, 103, 126
whey proteins, 30–31
whey starters
    active phase of, 68
    keeping of, 400–401
    mesophilic vs. thermophilic, 69, 410
    overview, 8, 56, 60–61
    for pickles, 275
    use of, 99, 401
    in wooden vats, 60
whey vinegar, 121
whisks, for cutting curd, 127
white-rinded cheeses. *See also Geotrichum candidum*
    leaf wrapping of, 175
    orange rind development on, 431–32
    unwanted *Mucor* growth on, 430
    washing rinds of, 166
wine, natural
    cheese on lees from, 356, 357
    making of, 347–49, *348*
    starters for, 347–48, 421–22, 424
Winkler AG, 414
winter cheeses, 8, 359. *See also specific types*
wood
    for boards in caves, 157–58
    for cheesemaking tools, *127*, 128, 134
    in commercial cheesemaking, 413–14

wooden vats
    for caciocavallo, 22
    in commercial cheesemaking, 414
    description of, 124–25
    drying off and re-initiation of, 408
    for Gouda, 300
    keeping of, 406–8
    as mesophilic culture, 69
    original cheese make in, 97–98
    overview, 8
    renneting milk in, 99
    research papers on, 412
    as starter, 56, 58, 60, 98, 125, 406–8
    types of cheeses made in, *124*
wool, as cushioning for cheeses, 159
wrapping cheeses
    cheesecloth for, 110, 129–130, *130*
    leaves for, 174–75
    spruce bands for, 175, 376, *376*, 377
wrinkly-rinded lactic cheeses
    crottin cheeses, 234
    Époisses cheeses, 378
    hastening spaces for, 160, 170

Y

yak's milk, 42
yeasts
    for ambarees, 354
    milk fermented by, 352
    for natural wine, 347–48
    overview, 8
    in sourdough starters, 417, *418*, 419
    starter culture ratio effects on, 98
    unwanted growth of, 68, 420, 429
    in wooden vats, 60
yogurt
    goat's and sheep's butter from, 325, 326
    from goat's milk, 39
    as kaymak by-product, 366
    lack of sourness when curdling, 429
    making of, 206, 360–61
    for paneer, 395
    thermophilic cultures for, 69, 70, 409
    in winter, 359
yogurt cheeses
    boiling milk for, 118, 119
    making of, 207, 368, 369
    preservation methods, 370
    temperature guidelines, 112

Z

za'atar, 370
zizzona cheeses, 266

ABOUT THE AUTHOR

CHLOE GIRE

David Asher is a natural cheesemaker, bringing the traditions of dairying, fermentation, and coagulation back into this age-old craft. A former farmer and goatherd from the west coast of Canada, David now travels widely, sharing a very old but also very new approach to cheese.

Through teaching about the use of animal-warm milk, natural starter cultures, and rennet from calves and kids, David helps cheesemakers around the world reclaim their traditional cheeses. He also explores the relations of all food fermentations, and the important role of small-scale and traditional food production in our modern world. David is the author of *The Art of Natural Cheesemaking*.

the politics and practice of sustainable living

CHELSEA GREEN PUBLISHING

Chelsea Green Publishing sees books as tools for effecting cultural change and seeks to empower citizens to participate in reclaiming our global commons and become its impassioned stewards. If you enjoyed reading *Milk Into Cheese*, please consider these other great books related to cheesemaking and fermentation.

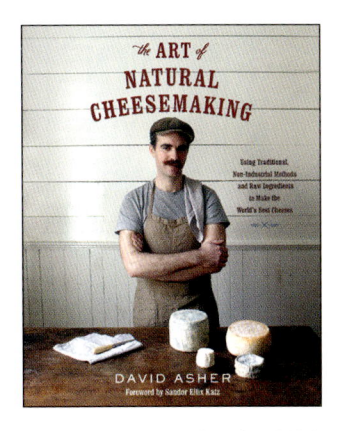

THE ART OF NATURAL CHEESEMAKING
Using Traditional, Non-Industrial Methods and Raw Ingredients to Make the World's Best Cheeses
DAVID ASHER
9781603585781
Paperback

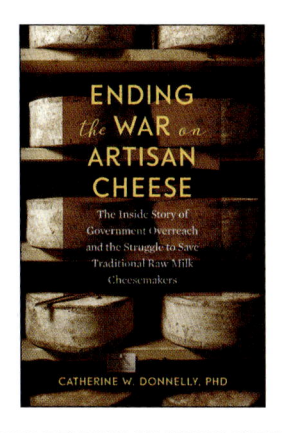

ENDING THE WAR ON ARTISAN CHEESE
The Inside Story of Government Overreach and the Struggle to Save Traditional Raw Milk Cheesemakers
CATHERINE DONNELLY
9781603587853
Paperback

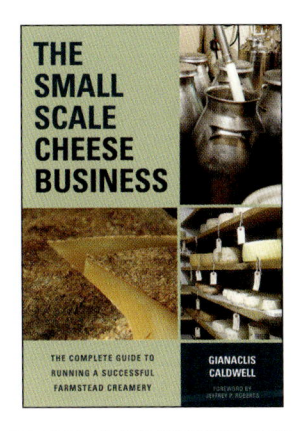

THE SMALL-SCALE CHEESE BUSINESS
The Complete Guide to Running a Successful Farmstead Creamery
GIANACLIS CALDWELL
9781603585491
Paperback

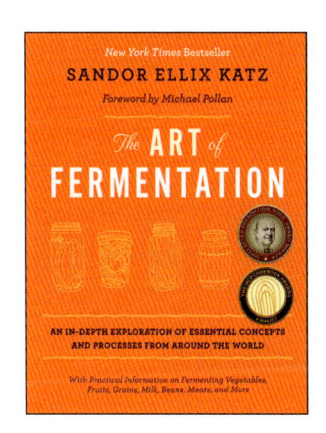

THE ART OF FERMENTATION
An In-Depth Exploration of Essential Concepts and Processes from around the World
SANDOR ELLIX KATZ
9781603582865
Hardcover

For more information,
visit **www.chelseagreen.com**.